Advanced Sensors Technologies Applied in Mobile Robot

Advanced Sensors Technologies Applied in Mobile Robot

Editors

Gregor Klančar
Marija Seder
Sašo Blažič

MDPI • Basel • Beijing • Wuhan • Barcelona • Belgrade • Manchester • Tokyo • Cluj • Tianjin

Editors
Gregor Klančar
University of Ljubljana,
Faculty of Electrical
Engineering
Slovenia

Marija Seder
University of Zagreb, Faculty
of Electrical Engineering and
Computing
Croatia

Sašo Blažič
University of Ljubljana,
Faculty of Electrical
Engineering
Slovenia

Editorial Office
MDPI
St. Alban-Anlage 66
4052 Basel, Switzerland

This is a reprint of articles from the Special Issue published online in the open access journal *Sensors* (ISSN 1424-8220) (available at: https://www.mdpi.com/journal/sensors/special_issues/Mobile_Robot).

For citation purposes, cite each article independently as indicated on the article page online and as indicated below:

LastName, A.A.; LastName, B.B.; LastName, C.C. Article Title. *Journal Name* **Year**, *Volume Number*, Page Range.

ISBN 978-3-0365-7238-3 (Hbk)
ISBN 978-3-0365-7239-0 (PDF)

Contents

About the Editors

Gregor Klančar

Gregor Klančar (Assoc. Prof.) received B.Sc. and Ph.D. degrees in Electrical Engineering from the University of Ljubljana, Faculty of Electrical Engineering, Slovenia, in 1999 and 2003, respectively. He is currently an Associate Professor with the Faculty of Electrical Engineering, University of Ljubljana. He lectures on autonomous mobile systems at graduate level and on the advanced control of autonomous systems at postgraduate level. His current research interests include autonomous mobile robots, motion control, trajectory tracking, path planning, localization, and agent-based behaviour systems.

Marija Seder

Marija Seder (Assoc. Prof.) received M.Sc. and Ph.D. degrees in Electrical Engineering from the University of Zagreb, Zagreb, Croatia, in 2004 and 2010, respectively. She is currently an Associate Professor with the Faculty of Electrical Engineering and Computing, Department of Control and Computer Engineering, University of Zagreb. She lectures on computer-controlled systems at graduate level and on the control of autonomous systems at postgraduate level. Her current research interests include mobile robotics, especially motion planning, path planning, coverage planning, obstacle avoidance, and environment exploration.

Sašo Blažič

Sašo Blažič (Prof.) received B.Sc., M.Sc., and Ph.D. degrees in Electrical Engineering from the Faculty of Electrical Engineering, University of Ljubljana, Ljubljana, Slovenia, in 1996, 1999, and 2002, respectively. He is currently a Professor with the Faculty of Electrical Engineering, University of Ljubljana, where he has been the Vice-Dean for Research since 2017. His current research interests include the adaptive, fuzzy, and predictive control of dynamical systems, modelling of nonlinear and evolving systems, autonomous mobile systems, trajectory tracking, and path planning of wheeled mobile robots.

sensors MDPI

Editorial

Advanced Sensors Technologies Applied in Mobile Robot

Gregor Klančar [1,*], Marija Seder [2] and Sašo Blažič [1]

1 Faculty of Electrical Engineering, University of Ljubljana , Tržaška 25, 1000 Ljubljana, Slovenia
2 Faculty of Electrical Engineering and Computing, University of Zagreb, Unska 3, 10000 Zagreb, Croatia
* Correspondence: gregor.klancar@fe.uni-lj.si

This special issue focuses on mobile robotic systems, where we are seeing a widespread increase in current applications as well as promising future applications enabled by the latest technologies in sensor development. Mobile robots are already making their way into our homes, modern manufacturing and warehousing systems are hard to imagine without driverless transport systems. Self-driving cars are already driving on normal roads, flying taxis are about to take off and offer new travel experiences and drones already have applications in delivery and remote sensing. There are also applications in agriculture, construction, medical care, surveillance, entertainment and other areas, some of which will also develop in unforeseen ways, all of which offer an emerging market with great potential. Advanced sensor technologies are critical in mobile robotics—a multidisciplinary field of research—to achieve automated or autonomous operation of mobile robots in these applications. They play a role in any navigation, motion control, action planning, decision making, environmental recognition, localisation, perception, object recognition, target tracking or object manipulation.

This special issue on advanced sensor technologies contains contributions on the latest developments in mobile robotic systems and related research. Various topics with different ideas and applications from mobile robotics have found their place in this special issue. They can be grouped into the three main areas of localisation and situational awareness, path planning and control algorithms. The three areas fit well with the Sense-Plan-Act architecture, which describes the most important basic activities required for the implementation of cognitive autonomous systems.

1. Sensing for Localisation

In [1], a novel method for calculating the odometry of a 3D LiDAR range image in real time is presented. Ego-motion is computed by iteratively imposing a coplanarity constraint between pairs of detected planner objects in the first step and their associated keypoints in the second step. In [2], humanoid robot control is reported using state-of-the-art motion capture systems in the high-frequency feedback control loop of humanoid robots. This can be an alternative in cases where state estimation is not reliable. Such external estimators can serve as a reference for the internal estimators, as presented in this work. Fingerprinting-based indoor 2D positioning method is proposed in [3], which utilizes the fusion of RSSI and magnetometer measurements. Autonomous navigation in mining tunnels based on artificial passive landmarks is addressed in [4]. The geometry has been optimized in order to ensure drift-free localization of mobile units equipped with LiDAR scanners. Computationally efficient high-level B-spline features extraction from 2D LiDAR is proposed in [5] with application to mapping problems. This work also provides a new benchmark for evaluating and comparing different feature generators.

2. Sensing for Situation Awareness

The identification and classification of attention deficit hyperactivity disorder (ADHD) in children is outlined in [6]. This is done through a game in which a mobile robot animates a child who must follow the robot's path. Using five Azure Kinect units equipped with depth

Citation: Klančar, G.; Seder, M.; Blažič, S. Advanced Sensors Technologies Applied in Mobile Robot. *Sensors* **2023**, *23*, 2958. https://doi.org/10.3390/s23062958

Received: 22 February 2023
Revised: 24 February 2023
Accepted: 3 March 2023
Published: 8 March 2023

sensors, the recorded skeletal data is analysed and classified using deep neural networks to output a diagnosis while the child carelessly plays the game. Article [7] presents an innovative strategy for collecting dirt samples for cleaning robots by combining geometric feature extraction and swarm algorithms. This combined approach generates an efficient optimal path that covers all identified dirt locations for an efficient cleaning mission. In addition, article [8] provides an annotated comprehensive dataset for dirt analysis. Nine classes of common domestic dirt and a labelled dataset of 3000 microscope dirt images taken from a semi-domestic environment. In [9], an AI-assisted system for predictive maintenance of mobile cleaning robots is presented that uses vibration signals to detect performance degradation and operational safety issues. A four-layer 1D convolutional neural network framework was developed and trained on a dataset of vibration signals generated by a self-developed autonomous steam mopping robot with different levels of degradation and hazardous operating environments. In [10], an approach is implemented to enable a drone to autonomously clean insulators on a power line. The algorithm for detecting and tracking dirty insulators is implemented and a special cleaning hardware is developed. In [11], a framework for false ceiling deterioration detection and mapping using deep neural network based object detection algorithm and teleoperated robot is presented. The object detection algorithm was trained on our custom false ceiling damage detection dataset consisting of four classes: structural defects (spalling, cracks, pitted surfaces and water damage), HVAC system damage (corrosion, mould and pipe damage), electrical damage (frayed wires) and pest infestation (termites and rodents).

3. Path Planning

Complete coverage path planning algorithm that generates smooth paths based on clothoids that allow a non-holonomic mobile robot to move in optimal time while following the path is described in [12]. This algorithm significantly reduces the coverage time, path length and overlap area, and increases the coverage rate compared to state-of-the-art full coverage algorithms. In [13], a novel solution for a spline path of a 5th order Bézier curve is proposed to obtain smooth trajectory planning with minimum time for wheeled mobile robots. The proposed trajectory optimisation considers constraints on the environmental space and constraints on the velocity, acceleration and jerk. In [14], a smooth navigation function combining Dijkstra-based discrete static potential field evaluation with bilinear interpolation is proposed. Modifications of the bilinear interpolation method are proposed to make it applicable to path-planning applications. In [15], a method based on gait biomechanics is presented for short-term prediction of pedestrian trajectories for real-time applications. This method relies on a single biomechanical variable and has a low computational cost, making it a viable solution for implementation in low-cost wearable devices.

4. Control Algorithms

An adaptive manipulator prescribed performance tracking motion control with global finite-time stability guarantees is proposed in [16]. In [17], a visual servo control approach is presented that enables an unmanned aerial vehicle to land autonomously on a fast-moving platform of another vehicle. In [18], the modelling and control of a spherical robot are proposed and tested with different control strategies. The model and examples with different control scenarios are available online. In [19], a global navigation function for model predictive control (MPC) of autonomous mobile robots with application to warehouse automation is proposed. The navigation function is based on a potential field derived from an E* graph search algorithm on a discrete occupancy grid and by bicubic interpolation.

This special issue brings innovative ideas that apply sensor technologies in mobile robotics in their own way. New ideas are presented for mobile robots that specialise in cleaning floors, power lines and HVAC systems. We also find innovative approaches to navigation path planning using local-minima-free potential fields, novel path primitives and/or their parameterisation for minimum-time planning, and various control approaches ranging from visual servoing to model predictive and adaptive trajectory tracking, applied to wheeled robots, humanoid manipulators and flying robots. Localisation approaches using LiDAR, motion capture systems, fingerprint-based and biomechanical gait systems are also discussed. In addition to advances in methodology, applications in healthcare, mining tunnels, cleaning, warehouses and other areas are mentioned. We believe that the works collected in the special issue Advanced Sensors Technologies Applied in Mobile Robot and its results will inspire other researchers in solving future research questions and applications in mobile robotics.

Funding: This research received no external funding.

Institutional Review Board Statement: Not applicable.

Informed Consent Statement: Not applicable.

Data Availability Statement: Not applicable.

Conflicts of Interest: The authors declare no conflict of interest.

References

1. Galeote-Luque, A.; Ruiz-Sarmiento, J.R.; Gonzalez-Jimenez, J. Efficient 3D Lidar Odometry Based on Planar Patches. *Sensors* **2022**, *22*, 6976. [CrossRef] [PubMed]
2. Popescu, M.; Mronga, D.; Bergonzani, I.; Kumar, S.; Kirchner, F. Experimental Investigations into Using Motion Capture State Feedback for Real-Time Control of a Humanoid Robot. *Sensors* **2022**, *22*, 9853. [CrossRef] [PubMed]
3. Sarcevic, P.; Csik, D.; Odry, A. Indoor 2D Positioning Method for Mobile Robots Based on the Fusion of RSSI and Magnetometer Fingerprints. *Sensors* **2023**, *23*, 1855. [CrossRef] [PubMed]
4. Torres-Torriti, M.; Nazate-Burgos, P.; Paredes-Lizama, F.; Guevara, J.; Auat Cheein, F. Passive Landmark Geometry Optimization and Evaluation for Reliable Autonomous Navigation in Mining Tunnels Using 2D Lidars. *Sensors* **2022**, *22*, 3038. [CrossRef] [PubMed]
5. Usman, M.; Ali, A.; Tahir, A.; Rahman, M.Z.U.; Khan, A.M. Efficient Approach for Extracting High-Level B-Spline Features from LIDAR Data for Light-Weight Mapping. *Sensors* **2022**, *22*, 9168. [CrossRef] [PubMed]
6. Lee, W.; Lee, D.; Lee, S.; Jun, K.; Kim, M.S. Deep-Learning-Based ADHD Classification Using Children's Skeleton Data Acquired through the ADHD Screening Game. *Sensors* **2023**, *23*, 246. [CrossRef] [PubMed]
7. Pathmakumar, T.; Muthugala, M.A.V.J.; Samarakoon, S.M.B.P.; Gómez, B.F.; Elara, M.R. A Novel Path Planning Strategy for a Cleaning Audit Robot Using Geometrical Features and Swarm Algorithms. *Sensors* **2022**, *22*, 5317. [CrossRef] [PubMed]
8. Pathmakumar, T.; Elara, M.R.; Soundararajan, S.V.; Ramalingam, B. Toward a Comprehensive Domestic Dirt Dataset Curation for Cleaning Auditing Applications. *Sensors* **2022**, *22*, 5201. [CrossRef] [PubMed]
9. Pookkuttath, S.; Rajesh Elara, M.; Sivanantham, V.; Ramalingam, B. AI-Enabled Predictive Maintenance Framework for Autonomous Mobile Cleaning Robots. *Sensors* **2022**, *22*, 13. [CrossRef] [PubMed]
10. Lopez Lopez, R.; Batista Sanchez, M.J.; Perez Jimenez, M.; Arrue, B.C.; Ollero, A. Autonomous UAV System for Cleaning Insulators in Power Line Inspection and Maintenance. *Sensors* **2021**, *21*, 8488. [CrossRef] [PubMed]
11. Semwal, A.; Mohan, R.E.; Melvin, L.M.J.; Palanisamy, P.; Baskar, C.; Yi, L.; Pookkuttath, S.; Ramalingam, B. False Ceiling Deterioration Detection and Mapping Using a Deep Learning Framework and the Teleoperated Reconfigurable 'Falcon' Robot. *Sensors* **2022**, *22*, 262. [CrossRef] [PubMed]
12. Šelek, A.; Seder, M.; Brezak, M.; Petrović, I. Smooth Complete Coverage Trajectory Planning Algorithm for a Nonholonomic Robot. *Sensors* **2022**, *22*, 9269. [CrossRef] [PubMed]
13. Benko Loknar, M.; Klančar, G.; Blažič, S. Minimum-Time Trajectory Generation for Wheeled Mobile Systems Using Bézier Curves with Constraints on Velocity, Acceleration and Jerk. *Sensors* **2023**, *23*, 1982. [CrossRef] [PubMed]
14. Klančar, G.; Zdešar, A.; Krishnan, M. Robot Navigation Based on Potential Field and Gradient Obtained by Bilinear Interpolation and a Grid-Based Search. *Sensors* **2022**, *22*, 3295. [CrossRef] [PubMed]
15. González, L.; López, A.M.; Álvarez, J.C.; Álvarez, D. Real-Time Short-Term Pedestrian Trajectory Prediction Based on Gait Biomechanics. *Sensors* **2022**, *22*, 5828. [CrossRef] [PubMed]
16. Vo, A.T.; Truong, T.N.; Kang, H.J. An Adaptive Prescribed Performance Tracking Motion Control Methodology for Robotic Manipulators with Global Finite-Time Stability. *Sensors* **2022**, *22*, 7834. [CrossRef] [PubMed]
17. Keipour, A.; Pereira, G.A.S.; Bonatti, R.; Garg, R.; Rastogi, P.; Dubey, G.; Scherer, S. Visual Servoing Approach to Autonomous UAV Landing on a Moving Vehicle. *Sensors* **2022**, *22*, 6549. [CrossRef] [PubMed]

18. Montenegro, G.; Chacón, R.; Fabregas, E.; Garcia, G.; Schröder, K.; Marroquín, A.; Dormido-Canto, S.; Farias, G. Modeling and Control of a Spherical Robot in the CoppeliaSim Simulator. *Sensors* **2022**, *22*, 6020. [CrossRef] [PubMed]
19. Klančar, G.; Seder, M. Coordinated Multi-Robotic Vehicles Navigation and Control in Shop Floor Automation. *Sensors* **2022**, *22*, 1455. [CrossRef] [PubMed]

sensors

MDPI

Article

Efficient 3D Lidar Odometry Based on Planar Patches

Andres Galeote-Luque, Jose-Raul Ruiz-Sarmiento * and Javier Gonzalez-Jimenez

Machine Perception and Intelligent Robotics Group (MAPIR-UMA), Malaga Institute for Mechatronics Engineering and Cyber-Physical Systems (IMECH.UMA), University of Malaga, 29071 Malaga, Spain
* Correspondence: jotaraul@uma.es

Abstract: In this paper we present a new way to compute the odometry of a 3D lidar in real-time. Due to the significant relation between these sensors and the rapidly increasing sector of autonomous vehicles, 3D lidars have improved in recent years, with modern models producing data in the form of range images. We take advantage of this ordered format to efficiently estimate the trajectory of the sensor as it moves in 3D space. The proposed method creates and leverages a flatness image in order to exploit the information found in flat surfaces of the scene. This allows for an efficient selection of planar patches from a first range image. Then, from a second image, keypoints related to said patches are extracted. This way, our proposal computes the ego-motion by imposing a coplanarity constraint between pairs <point, plane> whose correspondences are iteratively updated. The proposed algorithm is tested and compared with state-of-the-art ICP algorithms. Experiments show that our proposal, running on a single thread, can run 5× faster than a multi-threaded implementation of GICP, while providing a more accurate localization. A second version of the algorithm is also presented, which reduces the drift even further while needing less than half of the computation time of GICP. Both configurations of the algorithm run at frame rates common for most 3D lidars, 10 and 20 Hz on a standard CPU.

Keywords: mobile robots; localization and SLAM; 3D lidar; range sensing

Citation: Galeote-Luque, A.; Ruiz-Sarmiento, J.-R.; Gonzalez-Jimenez, J. Efficient 3D Lidar Odometry Based on Planar Patches. *Sensors* **2022**, *22*, 6976. https://doi.org/10.3390/s22186976

Academic Editor: Carlo Alberto Avizzano

Received: 8 August 2022
Accepted: 13 September 2022
Published: 15 September 2022

1. Introduction

For the safe navigation of autonomous vehicles it is essential to have an accurate estimation of their pose, that is, their position and orientation with respect to a given reference frame. These vehicles are typically equipped with sensors such as Inertial Measurement Units (IMUs), lidars or cameras, that allow them to perceive the vehicle and environment states and estimate their ego-motion. Regarding cameras, the most common approach for addressing this issue is *Visual Odometry* (VO), which consists of estimating the relative pose of the sensor along a sequence of images [1]. Similarly, *3D lidar odometry* permits the recovery of the vehicle motion along consecutive scans provided by a 3D lidar.

Particularly, 3D lidars stand out for yielding accurate geometric information from the scene in the form of point clouds, also known as *scans*, while performing reliably under changes in lighting conditions, which frequently occur in real-world scenarios [2]. The development of cheaper and more advanced 3D lidars has led to an increase in the popularity of these sensors for mobile platforms, especially autonomous vehicles. Examples of this are the autonomous driving platforms found in state-of-the-art projects such as Waymo [3,4] and Argoverse [5,6], as well as in datasets such as KITTI [7] or KAIST Urban [8].

The data points provided by traditional 3D lidars come in the form of unordered point clouds (i.e., a set of points defined by their Cartesian <X, Y, Z> coordinates, stored in an arbitrary manner). This brings 3D lidar odometry closer to the 3D registration problem: the issue of finding the rigid transformation between two consecutive point clouds, which, assuming a static world, corresponds to the sensor motion. In this regard, the Iterative Closest Point (ICP) algorithm [9], and its variants [10,11], are the most widespread methods mainly because they are effective and simple to implement. However popular, ICP algorithms are computationally expensive given the way the required calculations scale with the size of the point clouds. In a nutshell, ICP is based on finding, for every point in the source point cloud, the nearest point in the target point cloud, which requires calculating the distance between each possible pair of points. Thus, denser point clouds result in more distance calculations and, hence, a greater running time [12]. In order to obtain a correct convergence of the method, a high amount of points is needed, as the error yielded by each pair of points is averaged over the whole point cloud. This usually results in ICP running slower than the sensor frame rate, and therefore being forced to register a pair of non-consecutive scans with a greater relative motion. This becomes more of a limitation because of the serious chances of falling in local minima if the motion to recover is large, and/or no prior information about the motion is available for a good method initialization.

3D lidar sensors have seen an important improvement in recent years, with some modern models, especially those employing solid state technology, being able to produce an ordered point cloud, also referred to as ordered scan. These are composed of 3D points stored in a structured and consistent fashion. For example, the horizontal and vertical angles (polar and azimuth, respectively) of the spherical representation of the points can be equally distributed along the field of view. This enables the ordered scan to be represented as a range image, where the row and column relate to the azimuth (θ) and polar (ϕ) angles, respectively, and the value of each pixel represents the distance to the sensor. Having access to the geometric information of the scene with this new format paves the way for the development of novel odometry methods, which can capitalize on ordered point clouds to achieve a higher efficiency. In addition to this, nowadays the move of autonomous vehicles towards optimizing Size, Weight and Power (SWaP) [13] is clear, resulting in systems with scarce computing resources available to share between the different processes running on the platform. As a consequence, the usage of fast and efficient algorithms becomes a strong requirement.

In this paper we propose an efficient 3D lidar odometry method that performs at a high frame rate by taking advantage of the structure of ordered scans provided by modern 3D lidars. These data come in the form of range images, allowing us to leverage techniques popular in traditional Visual Odometry, especially in depth VO, which tend to be more efficient than ICP-like 3D registration methods. In the proposed method, the motion is recovered by imposing the coplanarity constraint [14] between pairs <point, plane> extracted from the input images. This reduces to finding the rigid transformation that minimizes the point-to-plane distance between pairs of features, similar to *indirect/feature-based* VO methods [15,16]. However, we skip solving for correspondences (e.g., using a minimum distance criterion) since the matches are iteratively established based on the projection function of the sensor, similar to what is adopted in *direct* methods such as [17,18]. The result is a hybrid algorithm that combines techniques from both *direct* and *indirect* VO methods to obtain the odometry of a 3D lidar by exploiting the flatness of surfaces present in most scenes, typically in man-made environments, both indoors and outdoors. Because we are imposing the coplanarity constraint to pairs <point, plane>, we are assuming that each point comes from its counterpart planar patch. This holds when the planar patches are sufficiently flat to be approximated by a plane, and their supporting regions are large enough with respect to the sensor relative pose. Thus, selecting planar patches that satisfy these requirements is essential in our method, and occupies much of the computation time. Also notice that, as the method approaches convergence, the motion to recover becomes smaller, making it more probable for the coplanarity constraint to be fulfilled. With each iteration of the algorithm, each point slides along their corresponding planar patch towards its centroid. In an ideal case, the points becomes coincidental with the centroid of their matched planar patch at convergence. Figure 1 shows a representation of the proposed technique applied to a simple scene, with consecutive iterations of the algorithm being shown.

It is worth mentioning that the applicability of the method could be extended to any sensor that delivers any kind of point cloud, such as traditional 3D lidar and even depth cameras, by creating a range image representation through the projection of the point cloud onto a spherical surface.

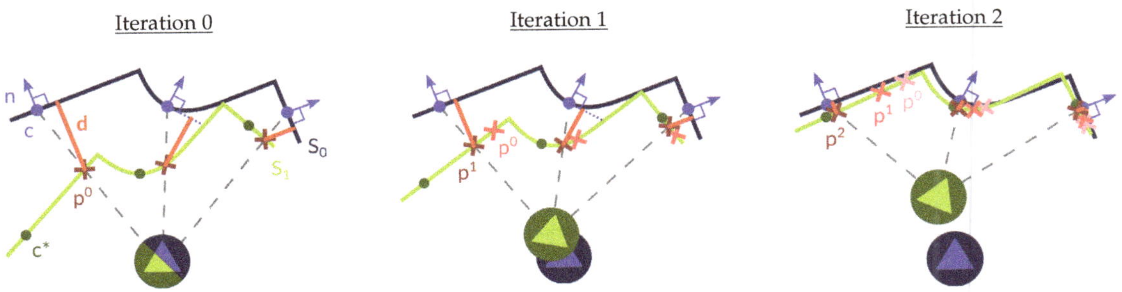

Figure 1. Consecutive iterations of the proposed algorithm. First, certain areas of the ordered scan S_0 (dark blue line) are selected based on their flatness. Planar patches defined by their center (**c**, blue dot) and normal (**n**, blue arrow) are fitted to the selection. These centers are projected (gray dashed lines) based on the current motion estimation onto S_1 (light green line), where the point ($\mathbf{p}^k$, red cross) related to said plane can be extracted (with k representing the iteration). By imposing the coplanarity constraint [14], the relative pose of the sensor can be calculated as the rigid transformation that minimizes the point-to-plane distances (d, red line) between each pair <point, plane>. In the next iteration, the estimated motion is used to update the point to which each plane is matched. This improves the probability of each pair fulfilling the coplanarity constraint, and thus results in a more accurate estimation of the movement. Notice how the selected point becomes closer and closer to its corresponding centroid ($\mathbf{c}^*$, dark green dots) with each iteration.

To validate the presented method, we have compared different configurations of it with state-of-the-art techniques over the dataset presented in VelodyneSLAM [19]. The results show how our method is capable of running in a single thread at real-time (∼30 Hz), five times faster than a multi-threaded GICP competitor, also outperforming the accuracy of other methods at a fraction of the computational cost. An alternative multi-threaded configuration has been also studied, reporting a higher accuracy while halving the computation time of GICP.

2. Related Work

For the purpose of clarity this section will be divided into two subsections: *Depth Visual Odometry* and *3D Lidar Odometry*. The proposed method leverages techniques found in depth VO, and in order to contextualize said techniques the state of the depth VO literature will be presented first. We will then proceed to review the state of the art regarding 3D lidar odometry, to correctly establish the place of the proposed method in the literature.

2.1. Depth Visual Odometry

Depth visual odometry applied to RGB-D cameras has been broadly researched since the appearance of these sensors. The approaches found in the literature can be divided into two main categories: *indirect* and *direct* methods. The former, also known as *feature-based*, are characterized for reducing the input images to a set of features which are matched by solving correspondences between them. The motion of the sensor is estimated by minimizing the distance between pairs of features. Examples of features studied in the literature are points [20], a combination of points and lines [21], and planes [15,16]. The correct finding of correspondences among features is essential in indirect methods; hence, most of the computation time is spent in this step. In particular for planar features, a plane segmentation algorithm is usually applied, which tends to slow the system down due to it being time-consuming.

On the contrary, direct methods avoid extracting features and instead minimize an energy function that is usually related to the difference between consecutive depth and/or RGB images when one of them is reprojected back against the other. Steinbrücker et al. [22] presented this concept applied to both RGB and depth images, where the photometric error is minimized, while the only purpose of the geometric information is to reproject one image against the other. This work was extended in [17] by introducing a new probabilistic formulation that produced better results. The work presented by Jaimez and Gonzalez-Jimenez [18] only utilizes the geometric information found in the depth image, estimating the motion by applying the range flow constraint densely. The result is a very fast and accurate depth odometry method, outperforming similar approaches found in the literature at the time.

Although not an odometry method, the Iterative Closest Point [9] has served as the comparison baseline to most of the depth odometry approaches. ICP was formulated for the registration of two unordered point clouds. Since depth images are a representation of an ordered point cloud, ICP can be applied to obtain the relative pose of the sensor given a pair of frames. Different versions of the ICP algorithm have been developed through the years: in [10] the point-to-plane distance is minimized instead of the original point-to-point; Generalized ICP [11] combines the two previous methods into a single probabilistic framework. The same idea was then expanded by Servos et al. [12] by adding more information in the form of channels, such as color or intensity, to reduce the ambiguity of the solution.

The method proposed in this paper could be categorized as *indirect* because it makes use of features (points and planes). As explained before, solving for correspondences is an essential procedure in traditional *feature-based* methods, because having features matched incorrectly usually results in a detriment in accuracy. We circumvent this critical step by iteratively matching the features based on the current relative pose, as in *direct* methods, thus allowing for more leniency in the pairing of the features and a faster execution time.

2.2. 3D Lidar Odometry

Despite 3D lidars producing geometric information too, such as RGB-D cameras, the unordered point clouds provided by early technology could not be represented as an image, rendering the previously mentioned VO methods unusable. The localization problem related to 3D lidar was seen closer to 3D registration, and thus the ICP algorithm and its variations dominated the literature. Nonetheless, over the years, other approaches have been developed, such as the 3D Normals Distributions Transform (NDT) [23], which stands out for offering an alternative to ICP. NDT was originally conceived for 2D localization [24], where the model is represented by a combination of normal distributions, describing the probability of finding a surface point at a certain position. Similar to the previously mentioned indirect methods, the proposal of Velas et al. [25] matches line features to estimate the odometry of a 3D lidar sensor.

One of the most common applications of 3D lidar odometry is as the front-end of SLAM systems, and even though it has not been as studied as depth odometry, the literature displays a wide collection of 3D lidar SLAM algorithms. In spite of the fact that there are SLAM methods where a custom feature-based odometry has been implemented [26,27], most of the 3D lidar SLAM systems perform the scan-to-scan alignment with some variation of ICP (see [19,28–31]).

In recent years, 3D lidars have been developed to produce ordered point clouds which can be represented as a range image, specially with the introduction of solid state lidar. This new representation of the geometric information of the scene fosters the creation of new methods that exploit its advantages. Given the novelty of these 3D lidars capable of generating ordered scans, the literature does not yet reflect this trend. In our proposal, we leverage the image representation of the ordered point clouds by employing techniques found in depth VO. This allows it to be a fast 3D lidar odometry method, which is desirable for various reasons. First, as the front-end of a SLAM system, having the localization available at a higher frame rate means the SLAM algorithm has more information to work with, while leaving more computational resources free for other processes running alongside. Additionally, the development of fast odometry methods suits the advance in robotics and autonomous vehicles towards optimizing Size, Weight and Power (SWaP).

3. Method Overview

This section summarizes how the proposed method estimates the motion of a 3D lidar over a sequence of range images. Thus, the objective is to find the relative pose of the sensor between two consecutive frames, defined by the rotation matrix and translation vector $(R, \mathbf{t})$, with $R \in \mathrm{SO}(3)$ and $\mathbf{t} \in \mathbb{R}^3$. The input is a pair of ordered scans, represented as range images, S_0 and S_1, where $S_i : (\Omega \in \mathbb{N}^2) \to \mathbb{R}$ is defined on the image domain Ω. Figure 2 shows the general workflow of our method.

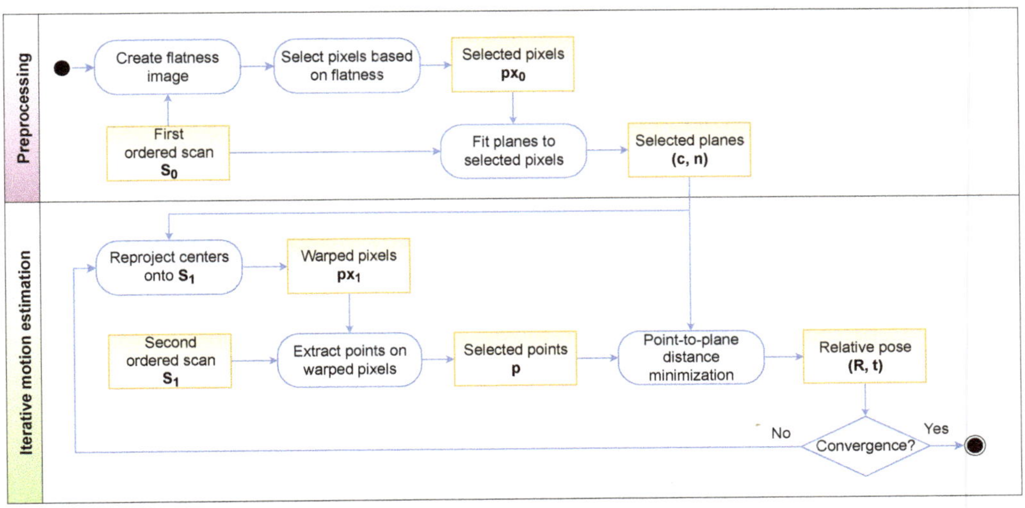

Figure 2. Overview graph showing the workflow of the proposed method. As a preprocessing step, a flatness image is computed from the first ordered scan S_0 to find the pixels most likely located in flat surfaces of the scene. Once planes have been fit to the selected pixels, the iterative motion estimation process can begin. It consists of reprojecting the selected planes onto S_1 based on the current motion estimation, which yields their corresponding pixel coordinates. From these image coordinates, the point related to the selected plane can be extracted, completing the feature pair <point, plane>. The relative pose estimation is then obtained by minimizing the point-to-plane distance of feature pairs. This is repeated until convergence.

The main idea of this method is to create pairs of planar and point features from S_0 and S_1 respectively, to then impose the coplanarity constraint to each pair. When this hypothesis holds true, the rigid transformation that minimizes the point-to-plane distance between them corresponds to the sensor motion [14]. Note that the supporting flat surface from which the planar patch is extracted does not need to be completely flat for the coplanarity constraint to apply. The bigger the supporting flat surface of the planar features, the higher the probability of fulfilling the hypothesis. Thus, a selection algorithm is employed to choose the pixels more probable of belonging to large flat surfaces.

The approach followed in the proposed algorithm consists of creating a flatness image (as seen in Figure 3) that holds the information of how flat the neighborhood around each point is, so pixels can be selected based on this score. Once the pixels from S_0 located in flat surfaces have been selected, a planar patch is fitted to the neighborhood of each one. The details of this selection process will be further explained in Section 3.1.

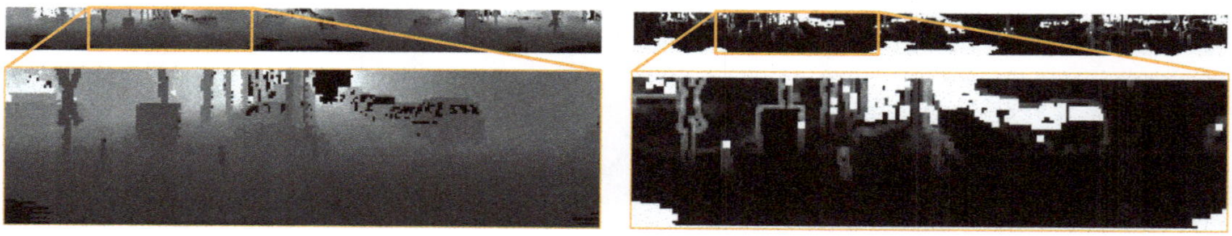

Figure 3. Flatness image (**right**) generated from S_0 (**left**) applying the LSF method. Flat surfaces such as the ground and walls appear darker, while non-planar pixels are displayed brighter. Images below are zoomed in areas of the panoramic images above.

The next step consists of finding the 3D points from S_1 related to each selected plane. This is an iterative process, since the points are extracted from the pixel coordinates obtained by reprojecting the planes onto S_1 based on the current motion estimation. For the first estimation, the movement is assumed to be zero, resulting in matching each planar feature with the 3D point in S_1 located in the same pixel coordinates. This will be expanded on Section 3.2.

This reprojection-based matching results in a set of planes from S_0 defined by their center and normal vector $(\mathbf{c}_i, \mathbf{n}_i)$, and their corresponding points $\mathbf{p}_i$ from S_1, with $\mathbf{c}_i \in C$, $\mathbf{n}_i \in N$ and $\mathbf{p}_i \in P$. The relative pose $(R, \mathbf{t})$ that minimizes the distance between the pairs of features also represents the motion between the input scans. It is possible, as long as the coplanarity constraint is fulfilled and each point belongs to the corresponding plane, to register both scans this way. Section 3.3 will delve into this estimation process.

As explained before, this is an iterative process, so the estimated motion is fed back to recalculate the points associated to each plane. This increases the probability of the paired features fulfilling the coplanarity constraint, which in turn improves the estimation of the relative pose. This can be repeated until convergence.

3.1. Flatness-Based Selection

This section describes the preprocessing step which reduces the input frame S_0 to a set of selected planar patches defined by their centers C and normal vectors N. These planar patches will be, at a later stage, matched with points P from S_1. This pairing will be considered valid if both features belong to the same surface of the scene, satisfying the coplanarity constraint. The fulfillment of this hypothesis depends on two main factors: the size of the selected planar patches and the movement to recover. Note that the latter is given by the speed of the sensor and cannot be controlled, unlike the former. To increase the probability of planar features being correctly matched, they are selected based on how flat their neighborhoods are. The bigger the supporting region of the selected plane $(\mathbf{c}_i,\mathbf{n}_i)$, the higher the probability for the point $\mathbf{p}_i$ to lie within it in the next frame, even after big movements. Some form of plane segmentation could be applied here to improve the selection, but these algorithms tend to be computationally expensive, preventing the method from running at real time. However, as previously discussed, opting for a faster alternative has the advantages of working with smaller movements, having the solution (pose estimation) available more often.

Considering this, the proposed selection approach is based on obtaining a flatness image, which holds information about how planar the area around each pixel of S_0 is. An example of a flatness image is shown in Figure 3. In this article, two ways of obtaining the flatness images are presented, each one prioritizing accuracy or efficiency, respectively:

- **Least Squares Fitting (LSF)**: For each pixel, a plane is fitted to its neighborhood using LSF. Then, the error between the fit plane and the surrounding points is used as the value in the flatness image. This technique is very accurate at the cost of being computationally more expensive.
- **Curvature-based (Fast method)**: This approach exploits the function obtained by representing a 3D plane in the spherical coordinates of the range image. These images describe each point in spherical coordinates, where the value of each pixel is the distance (r) from the point to the sensor, and the image row and column relate to the azimuthal (θ) and polar (ϕ) angles, respectively. When a simple plane such as (1) is represented in said spherical coordinates system, the distance r of a point in the plane can be described as a function of ϕ and θ as in (2).

$$ax + by + cz = 1 \quad (1)$$

$$r(\phi, \theta) = \frac{1}{a \cos\phi \cos\theta + b \sin\phi \cos\theta + c \sin\theta} \quad (2)$$

Thus, the inverse of the distance function is a sine wave (3).

$$d(\phi, \theta) = \frac{1}{r} = a \cos\phi \cos\theta + b \sin\phi \cos\theta + c \sin\theta \quad (3)$$

Instead of trying to fit a sine wave to check for flat areas, the curvature of the neighborhood can be tested. The curvature of a function can be approximated by its second derivative [32], which means that the second derivative of the inverse of the distance d along both ϕ and θ should be low for pixels located in flat surfaces. However, non-planar areas, such as the intersection of two planes, will result in an abrupt change in curvature. The derivatives along both image axes can be efficiently calculated with the use of 2D convolutions, applying the Sobel operator. The value stored in the flatness image is the curvature (k) of each pixel, calculated as the sum of the absolute value of the second derivative of d along each axis (4).

$$k = |\frac{\partial^2 d}{\partial \phi^2}| + |\frac{\partial^2 d}{\partial \theta^2}| \quad (4)$$

Regardless of the technique used to obtain the flatness image, a Gaussian filter is applied to blur it afterwards, which helps to deal with pixels located near non-planar areas by leveraging the score of their neighborhoods.

Once the flatness image has been created, it is then used to select pixels from S_0 based on their score. To ensure a good distribution of pixels around the scene, the image is divided into blocks, and the best pixels from each block are selected. This also serves as a non-maximum suppression step.

A planar patch ($\mathbf{c}_i$, $\mathbf{n}_i$) is fitted to each neighborhood around the selected pixels in the image S_0, with the centroid of the group being the center of the plane. For the normal vector, the singular value decomposition is applied to the matrix created by stacking the difference of each point of the neighborhood and the centroid. The normal vector is then the left singular vector corresponding to the least singular value. A measurement of the fitness f_i, how planar the group of points is, is also estimated as the least singular value. Low values of this variable mean the group of points actually form a planar surface.

Take into account that it is no guarantee that the selected pixels are located in a planar surface just from the flatness image, but once the planes are fit into their neighborhood, the fitness f_i can be used to downweight selected pixels found near occlusions or in curved surfaces (see Section 3.3).

3.2. Iterative Projection-Based Matching

After the planar patches have been obtained from S_0, their corresponding points $\mathbf{p}_i$ have to be extracted from S_1. In traditional indirect methods, the features from both frames are matched based on some descriptor. Instead, in the proposed method, an iterative approach is implemented, inspired by direct methods. It consists of reprojecting the centers of the selected planes onto S_1 based on the current motion estimation, to find the pixels in S_1 that are more probable of representing the same scene surface as their corresponding planar region. This process works as follows:

1. The center of each plane $\mathbf{c}_i$ is transformed using the motion estimation $(R, \mathbf{t})$ to obtain $\mathbf{c}_i^*$:
$$\mathbf{c}_i^* = [x_c, y_c, z_c]^T = R^{-1}(\mathbf{c}_i - \mathbf{t}) \tag{5}$$
2. The spherical coordinates of the transformed center are then calculated:
$$\begin{cases} r_i &= \sqrt{x_c^2 + y_c^2 + z_c^2} \\ \phi_i &= \text{arctan2}(y_c, x_c) \\ \theta_i &= \arcsin(z_c / r_i) \end{cases} \tag{6}$$
3. The row (v) and column (u) on the image are obtained based on the spherical coordinates:
$$\begin{cases} v_i = \arg\min_j |\theta_{vec}(j) - \theta_i| \\ u_i = [(\text{columns} - 1)(\pi - \phi_i)/(2\pi)] \end{cases} \tag{7}$$
where θ_{vec} represents the vector of the azimuthal angles related to the row numbers, and [ ] is the rounding operator, which returns the integer number closest to the input value.
4. These pixel coordinates can then be used to obtain the points $\mathbf{p}_i$ from S_1. A Gaussian filter is applied over the neighboring points to reduce negative effect of noise.

During the first iteration, the relative pose between both range images is considered to be composed by the identity rotation $R = I$ and no translation $\mathbf{t} = 0$, which means that the points $\mathbf{p}_i$ are located in the same image coordinates as the centers $\mathbf{c}_i$ of the selected planes, but in the next image. Note that by imposing the coplanarity constraint the only requisite is that the points and planes selected belong to the same surface of the 3D scene, so there is no need for a perfect point-to-point matching. By selecting a pixel in S_0 located in a planar surface as explained in the previous section, there is a high probability that the point located at the same image coordinates in both frames lies on the same surface.

3.3. Relative Pose Estimation

This section describes how the relative pose between the two range images is calculated given the set of matching <point, plane> pairs, assuming they satisfy the coplanarity constraint. The objective is to find the rigid transformation $(R, \mathbf{t})$ that minimizes the point-to-plane distance of each pair of features, as shown in Equation (8). This is optimized by the Levenberg–Marquardt implementation from the well-known Ceres Solver [33]:

$$(R, \mathbf{t}) = \arg\min_{R,\mathbf{t}} \sum_i \rho\left(||\alpha_i \mathbf{n}_i \cdot (R\mathbf{p}_i + \mathbf{t} - \mathbf{c}_i)||^2\right) \tag{8}$$

where $\alpha_i = (1 - f_i)$ serves as a weight of the residual based on the fitness of the selected plane. This way, selected pixels located in surfaces with lower flatness have less impact on the final result. The Huber loss function ρ is applied to the residual to deal with outliers, which are pixels that do not fulfill the hypotheses, whether because they move with respect to the static scene or because they do not lie in the same surface in both frames. Along with the relative pose, the covariance of the solution is calculated, which holds the information of the uncertainty of each parameter of the rigid transformation.

The motion estimation is part of the iterative process previously mentioned (recall Figure 2), so the resulting transformation is fed back to update the points P in S_1 corresponding to each planar region. This improves the matching of the features by making it more probable that both of them belong to the same surface of the scene, improving in turn the motion estimation. This procedure is repeated until the convergence of the solution.

As a final process, a motion filter is applied to the resulting relative pose, similar to the one found in [18], in which the motion estimation from the previous pair of frames is leveraged if the covariance of the current solution is high. This has proven to be helpful in scenes with low planar information, improving the accuracy of the localization as will be proven in Section 4.4.

4. Experiments and Results

In this section, we evaluate the performance of our method resorting to sequences of 3D lidar scans from the widely-used Velodyne SLAM dataset [19] (see Section 4.1). The obtained results are compared with different implementations of the state-of-the-art GICP algorithm [11], including single-thread and multi-thread alternatives (Sections 4.2 and 4.3). For all the experiments, the test platform used is a standard PC with 8GB of RAM, running Ubuntu 20.04 with an Intel Core i7-7700HQ CPU with four cores and eight execution threads at 2.8 GHz. We also present an ablation study exploring different configurations of the proposed method (Section 4.4) and discuss the limitations of the proposal (see Section 4.5).

4.1. Testbed

The proposed method was evaluated on the two outdoor sequences provided by the Velodyne SLAM dataset [19], captured by a vehicle driving around town in a closed loop over a kilometer in length. The Velodyne HDL-64E sensor, running at 10 Hz, provides the geometric information in two formats: point clouds and 870×64 range images.

The error metric employed to compare the different methods is the one proposed in [34], known as relative pose error (RPE). Although the RMSE of the error will be shown because of its popularity in the odometry literature, the mean RPE will be used to compare the methods based on the arguments presented in [35].

The proposed method is configured with the following parameters:

- Neighborhoods are 3×3 blocks of pixels.
- The flatness image is divided into 16×16 blocks, and the four best pixels are selected from each block.

4.2. Time Performance

The performance of our method is compared to the widely popular GICP algorithm [11] as implemented in two different C++ libraries: PCL [36] and Open3D (O3D) [37]. The former is a basic single-thread (ST) implementation, while the latter applies a loss function to reduce the effect of outliers, and performs at a higher rate because of multi-threading.

The proposed method is tested with both versions of the flatness image introduced in Section 3.1, *fast* and *LSF*, with the second method being more accurate at the cost of computation time. For a fair comparison with the GICP implementation of Open3D, a multi-threaded (MT) implementation of the LSF variant is included. A breakdown of the elapsed times for every part of the proposed algorithm is shown in Table 1 for both versions. It is clear how the use of 2D convolutions in the *fast* version of the method greatly reduces the computational cost. Regarding the threaded implementation of our method, the only process worth parallelizing is the creation of the flatness image, resulting in a 59% reduction of computation time.

With the goal of performing a balanced comparison, a pair of configurations were employed in the GICP calculations, each one prioritizing accuracy or speed over the other one. In summary, seven odometry methods are compared in this section, four of them being a variation of GICP (PCL-fast, PCL-acc., Open3D-fast, Open3D-acc.), and three of them being the different versions of the proposed method (Ours-fast, Ours-LSF-ST, Ours-LSF-MT). The Open3D implementation of GICP and Ours-LSF-MT are multi-threaded, while the rest run on a single thread. Table 2 shows the computation time and frequency of each method, where we can see how the slower implementation of our method, using the LSF, outmatches the frame rate of the faster GICP (Open3D-fast), while the fast version achieves 4× its speed. Note that 3D lidar sensors usually run at 10–20 Hz, so the presented configurations of the proposed method work in real time. Figure 4a shows this information in a more visual way.

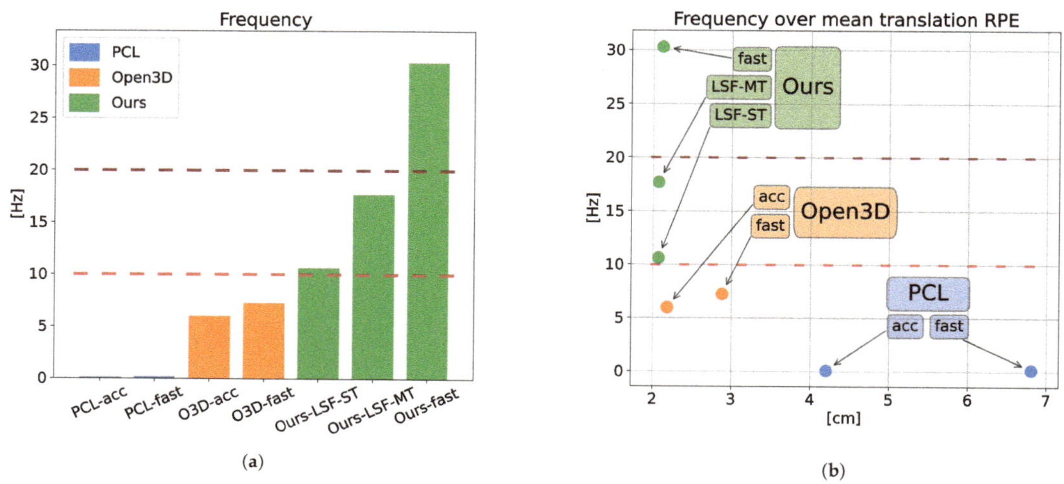

Figure 4. Frequency comparison of the different methods. (**a**) Frame rate of the different methods. The usual working frequencies of 3D lidar, 10 and 20 Hz, are displayed along the data. (**b**) Frequency of the different methods over the translational RPE. The usual working frequencies of 3D lidar, 10 and 20 Hz, are displayed along the data.

Table 1. Computation time breakdown of the proposed method.

Process	Fast (ST)	LSF (ST)	LSF (MT)
Flatness image	2.5 ms	63.5 ms	26 ms
Plane select. and fitting	3ms	3ms	3ms
Matching points extraction	2.5 ms	2.5 ms	2.5 ms
Motion estimation	25 ms	25 ms	25 ms
Total	33 ms	94 ms	56.5 ms

Table 2. Computation time comparison between the different methods.

Threaded	Method	Time per Frame	Frame Rate
No	PCL (fast)	5300 ms	0.188 Hz
	PCL (acc.)	8800 ms	0.114 Hz
	Ours (fast)	33 ms	30.30 Hz
	Ours (LSF, ST)	94 ms	10.64 Hz
Yes	Open3D (fast)	138 ms	7.24 Hz
	Open3D (acc.)	167 ms	5.98 Hz
	Ours (LSF, MT)	56.5 ms	17.69 Hz

4.3. Accuracy Results

Regarding the accuracy, Table 3 displays the translational and rotational error of the six methods being compared along the two sequences of the dataset. Note that the results obtained from Ours-LSF are the same whether or not it is multi-threaded. In said table we can see how, when comparing the LSF version of our method with the competitor showing the lowest error (Open3D-acc), our proposal is able to achieve a reduction in the translational error of 5% and 19% of the mean and median value of the RPE, respectively, while speeding up its frame rate by a factor of 2.9×. Even the proposed Ours-fast version, which works five times faster than Open3D-acc., reaches a similar accuracy improvement. Regarding the rotation error, a similar performance is achieved, with the difference between GICP and our method being below 3%. Note that there is a notable difference between comparing the RMSE, mean error, and median error of the different methods. We agree with [35] that the RMSE is not able to correctly capture the general accuracy of the data since it severely penalizes instances of big errors, however scarce they are. The mean absolute error on the other hand allows for a fairer comparison of the accuracy of the presented methods. In addition, the median absolute error conveys the accuracy obtained in most of the frames along each sequence. Figure 5 shows a visual representation of this evaluation, where we can check the superior performance of our proposal regarding accuracy. Moreover, for a qualitative comparison, the trajectories computed by each method are displayed in Figure 6.

Table 3. RPE comparison between the different methods.

(a) Translational RPE.

[cm/frame]		PCL fast	PCL acc.	O3D fast	O3D acc.	Ours fast	Ours LSF
RMSE	Scen. 1	11.55	8.54	5.68	**5.14**	6.95	7.04
	Scen. 2	14.68	9.22	**7.47**	7.54	7.98	7.83
	Avg	13.12	8.88	6.58	**6.34**	7.47	7.44
Mean	Scen. 1	6.55	4.31	2.97	2.30	2.22	**2.18**
	Scen. 2	7.07	4.09	2.79	2.06	2.01	**1.96**
	Avg	6.81	4.20	2.88	2.18	2.12	**2.07**
Median	Scen. 1	3.79	2.58	2.31	1.77	1.47	**1.44**
	Scen. 2	3.61	2.38	2.79	1.55	**1.23**	**1.23**
	Avg	3.70	2.48	2.55	1.66	1.35	**1.34**

(b) Rotational RPE.

[deg/frame]		PCL fast	PCL acc.	O3D fast	O3D acc.	Ours fast	Ours LSF
RMSE	Scen. 1	0.230	0.198	0.189	**0.184**	0.212	0.203
	Scen. 2	0.228	0.177	0.168	**0.163**	0.257	0.324
	Avg	0.229	0.188	0.179	**0.174**	0.235	0.263
Mean	Scen. 1	0.177	0.154	0.147	**0.142**	0.145	0.144
	Scen. 2	0.171	0.140	0.131	**0.126**	0.128	0.129
	Avg	0.174	0.147	0.139	**0.134**	0.137	0.136
Median	Scen. 1	0.134	0.119	0.112	0.109	0.108	**0.107**
	Scen. 2	0.133	0.111	0.102	0.097	0.093	**0.091**
	Avg	0.134	0.115	0.107	0.103	0.100	**0.099**

Figure 4b shows the running frequency of each method over their mean translation RPE, allowing a comparison of accuracy and speed combined. For example, it can be seen how Ours-fast only performs 3% better than O3D-acc in translation RPE, but runs 5× faster. Compared to the faster version of GICP, Ours-fast works at 4x the speed, and the accuracy improvement goes up to 26%. Regarding the *LSF* version of the proposed method, it yields a better accuracy, with a 5% and 28% improvement over the accurate and fast versions of Open3D, respectively, while running 2.9× and 2.4× faster than them.

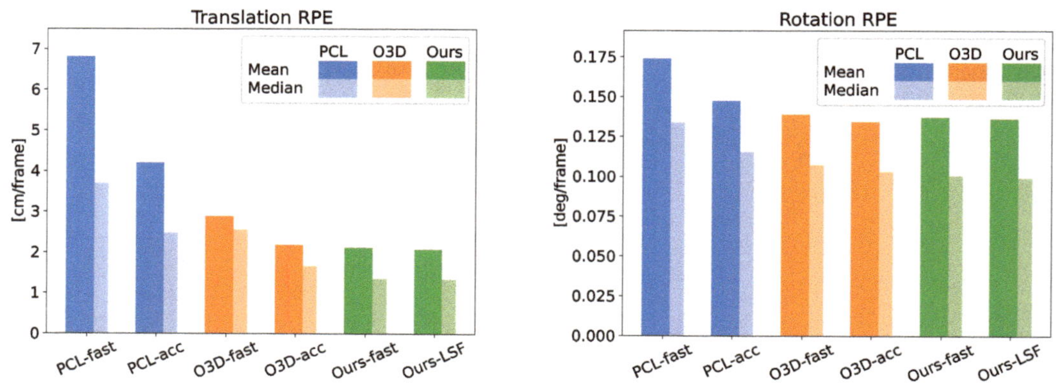

Figure 5. Comparison of the mean and median RPE of the different methods. Translational and rotational error are displayed on the left and right, respectively.

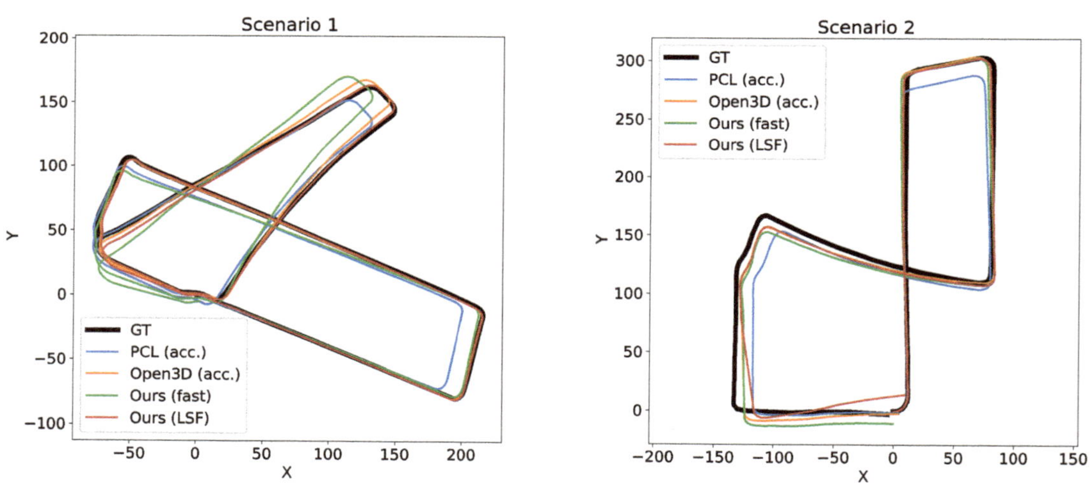

Figure 6. Top view of the resulting trajectories of the tested methods over each sequence of the dataset, along with the ground truth.

4.4. Ablation Study

The purpose of this section is to validate the contribution of the different processes employed in our proposal to the resulting performance. Particularly, three procedures are analyzed: the iterative update of the correspondences, the motion filter (see Section 3.3), and the blurring of the flatness image (Section 3.1). Thus, the baseline for this test is the base algorithm running a single iteration.

Based on Table 4, the effect of a second iteration proves to be worth it, reducing the translation error over 70%. Allowing the system to iterate until convergence improves the accuracy even further. Regarding the motion filter, it helps to mitigate the effect of scenes where not enough information is found (see Section 4.5), which happens sparsely throughout the sequences, resulting in an improvement of the accuracy of 3%. Similarly, applying a Gaussian filter to the flatness image helps with selecting pixels located far from non-planar areas, reducing the error 2.6% further.

Table 4. Improvements in accuracy of the different versions of the proposed method.

Version	Translation: Mean RPE (cm/frame)	Improvement over Previous Version
Base (1 iteration)	11.27	-
Base (2 iterations)	3.24	71.25%
Iterations until convergence	2.31	28.70%
Prev. and motion filter	2.24	3.03%
Prev. and flatness image blurring	2.18	2.68 %

4.5. Limitations

The formulation of the presented method entails the limitation caused by the aperture problem [38], meaning that certain scenarios exist in which the motion cannot be estimated caused by the lack of planes oriented in different directions. A clear example of these undetermined scenes would be a corridor, in which the translation along it cannot be recovered by imposing the coplanarity constraint. To solve this issue, in addition to the motion filter, information from another sensor, like IMUs or RGB cameras, could be leveraged to constrain the solution.

5. Conclusions

In this paper we have introduced a novel method for 3D lidar odometry that exploits the structure of the range images that modern sensors provide, as well as information from flat surfaces to achieve run-time operation and a low drift. These flat surfaces are leveraged to extract planar patches, described by their centroid and normal vector, from the first range image, which in turn are used to retrieve features in the second image in the form of points. The utilization of a flatness image is proposed to efficiently select planar patches, and two different ways to estimate it have been presented, each one prioritizing speed or accuracy over the other. The coplanarity constraint is imposed to each <point, plane> pair. This way the motion is estimated by minimizing the point-to-plane distance between pairs of features. Their correspondences are updated iteratively based on the current estimation of the relative pose, by employing the projection model of the sensor. When the point related to each plane is updated based on the current pose estimation, the coplanarity constraint becomes more probable to be fulfilled, thus improving the accuracy of the solution. Gaussian and motion filters are used to reduce the negative influence of noise and scenarios with limited information, respectively.

To test its viability, our approach has been compared with GICP, widely used in 3D lidar SLAM systems. Two GICP configurations have been used in the comparison, one focusing on accuracy and the other on speed, to match the two versions of the proposed method. When comparing the fast version of both approaches, our method turns out to be capable of achieving four times the frame rate using a single thread, while improving the accuracy by 26%. Regarding the implementations where the goal is to maximize the accuracy, our approach runs 2.95× faster than GICP, and yields 5% more accuracy. This reduction in computation time allows the resources of the system to be shared between the different processes running alongside.

In future works, the adaptability of the method parameters will be studied. For example, the size of the neighborhood can change depending on the distance to the point, or quadtrees can be used instead of fixed-size blocks in the selection. Being adaptive to the peculiarities of the scene at hand would permit the method to properly operate with a wider range of real-world conditions.

Author Contributions: Conceptualization, A.G.-L., J.-R.R.-S. and J.G.-J.; methodology, A.G.-L. and J.-R.R.-S.; software, A.G.-L.; validation, A.G.-L.; formal analysis, A.G.-L. and J.-R.R.-S.; investigation, A.G.-L., J.-R.R.-S. and J.G.-J.; resources, J.G.-J.; data curation, A.G.-L.; writing—original draft preparation, A.G.-L.; writing—review and editing, J.-R.R.-S. and J.G.-J.; visualization, A.G.-L. and J.-R.R.-S.; supervision, J.-R.R.-S. and J.G.-J.; project administration, J.-R.R.-S. and J.G.-J.; funding acquisition, J.G.-J. All authors have read and agreed to the published version of the manuscript.

Funding: This work has been supported by the grant program PRE2018-085026 and the research projects *HOUNDBOT* (P20_01302), funding by Andalusian Regional Government, and *ARPEGGIO* (PID2020-117057GB-I00), funded by Spain National Research Agency. The publication of this paper has been funded by the University of Malaga.

Institutional Review Board Statement: Not applicable.

Informed Consent Statement: Not applicable.

Data Availability Statement: Not applicable.

Conflicts of Interest: The authors declare no conflict of interest.

Abbreviations

The following abbreviations are used in this manuscript:

ICP	Iterative Closest Point
GICP	Generalized Iterative Closest Point
VO	Visual Odometry
SWaP	Size, Weight and Power
NDT	Normals Distribution Transform
SLAM	Simultaneous Localization and Mapping
LIDAR	Light Detection and Ranging
LSF	Least Squares Fitting
RPE	Relative Pose Error
RMSE	Root Mean Square Error
ST	Single Thread
MT	Multi-Threaded
IMU	Inertial Measurement Unit

References

1. Nistér, D.; Naroditsky, O.; Bergen, J. Visual odometry. In Proceedings of the 2004 IEEE Computer Society Conference on Computer Vision and Pattern Recognition, Washington, DC, USA, 27 June–2 July 2004; Volume 1, p. I-I.
2. Adis, P.; Horst, N.; Wien, M. D3DLO: Deep 3D LiDAR Odometry. In Proceedings of the 2021 IEEE International Conference on Image Processing (ICIP), Anchorage, AK, USA, 19–22 September 2021; pp. 3128–3132.
3. Sun, P.; Kretzschmar, H.; Dotiwalla, X.; Chouard, A.; Patnaik, V.; Tsui, P.; Guo, J.; Zhou, Y.; Chai, Y.; Caine, B.; et al. Scalability in perception for autonomous driving: Waymo open dataset. In Proceedings of the IEEE/CVF Conference on Computer Vision and Pattern Recognition, Seattle, WA, USA, 13–19 June 2020; pp. 2446–2454.
4. Ettinger, S.; Cheng, S.; Caine, B.; Liu, C.; Zhao, H.; Pradhan, S.; Chai, Y.; Sapp, B.; Qi, C.R.; Zhou, Y.; et al. Large scale interactive motion forecasting for autonomous driving: The waymo open motion dataset. In Proceedings of the IEEE/CVF International Conference on Computer Vision, Montreal, QC, Canada, 11–17 October 2021; pp. 9710–9719.
5. Chang, M.F.; Lambert, J.; Sangkloy, P.; Singh, J.; Bak, S.; Hartnett, A.; Wang, D.; Carr, P.; Lucey, S.; Ramanan, D.; et al. Argoverse: 3d tracking and forecasting with rich maps. In Proceedings of the IEEE/CVF Conference on Computer Vision and Pattern Recognition, Long Beach, CA, USA, 15–20 June 2019; pp. 8748–8757.
6. Wilson, B.; Qi, W.; Agarwal, T.; Lambert, J.; Singh, J.; Khandelwal, S.; Pan, B.; Kumar, R.; Hartnett, A.; Pontes, J.K.; et al. Argoverse 2: Next Generation Datasets for Self-Driving Perception and Forecasting. In Proceedings of the Thirty-Fifth Conference on Neural Information Processing Systems Datasets and Benchmarks Track (Round 2), Virtually, 6–14 December 2021.
7. Geiger, A.; Lenz, P.; Stiller, C.; Urtasun, R. Vision meets Robotics: The KITTI Dataset. *Int. J. Robot. Res.* **2013**, *32*, 1231–1237. [CrossRef]

8. Jeong, J.; Cho, Y.; Shin, Y.S.; Roh, H.; Kim, A. Complex Urban Dataset with Multi-level Sensors from Highly Diverse Urban Environments. *Int. J. Robot. Res.* **2019**, *38*, 642–657. [CrossRef]
9. Besl, P.J.; McKay, N.D. Method for registration of 3-D shapes. In Proceedings of the Sensor fusion IV: Control Paradigms and Data Structures. International Society for Optics and Photonics: Boston, MA, USA, 12–15 November 1992; Volume 1611, pp. 586–606.
10. Chen, Y.; Medioni, G. Object modelling by registration of multiple range images. *Image Vis. Comput.* **1992**, *10*, 145–155. [CrossRef]
11. Segal, A.; Haehnel, D.; Thrun, S. Generalized-ICP. In Proceedings of the Robotics: Science and Systems, Seattle, WA, 28 June–1 July 2009; Volume 2, p. 435.
12. Servos, J.; Waslander, S.L. Multi-Channel Generalized-ICP: A robust framework for multi-channel scan registration. *Robot. Auton. Syst.* **2017**, *87*, 247–257. [CrossRef]
13. Jones, K.H.; Gross, J.N. Reducing size, weight, and power (SWaP) of perception systems in small autonomous aerial systems. In Proceedings of the 14th AIAA Aviation Technology, Integration, and Operations Conference, Atlanta, GA, USA, 16–20 June 2014; p. 2705.
14. Ramalingam, S.; Taguchi, Y. A theory of minimal 3D point to 3D plane registration and its generalization. *Int. J. Comput. Vis.* **2013**, *102*, 73–90. [CrossRef]
15. Raposo, C.; Lourenço, M.; Antunes, M.; Barreto, J.P. Plane-based Odometry using an RGB-D Camera. In Proceedings of the BMVC. Citeseer, Bristol, UK, 9–13 September 2013.
16. Chen, B.; Liu, C.; Tong, Y.; Wu, Q. Robust RGB-D Visual Odometry Based on Planar Features. In Proceedings of the 2017 IEEE 7th Annual International Conference on CYBER Technology in Automation, Control, and Intelligent Systems (CYBER), Honolulu, HI, USA, 31 July–4 August 2017; pp. 449–453.
17. Kerl, C.; Sturm, J.; Cremers, D. Robust odometry estimation for RGB-D cameras. In Proceedings of the 2013 IEEE International Conference on Robotics and Automation, Karlsruhe, Germany, 6–10 May 2013; pp. 3748–3754.
18. Jaimez, M.; Gonzalez-Jimenez, J. Fast visual odometry for 3-D range sensors. *IEEE Trans. Robot.* **2015**, *31*, 809–822. [CrossRef]
19. Moosmann, F.; Stiller, C. Velodyne SLAM. In Proceedings of the IEEE Intelligent Vehicles Symposium, Baden-Baden, Germany, 5–9 June 2011; pp. 393–398.
20. Prakhya, S.M.; Bingbing, L.; Weisi, L.; Qayyum, U. Sparse depth odometry: 3D keypoint based pose estimation from dense depth data. In Proceedings of the 2015 IEEE International Conference on Robotics and Automation (ICRA), Seattle, WA, USA, 26–30 May 2015; pp. 4216–4223.
21. Lu, Y.; Song, D. Robust RGB-D odometry using point and line features. In Proceedings of the IEEE International Conference on Computer Vision, Santiago, Chile, 7–13 December 2015; pp. 3934–3942.
22. Steinbrücker, F.; Sturm, J.; Cremers, D. Real-time visual odometry from dense RGB-D images. In Proceedings of the 2011 IEEE International Conference on Computer Vision Workshops (ICCV Workshops), Barcelona, Spain, 6–13 November 2011; pp. 719–722.
23. Magnusson, M.; Lilienthal, A.; Duckett, T. Scan registration for autonomous mining vehicles using 3D-NDT. *J. Field Robot.* **2007**, *24*, 803–827. [CrossRef]
24. Biber, P.; Straßer, W. The normal distributions transform: A new approach to laser scan matching. In Proceedings of the 2003 IEEE/RSJ International Conference on Intelligent Robots and Systems (IROS 2003) (Cat. No. 03CH37453), Las Vegas, NV, USA, 27–31 October 2003; Volume 3, pp. 2743–2748.
25. Velas, M.; Spanel, M.; Herout, A. Collar line segments for fast odometry estimation from velodyne point clouds. In Proceedings of the 2016 IEEE International Conference on Robotics and Automation (ICRA), Stockholm, Sweden, 16–21 May 2016; pp. 4486–4495.
26. Liang, S.; Cao, Z.; Guan, P.; Wang, C.; Yu, J.; Wang, S. A novel sparse geometric 3-d lidar odometry approach. *IEEE Syst. J.* **2020**, *15*, 1390–1400. [CrossRef]
27. Shan, T.; Englot, B.; Meyers, D.; Wang, W.; Ratti, C.; Rus, D. Lio-sam: Tightly-coupled lidar inertial odometry via smoothing and mapping. In Proceedings of the 2020 IEEE/RSJ International Conference on Intelligent Robots and Systems (IROS), Las Vegas, NV, USA, 25–29 October 2020; pp. 5135–5142.
28. Zhang, J.; Singh, S. Visual-lidar odometry and mapping: Low-drift, robust, and fast. In Proceedings of the 2015 IEEE International Conference on Robotics and Automation (ICRA), Seattle, WA, USA, 26–30 May 2015; pp. 2174–2181.
29. Kuramachi, R.; Ohsato, A.; Sasaki, Y.; Mizoguchi, H. G-ICP SLAM: An odometry-free 3D mapping system with robust 6DoF pose estimation. In Proceedings of the 2015 IEEE International Conference on Robotics and Biomimetics (ROBIO), Zhuhai, China, 6–9 December 2015; pp. 176–181.
30. Agrawal, P.; Iqbal, A.; Russell, B.; Hazrati, M.K.; Kashyap, V.; Akhbari, F. PCE-SLAM: A real-time simultaneous localization and mapping using LiDAR data. In Proceedings of the 2017 IEEE Intelligent Vehicles Symposium (IV), Los Angeles, CA, USA, 11–14 June 2017; pp. 1752–1757.
31. Behley, J.; Stachniss, C. Efficient Surfel-Based SLAM using 3D Laser Range Data in Urban Environments. In Proceedings of the Robotics: Science and Systems, Pittsburgh, PA, USA, 26–30 June 2018; Volume 2018, p. 59.
32. Flynn, P.J.; Jain, A.K. On reliable curvature estimation. In Proceedings of the CVPR, San Diego, CA, USA, 4–8 June 1989; Volume 88, pp. 5–9.
33. Agarwal, S.; Mierle, K.; The Ceres Solver Team. Ceres Solver. Available online: http://ceres-solver.org (accessed on 23 July 2022).

34. Sturm, J.; Engelhard, N.; Endres, F.; Burgard, W.; Cremers, D. A benchmark for the evaluation of RGB-D SLAM systems. In Proceedings of the 2012 IEEE/RSJ International Conference on Intelligent Robots and Systems, Vilamoura-Algarve, Portugal, 7–12 October 2012; pp. 573–580.
35. Willmott, C.J.; Matsuura, K. Advantages of the mean absolute error (MAE) over the root mean square error (RMSE) in assessing average model performance. *Clim. Res.* **2005**, *30*, 79–82. [CrossRef]
36. Rusu, R.B.; Cousins, S. 3D is here: Point Cloud Library (PCL). In Proceedings of the IEEE International Conference on Robotics and Automation (ICRA), Shanghai, China, 9–13 May 2011.
37. Zhou, Q.Y.; Park, J.; Koltun, V. Open3D: A Modern Library for 3D Data Processing. *arXiv* **2018**, arXiv:1801.09847.
38. Shimojo, S.; Silverman, G.H.; Nakayama, K. Occlusion and the solution to the aperture problem for motion. *Vis. Res.* **1989**, *29*, 619–626. [CrossRef]

sensors

MDPI

Communication

Experimental Investigations into Using Motion Capture State Feedback for Real-Time Control of a Humanoid Robot

Mihaela Popescu [1,*], Dennis Mronga [2], Ivan Bergonzani [2], Shivesh Kumar [2] and Frank Kirchner [1,2]

[1] Robotics Group, Faculty of Mathematics and Computer Science, University of Bremen, 28359 Bremen, Germany
[2] Robotics Innovation Center, German Research Center for Artificial Intelligence (DFKI GmbH), 28359 Bremen, Germany
* Correspondence: popescu@uni-bremen.de

Abstract: Regardless of recent advances, humanoid robots still face significant difficulties in performing locomotion tasks. Among the key challenges that must be addressed to achieve robust bipedal locomotion are dynamically consistent motion planning, feedback control, and state estimation of such complex systems. In this paper, we investigate the use of an external motion capture system to provide state feedback to an online whole-body controller. We present experimental results with the humanoid robot RH5 performing two different whole-body motions: squatting with both feet in contact with the ground and balancing on one leg. We compare the execution of these motions using state feedback from (i) an external motion tracking system and (ii) an internal state estimator based on inertial measurement unit (IMU), forward kinematics, and contact sensing. It is shown that state-of-the-art motion capture systems can be successfully used in the high-frequency feedback control loop of humanoid robots, providing an alternative in cases where state estimation is not reliable.

Keywords: humanoid robot; state estimation; motion capture; Whole-Body Control

Citation: Popescu, M.; Mronga, D.; Bergonzani, I.; Kumar, S.; Kirchner, F. Experimental Investigations into Using Motion Capture State Feedback for Real-Time Control of a Humanoid Robot. *Sensors* **2022**, *22*, 9853. https://doi.org/10.3390/s22249853

Academic Editors: Sašo Blažič, Gregor Klancar and Marija Seder

Received: 31 October 2022
Accepted: 12 December 2022
Published: 15 December 2022

1. Introduction

Humanoid robots are complex systems, both in terms of modeling and control. Bipedal locomotion is particularly difficult due to the instability of the robot in walking phases with double or single ground contacts. Balance is highly dependent on the control approaches employed and the accuracy of the floating base state estimation. The latter is commonly achieved using onboard sensors, which are subject to drift and noise. In contrast, external tracking approaches provide a globally stable estimate of the robot's state, independent of inertial sensor drift and kinematic modeling errors such as leg flexibility.

Marker-based motion capture systems (MoCap) have been used for various robotic applications. One widely explored topic is human motion imitation. Motion data acquisition enables humanoid robots to perform human-like movement sequences such as walking and dancing. For instance, ref. [1] proposes a trajectory generation method for humanoid robots to imitate human walking gaits captured with a marker-based motion capture system. The human movements are adapted to match the kinematic structure, degrees of freedom, and joint limits of the humanoid robot. The work in [2] presents the use of human motion data to generate natural walking and turning motions on the humanoid robot HRP-4C, while considering dynamic balance. Moreover, the work of [3] addresses the topic of human–robot interaction by generating human-like locomotion trajectories for the humanoid robot TALOS. Motion capture is used to compare the computed robot trajectories with previously recorded human walking trajectories and to evaluate which walking pattern generation model is more realistic. Dancing motion generation is a challenging task as well, since it often requires quicker motions than walking. The work in [4] proposes a control approach, which enables the HRP-2 humanoid robot to perform human-like dancing based on motion capture data, while maintaining balance and enforcing actuation limits.

Another important application of motion capture systems is state estimation of mobile robots. Very often, MoCap is used as precise ground truth to determine the position and orientation of the floating base of a humanoid robot. For example, MoCap has been used to evaluate different state estimation approaches based on proprioceptive sensors [5], LiDAR and kinematic-inertial data fusion [6], as well as LiDAR fused with visual-inertial odometry [7]. Less frequently, external motion capture systems have been used to provide state feedback to the control loops of legged robots. The LittleDog quadruped robot [8] uses a set of retroreflective markers placed both on the robot body and legs as well as on the terrain surface to allow analysis of different locomotion strategies without robot perception. The motion capture runs at a low frequency of approximately 100 Hz since the speed of state estimation is not crucial for the balance of the robot with four contact points. The hopping leg Salto-1P [9] executes precise hopping by using motion capture data to estimate the position and orientation of the robot. However, the motion capture has a low frequency of 100 Hz, which is not sufficient for a fast feedback control loop and requires an additional Kalman filter for position estimation and fusion with the gyroscope readings from the onboard IMU for attitude estimation. Moreover, marker-based motion capture has been used to track the position and orientation of the HRP-2 humanoid robot's chest, while performing locomotion and pulling a fire hose, which acts as an external force on the robot's wrist [10]. The tracked robot pose is used as input to the walking pattern generator to correct the orientation drift and improve the robot's balance during locomotion. However, the success rate of the HRP-2 humanoid robot experiments is only 50%, and the motion capture system has a low acquisition frequency of 200 Hz.

In this work, we present an experimental study to demonstrate for the first time that it is possible to employ a high-frequency motion capture system for online stabilization of the humanoid robot RH5 [11]. We use motion capture as an alternative for proprioceptive state estimation. The whole-body controller in [12] is used to stabilize legged motions such as squats and balancing on one leg.

In particular, the paper presents (i) the usage of a high-frequency marker-based motion capture framework for robot floating base tracking, (ii) Whole-Body Control of the humanoid robot with motion capture position feedback, (iii) the experimental validation of the proposed approach with the humanoid robot RH5 performing squats and balancing on one leg and (iv) a comparison with a state estimation approach [13] based on internal sensor measurements, namely IMU, leg kinematics and foot contact sensors. We believe that motion capture can be a viable alternative to state estimation to address edge cases of humanoid locomotion where state estimation is not reliable. As an alternative, the two approaches could potentially be combined.

This paper is organized as follows. In Section 2, we describe the motion capture framework and the Whole-Body Control algorithm. Section 3 presents the experimental results of squatting and single leg balancing of the humanoid robot RH5. In Section 5, we draw the conclusions and propose future research directions.

2. Materials and Methods

First, we briefly describe the humanoid robot RH5 used in practical experiments. Second, we introduce the motion capture system and explain its application for tracking the position and orientation of the robot's floating base. Next, we describe the state estimation approach based on proprioceptive sensors used in this work for comparison with our motion capture system. Finally, we present the Whole-Body Control framework used on the humanoid robot RH5. The interaction between these components is depicted in Figure 1.

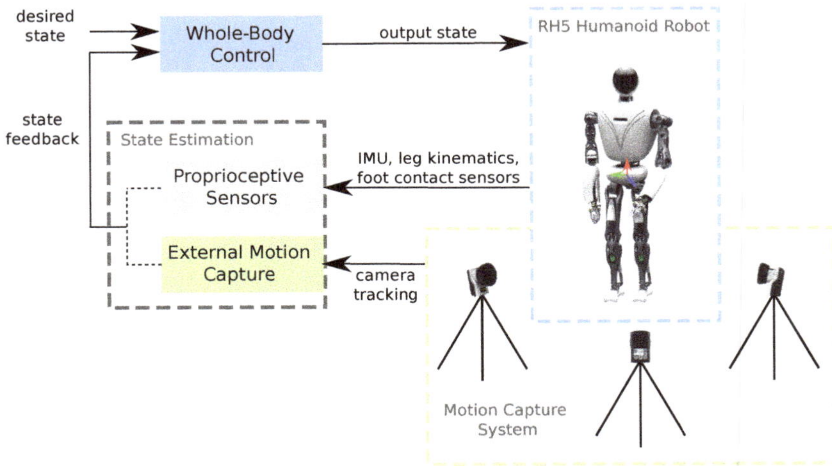

Figure 1. The control architecture of the humanoid robot RH5 includes a Whole-Body Controller that receives feedback from a state estimation module, based on either (i) external motion capture system or (ii) proprioceptive sensors.

2.1. Humanoid Robot RH5

The robot RH5 [11] is a 2 m tall, 62.5 kg humanoid driven by a hybrid combination of serial and parallel actuation modules. For example, the RH5 ankle joints are designed as parallel submechanisms with 2 degrees of freedom (DOF), which are arranged in series with the other leg actuation modules (see [14] for a comprehensive overview). The robot has 34 DOF and is equipped with various sensors, such as an inertial measurement unit, joint encoders, force-torque sensors, foot contact sensors and a stereo camera. For proprioceptive state estimation, we rely on the IMU sensor, joint encoders and foot contact sensors. The IMU model used here is part of the Xsens MTI-300 series of attitude and heading reference systems. The robot's foot contact with the ground consists of 4 contact sensors located at the corners of each foot. Additionally, there is a 6 DOF force/torque sensor on each foot. In the parallel submechanism modules, an absolute encoder is installed in the independent joints and a relative encoder in the linear actuators, ensuring correct forward and inverse geometric mappings.

2.2. Motion Capture System

The motion capture system used for rigid body tracking consists of 3 Oqus 300+ Qualisys cameras connected to a Windows 10 computer. The Qualisys Track Manager software allows tracking and streaming of rigid body poses over an Ethernet connection. We stream rigid body data in real time without further pre-processing steps to the RoCK software framework [15], which runs on the robot's main control PC (Ubuntu 18.04). The rigid body tracking has a frequency of 750 Hz and a variable latency of 2–4 ms.

System calibration is achieved by placing the calibration frame with the desired position and orientation in the cameras' field of view. The calibration process is fast and accurate, with average residual values of less than 0.5 mm. A new system calibration is only required after repositioning the cameras in the workspace, which means that recalibration between successive experiments is not required. The cameras can be placed in any configuration in the room as long as the markers are not occluded.

The motion capture system can be used to track any robotic platform and stream rigid body data in real time over the network. In our work, we use the motion capture system to track the robot {IMU} frame shown in Figure 2. In this way, we can retrieve the pose of the floating base of the humanoid and make a direct comparison with a state estimation approach based on IMU data.

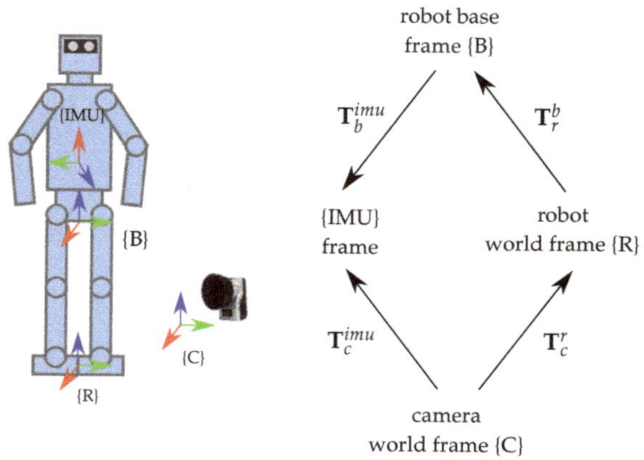

Figure 2. The coordinate frames used for robot floating base tracking are the camera world coordinate frame {C}, robot world coordinate frame {R}, robot base frame {B} and robot {IMU} frame. The corresponding transformation tree is depicted on the right hand side of the figure.

Three markers are required to determine the position and orientation of a rigid body. For redundancy, we place 4 reflective markers on the robot torso as shown in Figure 3. This ensures better tracking performance in case of occlusions and robustness against outliers caused by reflective robot surfaces. We define the tracked IMU rigid body as follows: (i) the origin corresponds to *Marker 1* placed on the center of the IMU sensor, and (ii) the orientation is aligned with the right-handed robot {IMU} frame (x-up, y-right, z-forward).

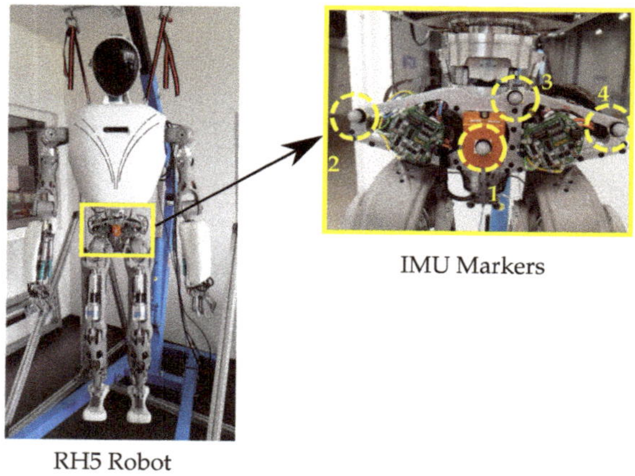

Figure 3. Four reflective markers are placed on the humanoid robot torso in order to track the robot IMU frame with a motion capture system.

Next, we apply a series of transformations to the IMU rigid body pose to (i) convert the tracked IMU pose to robot world coordinates using the camera-to-robot transformation $\mathbf{T}_c^r$ and (ii) obtain the robot's floating base pose $\mathbf{T}_r^b$ in the robot world coordinate system. The camera world frame {C} and robot world frame {R} are shown along with all other relevant transformations in Figure 2.

The camera world frame {C} is defined during the calibration procedure of the motion capture system. The transformation of the robot world frame $\mathbf{T}_r^{b,0}$ is defined as the projection of the base frame at the initial time $t = 0$ onto the ground plane. The z-position component is set to zero and the orientation of the world frame is set to the initial orientation of the base frame. Then, the transformation between camera and robot world frame $\mathbf{T}_c^r$ is computed using the transformation chain rule in Equation (1):

$$\mathbf{T}_c^r = \mathbf{T}_c^{imu,0}\,(\mathbf{T}_b^{imu})^{-1}\,(\mathbf{T}_r^{b,0})^{-1}, \tag{1}$$

where $\mathbf{T}_c^{imu,0}$ is the tracked IMU rigid body pose in camera coordinates at time $t = 0$ and $\mathbf{T}_b^{imu}$ is the fixed IMU frame transformation with respect to the robot base frame.

Finally, we can recover the tracked floating base pose of the robot $\mathbf{T}_r^{b,i}$ in robot world coordinates {R} at time $t = i$. For this purpose, we apply a series of transformations from (i) robot to camera frame $(\mathbf{T}_c^r)^{-1}$, (ii) camera to IMU frame $\mathbf{T}_c^{imu,i}$ at time $t = i$ and (iii) IMU to floating base frame of the robot $(\mathbf{T}_b^{imu})^{-1}$ as shown in Equation (2):

$$\mathbf{T}_r^{b,i} = (\mathbf{T}_c^r)^{-1}\,\mathbf{T}_c^{imu,i}\,(\mathbf{T}_b^{imu})^{-1}. \tag{2}$$

In Equation (2), the time index i denotes transformations of tracked rigid bodies, which are updated at every time step. The other transformations are fixed frames that are constant over time.

2.3. State Estimation

The proprioceptive state estimator uses the invariant extended Kalman filter (InEKF) proposed by [13]. The filter fuses sensor information from IMU, leg odometry and foot contact sensors. The IMU linear acceleration $\boldsymbol{a}_{\text{imu}}$ and angular velocity $\boldsymbol{\omega}_{\text{imu}}$ data are used as input to the prediction step of the InEKF. The update step is performed based on leg kinematics $\boldsymbol{q}, \dot{\boldsymbol{q}}$ and foot contact information $\boldsymbol{f}_{\text{ext}}$ as shown in Figure 4.

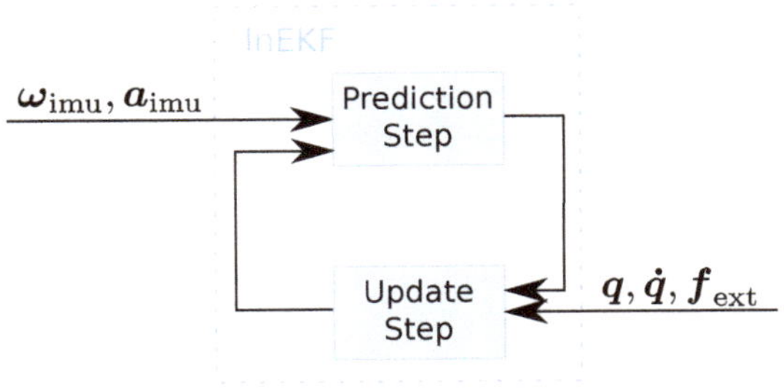

Figure 4. Prediction and update blocks of the InEKF proprioceptive state estimator.

The system state $\boldsymbol{X} \in \mathbb{R}^{(n+5)\times(n+5)}$ estimated by the InEKF is defined in Equation (3):

$$\boldsymbol{X} = \begin{bmatrix} \boldsymbol{R} & \boldsymbol{v} & \boldsymbol{p} & \boldsymbol{p}_{C_1} & \cdots & \boldsymbol{p}_{C_n} \\ \mathbf{0}_{1,3} & 1 & 0 & 0 & \cdots & 0 \\ \mathbf{0}_{1,3} & 0 & 1 & 0 & \cdots & 0 \\ \vdots & \vdots & \vdots & \vdots & \ddots & \vdots \\ \mathbf{0}_{1,3} & 0 & 0 & 0 & \cdots & 1 \end{bmatrix}_{(n+5),\,(n+5)}, \tag{3}$$

where $\boldsymbol{R} \in \mathbb{R}^{3\times3}, \boldsymbol{v} \in \mathbb{R}^3$ and $\boldsymbol{p} \in \mathbb{R}^3$ represent the orientation, velocity and position of the robot's floating base, and $\boldsymbol{p}_{C_i} \in \mathbb{R}^3$ represents the position of the n foot contact points.

In contrast to the standard EKF, the InEKF [16] takes advantage of Lie Group theory [17]. Lie Groups are collections of object symmetries, for instance, the collection of rotation matrices of a 3D object in space, known as the Special Orthogonal Group $SO(3)$.

Instead of using Jacobians to linearize the system, the InEKF operates on a linear vector space, namely the Lie algebra g of a given Lie Group $\mathcal{G}$. The Lie algebra is defined as the tangent to the Lie Group manifold at the identity element. The mapping from the Lie algebra to the Lie Group is given by the exponential map exp in Equation (4), while the reverse mapping is provided by the logarithmic map log from Equation (5):

$$exp : g \to \mathcal{G}; \hat{\tau} \mapsto \boldsymbol{X} = exp(\hat{\tau}) \tag{4}$$

$$log : \mathcal{G} \to g; \boldsymbol{X} \mapsto \hat{\tau} = log(\boldsymbol{X}), \tag{5}$$

where $\hat{\tau}$ is the estimated state on the Lie algebra and the $\boldsymbol{X}$ is the state represented on the matrix Lie Group manifold.

In the InEKF, the exponential map is used to update the state estimate and determine the exact error on the Lie Group manifold. The filter has strong convergence provided by the invariance properties of the Lie Group and allows linearization independent of the current system state. However, when the IMU accelerometer and gyroscope biases are added to the state matrix, it loses the group affine property required for a matrix Lie Group. This leads to an "Imperfect InEKF", and the estimation error cannot be exactly retrieved anymore. Moreover, the filter still suffers from inertial drift, yaw unobservability and uncertainties in forwards kinematics due to leg flexibility.

These shortcomings of the proprioceptive state estimator may hinder the execution of complex and dynamic motions required for bipedal locomotion and affect the robot stability during single leg support phases and in challenging environments. Hence, state feedback through motion capture is proposed as an alternative for developing and testing new controllers for the humanoid robot RH5.

2.4. Whole-Body Control

To stabilize the robot during motions such as squatting and balancing, we use a velocity-based Whole-Body Control (WBC) framework (https://github.com/ARC-OPT/wbc, accessed on 11 December 2022) [12], which solves the following instantaneous quadratic program (QP):

$$\begin{aligned} \min_{\dot{\mathbf{q}}} \quad & \| \textstyle\sum_i \mathbf{w}_i (\mathbf{J}^i \dot{\mathbf{q}} - \mathbf{v}_d^i) \|_2 \\ \text{s.t.} \quad & \mathbf{J}_c^j \dot{\mathbf{q}} = 0, \quad j = 1, \ldots, K \\ & \dot{\mathbf{q}}_m \leq \dot{\mathbf{q}} \leq \dot{\mathbf{q}}_M, \end{aligned} \tag{6}$$

where $\dot{\mathbf{q}} \in \mathbb{R}^{6+n}$ are the robot joint velocities, including 6-DOF virtual floating base, n number of robot joints, $\mathbf{v}_d^i \in \mathbb{R}^6$ is the desired spatial velocity for the i-th task, $\mathbf{J}^i \in \mathbb{R}^{6\times(6+n)}$ is the associated task Jacobian and $\mathbf{w_i} \in \mathbb{R}^6$ the task weights that control the priority of a task. The QP is subject to joint velocity limits $\dot{\mathbf{q}}_m, \dot{\mathbf{q}}_M \in \mathbb{R}^{6+n}$, as well as K rigid contact constraints, where $\mathbf{J}_c^j \in \mathbb{R}^{6\times(6+n)}$ is the contact Jacobian of the j-th contact point. The QP in Equation (6) can be solved using any standard QP solver, e.g., [18]. Robot tasks are specified by providing trajectories for $\mathbf{v}_d^i$, for instance, by means of feedback controllers designed around the QP in Equation (6). For full pose control, this can be achieved as follows:

$$\mathbf{v}_d = \mathbf{v}_r + \mathbf{K}_p \begin{pmatrix} \mathbf{x}_r - \mathbf{x} \\ \theta \hat{\omega}_r^a \end{pmatrix}, \tag{7}$$

where $\mathbf{v}_r \in \mathbb{R}^6$ is the feed forward spatial velocity, $\mathbf{K}_p \in \mathbb{R}^{6\times 6}$ the feedback gain matrix, $\mathbf{x}_r, \mathbf{x} \in \mathbb{R}^3$ the reference and actual position of the robot and $\theta\hat{\omega}_r^a \in \mathbb{R}^3$ the difference in orientation between actual and reference pose using a singularity-free representation [19].

The solution of Equation (6) is fed into an inverse dynamics solver [20] as shown in Figure 5, which not only computes the joint torques $\boldsymbol{\tau} \in \mathbb{R}^n$ for the entire robot, but also projects $\mathbf{q}, \dot{\mathbf{q}}, \boldsymbol{\tau}$ into the actuator space of the system, including all parallel kinematic mechanisms (PKM) of RH5. This avoids the usual mechanism-specific transformation of the solution to the actuators of each PKM, e.g., to the linear actuators in the RH5 ankle mechanism. As a result, we obtain the reference actuator positions, velocities and forces/torques $\mathbf{u}, \dot{\mathbf{u}}, \boldsymbol{\tau}_{u,r} \in \mathbb{R}^p$, where p is the number of actuators in the robot. On actuator level, these are stabilized using a PD position controller with force/torque feed forward, which runs at 1 kHz, as shown in Equation (8):

$$\boldsymbol{\tau}_{u,d} = \boldsymbol{\tau}_{u,r} + \mathbf{K}_{d,u}(\dot{\mathbf{u}}_r - \dot{\mathbf{u}}) + \mathbf{K}_{p,u}(\mathbf{u}_r - \mathbf{u}), \tag{8}$$

where $\mathbf{K}_{d,u}, \mathbf{K}_{p,u} \in \mathbb{R}^{p \times p}$ are again diagonal feedback gain matrices.

The controller in Equation (7) is used to control the center of mass (CoM) and the orientation of the upper body of the humanoid robot, where the state feedback $\mathbf{x}_{bf}$ of the robot's floating base is provided by either (i) the external motion capture framework described in Section 2.2 or by using (ii) an internal state estimation approach as described in Section 2.3.

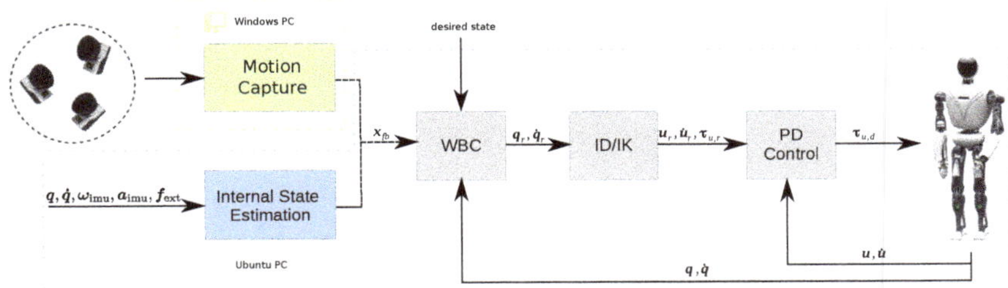

Figure 5. Control architecture to stabilize the desired robot motions.

3. Results

In this section, we present experimental results obtained with the humanoid robot RH5. The robot is supposed to perform two different whole-body motions, namely squatting and balancing on one leg (please find the Supplementary Video S1 under https://www.youtube.com/watch?v=bqiBvVHf2i0, accessed on 11 December 2022). The two sets of experiments are performed using the whole-body controller in Section 2.4 with state feedback from motion capture system and proprioceptive state estimation, respectively. In the squatting experiments, tracking of the vertical motion of the robot is evaluated at two execution speeds of 10 and 16 s per squat. In the single leg balancing experiments, CoM and foot tracking are evaluated when the robot raises one leg at two different heights of 10 cm and 15 cm, respectively. At the beginning of each experiment, the robot is placed in its initial joint configuration and stands freely on the floor with both feet in contact with the ground. For safety reasons, a movable cord is attached to the robot's torso, which is secured by a crane and neither restricts the robot's movement nor affects its stability.

3.1. Squatting Experiment

In the squatting experiments, the floating base of the robot performs a translation along the vertical z-axis with a height difference of 14 cm, as shown in Figure 6. Using the whole-body controller described in Section 2.4, we constrain the feet to be in contact with the ground during motion execution. In addition, we split the squatting motion into two control tasks, namely the root task and the CoM task, as shown in Table 1. The root task is used to constrain the floating base of the robot to follow the desired vertical motion on the z-axis and minimize the lateral motion on the y-axis. The position of the floating base on the x-axis is not constrained to allow the root frame to move forward and backward if necessary, similar to the human squat. Furthermore, the CoM task is used to keep the ground projection of the robot CoM centered in the support polygon to enforce balance.

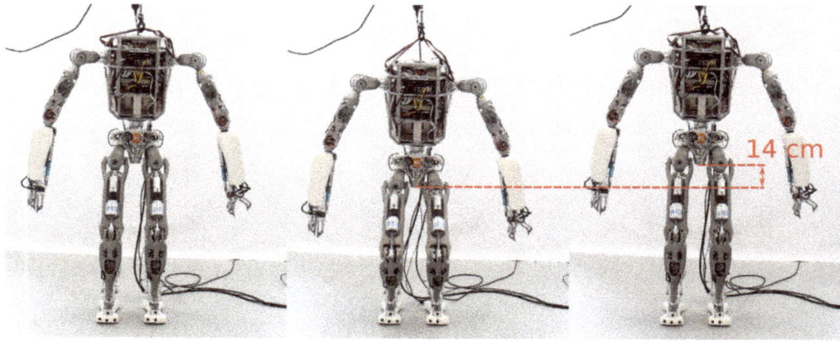

Figure 6. Time lapse of the humanoid robot RH5 performing squats with an amplitude of 14 cm.

We performed two experiments with five squat repetitions each. The execution speed was constant in both scenarios, while the time interval for a squat varies, namely (i) experiment S1 with one squat per 16 s and (ii) experiment S2 with one squat per 10 s. In this way, we can evaluate the squat movement with and without a stabilization break between the movement direction changes. Both sets of experiments were successfully conducted using state feedback from (a) external motion tracking and (b) proprioceptive state estimation.

Table 1. Task weights used in the whole-body controller for squatting and single leg balancing.

Experiment	Task	Weights					
		x	y	z	θ_x	θ_y	θ_z
Squatting	CoM	6	6	0	-	-	-
	Root	0	1	1	1	1	1
Balancing	CoM	6	6	1	-	-	-
	Feet	1	1	1	1	1	1
	Wrists	1	1	0	0	0	0

Due to the initial yaw angle unobservability of the proprioceptive state estimator, the reference frame of the estimator is arbitrarily rotated around the z-axis. We account for this rotation by adjusting the setpoints accordingly. To obtain a rotation-invariant error and compare the two state feedback approaches, i.e., MoCap and proprioceptive state estimation, we compute the 3D Euclidean space root-mean-square error (RMSE) between the desired and measured robot position, as shown in Equation (9):

$$\mathcal{E}_c = \sqrt{\mathcal{E}_{c,x}^2 + \mathcal{E}_{c,y}^2 + \mathcal{E}_{c,z}^2}\,, \tag{9}$$

where $\mathcal{E}_{c,x}$ and $\mathcal{E}_{c,y}$ represent the CoM tracking error on the x and y-axis, $\mathcal{E}_{c,z}$ represents the floating base tracking error on the z-axis, and $\mathcal{E}_c$ represents the RMSE tracking error in Euclidean space. Since the calibration residuals of the motion capture system are less than or equal to ± 0.5 mm, we evaluate the experimental data to an accuracy of 1 mm.

The tracking results of experiment S1 are shown in Figure 7. We observe slightly better stability using external motion capture feedback compared to proprioceptive state estimation. As shown in Table 2, proprioceptive feedback is responsible for larger errors when the time interval between squats decreases in experiment S2. This shows that state estimation becomes less accurate during fast movements, whereas performance does not change significantly during squatting with external motion capture feedback.

Table 2. Tracking error of the robot CoM position $\mathcal{E}_c$ along the three axes and foot position on the z-axis ($\mathcal{E}_p$) during the squatting and single leg balancing experiments. The highlighted values represent the smallest CoM and foot position tracking errors for every set of experiments.

Experiment	State Feedback	RMSE [m]				
		$\mathcal{E}_c$	$\mathcal{E}_{c,x}$	$\mathcal{E}_{c,y}$	$\mathcal{E}_{c,z}$	$\mathcal{E}_p$
S1 (16 s)	MoCap Tracking	**0.004**	0.004	0.001	0.001	-
	State Estimation	0.008	0.007	0.004	0.001	-
S2 (10 s)	MoCap Tracking	**0.004**	0.004	0.001	0.002	-
	State Estimation	0.027	0.010	0.025	0.001	-
B1 (10 cm)	MoCap Tracking	**0.025**	0.002	0.025	0.001	0.006
	State Estimation	0.026	0.018	0.018	0.001	**0.002**
B2 (15 cm)	MoCap Tracking	**0.023**	0.002	0.023	0.001	0.008
	State Estimation	0.026	0.017	0.019	0.001	0.008

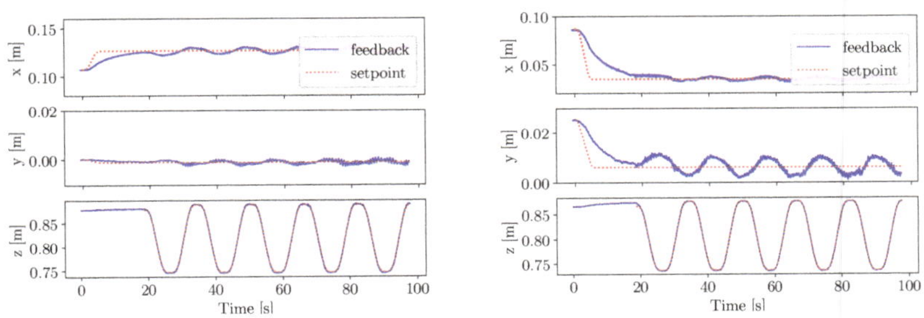

(**a**) (**b**)

Figure 7. Squatting experiments S1, where the robot CoM position on x and y-axis and the floating base position on the z-axis are tracked by the whole-body controller using (**a**) motion capture state feedback and (**b**) proprioceptive state estimation. (**a**) Squatting motion using external motion capture state feedback with the respective RMSE as follows: $\mathcal{E}_{c,x} = 0.004$, $\mathcal{E}_{c,y} = 0.001$ and $\mathcal{E}_{c,z} = 0.001$. (**b**) Squatting motion using proprioceptive state estimation feedback with the respective RMSE as follows: $\mathcal{E}_{c,x} = 0.007$, $\mathcal{E}_{c,y} = 0.004$ and $\mathcal{E}_{c,z} = 0.001$.

3.2. One Leg Balancing Experiment

During the balancing experiments, the robot starts with both feet in contact with the ground. From the double leg support phase, the robot switches to the single leg support phase by shifting the CoM to the right foot and raising the left foot to a 15 cm height, as shown in Figure 8.

We achieved this behavior using Whole-Body Control, with a Cartesian task for raising the left foot and a CoM task for constraining the position of the robot's center of mass. To achieve human-like motion, both wrists are constrained by Cartesian tasks to keep them in front of the torso. Moreover, the contact constraint of the left foot is dynamically disabled during the lift-off phase and re-enabled during touchdown.

The setpoints for the tasks are generated using a trajectory interpolator and executed at joint level using the PD position controller in [20]. To enforce static balance of the robot, larger weights have been chosen for the x and y-axes of the CoM position with respect to the CoM vertical axis, as shown in Table 1.

We successfully performed experiments on balancing on one leg by tracking the robot's floating base using (i) a motion capture system and (ii) proprioceptive pose estimation. We defined two scenarios, namely experiment B1 and B2, in which the vertical setpoint of the left foot reaches a height of 10 cm and 15 cm, respectively. In both experiments, the center of mass has been lowered by 4 cm on the z-axis to increase stability.

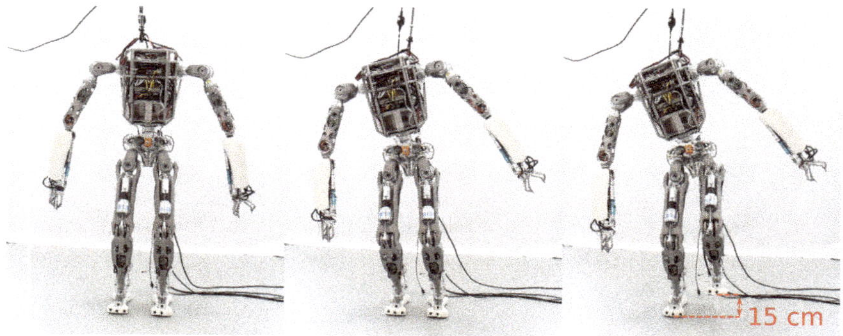

Figure 8. Time lapse of the humanoid robot RH5 balancing on the right leg, while raising the left leg at 15 cm above the ground.

The results of experiment B2 are shown in Figure 9. We notice larger oscillations on the x and y-axis when using proprioceptive state estimation feedback, as opposed to external motion capture feedback. In both balancing experiments with motion capture feedback, we observe stable single leg balancing, as summarized in Table 2.

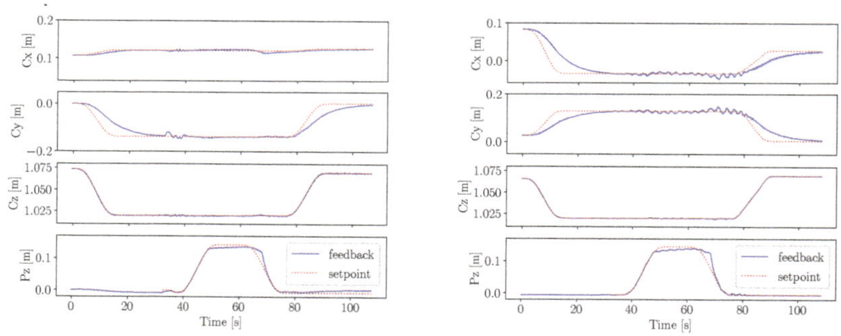

(**a**) (**b**)

Figure 9. Single leg balancing experiments B2, where the robot CoM position C_x, C_y, and C_z on the x, y and z-axis, respectively, as well as the foot position P_z on the z-axis are tracked by the whole-body controller using (**a**) motion capture state feedback and (**b**) proprioceptive state estimation. (**a**) One leg balancing using external motion capture state feedback with the respective RMSE as follows: $\mathcal{E}_{c,x} = 0.002$, $\mathcal{E}_{c,y} = 0.023$, $\mathcal{E}_{c,z} = 0.001$ and $\mathcal{E}_p = 0.008$. (**b**) One leg balancing using proprioceptive state estimation feedback with the respective RMSE as follows: $\mathcal{E}_{c,x} = 0.017$, $\mathcal{E}_{c,y} = 0.019$, $\mathcal{E}_{c,z} = 0.001$ and $\mathcal{E}_p = 0.008$.

4. Discussion

The experiments with the RH5 humanoid on squatting and single leg balancing compare two approaches for providing state feedback for Whole-Body Control, namely an external motion capture system and proprioceptive state estimation.

Proprioceptive state estimation provides fast state estimates relying only on proprioceptive sensors such as IMU, position readings from the joints and contact sensors. However, it suffers from yaw unobservability, and we apply an additional transformation to the desired COM trajectory to cope with the initial yaw estimation error. Moreover, the results show that both squatting and single leg balancing motions with proprioceptive state estimation feedback suffer from oscillations when the speed or complexity of the motion increases. This might be caused by the uncertainties in the "Imperfect InEFK" estimation, since the IMU biases from the state vector do not satisfy the matrix Lie group properties.

In contrast, the external motion capture system consists of cameras tracking reflective markers on the robot's IMU frame. The employed motion capture system is able to provide accurate and fast state feedback to the whole-body controller with minimal setup and calibration efforts. The results show that external motion capture feedback contributes to more stable motions in the squatting and single leg balancing experiments. Due to its robustness and suitability for high-frequency closed-loop control, this method could enable the robot to execute more complex motions in the future, such as walking, climbing stairs and multi-contact tasks. Thus, external motion capture feedback can contribute to the development and testing of robot capabilities and Whole-Body Control algorithms.

5. Conclusions

Floating base state estimation plays a key role in bipedal locomotion of a humanoid robot since state estimation errors can affect the robot's balance in double or single leg support phases. In this work, we show investigations on the use of external motion capture feedback for humanoid robot control and compare it with a state-of-the-art proprioceptive state estimation method. We perform two different whole-body motions with the humanoid robot RH5, namely squatting and single leg balancing and track the robot's floating base using external cameras. We demonstrate that high-frequency external motion capture feedback can be reliably used for Whole-Body Control of humanoid robots and shows better stability than proprioceptive sensing, which is subject to noise and drift. As possible applications, external motion capture could be used both in industrial workspaces such as factories and in research laboratories in parallel with the development of better proprioceptive state estimation approaches to improve Whole-Body Control algorithms and explore the capabilities of humanoid robots.

In future work, we consider addressing possible MoCap errors such as outlier rejection and marker placement in order to increase performance. Moreover, fusion of proprioceptive state estimation and real-time motion capture data could reduce the state estimation drift and enable the robot to perform robust bipedal locomotion or other multi-contact tasks.

Supplementary Materials: The following supporting information can be accessed at: https://www.youtube.com/watch?v=bqiBvVHf2i0, Video S1: Experimental Investigations into Using Motion Capture State Feedback for Real-Time Control of a Humanoid Robot.

Author Contributions: Conceptualization, S.K., F.K., D.M. and M.P.; methodology, D.M., S.K., M.P. and I.B.; software, S.K., M.P., D.M. and I.B.; validation, S.K. and D.M.; formal analysis, M.P. and I.B.; investigation, M.P., I.B. and D.M.; resources, S.K. and F.K.; data curation, M.P. and I.B.; writing—original draft preparation, M.P., S.K. and D.M.; writing—review and editing, F.K. and S.K.; visualization, M.P., I.B. and D.M.; supervision, F.K. and S.K.; project administration, S.K., M.P and F.K.; funding acquisition, F.K. and S.K. All authors have read and agreed to the published version of the manuscript.

Funding: This research was done in the M-RoCK project (FKZ 01IW21002) and VeryHuman project (FKZ 01IW20004) funded by the Federal Ministry of Education and Research (BMBF) and is additionally supported with project funds from the federal state of Bremen for setting up the Underactuated Robotics Lab (Grant Number: 201-001-10-3/2021-3-2).

Institutional Review Board Statement: Not applicable.

Informed Consent Statement: Not applicable.

Data Availability Statement: Not applicable.

Conflicts of Interest: The authors declare no conflict of interest.

References

1. Boutin, L.; Eon, A.; Zeghloul, S.; Lacouture, P. An auto-adaptable algorithm to generate human-like locomotion for different humanoid robots based on motion capture data. In Proceedings of the 2010 IEEE/RSJ International Conference on Intelligent Robots and Systems, Taipei, Taiwan, 18–22 October 2010; pp. 1256–1261. [CrossRef]
2. Miura, K.; Morisawa, M.; Nakaoka, S.; Kanehiro, F.; Harada, K.; Kaneko, K.; Kajita, S. Robot motion remix based on motion capture data towards human-like locomotion of humanoid robots. In Proceedings of the 2009 9th IEEE-RAS International Conference on Humanoid Robots, Paris, France, 7–10 December 2009; pp. 596–603. [CrossRef]
3. Maroger, I.; Stasse, O.; Watier, B. Walking Human Trajectory Models and Their Application to Humanoid Robot Locomotion. In Proceedings of the 2020 IEEE/RSJ International Conference on Intelligent Robots and Systems (IROS), Las Vegas, NV, USA, 25–29 October 2020; pp. 3465–3472. [CrossRef]
4. Ramos Ponce, O.E.; Mansard, N.; Stasse, O.; Benazeth, C.; Hak, S.; Saab, L. Dancing Humanoid Robots: Systematic use of OSID to Compute Dynamically Consistent Movements Following a Motion Capture Pattern. *IEEE Robot. Autom. Mag.* **2015**, *22*, 16–26. [CrossRef]
5. Ramadoss, P.; Romualdi, G.; Dafarra, S.; Chavez, F.J.A.; Traversaro, S.; Pucci, D. DILIGENT-KIO: A Proprioceptive Base Estimator for Humanoid Robots using Extended Kalman Filtering on Matrix Lie Groups. *arXiv* **2021**, arXiv:2105.14914.
6. Sushrutha Raghavan, V.; Kanoulas, D.; Zhou, C.; Caldwell, D.G.; Tsagarakis, N.G. A Study on Low-Drift State Estimation for Humanoid Locomotion, Using LiDAR and Kinematic-Inertial Data Fusion. In Proceedings of the 2018 IEEE-RAS 18th International Conference on Humanoid Robots (Humanoids), Beijing, China, 6–9 November 2018; pp. 1–8. [CrossRef]
7. Lasguignes, T.; Maroger, I.; Fallon, M.; Ramezani, M.; Marchionni, L.; Stasse, O.; Mansard, N.; Watier, B. ICP Localization and Walking Experiments on a TALOS Humanoid Robot. In Proceedings of the 2021 20th International Conference on Advanced Robotics (ICAR), Ljubljana, Slovenia, 6–10 December 2021; pp. 800–805. [CrossRef]
8. Murphy, M.P.; Saunders, A.; Moreira, C.; Rizzi, A.A.; Raibert, M. The LittleDog robot. *Int. J. Robot. Res.* **2011**, *30*, 145–149. [CrossRef]
9. Yim, J. Hopping Control and Estimation for a High-Performance Monopedal Robot, Salto-1P. Ph.D. Thesis, UC Berkeley, Berkeley, CA, USA, 2020.
10. Ramirez-Alpizar, I.G.; Naveau, M.; Benazeth, C.; Stasse, O.; Laumond, J.P.; Harada, K.; Yoshida, E. Motion Generation for Pulling a Fire Hose by a Humanoid Robot. In Proceedings of the 16th IEEE-RAS International Conference on Humanoid Robotics (HUMANOIDS 2016), Cancun, Mexico, 15–17 November 2016.
11. Eßer, J.; Kumar, S.; Peters, H.; Bargsten, V.; Fernandez, J.d.G.; Mastalli, C.; Stasse, O.; Kirchner, F. Design, analysis and control of the series-parallel hybrid RH5 humanoid robot. In Proceedings of the 2020 IEEE-RAS 20th International Conference on Humanoid Robots (Humanoids), Munich, Germany, 19–21 July 2021; pp. 400–407. [CrossRef]
12. Mronga, D.; Kumar, S.; Kirchner, F. Whole-Body Control of Series-Parallel Hybrid Robots. In Proceedings of the 2022 International Conference on Robotics and Automation (ICRA), Philadelphia, PA, USA, 23–27 May 2022; IEEE: Piscataway Township, MJ, USA, 2022; pp. 228–234. [CrossRef]
13. Hartley, R.; Jadidi, M.G.; Grizzle, J.; Eustice, R.M. Contact-Aided Invariant Extended Kalman Filtering for Legged Robot State Estimation. In Proceedings of the Robotics: Science and Systems XIV. Robotics: Science and Systems Foundation, Pennsylvania, PA, USA, 26–30 June 2018. [CrossRef]
14. Kumar, S.; Wöhrle, H.; de Gea Fernández, J.; Müller, A.; Kirchner, F. A survey on modularity and distributivity in series-parallel hybrid robots. *Mechatronics* **2020**, *68*, 102367. .: 10.1016/j.mechatronics.2020.102367. [CrossRef]
15. ROCK, the Robot Construction Kit. Available online: http://www.rock-robotics.org (accessed on 11 December 2022).
16. Barrau, A.; Bonnabel, S. The Invariant extended Kalman filter as a stable observer. *IEEE Trans. Autom. Control* **2017**, *62*, 1797–1812. [CrossRef]
17. Solà, J.; Deray, J.; Atchuthan, D. A micro Lie theory for state estimation in robotics. *arXiv* **2018**, arXiv:1812.01537.
18. Ferreau, H.; Kirches, C.; Potschka, A.; Bock, H.; Diehl, M. qpOASES: A parametric active-set algorithm for quadratic programming. *Math. Program. Comput.* **2014**, *6*, 327–363. [CrossRef]
19. Lynch, K.; Park, F. *Modern Robotics*; Cambridge University Press: Cambridge, MA, USA, 2017; pp. 85–89.
20. Kumar, S.; Szadkowski, K.A.v.; Mueller, A.; Kirchner, F. An Analytical and Modular Software Workbench for Solving Kinematics and Dynamics of Series-Parallel Hybrid Robots. *J. Mech. Robot.* **2020**, *12*. 021114. [CrossRef]

sensors

MDPI

Article

Indoor 2D Positioning Method for Mobile Robots Based on the Fusion of RSSI and Magnetometer Fingerprints

Peter Sarcevic [1,*], Dominik Csik [1,2] and Akos Odry [1]

1 Department of Mechatronics and Automation, Faculty of Engineering, University of Szeged, Moszkvai krt. 9, 6725 Szeged, Hungary
2 Doctoral School of Applied Informatics and Applied Mathematics, Óbuda University, Bécsi str. 96/b, 1034 Budapest, Hungary
* Correspondence: sarcevic@mk.u-szeged.hu

Abstract: Received signal strength indicator (RSSI)-based fingerprinting is a widely used technique for indoor localization, but these methods suffer from high error rates due to various reflections, interferences, and noises. The use of disturbances in the magnetic field in indoor localization methods has gained increasing attention in recent years, since this technology provides stable measurements with low random fluctuations. In this paper, a novel fingerprinting-based indoor 2D positioning method, which utilizes the fusion of RSSI and magnetometer measurements, is proposed for mobile robots. The method applies multilayer perceptron (MLP) feedforward neural networks to determine the 2D position, based on both the magnetometer data and the RSSI values measured between the mobile unit and anchor nodes. The magnetic field strength is measured on the mobile node, and it provides information about the disturbance levels in the given position. The proposed method is validated using data collected in two realistic indoor scenarios with multiple static objects. The magnetic field measurements are examined in three different combinations, i.e., the measurements of the three sensor axes are tested together, the magnetic field magnitude is used alone, and the Z-axis-based measurements are used together with the magnitude in the X-Y plane. The obtained results show that significant improvement can be achieved by fusing the two data types in scenarios where the magnetic field has high variance. The achieved results show that the improvement can be above 35% compared to results obtained by utilizing only RSSI or magnetic sensor data.

Keywords: position estimation; indoor positioning; localization; mobile robots; sensor fusion; RSSI; magnetometer; fingerprint

Citation: Sarcevic, P.; Csik, D.; Odry, A. Indoor 2D Positioning Method for Mobile Robots Based on the Fusion of RSSI and Magnetometer Fingerprints. *Sensors* **2023**, *23*, 1855. https://doi.org/10.3390/s23041855

Academic Editors: Gregor Klancar, Marija Seder and Sašo Blažič

Received: 19 December 2022
Revised: 31 January 2023
Accepted: 2 February 2023
Published: 7 February 2023

1. Introduction

Indoor positioning techniques are utilized in a large variety of applications, such as emergency management [1], smart energy management [2], heating, ventilation, and air conditioning (HVAC) control systems [3], occupancy detection [4], and industrial monitoring [5]. They can be used to determine the position of humans, mobile robots, and objects, etc. In mobile robot applications, the localization problem is the first critical task which needs to be solved in control algorithms, since it directly influences their success. Localization provides the robot pose estimate using a sensor fusion framework, where generally relative and absolute poses are fused with probabilistic approaches. The accurate estimation of the absolute position plays a crucial role in these sensor fusion-based localization methods. In an indoor environment, the GPS cannot provide reliable measurement data, so other technologies have to be considered. Various technologies can be used for this task, such as cameras, LiDAR [6], and radio communication modules [7], etc. The fusion of these technologies was also widely used in related research [8].

In wireless sensor networks (WSN), wireless signal-based techniques are widely used in indoor positioning [9]. These techniques can be based on different types of extracted parameters, such as time of arrival (ToA), time of flight (ToF), angle of arrival (AoA), time difference of flight (TDoF), time difference of arrival (TDoA), received signal strength indicator (RSSI), and channel state information (CSI) [10]. The used methods can be divided into two major groups: geometric and fingerprint-based methods [10]. Geometric approaches include trilateration, multilateration, and triangulation methods. Various measurement parameters can be used in the case of these methods, such as ToA, ToF, or AoA.

The most often applied data type is the RSSI, which can be read from the transceiver modules. In the case of both method families, the position is estimated based on RSSI measurements between multiple anchor nodes with a known position and a unit with an unknown position. In geometric-based algorithms, the RSSI value is converted to distance based on an appropriate model, and then the position is estimated using trilateration [11]. In the case of fingerprint-based methods, measurements are collected in a room at fixed points with a given resolution, then a pattern recognition or pattern-matching algorithm (e.g., artificial neural networks (ANN), k-nearest neighbor (k-NN), etc. [12]) is trained using the measurement data, which is later utilized to determine the position of the unknown point [10,13,14]. Deep learning-based methods also offer an alternative solution to the problem [15,16], but they require much more computational and memory capacity than other methods. This can affect the real-time operation of the system, which is particularly important in cases of their use on mobile robots, where the embedded system must perform other tasks also. RSSI-based technologies are affected by various reflections, interferences and noises, which can influence their localization performance significantly, especially in indoor applications [17]. Some methods even utilize the trilateration and fingerprint-based results together in the fusion algorithm to improve the performance [18], while other methods fuse fingerprinting and time-based techniques [19].

In an outdoor environment, magnetic sensors are mainly utilized as compasses, but indoors they are almost useless in such form, since there are several hard iron and soft iron sources that affect the measurements [20]. This geomagnetism-based technique has attracted considerable attention in recent years [21,22], and has been shown to be a promising technology for indoor localization, since the field is more stable with much lower random fluctuation as compared to other signals [23,24]. Each building has its own unique ambient magnetic field, and if these local anomalies have sufficient variability, then they can be utilized during indoor localization [25].

Various approaches were previously proposed for indoor localization using geomagnetism, but these mainly consider human localization with smartphone-based systems [21,26]. The proposed methods are usually based on a sequence of geomagnetic data, which are generally fused with other data, such as WiFi [23,27–30], CSI [31], inertial measurement units (IMU) [32], pedestrian dead reckoning (PDR) [28,30,33], or pedometers [34]. In [33], a fusion algorithm was proposed which combined PDR and matching in a magnetic fingerprint map. Some works utilized only the sequence of magnetic field measurements with pattern recognition techniques [35–39]. In [24], the authors utilized the fingerprints of the changes in the raw magnetic field during indoor localization, but the performance can be largely affected by local distortions. Subbu et al. proposed an indoor localization solution by classifying signatures based on their patterns, using a dynamic time warping (DTW)-based approach [35]. The evaluation was performed using fingerprints containing simulated signatures of different ferromagnetic objects. In [36], an indoor positioning technique was proposed, which utilizes the differences in 2D magnetic field measurements that are collected over a distance with a pattern-matching algorithm. A 3D accelerometer was used to find the vertical direction. The authors assumed the following: the attitude variation of the smartphone is small; the change in the user velocity occurs only locally and sparsely; and the indoor environment is considered with no significantly varying magnetic field. In [37], a 2D geomagnetic fingerprint-based method was proposed, which uses point-to-point fingerprint matching. The method does not require an indoor map since it generates an indoor geomagnetic map, including all the turning points and connecting links with a crowdsourcing data collection module. The method also includes step and turning detection based on inertial sensors. The results based on experiments in a five-floor office building showed an average error smaller than 1.5 m. Ashraf et al. used geomagnetic field patterns with convolutional neural networks to perform indoor localization [38]. The performance was evaluated using measurements in two buildings with different experimental environments and path geometry. The results demonstrated that the users could be localized within 1.01 m at 75%. In [39], the magnetic field positioning performance of different machine learning methods was tested using pre-processed raw magnetic data. The test area was a corridor, where magnetic fingerprints were captured for 30 points.

The magnetic sensor-based technology was also used for the localization of mobile robots. In [25], global self-localization was considered for a mobile robot in one dimension using only a three-axis magnetometer. The Monte Carlo localization (MCL)-based technique was tested in experiments conducted in the corridors of four buildings. Both the three-dimensional measurement vector and the magnitude were tested in the method. Lee et al. proposed an indoor positioning system that recognizes magnetic sequence patterns by using a deep neural network [40]. The location was estimated by detecting the patterns of landmarks, using features extracted from the magnetic sequences. The achieved accuracy was 0.8 m in a corridor and 2.3 m in an atrium. In [41], magnetic fingerprints collected using a vision-based motion capture system were used together with odometry data in the localization framework. A particle filter-based method was proposed and the reported average errors achieved were 9 cm. The authors of [42] proposed a deep neural network-based method, which utilizes magnetometer measurement sequences, and applied it for both human and mobile robot localization. The method was tested with two public datasets and provided a localization error of around 1 m for both setups.

The disturbances in the magnetic field measured at different points can be used to form a fingerprint equivalent which is used in RSSI-based methods and carries additional information that can improve the efficiency of the RSSI-based fingerprinting. This study deals with indoor positioning using RSSI and magnetometer data together in a fingerprinting-based system, and the evaluation of the improvement provided by the proposed data fusion approaches. The contributions of this work can be summarized as follows:

- A novel 2D fingerprint-based positioning method is proposed, which fuses both the RSSI and magnetometer fingerprints using multilayer perceptron (MLP) neural networks. The method utilizes fingerprints measured in one plane near to the ground, since the goal is to provide absolute position information for the sensor fusion framework of a mobile robot. To the authors' best knowledge, no such method was previously proposed.
- The proposed method is validated using measurements collected in two different indoor scenarios. Both scenarios are realistic since they include static obstacles.
- Three different combinations of magnetometer data are tested in the study. In the first version, the magnetometer measurements of the three sensor axes are tested. The second version utilizes the Z-axis measurements together with the magnetic field magnitudes in the X-Y plane, while the 3D magnitudes are utilized in the third version.
- The results obtained using the different fusion versions are compared with the results provided by utilizing only RSSI or magnetometer data to examine the improvement caused by the fusion of two data types.

The rest of the paper is organized as follows. Section 2 presents the proposed fingerprinting-based method. The applied measurement data are presented in Section 3. The experimental results are discussed in Section 4, while Section 5 summarizes the results of the paper.

2. Position Estimation Using the Fusion of RSSI and Magnetic Fingerprints

The aim of the proposed positioning method is to fuse RSSI and magnetometer data, thereby improving the overall positioning performance.

2.1. RSSI-Based Positioning

RSSI is the calculation of real signal power received by a receiver and is typically expressed in decibel milliwatts (dBm) [10]. This measure can be used to measure the distance between transmitter and receiver devices, based on the transmitted and received signal power differences. RSSI measurements can be used in two ways in positioning methods. In geometric approaches, distances are calculated from the RSSI values using an adequate model. Two propagation models have been generally used to convert the RSSI to distance: the free-space models and the log-normal models. The free-space models are simple, but they do not consider the obstacles between receivers and transmitters, so they are often limited in real applications. The log-normal models are more suitable for different applications due to their flexibility. In fingerprinting-based techniques, the measured RSSIs of access points (AP) collected at given points of the area of interest are used to form the fingerprint database. This fingerprint database is used with a pattern recognition or pattern-matching algorithm, which can be later used to determine the position of a point given by a new measurement vector. Although geometric approaches are largely affected by obstacles, they do not have an effect on fingerprint-based techniques if they are static.

2.2. Magnetic Fingerprints

Vector magnetometers measure the magnetic field in three dimensions, and they are mainly used as compasses based on Earth's magnetic field. The measurements of these sensors are affected by different factors, which can be classified into two groups. The first group includes deterministic errors, which occur due to manufacturing imperfections, and can be compensated for by calibrating the sensors. Such errors are the scale factor, the bias, and the non-orthogonality error. Magnetic sensors are also affected by external magnetic influences, which are caused by materials that generate or distort the magnetic field. These are called hard iron and soft iron effects [22,43,44]. Hard iron errors are time-invariant, undesired magnetic fields generated by ferromagnetic materials with permanent magnetism, which are additive to Earth's magnetic field. The hard iron distortion is modeled by a 3×1 vector ($\boldsymbol{b}_{HI}$). Soft iron distortion is the result of a material that influences or distorts a magnetic field, but does not necessarily generate a magnetic field itself, and is therefore not additive. The effect of the soft iron distortion is modeled by a 3×3 matrix ($\boldsymbol{A}_{SI}$). The sensor model supposing that the deterministic errors are calibrated can be given by Equation (1).

$$\boldsymbol{B} = \boldsymbol{A}_{SI}\boldsymbol{m} + \boldsymbol{b}_{HI}, \tag{1}$$

where $\boldsymbol{B} = \begin{bmatrix} B_x & B_y & B_z \end{bmatrix}^T$ is the measured output vector, while $\boldsymbol{m} = \begin{bmatrix} m_x & m_y & m_z \end{bmatrix}^T$ is the magnetic field vector.

In an indoor environment, the measured magnetic field affected by the distortions at different points can be used to construct a fingerprint of an indoor space, which can be applied during localization.

2.3. Proposed Positioning Method

The magnetic sensor readings cannot be used for distance calculation, but the provided measurements of points with a known position can be utilized to construct a multidimensional fingerprint. Thus, it is obvious that the RSSI and magnetometer data should be fused in a fingerprinting-based method. The proposed method, which can be seen in Figure 1, utilizes together the fingerprints of *RSSI* values and the fingerprints of magnetometer measurements to train offline the MLP neural network. This trained MLP can be used in real-time to compute the $(\hat{x}, \hat{y})$ position of an unknown point based on the measurements collected at the given point. The used *RSSI* values are measured between APs with a known position and the mobile robot, while the magnetic field measurements are provided by a magnetometer installed on the mobile robot. In Figure 1, the magnetometer is connected with a dashed line to the fingerprinting algorithm since different versions of listed data are tested.

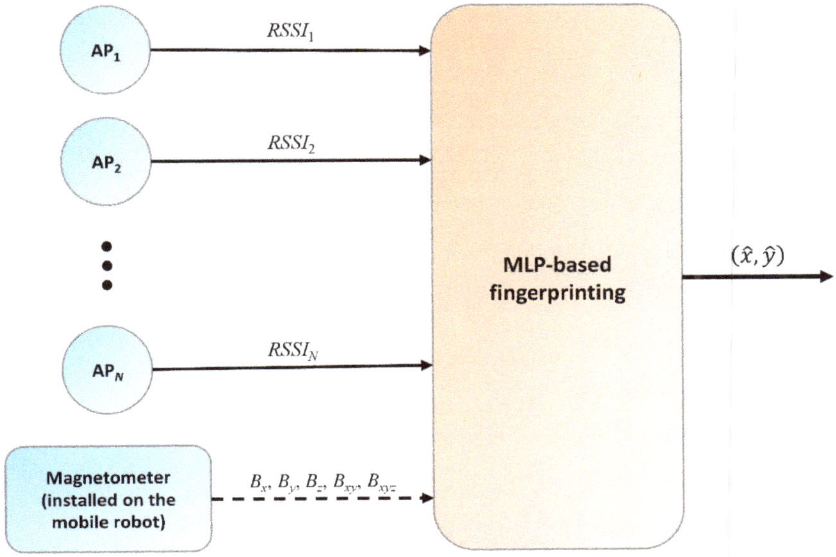

Figure 1. Architecture of the proposed method.

2.3.1. Magnetometer Data

In total, three different possibly usable versions of magnetometer data are tested in the analysis:

- The first version utilizes the measurements on the three sensor axes, i.e., the B_x, B_y, and B_z measurements. This kind of application of the magnetometer readings is only possible if the orientation of the sensor in the global coordinate frame is known for the given point, since different orientations at the same location result in different sensor readings. This version can be used to examine the most achievable positioning performance of the proposed method, since it contains the most information.
- In the second version, the 3D magnetic field magnitude (B_{xyz}), which can be computed from the sensor outputs using Equation (2), is utilized since it is orientation independent.

$$B_{xyz} = \sqrt{B_x^2 + B_y^2 + B_z^2}, \tag{2}$$

- If the mobile robot is moving on a flat surface, then the degree of freedom decreases to three, i.e., (X, Y, θ). This makes the measurements in the Z-axis directly usable. The magnetic field magnitude in the X-Y plane (B_{xy}), which can be calculated using Equation (3), can also be utilized without the knowledge of the θ angle. Thus, this version uses together the B_z and the B_{xy}.

$$B_{xy} = \sqrt{B_x^2 + B_y^2}, \tag{3}$$

2.3.2. Fingerprinting Algorithm

The proposed method applies MLP neural networks to perform positioning in 2D. The MLP is a widely used technique and it has been shown to provide the best results in RSSI-based fingerprint tasks compared to well-known methods, such as the weighted K-nearest neighbor (WKNN) and the random forest (RF) algorithms [45].

The method applies measurements collected at grid points of an indoor space to train the MLP. The measurement database of RSSI data used during the training process consists of WiFi $RSSI_{i,(x_c,y_c)}$ measurements for $i = 1, 2, \ldots, N$ APs. The measurements collected in a grid based on the (x_c, y_c) coordinates form a fingerprint for each AP. Similarly, the $B_{x,(x_c,y_c)}$, $B_{y,(x_c,y_c)}$, and $B_{z,(x_c,y_c)}$ magnetometer measurements taken at the grid points and the computed $B_{xy,(x_c,y_c)}$, and $B_{xyz,(x_c,y_c)}$ magnitudes also form fingerprints. These fingerprints for the different data types are also utilized during the MLP training.

The developed MLP contains three layers: an input layer, a hidden layer, and an output layer. An input vector is formed for each given location from the corresponding $RSSI_i$ values measured between the mobile node and the N APs, together with the magnetic field strengths used in different versions. The output layer has two neurons, which provide the X and Y coordinates. For a given point in the training datasets, the target values corresponding to the input vector are defined by the (x_c, y_c) coordinates. The optimal number of neurons in the hidden layer should be defined by testing different configurations. The proposed MLP uses tangent sigmoid activation functions in the hidden layer and linear transfer functions in the output layer.

3. Applied Measurement Data

The proposed method was validated using measurements collected in two different scenarios. In the case of both scenarios, RSSI values were measured between multiple anchor nodes with a known position and one mobile node. The WiFi was chosen to provide the RSSI measurements, since it is the most suitable and popular wireless standard, and it is widely used in indoor localization methods [10,46]. The WiFi has high bitrate, high scalability, and is relatively less affected by external factors compared to other wireless standards, such as Bluetooth low energy (BLE), ZigBee, LoRaWAN, radio frequency identification (RFID), and ultra-wideband (UWB). Other technologies also have their own advantages and disadvantages. BLE is also a widely used technology in positioning applications [47,48]. The advantages of this technology are its low power consumption and fast connection establishment between the modules, but the lower range is a big disadvantage in the case of this technology [49]. Passive RFID technology-based systems are also popular in localization applications. The main advantage of these systems is their low cost, but their operation range is limited [50], which results in a high number of used tags [51]. The UWB technology provides high precision and low loss [52], but the higher cost is a big disadvantage of these systems.

Magnetometer measurements were also collected in the case of the mobile node. The measurements were taken in one plane near to the ground, which would be the case if the system was used with a mobile robot. Also, the disturbance sources are closer to the sensor, so they provide greater effects.

3.1. Measurement System

The applied measurement system consists of five anchor nodes ($N = 5$) and one mobile node. All six units are based on a NodeMCU ESP-32S V1.1 microcontroller board, which is capable of WiFi communication. The mobile node is also equipped with a three-axis magnetic sensor.

The used magnetometer is an HMC5883L type three-axis digital compass IC, which utilizes Honeywell's anisotropic magnetoresistive (AMR) technology. The directional sensors feature precision in-axis sensitivity and linearity, and very low cross-axis sensitivity. The measurement range of the magnetic sensor is ± 810 µT in 12-bit resolution with a 160 Hz maximal sampling rate. It also contains an I2C interface, which can be used to both configure the sensors and read the measurement values.

Since the data were processed offline, a central unit was used to collect the measurement data. The mobile node forwarded the magnetometer readings and the RSSI values read from the transceiver of the ESP32 board during communication with the APs.

3.2. Data Acquisition

In the case of both scenarios, the measurements were collected at defined grid points. At all points, 10 RSSI measurements were recorded for all APs and 10 magnetometer measurements were also collected. The 10 samples were later averaged to decrease the effect of noise.

During the data acquisition process, the WiFi channel was set to 1 and the antenna Tx power was 19.5 dBm in the case of all wireless modules.

The version where the measurements of the three magnetometer axes are utilized required the knowledge of the mobile unit's orientation. To obtain usable magnetometer data for this version, the orientation of the mobile unit was set to as constant as possible at different points.

Besides the hard and soft iron effects, the magnetometer measurements are also affected by deterministic errors, such as scale factors, bias, and non-orthogonality errors. To compensate for these effects, the magnetometer was calibrated using an evolutionary algorithm-based calibration method, which utilizes measurements taken in multiple orientations to compute the calibration parameters [53]. This process was performed before data acquisition.

3.2.1. First Scenario

In the first scenario, a smaller room was chosen, which had enclosing dimensions of 6.6 m × 3.6 m. Figure 2 shows the measurement environment in the room and its schematic drawing with the position of the anchor nodes. To make the setup realistic, multiple static objects were left in the room, such as cabinets and tables. It is also important to note that one of the walls was made from glass, moreover a concrete column could also be found in the room. Some measurement points were none-line-of-sight (NLOS) due to the objects.

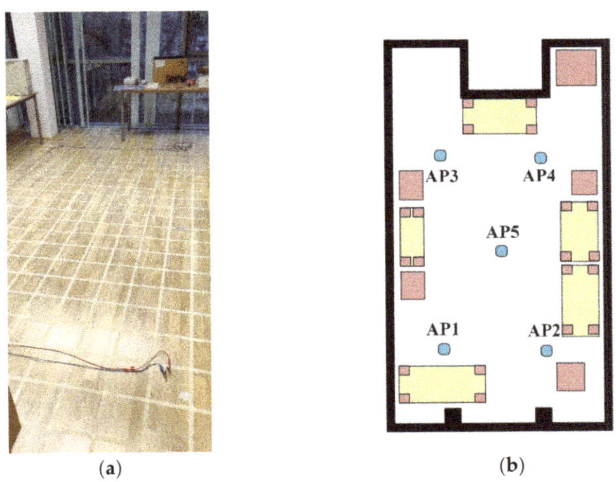

Figure 2. Measurement setup of the first scenario: (**a**) environment; (**b**) schematic drawing.

In the grid, which was used to determine the measurement points, squares with a side length of 20 cm were defined. The height of both the mobile unit and the APs was 4 cm. Altogether 426 points were used for data acquisition. The mobile node was moved manually between the measurement points and there was no one in the room during the measurements. The measurements were collected over the course of one day. The heatmaps of the collected RSSI measurements for the 5 APs and the magnetometer data are shown in Figure 3 and Figure 4, respectively. The white parts in the figures represent the places where measurements could not be taken due to the obstacles, while the black rectangles show the positions of the APs.

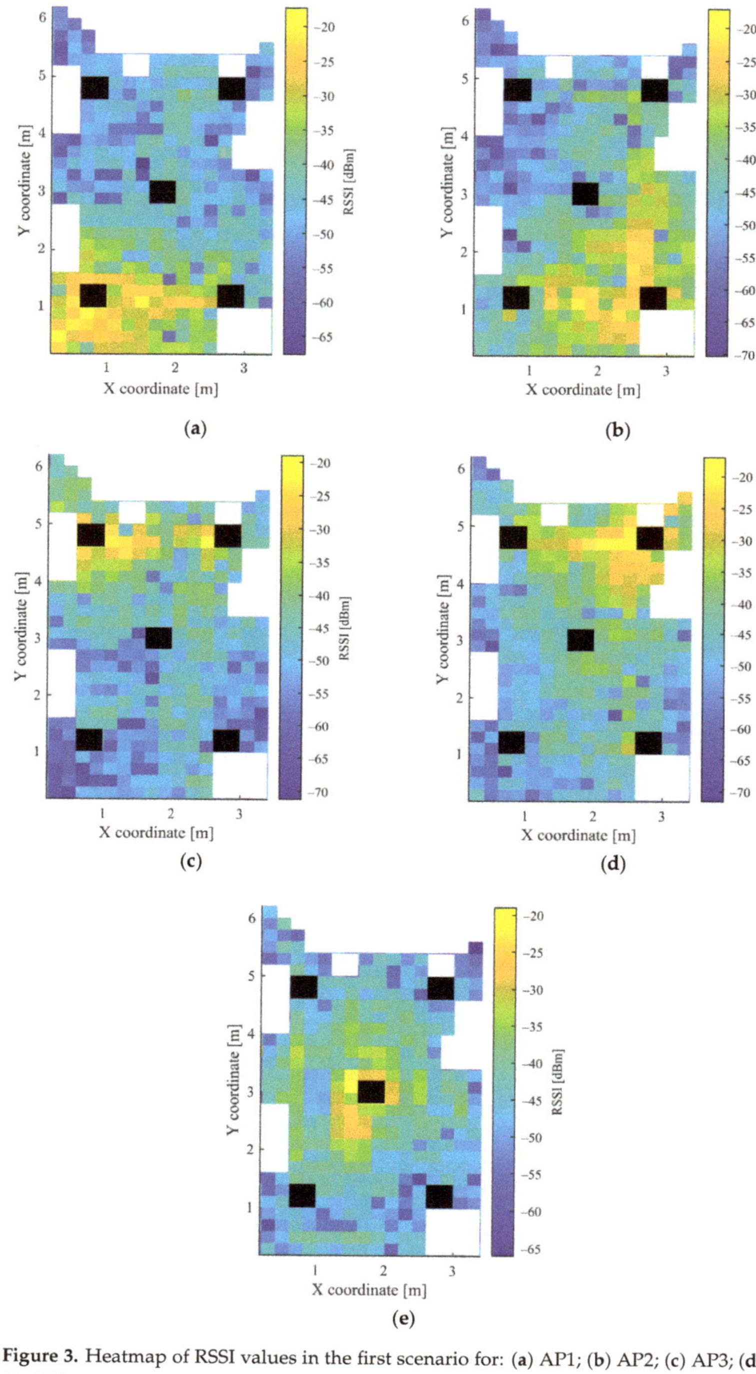

Figure 3. Heatmap of RSSI values in the first scenario for: (**a**) AP1; (**b**) AP2; (**c**) AP3; (**d**) AP4; and (**e**) AP5.

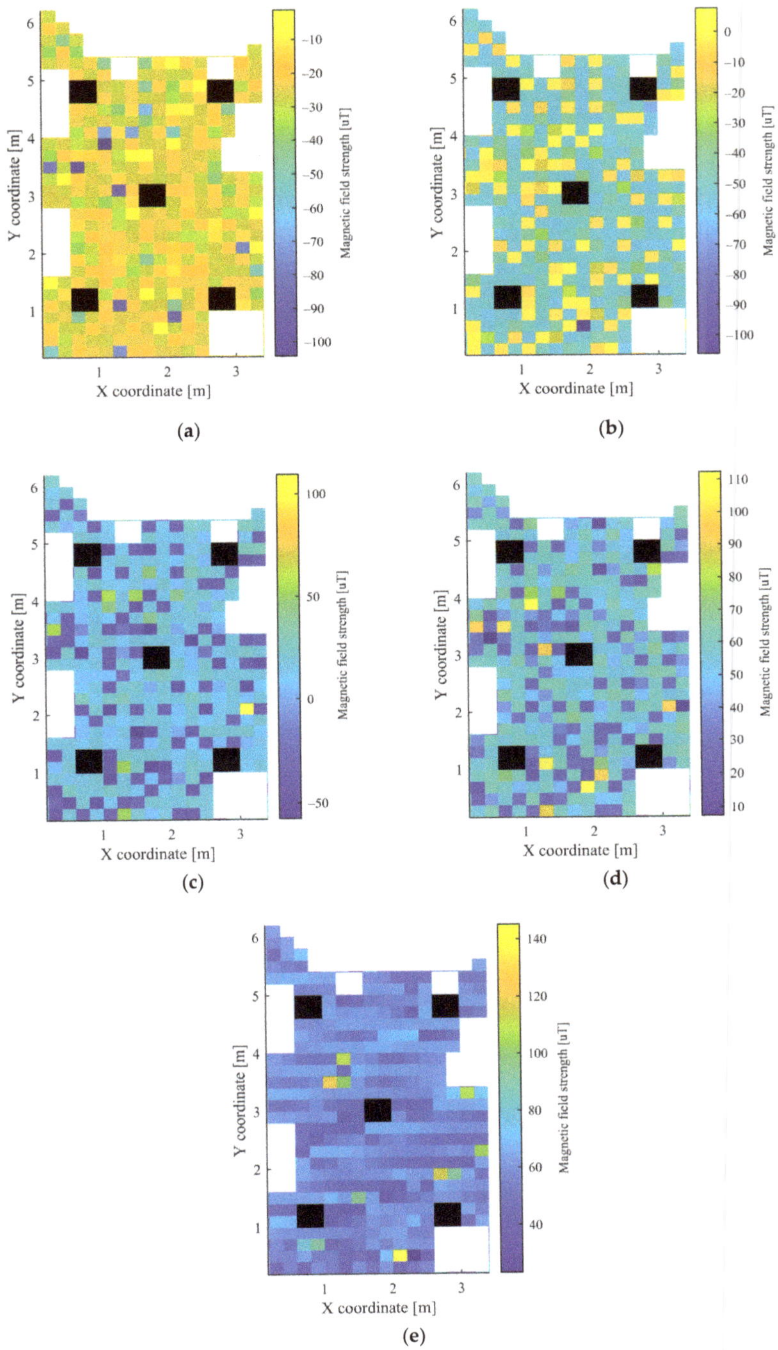

Figure 4. Heatmap of the magnetic field strength in the first scenario for: (**a**) B_x; (**b**) B_y; (**c**) B_z; (**d**) B_{xy}; and (**e**) B_{xyz}.

3.2.2. Second Scenario

The measurements of the second scenario were collected in a larger laboratory with enclosing dimensions of 12 m × 8 m. Both the measurement environment in the laboratory and its schematic drawing with the position of the APs can be seen in Figure 5. Similarly to the first scenario, static objects were present in the laboratory during the measurements, such as tables, robotic cells, and a linear rail with a robotic arm. Due to the objects, many points were NLOS to some APs.

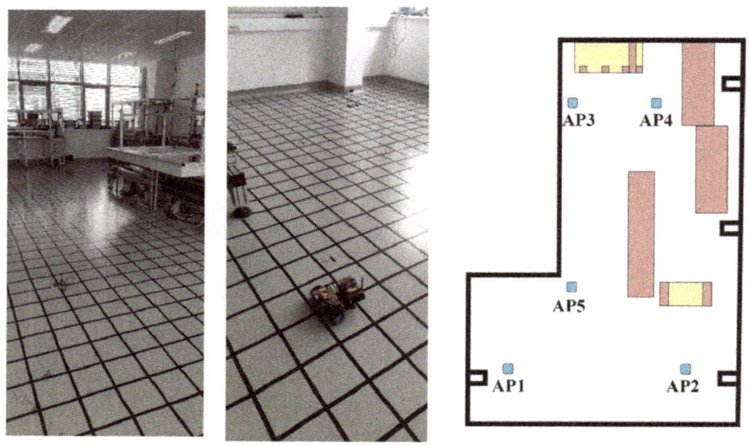

(a) (b) (c)

Figure 5. Measurement setup of the second scenario: (**a**) environment of the laboratory; (**b**) environment with the mobile robot; and (**c**) schematic drawing.

The grid was also defined using 20 cm distances, as in the first scenario. Due to the high number of points, the movement of the measurement unit between the grid points was realized using a mobile robot. The height of the mobile unit and the APs was 10.8 cm and 3.5 cm, respectively. Altogether, measurements were collected at 1408 points, defined by the grid. Further measurements were taken at 20 random positions, which were later used for the testing of the trained MLPs. These points were defined by random coordinates and were not identical with any grid point coordinates. The measurements were collected over the course of one day. Figure 6 shows the heatmaps of the RSSI values from the 5 APs, while the heatmaps of the magnetic sensor data can be seen in Figure 7. The white parts in both Figures 6 and 7 represent the places where measurements could not be taken due to the obstacles, while black parts show the places of the APs.

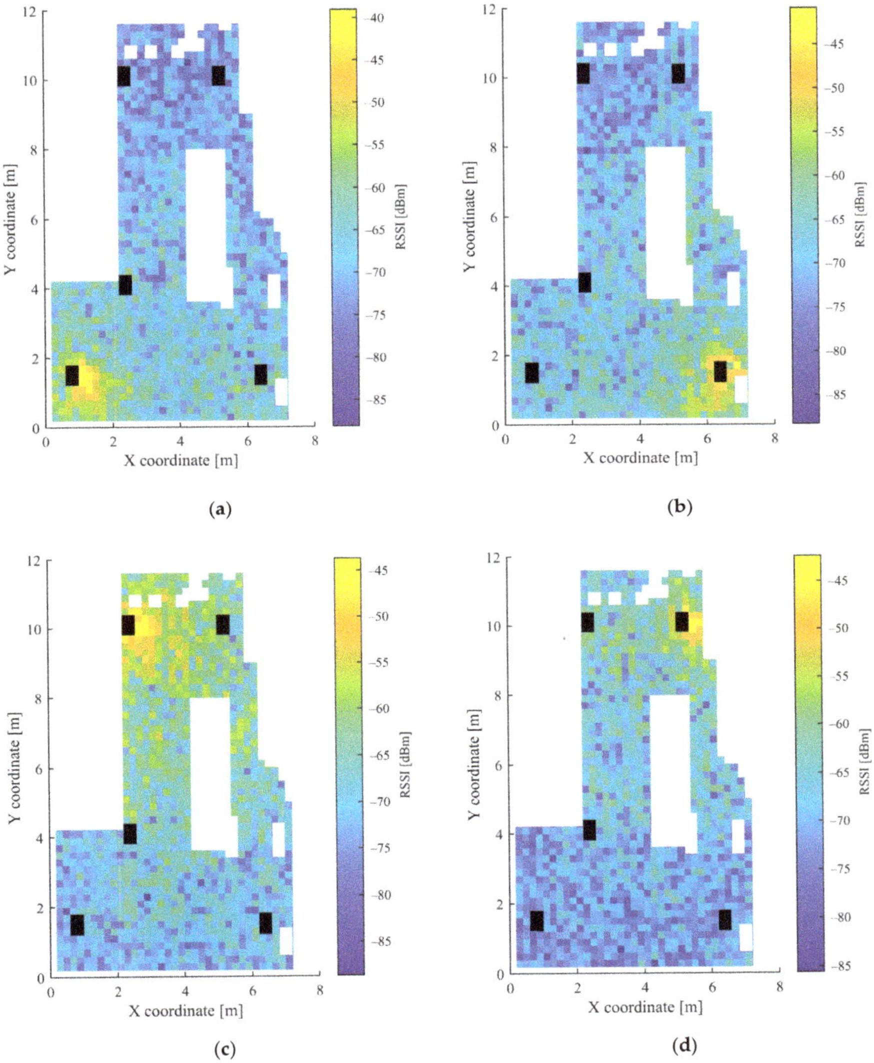

Figure 6. *Cont.*

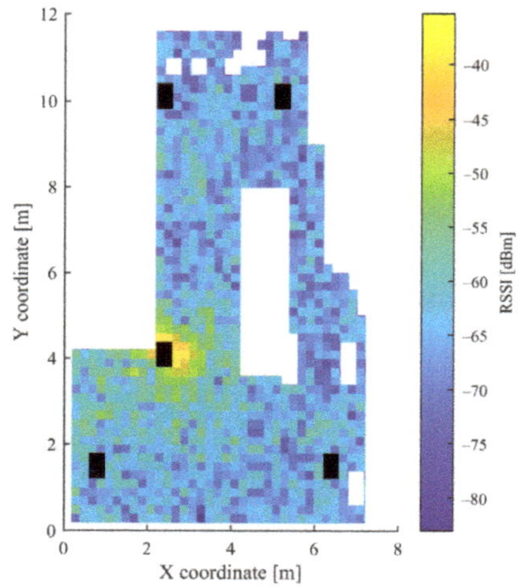

(e)

Figure 6. Heatmap of RSSI values in the second scenario for: (**a**) AP1; (**b**) AP2; (**c**) AP3; (**d**) AP4; and (**e**) AP5.

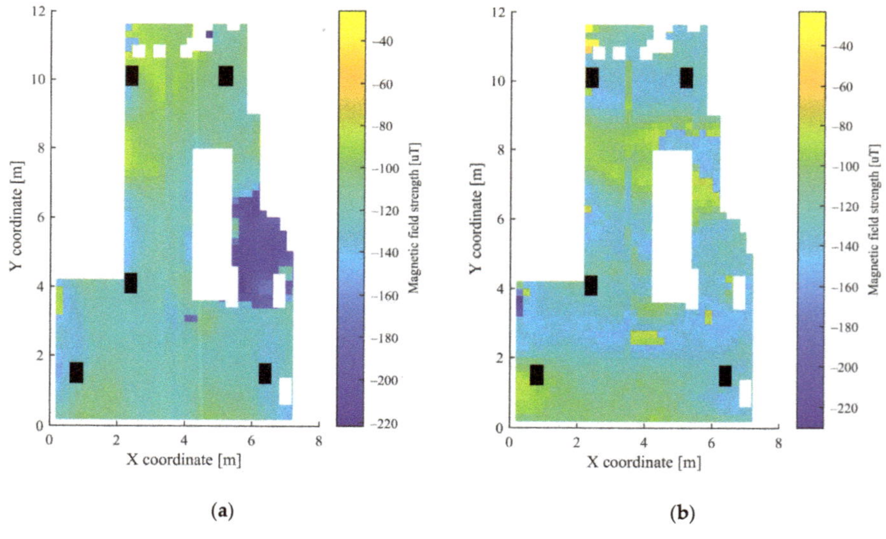

Figure 7. *Cont.*

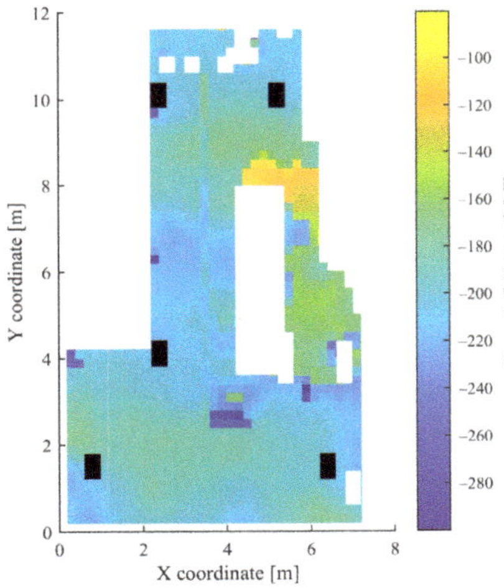

(c)

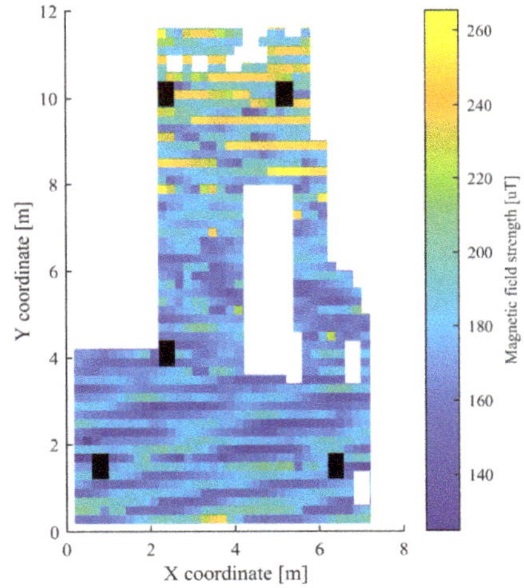

(d)

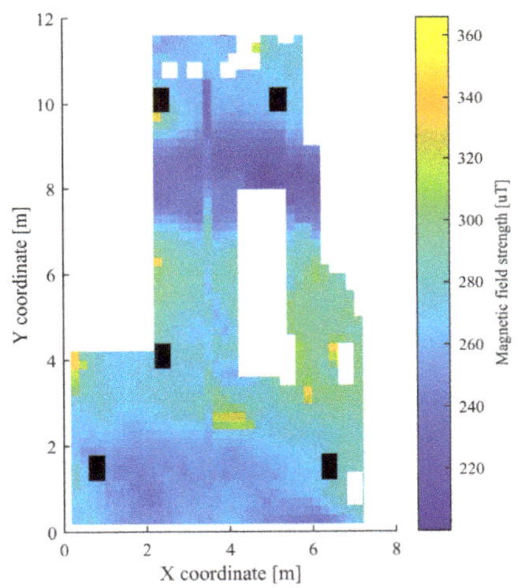

(e)

Figure 7. Heatmap of the magnetic field strength in the second scenario for: (**a**) B_x; (**b**) B_y; (**c**) B_z; (**d**) B_{xy}; and (**e**) B_{xyz}.

4. Experimental Results

It can be observed from Figures 3 and 4 for the first scenario and from Figures 6 and 7 for the second scenario, that the heatmaps of the RSSI data and magnetic field strengths carry additional information when compared to each other. The histogram of the B_{xyz} magnetic field magnitude for the two scenarios can be seen in Figure 8. The histograms show that the variance of the magnetometer measurements is much higher in the case of the second scenario, which is the laboratory.

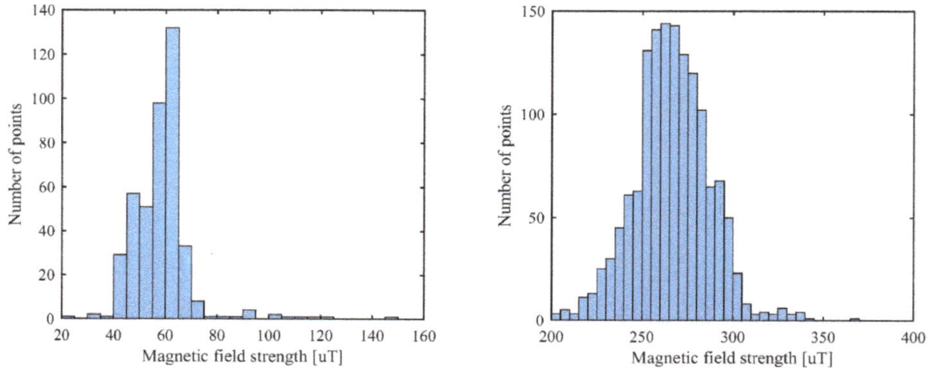

Figure 8. Histogram of the B_{xyz} magnetic field magnitude for: (**a**) first scenario; and (**b**) second scenario.

4.1. Datasets and MLP Training

The data measured at the grid points were utilized to train the MLP neural networks separately for the two scenarios. Since no test points were determined in the first scenario, two separate datasets were defined. In the first dataset, all points were utilized as training data without test data to examine the obtainable error levels at the grid points. In the second dataset, a subset of points containing every second point was used in the training process, while the remaining 50% of the points were applied for the testing of the trained MLPs.

Altogether, seven combinations of used data were tested to examine their performance in the two scenarios. The tested combinations are listed as follows:

1. *RSSI*
2. B_x, B_y, B_z
3. B_{xy}, B_z
4. B_{xyz}
5. *RSSI*, B_x, B_y, B_z
6. *RSSI*, B_{xy}, B_z
7. *RSSI*, B_{xyz}

The training of the MLP neural networks for all datasets was tested with 1–100 hidden layer neurons to find the necessary configuration. Since the achievable error rates largely depend on the initial weights, all configurations were tested 10 times and the results containing the lowest error rates were used later in the evaluation process. Multiple performance metrics were utilized in the evaluation process. The mean absolute error (MAE) was used as the main metric, which can be calculated using Equation (4),

$$\text{MAE} = \frac{1}{M}\sum_{i=1}^{M} \text{E}_i, \tag{4}$$

where M is the number of points, while E_i is the error corresponding to the ith point, which can be calculated using Equation (5).

$$\text{E}_i = \sqrt{(\hat{x} - x_c)^2 + (\hat{y} - y_c)^2} \tag{5}$$

In Equation (5), $\hat{x}$ and $\hat{y}$ are the determined coordinates, while x_c and y_c are the real position of the ith point.

Other metrics corresponding to the lowest achieved MAE were also computed. These were the standard deviation (STD) and the root mean squared error (RMSE), which can be calculated using Equations (6) and (7), respectively.

$$\mathrm{STD} = \sqrt{\frac{1}{M-1}\sum_{i=1}^{M}(\mathrm{E}_i - \mathrm{MAE})^2} \tag{6}$$

$$\mathrm{RMSE} = \sqrt{\frac{\sum_{i=1}^{M}\mathrm{E}_i^2}{M}} \tag{7}$$

The hyperparameters of the MLP training process can be found in Table 1. The training was done offline in MATLAB on a PC with the following specifications: Intel Core i5-12600K 3.69 GHz CPU, 16 GB DDR4 3600 MHz RAM, M2 SSD, ASUS GeForce RTX 3080 10GB GPU. The training process was time consuming since multiple hours were required, even for one version. Although it is a complex and time-consuming process to find the optimal network for a given version, it does not affect the real-time operation after implementation.

Table 1. Hyperparameters of the MLP training process.

Hyperparameter	Value
training function	Levenberg–Marquardt backpropagation
performance function	mean squared error (MSE)
normalization	minmax to range $[-1, +1]$
ratio of data used for training	70%
ratio of data used for validation	30%
maximum number of epochs to train	6000
performance goal	0
maximum validation failures	20
minimum performance gradient	10^{-7}
maximum time to train in seconds	inf

4.2. Evaluation of the First Scenario

The best achieved MAE results and the corresponding STD and RMSE results, using different versions of data for the two datasets in the case of the first scenario, are summarized in Table 2. It can be observed from the results that using the proposed method, the errors significantly decrease in the case of the training data. Using datasets where all points were used as training data, significant improvements were achieved using the versions that utilized $RSSI$, B_x, B_y, B_z and $RSSI$, B_{xy}, B_z, compared to results where only $RSSI$ was used. Here, the MAE decreased from 50.71 cm to 34.64 cm and 41.68 cm, respectively. In the datasets where every second point was utilized, the MAE using only $RSSI$ was more than 10 cm lower than when all points were used. The improvement caused by the fusion with magnetometer data was 7–10 cm, using any version. It should be noticed that when using the $RSSI$, B_{xy}, B_z version, an MAE more than 2 cm lower was achieved than with the other two fusion-based versions. In the case of test data, no improvement can be noticed since the $RSSI$ and the fusion-based versions provided nearly the same results. This can be caused by the low variance of the magnetic field measurements in this scenario, and the large distances between the points due to the usage of every second point. The achieved results using only magnetometer data in this scenario show a much lower performance than the other versions.

Table 2. Best achieved MAE and the corresponding STD and RMSE results for different datasets in the first scenario using different versions of data.

Used Data	All Points in the Grid (Training Data)		Every Second Point in the Grid (Training Data)		Every Second Point in the Grid (Test Data)	
	MAE ± STD [cm]	RMSE [cm]	MAE ± STD [cm]	RMSE [cm]	MAE ± STD [cm]	RMSE [cm]
RSSI	50.71 ± 44.05	67.13	40.04 ± 37.39	56.91	75.43 ± 58.12	97.06
B_x, B_y, B_z	96.88 ± 60.38	118.01	87.48 ± 54.95	104.86	141.48 ± 76.47	161.10
B_{xy}, B_z	119.01 ± 65.28	138.07	105.52 ± 65.93	126.61	145.12 ± 73.04	163.54
B_{xyz}	159.19 ± 74.25	175.72	149.67 ± 75.32	167.64	167.02 ± 76.12	183.48
RSSI, B_x, B_y, B_z	34.64 ± 32.87	51.86	32.44 ± 37.27	49.35	76.67 ± 56.20	96.81
RSSI, B_{xy}, B_z	41.68 ± 36.58	55.42	29.07 ± 41.18	52.58	75.03 ± 55.33	93.14
RSSI, B_{xyz}	49.86 ± 45.38	67.39	32.54 ± 39.17	50.85	77.79 ± 58.17	97.05

Figure 9 shows the best achieved MAE using different setups, based on the hidden layer neuron numbers with different versions of used data. It can be seen from the results that the MAE starts to converge around 40–50 neurons in the case of the training data. Only slight improvements can be noticed with higher neuron numbers. In the case of the test data, the MAE converges around 20 neurons.

The cumulative density functions (CDF) of errors for the different used versions of data can be seen in Figure 10. The errors larger than 300 cm were truncated. It can be noticed that the proposed method significantly improves the error distribution in the case of the training data. For example, when using all grid points for training, around 60% of points give below 50 cm errors with only *RSSI*, which increases to ~70% and ~80% using *RSSI*, B_{xy}, B_z and *RSSI*, B_x, B_y, B_z, respectively. With the test data, the fused versions provide almost identical curves, as with using only *RSSI*.

4.3. Evaluation of the Second Scenario

Table 3 presents the best achieved MAE and the corresponding STD and RMSE results in the second scenario for the different combinations of used data. Much higher improvements can be noticed with the proposed method compared to the results achieved in the first scenario, which is obviously the cause of the higher variance of the magnetic field readings. It is very important to note that approximately 20 cm better results were obtained in the case of both the training and test data using B_x, B_y, B_z, compared to the results achieved using only *RSSI*. It is also important to notice that the MAE for almost all versions was lower in the case of the test data than with the training data. This can be caused by lower error rates in the areas where the random locations were defined. The proposed method resulted in significantly lower error rates for all three versions compared to both *RSSI* and B_x, B_y, B_z, in the case of both the training and test data. The best results were provided by the version utilizing *RSSI*, B_x, B_y, B_z, for which the achieved MAE was 99.59 cm and 77.27 cm for the training and test data, respectively. The improvement with this combination of used data resulted in more than 35% improvement compared to the results obtained using only *RSSI* or B_x, B_y, B_z. The version using *RSSI*, B_{xy}, B_z, which can be used in the case of flat surfaces without knowledge of orientation, resulted in approximately 10 cm worse results than using *RSSI* and the measurements of the three sensor axes together. Compared to the results based on only *RSSI*, this version resulted in more than 35 % improvement. The worst results of the three fused versions were provided by utilizing *RSSI* with the 3D magnetic field magnitude together. The provided results with this orientation-independent version were 128.96 cm for the training data and 126.25 cm for the test data. Compared to the *RSSI*-based results, the improvement was 25.37% and 16.71%, respectively.

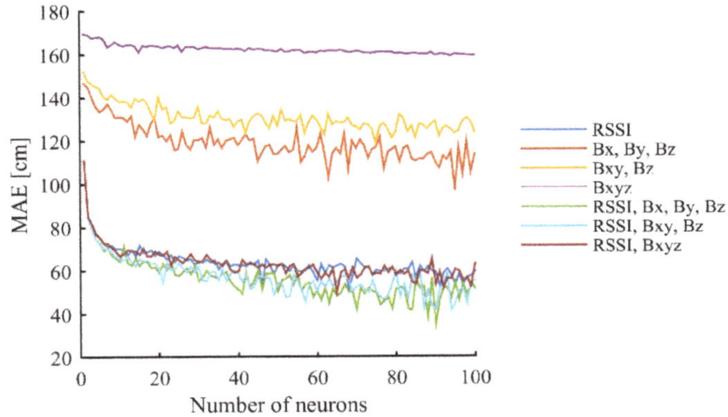

(**a**)

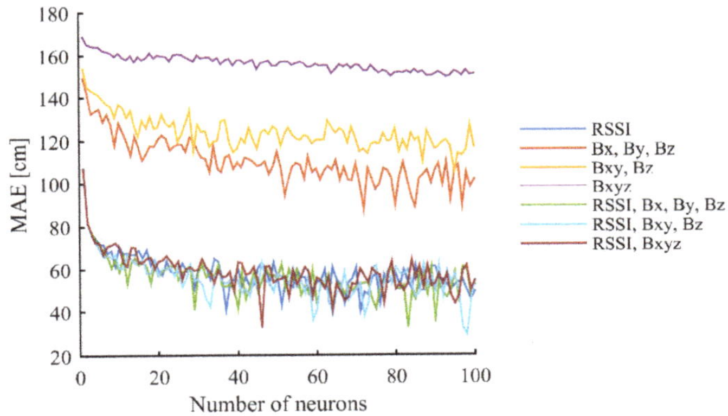

(**b**)

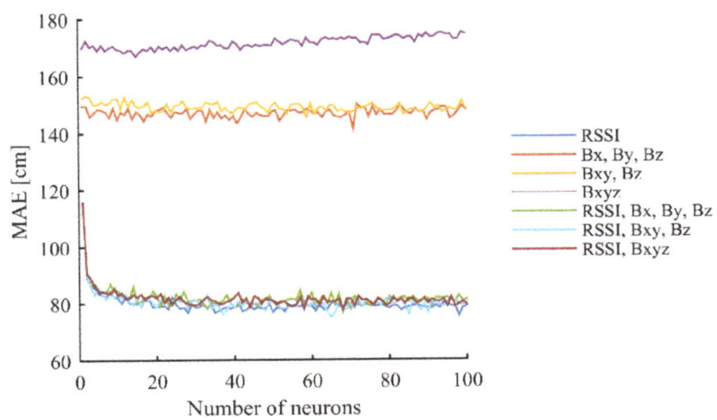

(**c**)

Figure 9. Results achieved in the first scenario using different number of hidden layer neurons with different versions of used data for: (**a**) all points in the grid (training data); (**b**) every second point in the grid (training data); and (**c**) every second point in the grid (test data).

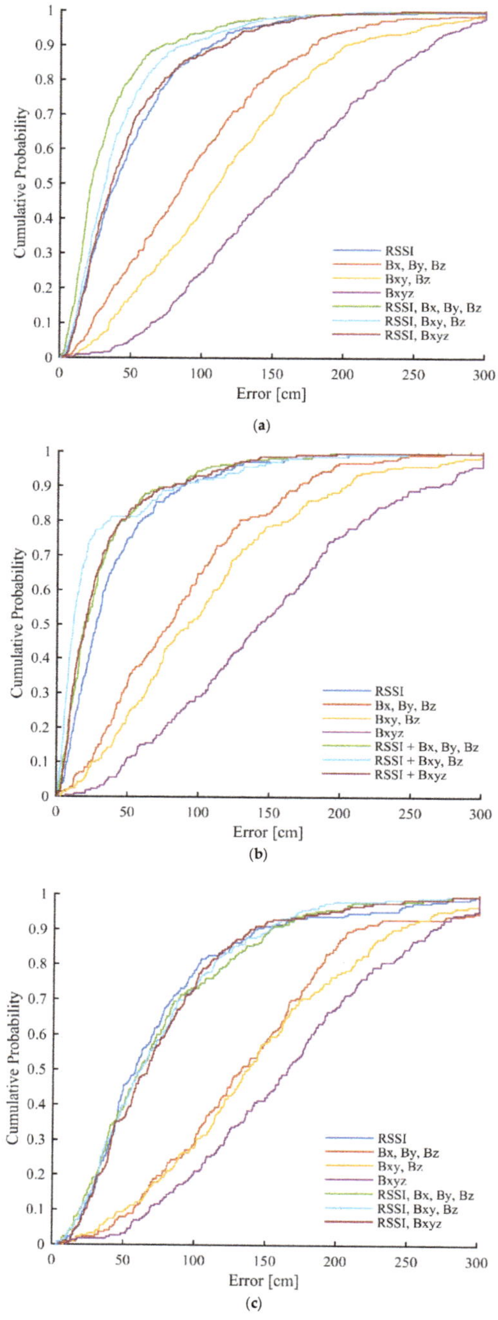

Figure 10. CDFs of errors achieved in the first scenario with different versions of used data for: (**a**) all points in the grid (training data); (**b**) every second point in the grid (training data); and (**c**) every second point in the grid (test data).

Table 3. Best achieved MAE and corresponding STD and RMSE results for different versions of used data in the second scenario.

Used Data	Grid Points (Training Data)		Test Points (Test Data)	
	MAE ± STD [cm]	RMSE [cm]	MAE ± STD [cm]	RMSE [cm]
$RSSI$	172.80 ± 111.14	205.43	151.58 ± 116.25	199.00
B_x, B_y, B_z	152.67 ± 129.81	200.37	130.49 ± 113.16	172.51
B_{xy}, B_z	228.28 ± 166.67	284.65	232.86 ± 172.64	295.42
B_{xyz}	285.20 ± 176.22	338.88	255.07 ± 193.52	317.23
$RSSI, B_x, B_y, B_z$	99.59 ± 78.93	127.06	77.27 ± 76.75	110.47
$RSSI, B_{xy}, B_z$	110.19 ± 86.44	140.03	87.63 ± 94.27	137.01
$RSSI, B_{xyz}$	128.96 ± 96.54	161.07	126.25 ± 115.66	175.58

The achieved MAE using a different number of neurons in the hidden layer, for the various versions of used data recorded in the second scenario, can be seen in Figure 11. It can be seen from the results that most versions converge with 30–40 neurons in the case of the training data and with 20 neurons using the test data, but slight improvements can also be noticed with a higher number of hidden layer neurons in both cases.

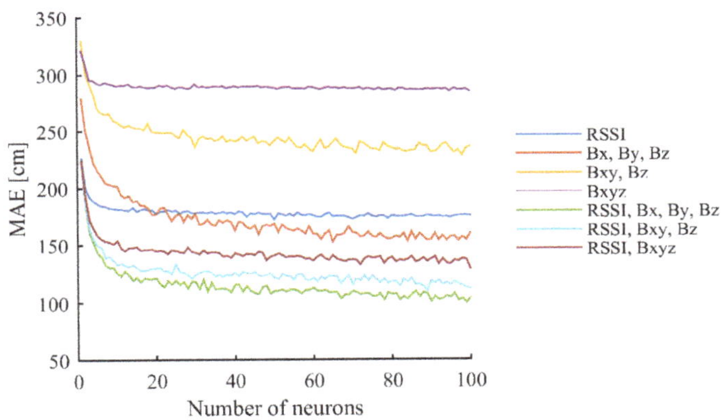

(**a**)

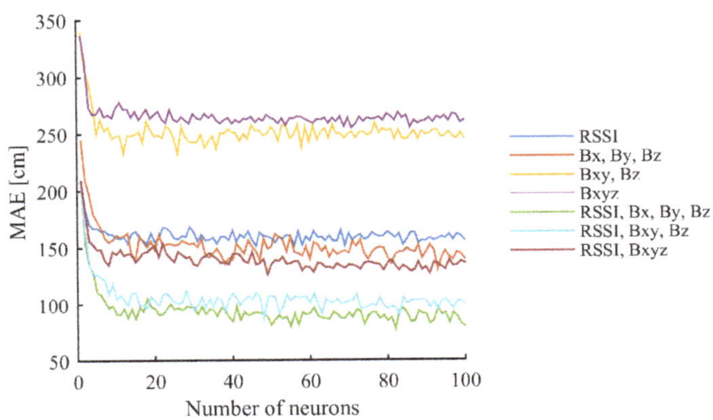

(**b**)

Figure 11. Achieved results in the second scenario using different number of hidden layer neurons with different versions of used data for: (**a**) grid points (training data); and (**b**) test points.

The CDFs of error for the best setups, using different versions of used data in the case of the second scenario, can be seen in Figure 12. The errors larger than 600 cm were truncated. It can be observed from the results that significant improvements can be achieved in the error distributions using the different versions of the proposed method.

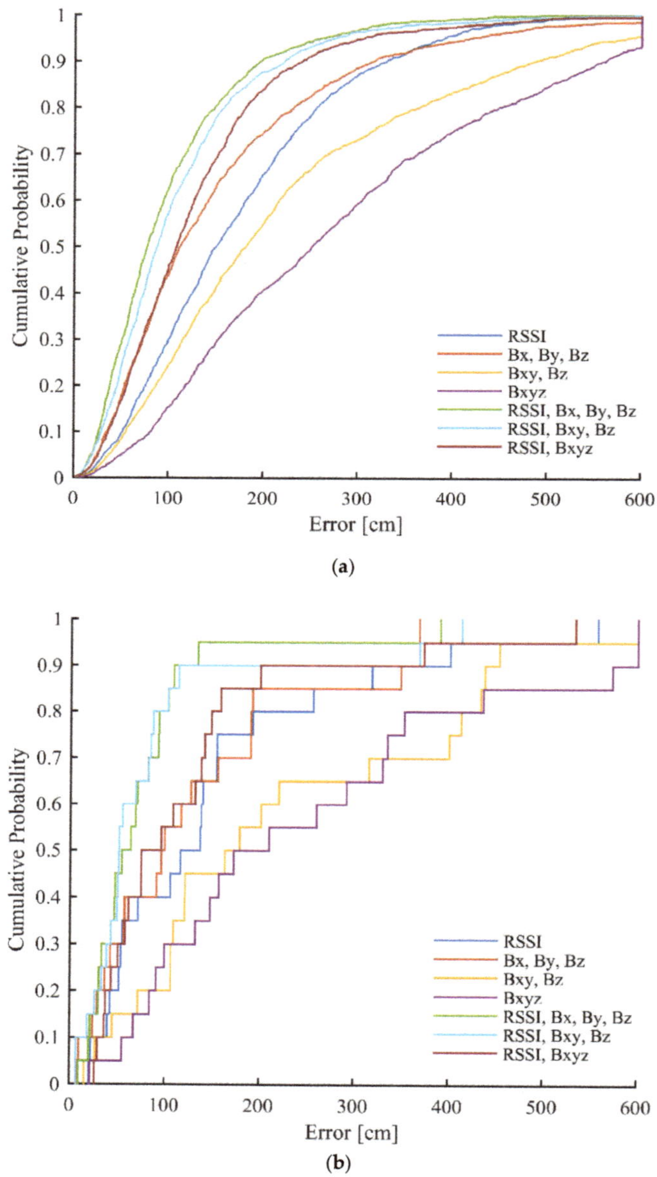

Figure 12. CDFs of errors achieved in the first scenario with different versions of used data for: (**a**) grid points (training data); and (**b**) test points.

5. Conclusions

In this work, a novel fingerprinting-based positioning method was proposed, which fuses the RSSI data measured between anchor nodes and a mobile node with the magnetic field measurements collected from the mobile unit. In total, three different versions of used magnetometer data were tested in the proposed method. The position is determined using three-layer MLP neural networks.

The method was validated using data collected in two different realistic indoor scenarios, i.e., in a smaller room and a larger laboratory. Multiple static objects were present in both scenarios. The MLP ANNs were tested with a various number of hidden layer neurons to find the optimal configuration.

The achieved results show that significant improvement, even above 35%, can be achieved using the proposed method, compared to the results achieved using only the RSSI or magnetic sensor data, if the variance of the magnetic field measurements is high in the observed indoor space. In the first scenario, where the variance of magnetometer readings was lower, improvement could be noticed only in the case of the training data. In the second scenario, where the variance in the magnetic field strength was much higher, significant improvements were achieved. The best results were obtained by fusing RSSI data with magnetometer measurements in the three axes, but this version requires knowledge of the orientation of the mobile unit in the global coordinate frame. The version using the RSSI data, together with the magnetometer measurements in the Z-axis and the magnitude in the X-Y plane, provided more than 35% better results compared to only RSSI-based error rates. This version can be used in the case of flat surfaces without knowledge of the orientation. Utilizing the RSSI data together with the 3D magnetometer magnitudes resulted in approximately 25% and 15% improvement in the case of the training and test data, respectively.

The limitations of the proposed approach include that the magnetic sensor needs to be close to the objects causing the disturbances. Another limitation is that some versions depend on the orientation, which should be solved by implementing an orientation estimation framework that can provide the necessary information. Further future goals include achieving further improvements in the error rates by incorporating other data and its application in a sensor fusion-based localization framework.

Author Contributions: Conceptualization, P.S.; methodology, P.S., D.C. and A.O.; software, P.S., D.C. and A.O.; validation, P.S.; formal analysis, P.S., D.C. and A.O.; investigation, P.S., D.C. and A.O.; resources, P.S., D.C. and A.O.; data curation, D.C.; writing—original draft preparation, P.S.; writing—review and editing, A.O.; visualization, P.S. and D.C.; supervision, P.S. and A.O.; project administration, P.S. and A.O.; funding acquisition, P.S. and A.O. All authors have read and agreed to the published version of the manuscript.

Funding: The work was supported by the National Research, Development, and Innovation Fund of Hungary through project no. 142790 under the FK_22 funding scheme.

Institutional Review Board Statement: Not applicable.

Informed Consent Statement: Not applicable.

Data Availability Statement: The data presented in this study are openly available at: https://github.com/petersarcevic/rssi_magnetometer_fingerprint_database (accessed on 30 January 2023).

Conflicts of Interest: The authors declare no conflict of interest.

References

1. Mukhopadhyay, A.; Mallisscry, A. TELIL: A Trilateration and Edge Learning based Indoor Localization Technique for Emergency Scenarios. In Proceedings of the International Conference on Advances in Computing, Communications and Informatics (ICACCI), Bangalore, India, 19–22 September 2018. [CrossRef]
2. Tekler, Z.D.; Low, R.; Yuen, C.; Blessing, L. Plug-Mate: An IoT-based occupancy-driven plug load management system in smart buildings. *Build. Environ.* **2022**, *223*, 109472. [CrossRef]

3. Balaji, B.; Xu, J.; Nwokafor, A.; Gupta, R.; Agarwal, Y. Sentinel: Occupancy based HVAC actuation using existing WiFi infrastructure within commercial buildings. In Proceedings of the ACM Conference on Embedded Networked Sensor Systems (SenSys), Rome, Italy, 11–15 November 2013. [CrossRef]
4. Tekler, Z.D.; Chong, A. Occupancy prediction using deep learning approaches across multiple space types: A minimum sensing strategy. *Build. Environ.* **2022**, *226*, 109689. [CrossRef]
5. Gogolák, L.; Fürstner, I. Wireless Sensor Network Aided Assembly Line Monitoring According to Expectations of Industry 4.0. *Appl. Sci.* **2021**, *11*, 25. [CrossRef]
6. Peng, Y.; Niu, X.; Tang, J.; Mao, D.; Qian, C. Fast Signals of Opportunity Fingerprint Database Maintenance with Autonomous Unmanned Ground Vehicle for Indoor Positioning. *Sensors* **2018**, *18*, 3419. [CrossRef] [PubMed]
7. Wang, J.; Park, J. An Enhanced Indoor Positioning Algorithm Based on Fingerprint Using Fine-Grained CSI and RSSI Measurements of IEEE 802.11n WLAN. *Sensors* **2021**, *21*, 2769. [CrossRef]
8. Kolakowski, M. Automated Calibration of RSS Fingerprinting Based Systems Using a Mobile Robot and Machine Learning. *Sensors* **2021**, *21*, 6270. [CrossRef]
9. Yang, T.; Cabani, A.; Chafouk, H. A Survey of Recent Indoor Localization Scenarios and Methodologies. *Sensors* **2021**, *21*, 8086. [CrossRef]
10. Singh, N.; Choe, S.; Punmiya, R. Machine Learning Based Indoor Localization Using Wi-Fi RSSI Fingerprints: An Overview. *IEEE Access* **2021**, *9*, 127150–127174. [CrossRef]
11. Qin, J.; Sun, S.; Deng, Q.; Liu, L.; Tian, Y. Indoor Trajectory Tracking Scheme Based on Delaunay Triangulation and Heuristic Information in Wireless Sensor Networks. *Sensors* **2017**, *17*, 1275. [CrossRef]
12. Roy, P.; Chowdhury, C. A Survey of Machine Learning Techniques for Indoor Localization and Navigation Systems. *J. Intell. Robot. Syst.* **2021**, *101*, 63. [CrossRef]
13. Luo, R.C.; Hsiao, T.J. Dynamic Wireless Indoor Localization Incorporating with an Autonomous Mobile Robot Based on an Adaptive Signal Model Fingerprinting Approach. *IEEE Trans. Ind. Electron.* **2019**, *66*, 1940–1951. [CrossRef]
14. Tian, X.; Shen, R.; Liu, D.; Wen, Y.; Wang, X. Performance Analysis of RSS Fingerprinting Based Indoor Localization. *IEEE Trans. Mob. Comput.* **2016**, *16*, 2847–2861. [CrossRef]
15. Alhomayani, F.; Mahoor, M.H. Deep learning methods for fingerprint-based indoor positioning: A review. *J. Locat. Based Serv.* **2020**, *14*, 129–200. [CrossRef]
16. Yoo, J. Multiple Fingerprinting Localization by an Artificial Neural Network. *Sensors* **2022**, *22*, 7505. [CrossRef]
17. Xue, W.; Qiu, W.; Hua, X.; Yu, K. Improved Wi-Fi RSSI Measurement for Indoor Localization. *IEEE Sens. J.* **2017**, *17*, 2224–2230. [CrossRef]
18. Poulose, A.; Kim, J.; Han, D.S. A Sensor Fusion Framework for Indoor Localization Using Smartphone Sensors and Wi-Fi RSSI Measurements. *Appl. Sci.* **2019**, *9*, 4379. [CrossRef]
19. Rizk, H.; Elmogy, A.; Yamaguchi, H. A Robust and Accurate Indoor Localization Using Learning-Based Fusion of Wi-Fi RTT and RSSI. *Sensors* **2022**, *22*, 2700. [CrossRef]
20. Li, B.; Gallagher, T.; Dempster, A.G.; Rizos, C. How feasible is the use of magnetic field alone for indoor positioning? In Proceedings of the International Conference on Indoor Positioning and Indoor Navigation (IPIN), Sydney, Australia, 13–15 November 2012. [CrossRef]
21. He, S.; Shin, K.G. Geomagnetism for Smartphone-Based Indoor Localization: Challenges, Advances, and Comparisons. *ACM Comput. Surv.* **2018**, *50*, 1–37. [CrossRef]
22. Ouyang, G.; Abed-Meraim, K. A Survey of Magnetic-Field-Based Indoor Localization. *Electronics* **2022**, *11*, 864. [CrossRef]
23. Wu, H.; Mo, Z.; Tan, J.; He, S.; Chen, S.-H.G. Efficient Indoor Localization Based on Geomagnetism. *ACM Trans. Sens. Netw.* **2019**, *15*, 42. [CrossRef]
24. Chen, Y.; Zhou, M.; Zheng, Z. Learning Sequence-Based Fingerprint for Magnetic Indoor Positioning System. *IEEE Access* **2019**, *7*, 163231–163244. [CrossRef]
25. Haverinen, J.; Kemppainen, A. Global indoor self-localization based on the ambient magnetic field. *Rob. Auton. Syst.* **2009**, *57*, 1028–1035. [CrossRef]
26. Ashraf, I.; Zikria, Y.B.; Hur, S.; Park, Y. A Comprehensive Analysis of Magnetic Field Based Indoor Positioning with Smartphones: Opportunities, Challenges and Practical Limitations. *IEEE Access* **2020**, *8*, 228548–228571. [CrossRef]
27. Shu, Y.; Bo, C.; Shen, G.; Zhao, C.; Li, L.; Zhao, F. Magicol: Indoor Localization Using Pervasive Magnetic Field and Opportunistic WiFi Sensing. *IEEE J. Sel. Areas Commun.* **2015**, *33*, 1443–1457. [CrossRef]
28. Ban, R.; Kaji, K.; Hiroi, K.; Kawaguchi, N. Indoor Positioning Method Integrating Pedestrian Dead Reckoning with Magnetic Field and WiFi Fingerprints. In Proceedings of the International Conference on Mobile Computing and Ubiquitous Networking (ICMU), Hakodate, Japan, 20–22 January 2015. [CrossRef]
29. Li, Y.; He, Z.; Nielsen, J.; Lachapelle, G. Using Wi-Fi/magnetometers for indoor location and personal navigation. In Proceedings of the International Conference on Indoor Positioning and Indoor Navigation (IPIN), Banff, AB, Canada, 13–16 October 2015. [CrossRef]
30. Li, Y.; Zhuang, Y.; Lan, H.; Zhou, Q.; Niu, X.; El-Sheimy, N. A Hybrid WiFi/Magnetic Matching/PDR Approach for Indoor Navigation with Smartphone Sensors. *IEEE Commun. Lett.* **2016**, *20*, 169–172. [CrossRef]

31. Huang, X.; Guo, S.; Wu, Y.; Yang, Y. A fine-grained indoor fingerprinting localization based on magnetic field strength and channel state information. *Pervasive Mob. Comput.* **2017**, *41*, 150–165. [CrossRef]
32. Ashraf, I.; Hur, S.; Park, Y. mPILOT-Magnetic Field Strength Based Pedestrian Indoor Localization. *Sensors* **2018**, *18*, 2283. [CrossRef]
33. Wang, G.; Wang, X.; Nie, J.; Lin, L. Magnetic-Based Indoor Localization Using Smartphone via a Fusion Algorithm. *IEEE Sens. J.* **2019**, *19*, 6477–6485. [CrossRef]
34. Wu, H.; He, S.; Chen, S.-H.G. A Graphical Model Approach for Efficient Geomagnetism-Pedometer Indoor Localization. In Proceedings of the IEEE International Conference on Mobile Ad Hoc and Sensor Systems (MASS), Orlando, FL, USA, 22–25 October 2017. [CrossRef]
35. Subbu, K.P.; Gozick, B.; Dantu, R. LocateMe: Magnetic-Fields-Based Indoor Localization Using Smartphones. *ACM Trans. Intell. Syst. Technol.* **2013**, *4*, 73. [CrossRef]
36. Kim, B.; Kong, S.-H. A Novel Indoor Positioning Technique Using Magnetic Fingerprint Difference. *IEEE Trans. Instrum. Meas.* **2016**, *65*, 2035–2045. [CrossRef]
37. Fan, X.; Wu, J.; Long, C.; Zhu, Y. Accurate and Low-cost Mobile Indoor Localization with 2-D Magnetic Fingerprints. In Proceedings of the ACM Workshop on Mobile Crowdsensing Systems and Applications (CrowdSenSys), Delft, The Netherlands, 6–8 November 2017. [CrossRef]
38. Ashraf, I.; Kang, M.; Hur, S.; Park, Y. MINLOC: Magnetic Field Patterns-Based Indoor Localization Using Convolutional Neural Networks. *IEEE Access* **2020**, *8*, 66213–66227. [CrossRef]
39. Ouyang, G.; Abed-Meraim, K. Analysis of Magnetic Field Measurements for Indoor Positioning. *Sensors* **2022**, *22*, 4014. [CrossRef]
40. Lee, N.; Ahn, S.; Han, D. AMID: Accurate Magnetic Indoor Localization Using Deep Learning. *Sensors* **2018**, *18*, 1598. [CrossRef]
41. Frassl, M.; Angermann, M.; Lichtenstern, M.; Robertson, P.; Julian, B.J.; Doniec, M. Magnetic Maps of Indoor Environments for Precise Localization of Legged and Non-legged Locomotion. In Proceedings of the IEEE/RSJ International Conference on Intelligent Robots and Systems (IROS), Tokyo, Japan, 3–7 November 2013. [CrossRef]
42. Antsfeld, L.; Chidlovskii, B. Magnetic Field Sensing for Pedestrian and Robot Indoor Positioning. In Proceedings of the International Conference on Indoor Positioning and Indoor Navigation (IPIN), Lloret de Mar, Spain, 29 November–2 December 2021. [CrossRef]
43. Secer, G.; Barshan, B. Improvements in deterministic error modeling and calibration of inertial sensors and magnetometers. *Sens. Actuator A Phys.* **2016**, *247*, 522–538. [CrossRef]
44. Hadjigeorgiou, N.; Asimakopoulos, K.; Papafotis, K.; Sotiriadis, P.P. Vector Magnetic Field Sensors: Operating Principles, Calibration, and Applications. *IEEE Sens. J.* **2021**, *21*, 12531–12544. [CrossRef]
45. Csik, D.; Odry, A.; Sarcevic, P. Comparison of RSSI-Based Fingerprinting Methods for Indoor Localization. In Proceedings of the IEEE International Symposium on Intelligent Systems and Informatics (SISY), Subotica, Serbia, 15–17 September 2022.
46. He, S.; Chan, S.-H.G. Wi-Fi Fingerprint-Based Indoor Positioning: Recent Advances and Comparisons. *IEEE Commun. Surv. Tutor.* **2016**, *18*, 466–490. [CrossRef]
47. Giovanelli, D.; Farella, E.; Fontanelli, D.; Macii, D. Bluetooth-Based Indoor Positioning Through ToF and RSSI Data Fusion. In Proceedings of the International Conference on Indoor Positioning and Indoor Navigation (IPIN), Nantes, France, 24–27 September 2018. [CrossRef]
48. Tekler, Z.D.; Low, R.; Gunay, B.; Andersen, R.K.; Blessing, L. A scalable Bluetooth Low Energy approach to identify occupancy patterns and profiles in office spaces. *Build. Environ.* **2020**, *171*, 106681. [CrossRef]
49. Mackensen, E.; Lai, M.; Wendt, T.M. Bluetooth Low Energy (BLE) based wireless sensors. In Proceedings of the IEEE SENSORS, Taipei, Taiwan, 28–31 October 2012. [CrossRef]
50. Hahnel, D.; Burgard, W.; Fox, D.; Fishkin, K.; Philipose, M. Mapping and localization with RFID technology. In Proceedings of the IEEE International Conference on Robotics and Automation (ICRA), New Orleans, LA, USA, 26 April–1 May 2004. [CrossRef]
51. Tesoriero, R.; Tebar, R.; Gallud, J.A.; Lozano, M.D.; Penichet, V.M.R. Improving location awareness in indoor spaces using RFID technology. *Expert Syst. Appl.* **2010**, *37*, 894–898. [CrossRef]
52. Ma, J.; Duan, X.; Shang, C.; Ma, M.; Zhang, D. Improved Extreme Learning Machine Based UWB Positioning for Mobile Robots with Signal Interference. *Machines* **2022**, *10*, 218. [CrossRef]
53. Sarcevic, P.; Pletl, S.; Kincses, Z. Evolutionary algorithm based 9DOF sensor board calibration. In Proceedings of the IEEE International Symposium on Intelligent Systems and Informatics (SISY), Subotica, Serbia, 11–13 September 2014. [CrossRef]

sensors

MDPI

Article

Passive Landmark Geometry Optimization and Evaluation for Reliable Autonomous Navigation in Mining Tunnels Using 2D Lidars

Miguel Torres-Torriti [1,*,†], Paola Nazate-Burgos [1,†], Fabián Paredes-Lizama [1,†], Javier Guevara [2] and Fernando Auat Cheein [2]

1 Department of Electrical Engineering, Pontificia Universidad Católica de Chile, Santiago 782-0436, Chile; pjnazate@uc.cl (P.N.-B.); ftparede@uc.cl (F.P.-L.)
2 Department of Electronic Engineering, Universidad Técnica Federico Santa María, Valparaíso 239-0123, Chile; dario.guevara@sansano.usm.cl (J.G.); fernando.auat@usm.cl (F.A.C.)
* Correspondence: mtorrest@ing.puc.cl
† These authors contributed equally to this work.

Abstract: Autonomous navigation in mining tunnels is challenging due to the lack of satellite positioning signals and visible natural landmarks that could be exploited by ranging systems. Solutions requiring stable power feeds for locating beacons and transmitters are not accepted because of accidental damage risks and safety requirements. Hence, this work presents an autonomous navigation approach based on artificial passive landmarks, whose geometry has been optimized in order to ensure drift-free localization of mobile units typically equipped with lidar scanners. The main contribution of the approach lies in the design and optimization of the landmarks that, combined with scan matching techniques, provide a reliable pose estimation in modern smoothly bored mining tunnels. A genetic algorithm is employed to optimize the landmarks' geometry and positioning, thus preventing that the localization problem becomes ill-posed. The proposed approach is validated both in simulation and throughout a series of experiments with an industrial skid-steer CAT 262C robotic excavator, showing the feasibility of the approach with inexpensive passive and low-maintenance landmarks. The results show that the optimized triangular and symmetrical landmarks improve the positioning accuracy by 87.5% per 100 m traveled compared to the accuracy without landmarks. The role of optimized artificial landmarks in the context of modern smoothly bored mining tunnels should not be understated. The results confirm that without the optimized landmarks, the localization error accumulates due to odometry drift and that, contrary to the general intuition or belief, natural tunnel features alone are not sufficient for unambiguous localization. Therefore, the proposed approach ensures grid-based SLAM techniques can be implemented to successfully navigate in smoothly bored mining tunnels.

Keywords: underground mining robots; scan matching; localization and SLAM in tunnels; 2D lidar navigation; GPS-denied environment

Citation: Torres-Torriti, M.; Nazate-Burgos, P.; Paredes-Lizama, F.; Guevara, J.; Auat Cheein, F. Passive Landmark Geometry Optimization and Evaluation for Reliable Autonomous Navigation in Mining Tunnels Using 2D Lidars. *Sensors* **2022**, *22*, 3038. https://doi.org/10.3390/s22083038

Academic Editors: Sašo Blažič, Gregor Klancar and Marija Seder

Received: 9 February 2022
Accepted: 6 April 2022
Published: 15 April 2022

1. Introduction

Improving underground mining productivity requires loaders with increased levels of autonomy in hauling and excavation tasks [1,2]. The capability of a machine to solve its pose both locally and globally in the network of tunnels is essential to achieve high autonomy levels.

The pioneering work [3] shows the feasibility of using simultaneous localization and mapping (SLAM) techniques to map old abandoned mines. At the heart of the approach in [3] is the scan matcher [4], which delivers locally consistent maps and estimates of the robot's motion relying on the existence of structural elements, such as pillars and beams, that facilitate the data association. However, modern tunnel boring techniques produce smooth tunnels in hard rock or soft ground, with new techniques that even avoid traditional support pillars [5]. This makes correct scan matching infeasible without landmarks, unlike cases such as that of abandoned coal mines [3] or silver mines [6] several decades old. Hence, an approach is proposed here that employs optimized artificial *passive* landmarks that ensure correct data association for consistent mapping and can be cheaply manufactured, installed, require little or no maintenance, and may be easily replaced if damaged. Our results show that in environments such as smooth mining tunnels without landmarks, the localization problem becomes the ill-posed environments, and SLAM cannot be solved without the aid of landmarks or beacons, as noted by [7].

The design of a landmark's optimized geometry and spacing is essential to ensure its identifiability, as well as an accurate and reliable localization of existing semi-autonomous tele-operated mining loaders equipped with standard 2D lidar scanners. Two landmark models are proposed and evaluated, one which employs shape primitives and another with a completely free shape described by a piecewise linear function. Hence, the main contributions of this work are in the landmark parameters' optimization using a genetic algorithm, identifying suitable landmark shapes, and validating the localization approach for loaders in underground tunnels. The methodology was implemented in a simulated environment and validated experimentally using the semi-autonomous industrial CAT 262C skid-steer loader developed by the authors for research purposes [2]. The machinery used is shown in Figure 1a, and the mock-up tunnel used for experimental validation can be appreciated in Figure 1b,c.

Figure 1. Experimental setup showing the semi-autonomous CAT 262C and the mock-up tunnel: (**a**) CAT 262C and the front lidar Sick LMS 511; (**b**) Tunnel mock-up without landmarks; (**c**) Tunnel mock-up with landmarks.

The applicability of the landmark-based localization approach and SLAM in real mining environments is demonstrated in Section 5.3 using an accurate 3D model of a 100 m tunnel segment from the El Teniente mine located in Chile [8], which is the largest underground copper mine in the world.

This paper is organized as follows. Section 2 discusses the related work concerning autonomous navigation for robots in mining tunnels. Section 3 explains the preliminary mathematical notions of localization by scan matching. Section 4 presents the proposed approach, including the landmark models and their parametric description for optimization by means of a search strategy based on a genetic algorithm. The fifth section discusses the results obtained in simulations and experiments with an industrial skid-steer compact excavator CAT 262C. Finally, the discussion and conclusions are mentioned in Sections 6 and 7, respectively.

2. Related Work

Since the first Automated Guided Vehicles (AGVs) were introduced to the market in 1950s, AGVs have evolved from pure wire or magnetic tape following mobile platforms into more sophisticated laser-guided vehicles (LGVs) or Autonomous Mobile Robots (AMRs). Navigation strategies have become particularly important because they enable the localization of mobile platforms by determining position and orientation. Early approaches used beacons, barcodes, and a combination of sensors for wheel odometry and evolving into more sophisticated laser scanning and vision systems for environmental recognition. Landmarks are useful for mobile robot navigation because they provide references for localization strategies that determine the position and orientation of mobile platforms. Humans use landmarks as a spatial representation of the environment to locate themselves and provide references to others. It is common to use landmarks to define the location of other objects/regions or in the creation of maps [9]. A landmark is a building, place, or object that is used for location and can be easily recognized. Robots, like humans, can recognize landmarks from their environment, and according to Thrun [10] landmarks can be those artificially placed in the environment or natural landmarks, which need to be discovered through algorithms to detect walls, corners, colors, etc. The first studies in the literature concerning localization methods for load-haul and dump (LHD) vehicles considered dead reckoning and artificial beacons [11]. The beacons used to correct the cumulative error from the odometry were retro-reflective markers in the roof detected using passive optical switches or passive LC (inductive-capacitive) resonators detected with simple antennas. Later developments included gyro sensors and laser-based guidance control systems (GCS). Gyros provide heading information, while GCS laser sensors allow locating the vehicles relative to the beacons using the beacons' bearing angle [12]. This approach requires the beacons' position to be known in the map the environment. An approach that combines RFID beacons as landmarks [13] combined with 2D lidar scanners for accurate mapping with the aim of using the maps in later localization technology.

Recent publications [14,15] propose the use of lidar-based SLAM approaches for autonomous navigation of LHDs. SLAM with laser scanners requires distinctive features in the environment that could serve as natural landmarks or anchor points without which the correct alignment of scans is not possible [16,17]. Unfortunately, unlike natural caves, mining tunnels are relatively straight or slowly curving, with walls that do not have distinctive uniquely identifiable elements, either because the walls have a repetitive coarse texture or are relatively smooth because modern tunnel boring machines and shotcrete spraying are employed in their construction.

Androulakis [18] uses a 2D lidar scanner to extract two types of features from pillar coal mine: linear segments for modeling entrance ribs and significant points for modeling intersection corners; however, between mine corners or entrances may be hundreds of meters where platforms will not have a robust reference or landmark. Therefore, landmarks, beacons, or some other kind of marker must be introduced in order to make SLAM-based approaches truly feasible. Based on the above, artificial landmarks can be justified for practical application in the navigation of mobile robots in harsh environments, and although the approach may be considered somewhat old and rudimentary, there are some basic issues in the design of optimized artificial landmarks that have not yet been resolved. In particular, the design of the optimized shape and spacing of landmarks must be studied to ensure their identifiability, as well as accurate and reliable localization of mobile robotic platforms. It is worth mentioning that the localization problem by means of scan matching can also be solved using different registration algorithms, such as the Iterative Closest Point (ICP) algorithm [19] or the Normal Distribution Transform (NDT) [20], but its analysis is out of the scope of this work. For a comparison of scan-matching approaches applied to the 3D mapping of underground mines, see, for example, the in-depth study by [21].

Others fields where landmarks are applied for robot localization include:

- Navigation in fruit groves with large tree canopies that create tunnel-like conditions [22];
- Visual place recognition (VPR) in changing environments for autonomous navigation exploiting landmarks to improve matching between images [23–25];
- Robotic guidance using sonar systems aided by landmarks that are inspired by the shape of flowers that act as sonar reflectors to improve localization and navigation [26];
- Indoor navigation [10,27];
- Outdoor urban navigation [28–30].

Methods using landmarks, such as poles or trees, for localization in GPS-denied agricultural and urban environments can be found in previous work by the authors [22,31] and references therein.

3. Preliminary Notions of Localization by Scan Matching

Let $\mathbf{q}_t = (x_t, y_t, \theta_t)$ denote the pose vector of the mobile platform, composed by the position coordinates (x_t, y_t) in the 2D plane and its orientation θ_t measured with respect to the horizontal axis of the world reference frame at time t. Denote by $\mathcal{Z}_t = \{(r^s_{t,k}, \theta^s_{t,k})$: $k = 1, 2, \ldots, m\}$ the set of lidar sensor measurements at time t expressed as a set of points with polar coordinates of the sensor's frame of reference, which, for simplicity of exposition, is assumed to be coincident with the mobile platform's frame of reference. Thus, $r^s_{t,k}$ represents the k-th measured distance from the mobile robot's center of mass to the object reflecting the laser's beam and $\theta^s_{t,k}$ the direction of the corresponding beam. For a platform that is estimated to be located at $\hat{\mathbf{q}} = (\hat{x}_t, \hat{y}_t, \hat{\theta}_t)$ at time t, the measurements in $\mathcal{Z}_t$ imply that that the estimated locations of objects in the map have Cartesian coordinates in the world reference frame are given by:

$$\hat{m}(\hat{\mathbf{q}}, \mathcal{Z}_{t,k}) = \begin{bmatrix} (r^s_{t,k} + n_r)\cos(\hat{\theta}_t + \theta^s_{t,k} + n_\theta) + \hat{x}_t + n_x \\ (r^s_{t,k} + n_r)\sin(\hat{\theta}_t + \theta^s_{t,k} + n_\theta) + \hat{y}_t + n_y \end{bmatrix}$$

where n_r, n_θ represent measurement noises in range and bearing, while n_x and n_y are position estimate uncertainties. Thus, the set of coordinates corresponding to the estimated location of objects in the world surrounding the robot is $\mathcal{M}_t(\hat{\mathbf{q}}, \mathcal{Z}_t) = \{\hat{z}(\hat{\mathbf{q}}, \mathcal{Z}_{t,k})$: $k = 1, 2, \ldots, m\}$. The robot localization problem can now be formulated as the problem of finding an estimated pose vector $\hat{\mathbf{q}}^*$ that minimizes the matching error between the true object locations in the map in the set $\mathcal{M} = \{(x_{m,i}, y_{m,i}) : i = 1, 2, \ldots, p\}$ and the estimated location of the objects in set $\mathcal{M}_t(\hat{\mathbf{q}}, \mathcal{Z}_t)$:

$$\hat{\mathbf{q}}^* = (\hat{x}^*, \hat{y}^*, \hat{\theta}^*) = \arg\min_{(\hat{x},\hat{y},\hat{\theta})} \bar{h}_K(\mathcal{M}, \mathcal{M}_t(\hat{\mathbf{q}}, \mathcal{Z}_{t,k})) = \frac{1}{K}\sum_{i=1}^{K} h_i(\mathcal{M}, \mathcal{M}_t(\hat{\mathbf{q}}, \mathcal{Z}_t)) \quad (1)$$

where $\bar{h}_K(\mathcal{A}, \mathcal{B})$ is the modified Hausdorff distance computed with the K best-matching object coordinates between sets of coordinates $\mathcal{A}$ and $\mathcal{B}$. For further details, please see [4].

Scan matching (1) based on the Iterative Closest Point (ICP) algorithm [32] as a metric for the distance between points in the map and the measurements set can also be employed instead of the modified Hausdorff distance to solve the localization problem.

4. Proposed Approach for Reliable Localization

A key aspect of the navigation in tunnels using the Hausdorff-based localization approach is the adequate definition of landmarks that ensure that the scan-matching problem can be solved unambiguously. To this end, the optimization of the landmarks' geometry and positioning requires adequate parametrization. Two landmark models were considered and evaluated. Their characteristics are explained in the next subsections, together with the implementation aspects concerning the genetic algorithm search strategy, the simultaneous localization and mapping strategy, and the optimized landmark search process.

The optimization of the parameters that define a landmark's geometrical characteristics for a given model is carried out using a genetic algorithm search strategy. The parameters of each landmark model, such as height, width, spacing of shape primitives, or the steps of a piecewise linear function (see Section 4.1 for specific details) are treated like genes that characterize an individual in a population of living organisms. The goal is to find a set of parameters (genes), which define the optimized genome or chromosome in the sense that the optimized genome is the one that delivers the best value of a fitness function (a performance or objective function of the optimization problem). In the context of robot localization, the fitness function can be defined as the total localization mean square error (MSE):

$$E = \frac{1}{N}\sum_{i=1}^{N}(x_i^g - \hat{x}_i^*)^2 + (y_i^g - \hat{y}_i^*)^2 + (\theta_i^g - \hat{\theta}_i^*)^2 \tag{2}$$

where $(x_i^g, y_i^g, \theta_i^g)$ are the ground truth values, and $(x_i^*, y_i^*, \theta_i^*)$ are the estimated pose values for samples $i = 1, 2, \ldots N$ of the robot's trajectory along the tunnel with landmarks whose optimized geometry was found by the genetic algorithm search strategy. The ground truth values are available in simulation. In the validation experiments, the ground truth data are generated with an RTK-DGPS (real-time kinematic differential GPS) that delivers centimeter-level positioning accuracy (see Section 5 for specification details).

4.1. Landmark Parametrization

Two approaches for landmark generation are considered. The first one employs shape primitives as the basis for the definition of the genome. With this approach, the genetic algorithm seeks a combination of shape primitive that minimize the localization error in the solution of (1). The second approach defines the landmark as a piecewise linear function with points of varying heights. In this case, the genetic algorithm finds the height values that minimize the localization error. A difference with the approach based on shape primitives is that, in this second approach, the shape of the landmark is initially completely free and not conditioned by the selection of primitives.

Before introducing the landmark models, it is convenient to introduce the following definitions and notation. Let $H(x) \stackrel{def}{=} \{0\,\forall x < 0; 1\,\forall x \geq 0\}$ denote the Heaviside step function, the boxcar function of height h and width δ centered at x can then be defined as: $\sqcap_{h,\delta}(x) \stackrel{def}{=} h \cdot (H(x + \delta/2) - H(x - \delta/2))$. Define the linear segment over interval $(-\delta/2, \delta/2]$ starting at a height a and ending at height b as $l_\delta(x; a, b) = [((b-a)/\delta)(x - \delta/2) + a] \sqcap_{1,\delta}(x)$.

4.1.1. Landmarks Based on Shape Primitives

To model a landmark using shape primitives the following four functions were considered:

- Shape 1 (triangular): $s_1(x; W, H) \stackrel{def}{=} \wedge(x; W) = (1 - |2x/W|) \sqcap_{H,W}(x)$;
- Shape 2 (rectangular): $s_2(x; W, H) \stackrel{def}{=} \sqcap_{H,W}(x)$;
- Shape 3 (parabolic): $s_3(x; W, H) \stackrel{def}{=} \cap(x; W) = (1 - (2x/W)^2) \sqcap_{H,W}(x)$;
- Shape 4 (linear): $s_4(x; W, H) \stackrel{def}{=} l_W(x; 0, 1) H \sqcap_{H,W}(x)$

The chromosome (sometimes called genotype or genome in the genetic algorithms literature) that represents an individual (a realization of a particular landmark) is defined in terms of genes that characterize the individual. A landmark model built using shape primitives has the following genome:

$$\mathcal{G}_P = \{W, H, D\} \tag{3}$$

where the genes correspond to the landmark's width W, its height H, and the separation distance D between consecutive landmarks. The graphic representation of the width, height, and separation between landmarks is shown in Figure 2. Regardless of the shape geometry, whether it corresponds to a triangle, rectangle, parabola, or line, the shape can be bounded by a box of width W and height H, as shown in the case of the triangle in Figure 2a. Additionally, the separation between landmarks is defined by the reference distance D, which can be the same between all landmarks, as shown in Figure 2b or defined to randomly vary in a interval $[D^* - 0.625, D^* + 0.625]$ m, where D^* is the optimized separation distance found with the search strategy explained in Section 4.4. In contrast, piecewise linear landmarks have six positions whose heights must be optimized together, with the overall width W and height H parameters, to produce an optimized shape with more degrees of freedom as illustrated in Figure 2c and explained in more detail in the next subsection.

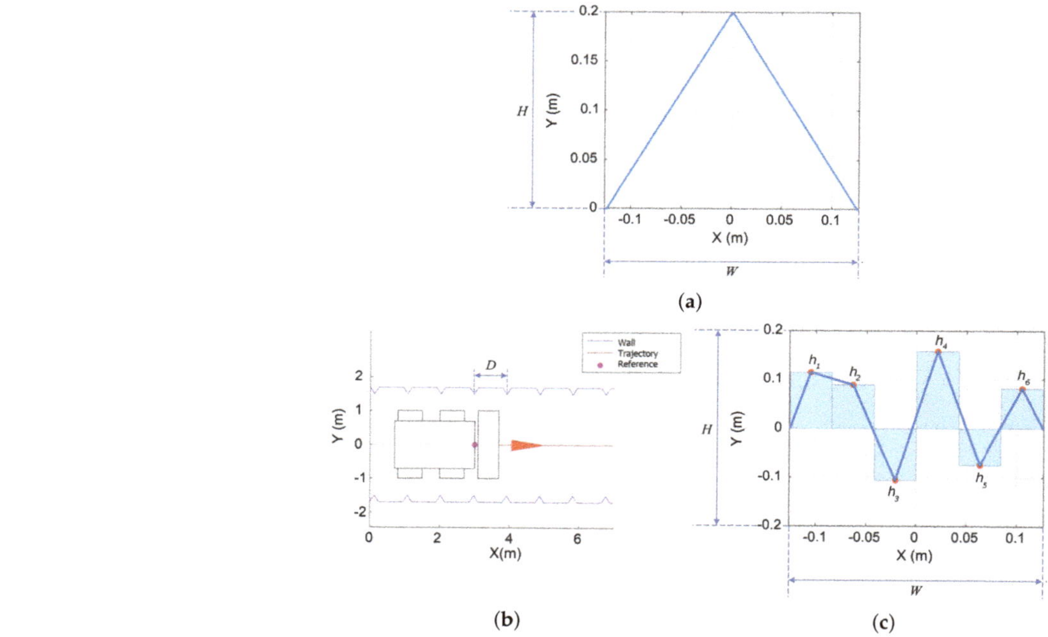

Figure 2. General size parameters of landmarks in terms of height H, width W in (**a**) and separation between landmarks D in (**b**). Shape parameters $h_i, i = 1, 2, \ldots, 6$ of the piecewise linear free shape landmarks define segment height (**c**).

4.1.2. Landmarks Based on Piecewise Linear Functions

The n-segments landmark of width W can be defined as a piecewise linear function with points of height $h_i, i = 1, 2, \ldots, n$ as:

$$L(x; W, H, h_1, h_2, \ldots, h_n) \stackrel{def}{=} \sum_{k=1}^{n} H\, l_{(W/n)}(x - (k-1)W/n + W/2; h_{k-1}, h_k)$$

where $h_0 = 0$. It is to be noted that the landmark model $L(x; W, H, h_1, h_2, \ldots, h_n)$ is centered at $x = 0$.

A landmark model built using a piecewise linear function has the following genome:

$$\mathcal{G}_L = \{W, H, D, h_1, h_2, h_3, h_4, h_5, h_6\} \tag{4}$$

where the genes correspond to the landmark's height points h_i, $i = 1, 2, \ldots, 6$, its width scaling W, its height scaling H, and the separation distance D between consecutive landmarks. A graphical representation of the piecewise linear landmark model is shown in Figure 2c. The range of values for each gene is summarized in Table 1 of the next section. The number of sections in the piecewise linear function considers the fact that polygonal shapes with more points for a given width W can be made smoother, but given sensor noise, additional smoothness does not provide additional distinctiveness. Using two or three segments would result in linear, triangular, or rectangular shapes already considered as part of the shape primitives. Therefore, to determine whether two consecutive shapes, e.g., two triangles or a triangle and a rectangle, offer an additional advantage, the piecewise linear free shape must have at least six parameters. Of course, it is possible to explore even more intricate geometries at the expense of an increased computational burden. Here, it was decided to limit the number of segments to six, but to compensate this limitation by also testing variants that can be easily computed, such as the horizontal symmetry and the vertically inverted landmark variations, as will be shown in the numerical computations and simulation Section 5.1.

Table 1. Minimum and maximum parameter values for the landmark models.

			Parameter Values (Genes)								
Landmark Model		**Units**	H	W	D	h_1	h_2	h_3	h_4	h_5	h_6
Primitive shape	Min.	m	0.01	0.01	0	-	-	-	-	-	-
	Max.	m	0.30	0.60	100	-	-	-	-	-	-
Piecewise linear	Min.	m	0.01	0.05	0	−2.00	−2.00	−2.00	−2.00	−2.00	−2.00
	Max.	m	0.35	0.150	750	2.00	2.00	2.00	2.00	2.00	2.00

4.2. Genetic Algorithm Implementation

Once N individuals characterized by chromosomes $\mathcal{G}_i = \{g_{i,1}, g_{i,2}, \ldots, g_{i,n}\}$, $i = 1, 2, \ldots, N$ of the form (3) or (4) have been initially created by sampling from a uniform distribution $U[g_j^{min}, g_j^{max}]$ with lower and upper bound values g_j^{min}, g_j^{max} from Table 1 for each gene $g_{i,j}$ to build an initial population, the genetic algorithm implemented iterates over the standard steps of fitness evaluation of each individual, selection of individuals, crossover (recombination) of individuals, mutation individuals, and insertion of offspring into the new generation as explained in [33,34]. The population size employed was of $N = 100$ individuals. This number of individuals was empirically found to provide a good trade-off between ensuring a sufficiently large population for convergence while, at the same time, keeping computation time as low as possible.

The fitness evaluation function is the total localization MSE (2). The fitness score of each individual is employed to rank individuals, i.e., sort them in terms of ascending MSE. The selection of individuals employs a stochastic sampling known as stochastic universal sampling or systematic resampling [35,36], in which an initial random number $p_0 \in U[0, 1/N]$ is generated. Individuals laying along a line in which each one has a length proportional to its fitness value are selected by a pointer that takes constant size steps according to $p_k = (k-1)F/N + p_0$, where F is the total fitness (the sum of each individuals' fitness) [35]. The reproduction step selects the best 5% of the total population and employs 75% of the remaining population for crossover. The crossover rule selects the genes (parameters) of consecutive parents according to a selection function in which a random binary vector of the length of the chromosome containing 0's and 1's is generated to indicate whether the gene value must be taken from one parent or the other. Next, the mutation step generates small random variations $\delta_{i,j}$ of the i-th child gene j by sampling a normal distribution $\delta_{i,j} \sim N(0, \sigma^2)$, where $\sigma = (g_j^{max} - g_j^{min})/2$, and sets the gene $g_{i,j}$ of offspring i to a new value $g_{i,j} + \delta_{i,j}$. The reinsertion step simply creates a new population which includes the best 5% individuals and the remaining reproduced population. The number of offsprings generated in each iteration by crossover and mutation is such that the total amount of individuals N is kept constant from one iteration to the other. The stopping condition included a maximum number of iterations of 100, which was never reached because the condition on average relative decrease of the fitness function of 0.1% was met first, as shown in the results Section 5.

4.3. Simultaneous Localization and Mapping

For fast online computation, we employ an Extended Kalman filter and a likelihood field for map probability; see [37,38] for further details. The approach in [37], known as GMapping, is a popular algorithm that employs a Rao–Blackwellized particle filter to estimate the joint posterior. Our approach is similar to that of [38] in that it combines the scan matching and an adaptive update of the likelihood field instead of particle filters proposed in [37] to achieve similar results in terms of the root mean square (RMS) error and low execution time for practical real-time implementation.

In order to make the localization more efficient and accurate, the tunnel walls are removed in order to extract the landmarks and improve the localization's accuracy. Tunnel walls may have some variability or roughness, but this variability is insufficient for unambiguous localization because the magnitude of the variability is comparable to the accuracy of commercially available lidars. Thus, tunnel walls are perceived as practically smooth straight or gradually curving walls. The background removal for landmark extraction is performed using the Random Sample and Consensus (RANSAC) algorithm [27,39]. To account for the possible curvature of the tunnel trajectory, tunnel walls are modeled as a cubic polynomial [40]. All points in the measurements set that do not fit the cubic polynomial within a tolerance margin are labeled as landmark points, as shown in Figure 3b, and are employed in the solution of the localization problem (1).

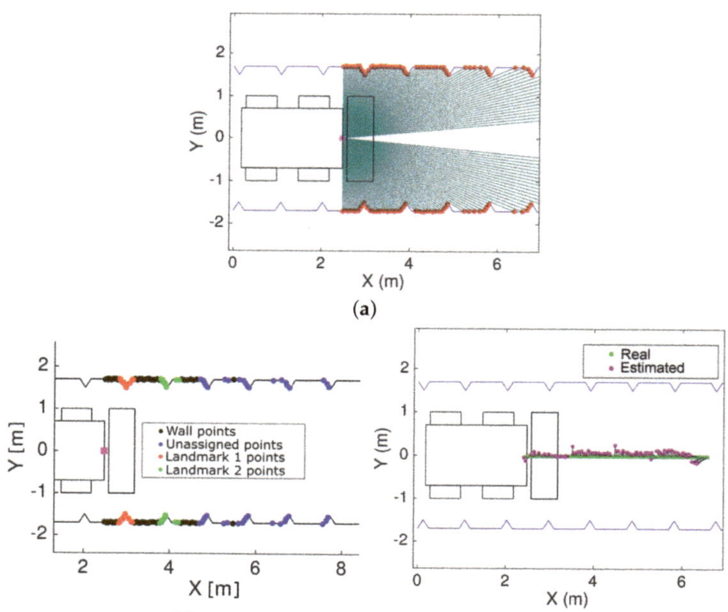

Figure 3. Top view of an ideal tunnel with triangular landmarks showing the matching of noisy lidar measurements to the map (**a**), the classification of landmark and wall points (**b**), and the estimated position using the proposed methodology (**c**).

A top view of a machine moving along a tunnel is presented in Figure 3, which shows the matching of lidar measurements to the triangular landmarks in the map, and the resulting pose estimated using the Hausdorff-based scan-matching approach. To illustrate the matching process, Figure 3a presents an ideal predefined map consisting of lateral tunnel walls with symmetrically and equally spaced triangular landmarks. Despite these landmarks not being optimized in shape and separation, the simulated lidar measurements, including noise in range, are matched, minimizing the modified Hausdorff distance (1). Once the matching has been carried out, the landmarks are classified into wall and landmark measurements. The black points in Figure 3b correspond to wall points as detected by the RANSAC algorithm. The remaining points are treated as landmarks. The matching process considering only the landmarks yields the pose, i.e., position and orientation, thus solving the localization of the machine relative to the landmarks. The pose measurements obtained with the matching procedure can be filtered to generate position and orientation estimates, which are compared in Figure 3c.

4.4. Optimized Landmark Search Process

The process implemented to find the best landmark shapes and spacing is illustrated in Figure 4. The process starts by considering a reference tunnel without landmarks $\mathcal{T}$, a known state trajectory of the robot moving along the tunnel $\mathbf{x}$, and a set $\mathcal{G}_1$ containing N individuals whose chromosomes or genomes define N tentative geometries and distances between landmarks. The initial set of chromosomes $\mathcal{G}_1$ is employed to generate a first set $\mathcal{M}_1$ containing N variations of tunnel $\mathcal{T}$ populated with landmarks according to the separation distance parameter. When creating the map, the spacing between consecutive landmarks $d_\ell, \ell = 1,2,3,\ldots$, is drawn from a uniform distribution $d_\ell \sim \mathcal{U}[D^* - 0.625, D^* + 0.625]$ m, where the value D^* is the value of the optimized landmark separation found in the previous iteration. Hence, the position of landmark $\ell = 1,2,3,\ldots,$ is defined as $\mathbf{p}_\ell = \mathbf{p}_{\ell-1} + d_\ell$ with $\mathbf{p}_0 = 0$. The randomly varying distance in a bounded interval is important in order to avoid ambiguous matching of consecutive landmarks due to repeating landmark separations. Then, the SLAM problem is solved for the simulated robot following trajectory $\mathbf{x}$ in the N maps in $\mathcal{M}_1$. The fitness function for the pose error (2) is evaluated for the N maps. Unless the stopping conditions explained in the subsection concerning the genetic algorithm implementation are met, the genetic algorithm must select the best candidates, produce crossover, and iterate until a chromosome $\mathcal{G}^*$ defining the optimized landmark geometry is returned.

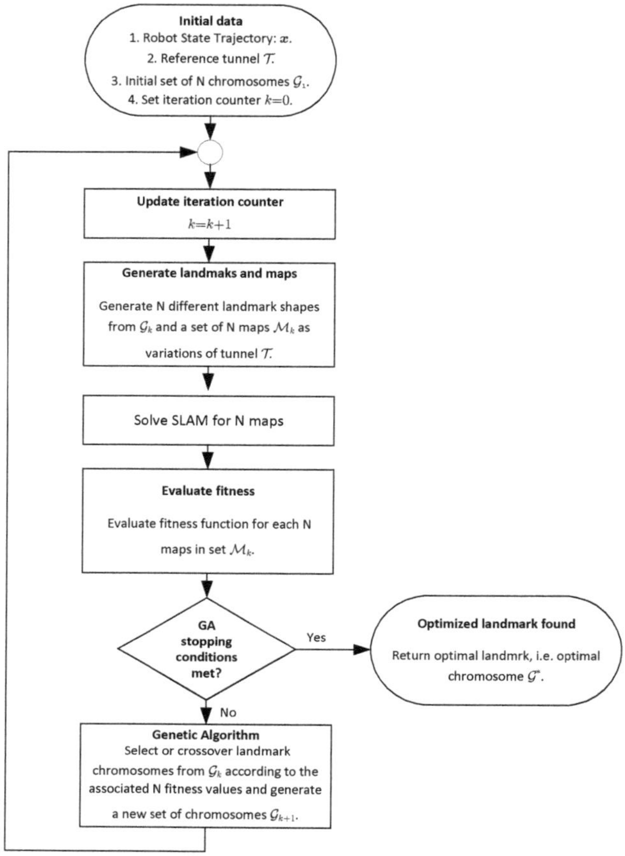

Figure 4. Implemented search scheme for optimized landmark geometries and spacing.

5. Results

The proposed approach is evaluated both in simulations and experimentally. The implementation of the robot trajectory simulation and SLAM, as well as the genetic algorithm, were implemented in Python without using other libraries than the standard mathematical function libraries NumPy and SciPy for numerical computations with arrays and matrices, integration of the ordinary differential equation of the robot's dynamics using the `odeint` function. The motion model equations are explained in detail in [2] and describe motion dynamics of a semi-autonomous industrial compact skid-steer loader CAT 262C employed in the experiments. The simulations use a grid map with a 1 cm^2 per pixel resolution and a position sensor model with a distance RMS error of 5 cm, which means 95% of the measurements are contained in an 8.65 m radius circle. The sampling frequency of the simulated system is 1 kHz and it is assumed that the same clock rate is employed for all sensors and the control loop. For the visualization of results, we use PyGame and Matplotlib libraries.

The experiments employ a semi-autonomous industrial compact skid-steer loader CAT 262C equipped with one Sensor STIM300 inertial measurement unit (IMU), two VectorNav IMU's, one Piksi SwiftNav RTK-DGPS, two torque sensors by Manner Sensortelemetrie, four Sick LMS 511 lidars, wheel encoders, TE Connectivity MEAS inclination sensors, control, and navigation computer (running ROS Melodic, sampling sensors at 100 Hz) and wireless communication interfaces. The Sick LMS 511 lidar is designed for industrial operation outdoors even with dust or rain, allowing for multiple echoes and materials with different absorption/reflectance levels. The reflectance of soil/rocks is typically in the range of 50–60% [41,42], and given the laser beam power employed by LMS 511 and the manufacturer specifications [43], this lidar can scan soil or rocks up to 60–65 m without the aid of retro-reflective markers. In the experiments, we used common cardboard landmarks, which have a reflectance in the operating wavelength of lidar equivalent to that of soil/rocks [41].

The skid-steer loader in the test site is shown in Figure 1. The experiments were carried out in a mock-up of the tunnel without and with the optimized landmarks found in simulation to validate the approach. Following previously published work [44], we have selected the RMS error to assess the localization error.

The Hausdorff scan matching implemented in this work considered 80% of the best matching points that minimize the modified directed Hausdorff distance with respect to the reference model in order to improve the data association following the tuning recommendations in [4], i.e., K in (1) is set $K = 0.8m$, where m is the total number of measurements. Since the scan matching procedure sorts the lidar measurements starting with the best fitting points, discarding the worst 20% of the matched points removes the matching bias and ensures sufficient measurements are available so that the matching does not become an ill-posed problem. An adaptive threshold K may be implemented in terms of an expectation-maximization strategy, but this aspect necessitates new theoretical developments beyond the scope of the current work to ensure the optimality of a dynamically adjusted threshold. To show that the choice of the fixed 80% threshold is adequate for practical applications, consider Figure 5, which presents the outcome of a simulation experiment in which 300 noisy lidar measurement points must be aligned to a reference model. The lidar ranging error is considered to have zero-mean Gaussian distribution with standard deviation $\sigma = 0.05$ cm, which is a typical value for the Sick LMS 511 employed in our experimental validation. The Sick LMS 511 can deliver 720 scan points with an angular resolution of 0.25° covering a 180° field-of-view. Here, we are using less than half the points that may be obtained using Sick LMS 511 for testing purposes. In practice, the number of scan points covering a landmark will depend on the distance to the landmark and scanning angular resolution, which can be adjusted to different values between 0.042° and 1° in the case of Sick LMS 511. As shown in Figure 5a, when a 100% of the lidar measurements are employed, there exists a bias in the final alignment due to spurious measurements. On the other hand, when the 80% best-matching points are selected in the computation of the modified Hausdorff distance (1), the noisy point cloud is fitted more accurately to the reference model, as shown in Figure 5b.

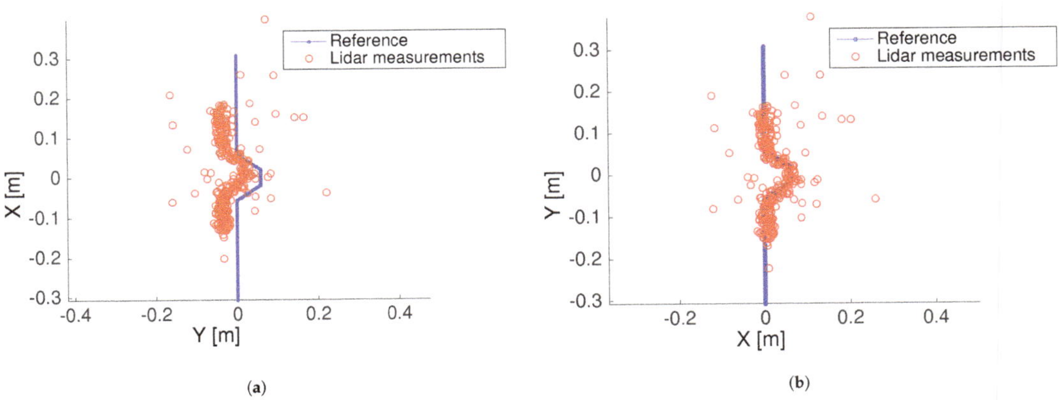

Figure 5. Scan-matching results using Hausdorff distance considering all measurements (**a**) and 80% of the best matching points (**b**).

5.1. Numerical Computation and Simulation Results

The robot simulation and the genetic algorithm to find the optimized landmarks were implemented in Python. The following trials were considered: P1—triangular primitive; P2—rectangular primitive; P3—parabolic primitive; P4—linear primitive; F1—piecewise linear free shape; F2—piecewise linear inverted free shape; F3—piecewise linear symmetric free shape; and F4—piecewise linear symmetric inverted free shape. If $s(x)$ is a shape, then the inverted shape is $1 - s(x)$. A symmetric shape is a shape that is an even function, i.e., $s(x)$ is symmetric if $s(x) = s(-x)$. In the implementation of the genetic algorithm, the parameters (genes of each individual's chromosome) were allowed to take values in an interval whose lower and upper bounds are summarized in Table 1.

The convergence of the RMS position error component of fitness function for each iteration of the genetic algorithm while searching for an optimized landmark geometry and separation is shown in Figure 6. The resulting piecewise linear models are shown in Figure 7. The different curves that are shown in each graph of Figure 7 represent a realization of the best individual's chromosome for a given generation. After several iterations, the best individuals of each generation evolve and converge to overlapping shapes that strongly coincide, thus confirming that an optimized geometry minimizing the pose error (2) exists. It is to be noted that the relative average decrease in the fitness function (2) becomes less than 0.1% for either the ICP or Hausdorff matching approaches after 20 iterations when using the shape primitives and at least 40 iterations when using the linear piecewise landmark model because it has more parameters. The optimized landmarks found in iteration 45 for the different shapes and models were selected when testing the localization performance to make a fair comparison and remove the differing amount of iterations as a possible advantage factor.

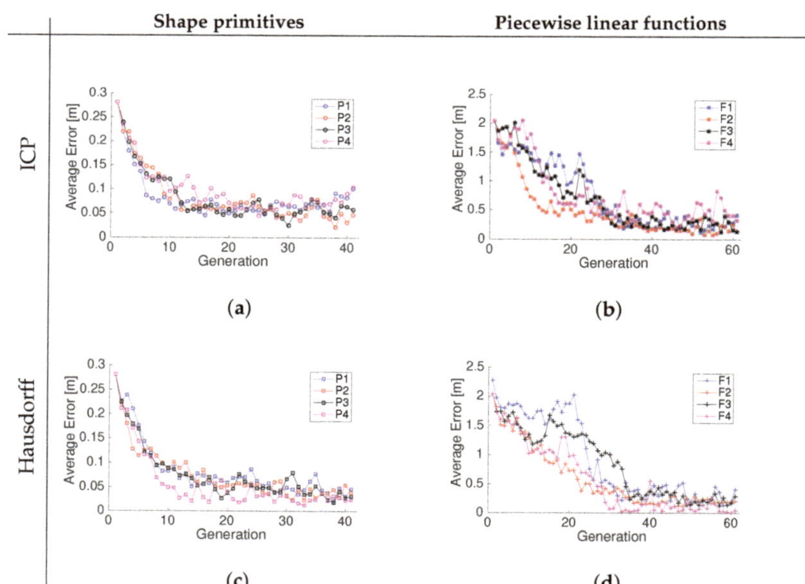

Figure 6. Convergence of the square root of the trace of the matching error covariance matrix for shape primitives (**a**,**c**) and piecewise linear functions (**b**,**d**) using ICP and Hausdorff matching, respectively.

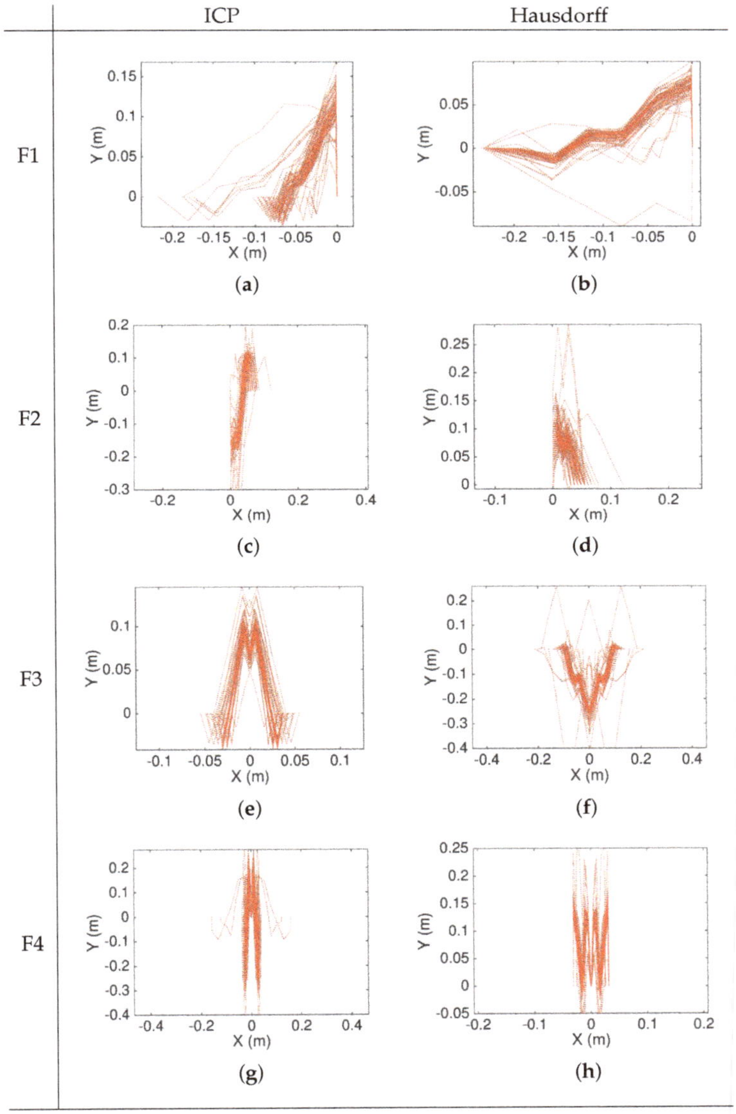

Figure 7. Optimized piecewise linear shapes found by the genetic algorithm after 45 iterations: F1—free (**a**,**b**); F2—inverted (**c**,**d**); F3—symmetric free shape (**e**,**f**); F4—symmetric inverted (**g**,**h**). The graphs show the evolution of multiple iterations superimposed showing the convergence to the optimized landmark geometry.

Regardless of the type of landmark, the results Figure 6 show that the Hausdorff matching converges with less variability than ICP. The genetic algorithm not only identified the best shapes for accurate matching, but also identified the optimal distance D between the landmarks, which was found to be between 8 and 10 m. Figure 8 shows that for the different landmarks, the initial proposed distance values D are approximately uniformly distributed. Regardless of the matching approach (ICP or Hausdorff), the distribution after 41 iterations of the *gene* associated to the separation D between landmarks concentrates around 9–10 m when using the shape primitives P1, P2, P3, or P4 landmarks, and they around 7–8 m when using the piecewise linear free shape landmarks F1, F2, F3, and F4. It is to be noted that in the case of landmarks F1, F2, F3, and F4, ICP tends to prefer closer landmarks with $D \approx 7$ m, while the Hausdorff matching produces lower RMS localization errors with landmarks separated by $D \approx 8$ m, as shown in Figure 8.

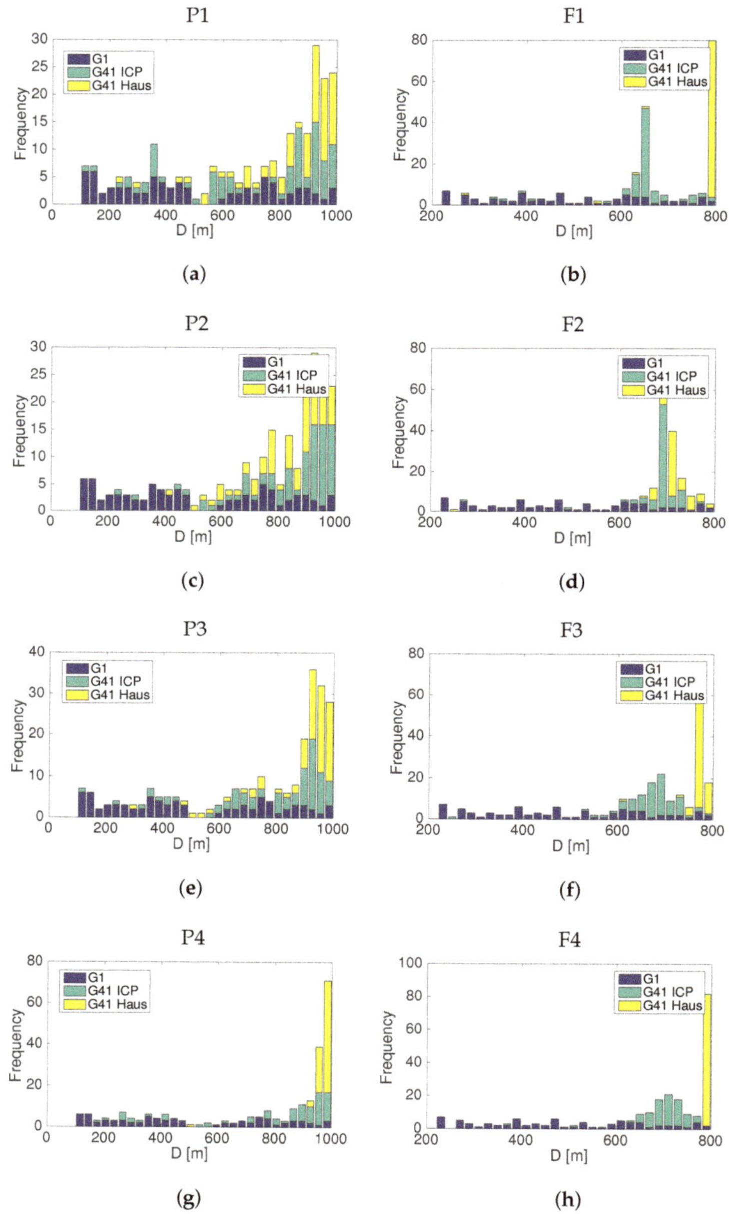

Figure 8. Histogram of the optimized distances between landmarks found by the genetic algorithm after 41 iterations (G41) compared to the initial distribution of distance genes (G1) for the primitive shapes P1, P2, P3, P4 (**a**,**c**,**e**,**g**), and the free piecewise linear shapes F1, F2, F3, F4 (**b**,**d**,**f**,**h**).

From the simulations presented in Table 2, it is possible to confirm that the triangular shape model (P1) yields the smallest RMS localization error for the robot in a simulated tunnel that was 10 m long, with an error of 22 mm using the Hausdorff matching strategy. The second best landmark is the symmetric inverted piecewise linear model (F4) resembling an inverted double triangular shape or "W" shown in Figure 7h, which yields an RMS localization error of 24 mm using the Hausdorff matching strategy. The results in Table 2 show that the best results are achieved with the Hausdorff matching strategy when compared to the ICP method. Even if ICP had a better performance than the Hausdorff matching strategy with two of the linear piecewise models, the Hausdorff matching technique delivers better results in all other cases because the RMS errors are 30–70% smaller.

Table 2. RMS localization error for each type of landmark found by the genetic algorithm.

	RMS Error [m]			RMS Error [m]	
Primitive Shape Models	**ICP**	**Hausdorff**	**Piecewise Linear Model**	**ICP**	**Hausdorff**
P1	0.042	0.022	F1	0.073	0.052
P2	0.091	0.026	F2	0.043	0.064
P3	0.063	0.028	F3	0.033	0.068
P4	0.065	0.026	F4	0.035	0.024

5.2. Experimental Validation

The experimental validation using the semi-autonomous CAT 262C skid-steer loader consisted of 15 repetitions each, first in a 10 m mockup tunnel without landmarks (experiment 1), then using the triangular shape primitive model P1 identified by the genetic algorithm (experiment 2), and, finally, the symmetric inverted piecewise linear landmark model F4 (experiment 3). The localization was solved with both the ICP and the Hausdorff matching strategy. The results in terms of average RMS localization error and 95% confidence intervals are summarized in Table 3. The experimental results reported in Table 3 employed the best landmarks evaluated in simulation as reported in Table 2, which are landmarks P1 (triangular shape primitive) and F4 (symmetric inverted piecewise linear free shape).

Table 3. RMS localization error using the optimized landmarks during experimental validation.

		RMS Localization Error [m]	
Experiment		**ICP**	**Hausdorff**
1	Without landmarks	20.765 ± 0.074	19.748 ± 0.113
2	P1—Triangular shape landmark	0.258 ± 0.046	0.235 ± 0.035
3	F4—Symmetric inverted landmark	0.206 ± 0.096	0.219 ± 0.093

The experimental results confirm that the symmetric inverted landmark F4 is slightly better compared to the triangular shape landmark model P1. However, the 26 mm difference on average is within the 95% confidence interval, which for the symmetric inverted landmark, is 93 mm. Compared to the case with no landmarks, which has an RMS localization error of almost twice the traveled distance (10 m) in the experiments, the localization approach with the proposed landmarks is very accurate and proves to be suitable for the localization of underground mining loaders and trucks. It is also to be noted that ICP performed better with an RMS localization error 13 mm smaller than the RMS localization error obtained with the Hausdorff matching strategy using the symmetric inverted landmark model F4. However, with the simpler triangular landmark model P1, ICP yields an RMS localization error that is 23 mm larger. Comparing the RMS errors presented in Tables 2 and 3, it is possible to observe that the experimental RMS positioning error is approximately 10 times larger than the RMS positioning error obtained in the simulations. This is mainly explained by the fact that the performance of the RTK-DGPS had, in practice, an RMS error of 8.3 cm, which means that about 95% of the measurements fall within a circle with a 14.4 cm radius. On the other hand, the clock rate of the different subsystems is different. The control loop was implemented at 100 Hz, but the RTK-DGPS provides measurements at 10 Hz, while the lidar and RTK-DGPS have 10 Hz sampling rates. Since the RTK-DGPS measurements are employed as ground truth, the practical RMS error includes the GPS error, but also the lidar's accuracy, which are approximately 5 cm.

5.3. Validation with an Underground Mine Dataset

A validation of the approach and the optimized landmarks is also carried out using the publicly available 3D point cloud dataset of the El Teniente copper mine located in Chile [8]. A 100 m section of one of the tunnels was extracted from the dataset and artificial landmarks P1 were added with randomly varying distances $D \sim U[9.25, 10.5]$ m around the optimized value found by the genetic algorithm to ensure non-uniform spacing between landmarks and thus avoid ambiguous matching of consecutive landmarks due to repeating landmark separations. The triangular landmark geometry P1 was chosen for validation with the data underground mine data set because it is a simpler geometry to manufacture and because it yielded an RMS localization error in the real-world experiments that is similar to that of the best landmark geometry F4 (see Table 3). Furthermore, the RMS localization error obtained in the runs of the Genetic Algorithm give a slight advantage to P1 over F4, when using the Hausdorff matching strategy, as shown in Table 2. A physically accurate model of the skid-steer loader developed in [2] was simulated to evaluate the effectiveness of the landmarks for SLAM using the scan matching procedure based on the modified Hausdorff distance [4]. The results are shown in Figure 9, which shows the traversed trajectory in Figure 9a, the matched point clouds in Figure 9b, the distance transform of the point clouds employed for matching using the modified Hausdorff distance in Figure 9c, and the resulting map and measured trajectory (red) compared to the trajectory ground truth (blue) in Figure 9d. The ground truth corresponds to the skid-steer loader's trajectory obtained by the model simulation assuming noise-free position sensors. On the other hand, the map considers a grid with a resolution of 10×10 cm^2 per pixel, while the measurement model considers the ranging error to be zero-mean Gaussian distributed with standard deviation $\sigma = 0.05$ cm, which is a typical value for the Sick LMS 511. An RMS localization error between the true position and the measured position of 0.163 ± 0.072 m was obtained after 15 repetitions, i.e., the simulation was repeated 15 times with a virtual machine driving in the tunnel considering the sensor noise parameters. The obtained localization error was registered to compute the RMS error across the 15 realizations and the 95% confidence interval of the RMS localization error. It is to be noted that without the landmarks, it is not possible to solve the SLAM problem correctly because the tunnel walls are almost smooth, thus causing the matching to diverge due to the lack of anchor points that could be used for reliable scan alignment. The RMS localization error in the dataset without landmarks obtained with the Hausdorff matching approach was 194.3 ± 0.22 m, while IPC resulted in an RMS localization error of 201.7 ± 0.15 m.

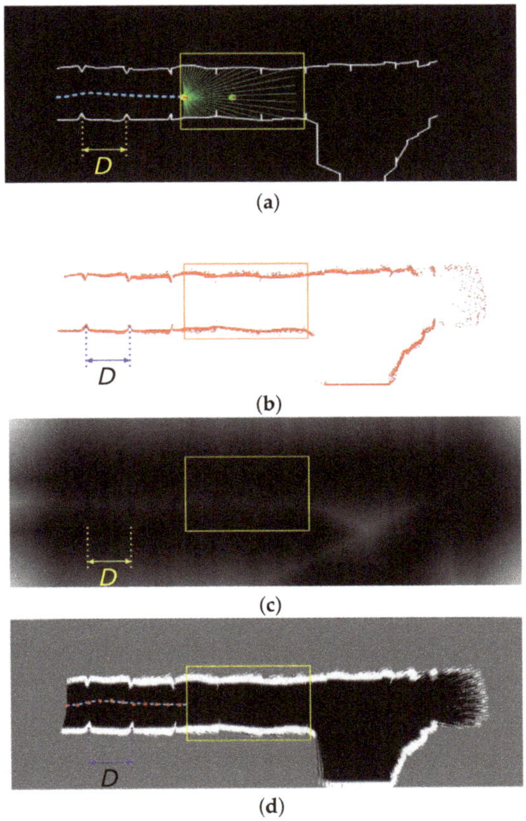

Figure 9. SLAM solution of the El Teniente tunnel showing the ground truth map, trajectory, and distance between landmarks (**a**); the laser rangefinder scan (**b**); the Voronoi distance transform of the scan (**c**); and the estimated grid map and trajectory results (**d**).

6. Discussion

The main findings after the experimental validation of the proposed strategy for navigation in mining tunnels are discussed as follows, considering both their significance and limitations:

1. The approach based on optimized artificial landmarks' geometry and spacing is suitable for localization and mapping in smoothly bored underground mining tunnels, where no GPS signal is available and where deploying and maintaining a network of active RF or optical beacons is costly and difficult.
2. Without landmarks, it is not possible to solve the localization problem using lidar information in smooth tunnels because the localization problem becomes ill-posed, as evidenced by the cumulative error of the positioning without landmarks reported in Table 3. Even if different SLAM techniques have been developed to reduce the well-known localization slip or drift problem [45], reliable underground localization and mapping requires accurate positioning drift-free strategies [7] to ensure industrial grade safety standards. Therefore, artificial landmarks are an essential part of the proposed solution for operation in adverse and challenging underground mining conditions. Other solutions, relying on SLAM algorithms and variants that employ natural landmarks may work partially and exhibit drift sporadically; thus, the use of natural landmarks is still not applicable for 24/7 working schedules required by the

mining industry. On the other hand, passive artificial landmarks may be cheaper to manufacture, install, and maintain compared to active RFID or IR beacons.

3. The optimization of landmark geometries for the different models (shape primitives and piecewise linear) yields expected positioning errors in the range 20–90 mm depending on the geometry. Considering the approximately 50 mm difference between the worst and best model, it is possible to conclude that adequate landmark design and optimization is worth the effort.
4. In addition to the development of an optimization scheme for the landmarks' geometry and spacing presented in Section 4.4) to improve localization, important contributions that are of practical relevance are the validation of: (i) the feasibility of the approach through experimental validation for localization in relatively smooth tunnels, in which traditional scan matching and visual features do not work due to the lack of sufficiently distinctive features that could be matched without ambiguity (see Table 3); (ii) the advantage of Hausdorff-based matching compared to the ICP method (see Table 2); and (iii) the gains in localization accuracy than can be achieved by optimizing the geometry and spacing of landmarks by means of a genetic algorithm search strategy (see Table 2).
5. The experiments were conducted with a mock-up of a smooth tunnel both with and without landmarks. Modern machine-bored tunnels are relatively smooth and lack features. Thus, the mock-up replicates a challenging geometry for matching and localization rather than visual appearance. In order to further validate the approach, an accurate 3D model of a 100 m section of one of the tunnels of the Chilean El Teniente copper mine from the dataset by [8] was employed. Fifteen iterations assuming typical motion disturbances and sensor noise, with magnitudes equivalent to those of the CAT 262C [2] and Sick LMS 511 lidar, were carried out to ensure statistical significance. Future work considers creating a new dataset and additional testing in different underground tunnels, which during this research has not been possible due to increased restrictions to access mining sites during the pandemic.
6. The RMS positioning error obtained in the experimental validation of Section 5.2 is influenced by the accuracy of the RTK-DGPS (RMS error of approximately 8.3 cm), which was employed as the ground truth. Another limitation arises from the accuracy and resolution of the lidar scanner, which is approximately 5 cm. We expect that the positioning accuracy measured in our experiments should improve with ongoing technological advances and the development of more accurate lidar and GPS sensors.
7. Concerning the practical implementation of the approach, two important aspects need to be considered: (i) the execution time and (ii) the environment's visibility conditions. The results presented in Section 5.3 show that the execution time is adequate for real-time implementation applicable to underground machines operating at standard speeds of 20 to 30 km/h. The effects of environmental visibility due to dust were not tested as part of this study. However, there exist laser range scanners and other vision systems that have been successfully employed in commercial collision avoidance systems for mining equipment, e.g., SICK's MINESIC100 EPS, MINESIC100 TCW or Visionary-B.
8. The accurate localization of artificial landmarks on the map does not need to be performed using accurate georeferencing or topographic stations since the landmark's location can be jointly estimated with the position. Once the landmarks have been deployed, practically no maintenance is required unless some are damaged and need to be replaced. The low-maintenance requirements are an advantage of the proposed solution compared to systems requiring energy supply and network connectivity for active optical and RF beacons.
9. Our ongoing research efforts are focused on improving the proposed approach with deep learning techniques and neural networks for different purposes, which include visual feature extraction, scene recognition, ego-motion estimation, and map matching. Techniques based on deep neural networks have shown promising results to improve lidar matching, e.g., OverlapNet by Chen et al. (2021) [46], and optical flow estimation,

e.g., Flownet by Fischer et al. (2015) [47], including RGB-D SLAM with convolutional neural networks [48] and 3D indoor scene mapping [49]. Hence, these techniques may improve the accuracy and robustness of lidar and visual matching, as well as motion estimation, which are essential for SLAM in underground tunnels. It is to be noted that an important challenge for the application of visual techniques in the harsh mining environments is the poor visibility in tunnels due to low illumination conditions and dust, as well as machine vibrations, which are typically not a problem in indoor or urban robotics.

7. Conclusions

An approach for reliable autonomous navigation in modern smoothly bored mining tunnels which ensures drift-free localization and consistent mapping has been developed and validated. The approach relies on the optimization landmark geometry and positioning (distance between landmarks). Finding the optimized parameters was achieved with a genetic algorithm search strategy. The results show that optimizing a free shape using a piecewise linear function leads to a inverted double triangular shaped landmark, while very similar results are obtained with the optimized triangular shape primitive. From a practical perspective, it may be more convenient to use simple optimized triangular-shaped landmarks because the positioning accuracy is on average around 22 cm, with a small difference of 2.6 cm, which is within the ± 9.3 cm confidence interval of the piecewise linear inverted double triangular shape. The experimental validation using a compact skid-steer excavator CAT 262C shows that without landmarks, the cumulative drift error steadily grows, and correct localization is not possible due to the ambiguity in lidar scan matchings. The experimental results thus confirm that using shape-optimized passive landmarks are a reliable alternative for localization and navigation in modern underground smoothly bored mining tunnels, for which electrically powered active optical or RF beacons are less likely to be accepted by the underground mining industry due to concerns on maintenance cost involved to prevent malfunctioning risks and ensure operational safety in case of a loss of power supply. The applicability of the localization approach for SLAM in real underground mines was verified using an accurate 3D model of a 100 m tunnel section of El Teniente mine in Chile, which is the largest underground copper mine in the world. Ongoing research is concerned with improving the accuracy and robustness of the proposed localization and mapping approach with deep learning techniques for ego-motion estimation, map matching, and the extraction of visual features that could be used as landmarks. An important challenge for the application of visual techniques in the harsh mining environments is the poor visibility in tunnels due to low illumination conditions and dust, as well as machine vibrations, which are typically not a problem in indoor or urban robotics. Our work in progress also considers improvements to the proposed approach for navigation in fruit groves and forests with large tree canopies that create tunnel-like conditions.

Author Contributions: M.T.-T., P.N.-B. and F.P.-L. contributed equally to the design, development and implementation, performing the experiments, and analyzing the results; J.G. and M.T.-T. edited and reviewed the original manuscript; M.T.-T. and F.A.C. provided initial conceptualization and advice along the research project. All authors contributed to the review of the state-of-the-art, the research, the writing of the article, and approved the submitted version. All authors have read and agreed to the published version of the manuscript.

Funding: This project has been supported by the National Agency of Research and Development (ANID, ex-Conicyt) under Fondecyt grant 1220140, Fondequip grant 120141, Basal grant FB0008 and PFCHA/ Doctorado Nacional/ 2020-21200700 and DGIIP-UTFSM.

Data Availability Statement: Mining tunnel 3D lidar data for "El Teniente" mine used in this paper are publicly available at [8].

Acknowledgments: We thank the RAL UC group for their help in setting up the mine mockup environment, Jorge Reyes and Daniel Calabi for their valuable suggestions during the experimental stage.

Conflicts of Interest: The authors declare that the research was conducted in the absence of any commercial or financial relationships that could be construed as a potential conflict of interest.

Abbreviations

The following abbreviations are used in this paper:

AGV	Automated Guided Vehicle
AMR	Autonomous Mobile Robot
DGPS	Differential GPS
EKF	Extended Kalman Filter
EM	Expectation-maximization algorithm
F1	Piecewise linear free shape
F2	Piecewise linear inverted free shape
F3	Piecewise linear symmetric free shape
F4	Piecewise linear symmetric inverted free shape
GCS	Guidance control system
GPS	Global Positioning System
ICP	Iterative Closest Point
IMU	Inertial measurement unit
LC	Inductive-capacitive
LGV	Laser-guided vehicle
LHD	Load-haul and dump
MSE	Mean square error
NDT	Normal Distribution Transform
P1	Triangular primitive
P2	Rectangular primitive
P3	Parabolic primitive
P4	Linear primitive
RANSAC	Random Sample and Consensus
RMS	Root mean square
RTK	Real time kinematic
SLAM	Simultaneous localization and mapping

References

1. Rigotti-Thompson, M.; Torres-Torriti, M.; Auat Cheein, F.; Troni, G. $\mathcal{H}_\infty$-based Terrain Disturbance Rejection for Hydraulically Actuated Mobile Manipulators with a Non-Rigid Link. *IEEE/ASME Trans. Mechatron.* **2020**, *25*, 2523–2533. [CrossRef]
2. Aguilera-Marinovic, S.; Torres-Torriti, M.; Auat-Cheein, F. General Dynamic Model for Skid-Steer Mobile Manipulators with Wheel–Ground Interactions. *IEEE/ASME Trans. Mechatron.* **2017**, *22*, 433–444. [CrossRef]
3. Thrun, S.; Thayer, S.; Whittaker, W.; Baker, C.; Burgard, W.; Ferguson, D.; Hahnel, D.; Montemerlo, D.; Morris, A.; Omohundro, Z.; et al. Autonomous exploration and mapping of abandoned mines. *IEEE Robot. Autom. Mag.* **2004**, *11*, 79–91. [CrossRef]
4. Donoso-Aguirre, F.; Bustos-Salas, J.P.; Torres-Torriti, M.; Guesalaga, A. Mobile robot localization using the Hausdorff distance. *Robotica* **2008**, *26*, 129–141. [CrossRef]
5. Zhang, X.; He, M.; Yang, J.; Wang, E.; Zhang, J.; Sun, Y. An Innovative Non-Pillar Coal-Mining Technology with Automatically Formed Entry: A Case Study. *Engineering* **2020**, *6*, 1315–1329. [CrossRef]
6. Lösch, R.; Grehl, S.; Donner, M.; Buhl, C.; Jung, B. Design of an Autonomous Robot for Mapping, Navigation, and Manipulation in Underground Mines. In Proceedings of the 2018 IEEE/RSJ International Conference on Intelligent Robots and Systems (IROS), Madrid, Spain, 1–5 October 2018; pp. 1407–1412. [CrossRef]
7. Thrybom, L.; Neander, J.; Hansen, E.; Landernas, K. Future Challenges of Positioning in Underground Mines. *IFAC-PapersOnLine* **2015**, *48*, 222–226. [CrossRef]
8. Leung, K.; Lühr, D.; Houshiar, H.; Inostroza, F.; Borrmann, D.; Adams, M.; Nüchter, A.; del Solar, J.R. Chilean underground mine dataset. *Int. J. Robot. Res.* **2017**, *36*, 16–23. [CrossRef]
9. Hu, H.; Gu, D. Landmark-based Navigation of Industrial Mobile Robots. *Ind. Robot Int. J.* **2004**, *27*, 458–467. [CrossRef]
10. Thrun, S. Finding landmarks for mobile robot navigation. In Proceedings of the 1998 IEEE International Conference on Robotics and Automation, Leuven, Belgium, 16–20 May 1998; Volume 2, pp. 958–963. [CrossRef]
11. Mäkelä, H.; Lehtinen, H.; Rintanen, K.; Koskinen, K. Navigation System for LHD Machines. *IFAC Proc. Vol.* **1995**, *28*, 295–300. [CrossRef]
12. Scheding, S.; Dissanayake, G.; Nebot, E.M.; Durrant-Whyte, H. An experiment in autonomous navigation of an underground mining vehicle. *IEEE Trans. Roboti. Autom.* **1999**, *15*, 85–95. [CrossRef]

13. Lavigne, N.; Marshall, J. A landmark-bounded method for large-scale underground mine mapping. *J. Field Robot.* **2012**, *29*, 861–879. [CrossRef]
14. Wu, D.; Meng, Y.; Zhan, K.; Ma, F. A LIDAR SLAM Based on Point-Line Features for Underground Mining Vehicle. In Proceedings of the 2018 Chinese Automation Congress (CAC), Xi'an, China, 30 November–2 December 2018; pp. 2879–2883. [CrossRef]
15. Ren, Z.; Wang, L.; Bi, L. Robust GICP-based 3D LiDAR SLAM for underground mining environment. *Sensors* **2019**, *19*, 2915. [CrossRef] [PubMed]
16. Guivant, J.E.; Masson, F.R.; Nebot, E.M. Simultaneous localization and map building using natural features and absolute information. *Robot. Auton. Syst.* **2002**, *40*, 79–90. [CrossRef]
17. Fairfield, N.; Kantor, G.; Wettergreen, D. Real-Time SLAM with Octree Evidence Grids for Exploration in Underwater Tunnels. *J. Field Robot.* **2007**, *24*, 3–21. [CrossRef]
18. Androulakis, V.; Sottile, J.; Schafrik, S.; Agioutantis, Z. Navigation system for a semi-autonomous shuttle car in room and pillar coal mines based on 2D LiDAR scanners. *Tunnell. Underground Space Technol.* **2021**, *117*, 104149. [CrossRef]
19. Donoso, F.; Austin, K.; McAree, P. Three new Iterative Closest Point variant-methods that improve scan matching for surface mining terrain. *Robot. Auton. Syst.* **2017**, *95*, 117–128. [CrossRef]
20. Magnusson, M.; Lilienthal, A.; Duckett, T. Scan registration for autonomous mining vehicles using 3D-NDT. *J. Field Robot.* **2007**, *24*, 803–827. [CrossRef]
21. Magnusson, M.; Nüchter, A.; Lörken, C.; Lilienthal, A.J.; Hertzberg, J. 3D mapping the Kvarntorp mine: A field experiment for evaluation of 3D scan matching algorithms. In Proceedings of the IEEE/RSJ International Conference on Intelligent Robots and Systems (IROS), Workshop on 3D Mapping, Nice, France, 22–26 September 2008.
22. Auat Cheein, F.; Torres-Torriti, M.; Rosell-Polo, J.R. Usability analysis of scan matching techniques for localization of field machinery in avocado groves. *Comput. Electron. Agricult.* **2019**, *162*, 941–950. [CrossRef]
23. Hou, Y.; Zhang, H.; Zhou, S.; Zou, H. Use of Roadway Scene Semantic Information and Geometry-Preserving Landmark Pairs to Improve Visual Place Recognition in Changing Environments. *IEEE Access* **2017**, *5*, 7702–7713. [CrossRef]
24. Holliday, A.; Dudek, G. Scale-Robust Localization Using General Object Landmarks. In Proceedings of the IEEE International Conference on Intelligent Robots and Systems, Madrid, Spain, 1–5 October 2018; pp. 1688–1694. [CrossRef]
25. Suenderhauf, N.; Shirazi, S.; Jacobson, A.; Dayoub, F.; Pepperell, E.; Upcroft, B.; Milford, M. Place Recognition with ConvNet Landmarks: Viewpoint-Robust, Condition-Robust, Training-Free. In Proceedings of the Robotics: Science and Systems, Rome, Italy, 13–17 July 2015. [CrossRef]
26. Simon, R.; Rupitsch, S.; Baumann, M.; Wu, H.; Peremans, H.; Steckel, J. Bioinspired sonar reflectors as guiding beacons for autonomous navigation. *Proc. Natl. Acad. Sci. USA* **2020**, *117*, 1367–1374. [CrossRef]
27. Nguyen, V.; Martinelli, A.; Tomatis, N.; Siegwart, R. A comparison of line extraction algorithms using 2D laser rangefinder for indoor mobile robotics. In Proceedings of the 2005 IEEE/RSJ International Conference on Intelligent Robots and Systems, Edmonton, Canada, 2–6 August 2005; pp. 1929–1934. [CrossRef]
28. Choe, Y.; Ahn, S.; Chung, M.J. Online urban object recognition in point clouds using consecutive point information for urban robotic missions. *Robot. Auton. Syst.* **2014**, *62*, 1130–1152. [CrossRef]
29. Javanmardi, M.; Javanmardi, E.; Gu, Y.; Kamijo, S. Precise mobile laser scanning for urban mapping utilizing 3D aerial surveillance data. In Proceedings of the 2017 IEEE 20th International Conference on Intelligent Transportation Systems (ITSC), Yokohama, Japan, 16–19 October 2017; pp. 1–8. [CrossRef]
30. Guevara, D.J.; Gené-Mola, J.; Gregorio Lopez, E.; Torres-Torriti, M.; Reina, G.; Auat Cheein, F. Comparison of 3D scan matching techniques for autonomous robot navigation in urban and agricultural environments. *J. Appl. Remote Sens.* **2021**, *15*, 024508. [CrossRef]
31. Veronese, L.d.P.; Auat Cheein, F.; Bastos-Filho, T.; Ferreira De Souza, A.; de Aguiar, E. A Computational Geometry Approach for Localization and Tracking in GPS-denied Environments. *J. Field Robot.* **2016**, *33*, 946–966. [CrossRef]
32. Rusinkiewicz, S.; Levoy, M. Efficient variants of the ICP algorithm. In Proceedings of the Third International Conference on 3-D Digital Imaging and Modeling, Quebec City, Canada, 28 May–1 June 2001; pp. 145–152. [CrossRef]
33. Chipperfield, A.J.; Fleming, P.J.; Pohlheim, H.; Fonseca, C.M. A genetic algorithm toolbox for Matlab. In Proceedings of the Tenth International Conference on Systems Engineering (ICSE '94), Conventry, United Kingdom, 6–8 September 1994; pp. 200–207.
34. Man, K.F.; Tang, K.S.; Kwong, S. Genetic algorithms: Concepts and applications [in engineering design]. *IEEE Trans. Ind. Electron.* **1996**, *43*, 519–534. [CrossRef]
35. Baker, J.E. Reducing Bias and Inefficiency in the Selection Algorithm. In Proceedings of the Second International Conference on Genetic Algorithms on Genetic Algorithms and Their Application, Cambridge, MA, USA, 28–31 July 1987; pp. 14–21.
36. Douc, R.; Cappé, O.; Moulines, E. Comparison of resampling schemes for particle filtering. In Proceedings of the 4th International Symposium on Image and Signal Processing and Analysis, ISPA 2005, Zagreb, Croatia, 15–17 September 2005; pp. 64–69. [CrossRef]
37. Grisetti, G.; Stachniss, C.; Burgard, W. Improved Techniques for Grid Mapping with Rao-Blackwellized Particle Filters. *IEEE Trans. Robot.* **2007**, *23*, 34–46. [CrossRef]
38. Pedrosa, E.; Lau, N.; Pereira, A. Online SLAM Based on a Fast Scan-Matching Algorithm. In *Progress in Artificial Intelligence*; Correia, L., Reis, L.P., Cascalho, J., Eds.; Springer: Berlin/Heidelberg, Germany, 2013; pp. 295–306.
39. Fischler, M.A.; Bolles, R.C. Random Sample Consensus: A Paradigm for Model Fitting with Applications to Image Analysis and Automated Cartography. *Commun. ACM* **1981**, *24*, 381–395. [CrossRef]

40. Tapia-Espinoza, R.; Torres-Torriti, M. Robust lane sensing and departure warning under shadows and occlusions. *Sensors* **2013**, *13*, 3270–3298. [CrossRef]
41. Speta, M.; Rivard, B.; Feng, J.; Lipsett, M.; Gingras, M. Hyperspectral imaging for the characterization of athabasca oil sands drill core. In Proceedings of the 2013 IEEE International Geoscience and Remote Sensing Symposium-IGARSS, Melbourne, Australia, 21–26 July 2013; pp. 2184–2187. [CrossRef]
42. Leblon, B.; Gallant, L.; Granberg, H. Effects of shadowing types on ground-measured visible and near-infrared shadow reflectances. *Remote Sens. Environ.* **1996**, *58*, 322–328. [CrossRef]
43. Sick AG. *LMS5xx Laser Measurement Sensors Operating Instructions*; Sick AG: Waldkirch, Germany, 2015.
44. Auat Cheein, F.; Torres-Torriti, M.; Hopfenblatt, N.B.; Prado, A.J.; Calabi, D. Agricultural service unit motion planning under harvesting scheduling and terrain constraints. *J. Field Robot.* **2017**, *34*, 1531–1542. [CrossRef]
45. Zhang, J.; Singh, S. Laser-visual-inertial Odometry and Mapping with High Robustness and Low Drift. *J. Field Robot.* **2018**, *35*, 1242–1264. [CrossRef]
46. Chen, X.; Läbe, T.; Milioto, A.; Röhling, T.; Behley, J.; Cyrill Stachniss, C. OverlapNet: A siamese network for computing LiDAR scan similarity with applications to loop closing and localization. *Auton. Robots* **2022**, *46*, 61–81. [CrossRef]
47. Dosovitskiy, A.; Fischer, P.; Ilg, E.; Häusser, P.; Hazirbas, C.; Golkov, V.; Smagt, P.V.D.; Cremers, D.; Brox, T. FlowNet: Learning Optical Flow with Convolutional Networks. In Proceedings of the 2015 IEEE International Conference on Computer Vision (ICCV), Santiago, Chile, 11–18 December 2015; pp. 2758–2766. [CrossRef]
48. Xu, G.; Li, X.; Zhang, X.; Xing, G.; Pan, F. Loop Closure Detection in RGB-D SLAM by Utilizing Siamese ConvNet Features. *Appl. Sci.* **2022**, *12*, 62. [CrossRef]
49. Weinmann, M.; Wursthorn, S.; Weinmann, M.; Hübner, P. Efficient 3D Mapping and Modelling of Indoor Scenes with the Microsoft HoloLens: A Survey. *PFG J. Photogramm. Remote Sens. Geoinform. Sci.* **2021**, *89*, 319–333. [CrossRef]

sensors

MDPI

Article

Efficient Approach for Extracting High-Level B-Spline Features from LIDAR Data for Light-Weight Mapping

Muhammad Usman [1,*], Ahmad Ali [1], Abdullah Tahir [1], Muhammad Zia Ur Rahman [1] and Abdul Manan Khan [2]

1 Department of Mechanical, Mechatronics, and Manufacturing Engineering, University of Engineering & Technology, Faisalabad Campus, Faisalabad 38000, Pakistan

2 Department of Mechanical Engineering, Hanbat National University, Deajeon 34158, Republic of Korea

* Correspondence: musman.me@gmail.com

Abstract: Light-weight and accurate mapping is made possible by high-level feature extraction from sensor readings. In this paper, the high-level B-spline features from a 2D LIDAR are extracted with a faster method as a solution to the mapping problem, making it possible for the robot to interact with its environment while navigating. The computation time of feature extraction is very crucial when mobile robots perform real-time tasks. In addition to the existing assessment measures of B-spline feature extraction methods, the paper also includes a new benchmark time metric for evaluating how well the extracted features perform. For point-to-point association, the most reliable vertex control points of the spline features generated from the hints of low-level point feature FALKO were chosen. The standard three indoor and one outdoor data sets were used for the experiment. The experimental results based on benchmark performance metrics, specifically computation time, show that the presented approach achieves better results than the state-of-the-art methods for extracting B-spline features. The classification of the methods implemented in the B-spline features detection and the algorithms are also presented in the paper.

Keywords: range sensing; feature detection; light-weight mapping; B-splines; localization

Citation: Usman, M.; Ali, A.; Tahir, A.; Rahman, M.Z.U.; Khan, A.M. Efficient Approach for Extracting High-Level B-Spline Features from LIDAR Data for Light-Weight Mapping. *Sensors* **2022**, *22*, 9168. https://doi.org/10.3390/s22239168

Academic Editor: Ali Khenchaf

Received: 11 October 2022
Accepted: 21 November 2022
Published: 25 November 2022

1. Introduction

The simultaneous localization and mapping of SLAM are already acknowledged as forming a discipline; therefore, the optimization of its specific modules (localization and mapping) is the focus of research these days. The main focus has been moved to the optimization and overall assimilation of these components. Localization tasks can be performed efficiently by using the extraction of distinctive regions or point features from the sensor data. The distinctive regions or points allow the computation of signatures for place recognition, which enables effective robot localization. Only a small group of point-features obtained from a local map was used in point-to-point associations for localization purposes. Three state-of-the-art point feature detectors designed for 2D laser scans are FLIRT [1], FALKO [2], and BID [3]. These low-level features are also used as landmarks for a lightweight map. The map represented using point features ignores the structural information of the surroundings and makes it challenging for robots to interact with that environment. The high-resolution structural information can be attained using occupancy grids [4]. However, to create a dense map using occupancy grids, a large state space is required. The high-level features also enable dense mappings with lesser state space requirements. The extraction of numerous high-level geometric features such as lines [5], polylines [6], circle features [7], and curve segments [8] has also been presented in the literature. Due to a single geometric feature extraction model, the approaches still require a more reliable and robust solution for environments with various shapes.

A comprehensive approach to express the curved and straight features has been presented in vision-based Curve SLAM [9]. A stereo camera image of paths was used to detect the left and right edges, and later, these edges were represented as Bezier curves features. The major drawback of using Bezier curves as features is that the degree of the polynomial is confined to the total number of control points. The degree of the polynomial is always one less than the number of control points. Therefore, the multifaceted shapes are hard to represent using a single Bezier curve. Another limitation of the Bezier curve is that displacing any of the control points changes the curve's shape completely, which makes it tough to manipulate the control point for curve modifications. A better alternative [10] is the utilization of B-splines in the description of straight and curved-shaped data obtained from a 2D laser scan.

The B-spline representation of a scan is found by the segmentation of the scanned data and then approximating the B-spline curve on the segments [10]. However, the segmentation methodology used in [10] results in the extreme breakdown and depletion of scan data. The curve-fitting method [10] also lacks the appropriate choice with respect to the total number of control points and the derived knot vector. In contrast, a better method for segmentation and the active B-spline curve-fitting approach with corner representations is presented in [3]. Although it results in a highly compact representation of the environment but the approach is computationally very expensive.

In this article, we proposed a novel methodology of B-spline feature extraction. In the existing state-of-the-art approach of B-spline feature extractions, the B-spline-based interest point detection (BID) is used for corner detection from laser scans. In BID, all points are tried as the candidates of the interest points by the active B-spline curve-fitting method using its neighbor points. The application of active B-spline approximation on every data point is a time-consuming practice, which greatly increases the time of the B-spline feature representation of the complete scan. The proposed approach is based on FALKO [2] interest point detection, which is a quicker method for corner detection. It reduces the overall execution time of the B-spline approximation of the complete scan. The performance of the novel approach was compared with the state-of-the-art methodology using the set of benchmark metrics presented in [3]. In the experimentation, the standard data sets of different indoor and outdoor environments are used. We not only suggested names for existing extraction methodologies based on the techniques used for the segmentation and the B-spline approximation but the algorithms of all approaches are also presented in our paper.

Contributions

The main contributions of this article are as follows:

- We proposed a novel method for the fast extraction of high-level B-spline features, which can be used for the accurate mapping of the environment suited for both curved and straight-line geometric shapes.
- The computation time of feature extraction is very crucial for mobile robots when they perform real-time tasks. Therefore, we added a new benchmark time metric in the existing set of metrics used for the evaluation of the B-spline feature extraction approach.
- The classification of B-spline feature extraction approaches based on the techniques used for segmentation, corner detection, and B-spline approximation is also presented in the paper.
- Finally, we also presented algorithms for all approaches that can be very effective when implementing the procedures.

The remaining paper is organized as follows: Section 2 provides the literature on low-level and high-level features detection approaches for LIDAR data, and Section 3 provides the problem's definition and methodology, Section 4 introduces B-Spline theory, Section 5 presents B-spline features extraction approaches, and Section 6 explains the experiments and the results. Finally, the conclusion is provided in Section 7.

2. Related Work

The robot's pose and the map are depicted by the modules of SLAM. Mapping is the process of estimating the map, and localization is referred to as the process of estimating the robot's pose. The localization of landmarks or features, which can be used to determine the exact location of a robot on a global map, is a problem that localization addresses. Environmental features, both artificial and natural, are shown on landmark maps. Smith et al. [11] came up with the ground-breaking solution of landmark-based EKF-SLAM. Although comprehensively studied [12] and more accurate visual keypoint-based place recognition techniques have their own set of difficulties regarding scalability, viewpoint, and slight variations. The object-based localization frameworks [13,14] provide a compromise between accurate keypoint localization and the capacity to take contextual and semantic information into account.

LiDAR-based Localization techniques can be classified and rely mostly on matching geometries. Numerous point registration techniques, the most well-known of which is iterative closest point (ICP) [15], involve a high pose prior and are, hence, unsuitable for global localization. Even though there are methods for global registration that operate outside of the local context [16,17], they still require keeping at least some of the point cloud data. This can be partially concentrated by simply collecting compact descriptors during map construction and localization. There are various low-level features or keypoints, such as the corner points, which have already been proposed specifically for the planer rangefinders. In [18], Bosse proposed the centroids of the clusters and the points of the large positive curvature as point features for the task of place recognition. Li and Olson in [19] suggested an approach that builds upon the method used in image processing: Kanade–Tomasi is a variant of the Harris corner detector. This method was applied to the laser scan data. The scan data were first converted into an image from which the point features were extracted. Later [20], they improved the point feature detector for LIDAR data, which was also based on the computer vision approach by computing the structure tensor of the surface's normal. The main concern is the proportional change in computational complexity of the image-processing-based approaches with the change in the size of the converted image from the scanned data. In FLIRT [1], a discrete framework of scale-space theory was used for re-scaling the scanned data; then, the point features were searched on different scales depending on the parameters such as the range, normal, or curvature. However, the single parameter-based detection of the point features may result in unstable points in FLIRT. Kallasi et al. proposed a point feature detector FALKO [2], which used the idea of edge intersection. The first step of the corner detection was the rough guess that was based on the evaluation of the parameters of the triangle generated by two boundary data points in a specific neighborhood and the corner point candidate. Therefore, it is likely to select the unstable candidates or will skip the possible candidates of the point features. In the B-spline-based interest point detector BID [3], the idea of neighborhood selection was taken from [2] but used in a better way by applying active B-spline approximation to the neighbor data points for the detection of point features.

Other than low-level feature extraction techniques, high-level feature extraction approaches such as lines and curve segments are also used for LIDAR data. High-level features have been included in many SLAM algorithms to address the limitations of point-based SLAM. Planes, image moments, line segments, office chairs, tables [21–23], objects, and rivers are a few examples of high-level features. The ability to compactly create a map of the environment is a desirable quality of the high-level structure. The outline of a door, the structure of an inside hallway, or other objects one may expect to find in the mapped place are represented by lines [24,25]. Additionally, lines give a sparse picture of the surroundings. For instance, each line is represented by only two points [23]: the start and end points of a line segment. The line features have also been implemented by [23] to map the environment using a vision-based sensor. The start and end points were represented by using line segments. High-level feature detectors presented for 2D laser data contain dynamic mappings based on circles and lines features [26], curve features based on adaptive curvature estimation [8], and weighted line fitting [27]. One of the foremost needed properties of high-level features is that they enable a compact method of mapping the structural details of the environment. A generalized method of representing the curve or line feature using the Bezier curves was explained in CurveSLAM [9] to create an efficient map of the stereo vision sensor's data. The methodology was used to create the straight and curved pathways as well as the yellow road lanes. A more comprehensive method of high-level features description for the varied and complex environments using B-splines was presented in [10]. However, the approach lacks smart segmentation procedures, choosing an initial number of control points and the parametrization for the B-spline's feature demonstration. In [3], active B-splines features extraction was implemented with efficient segmentation to reduce the wastage of data points and accurate representation of the line or curved segments. However, it is a computationally very expensive and slow approach.

3. Problem Definition and Methodology

In 2D LIDAR data, point features such as corners are taken as landmarks in the environment. These landmarks can also be used to represent a light-weight map of the environment. However, the structural information of the environment is not contained in the landmarks based on point features. Therefore, the robots are unable to interact with the actual environment by using the point-feature-based map. The occupancy grids used for map representations in [28] produced the structural details of the environment shown in Figure 1, but every grid requires dedicated memory to store each laser data point. The higher the resolution, the more space will be required to save the map.

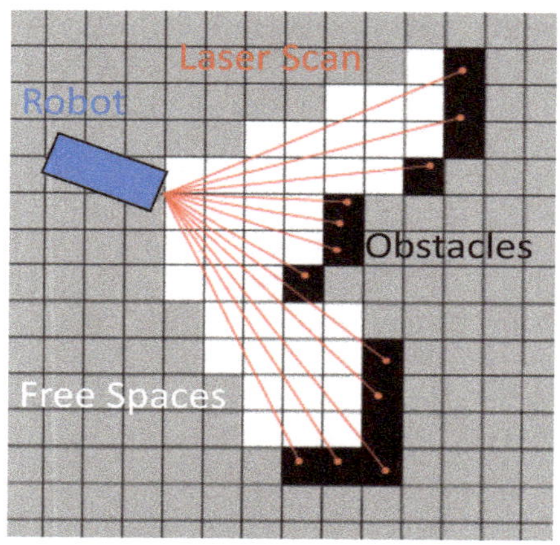

Figure 1. The occupancy grids used in [28] for a laser scan.

Various high-level features have been used in the literature to represent the map for using laser data points; however, these features are used for dedicated shapes such lines or curves. The B-splines are not only used for straight as well curved shapes to represent the map by fewer control points. However, in the literature, the methodologies used for B-spline representation still lacked an efficient approach.

In B-spline extractions, the two main steps are as follows: One is segmentation, and the second is the B-spline approximation of the segments. An additional step involving corner detection was presented in our previous work [3], which enabled not only light-weight representations of the map but also the corner representation. The detailed methodology [3] is presented in Section 5.3. The corner representation can also be used for B-spline associations. However, this [3] approach is computationally very expensive.

We proposed a faster method for representing laser data using B-splines. FALKO-based corner detection [2] significantly reduces the overall execution time of the B-spline representation and does not affect the other performance metrics of the approach. The three steps includes FALKO-based interest point detection, segmentation, and finally, B-spline curve fitting. It is represented in Figure 2.

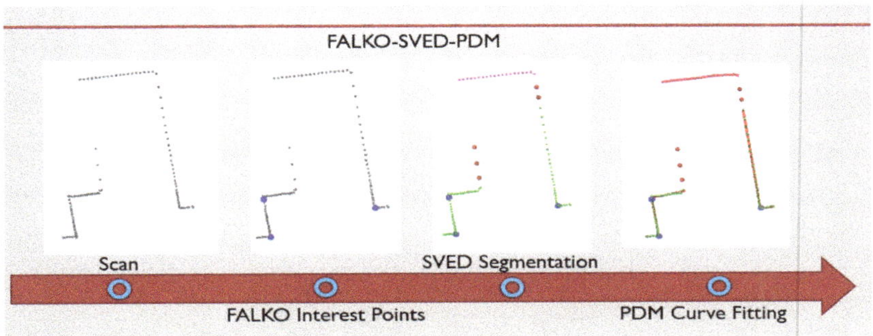

Figure 2. The three steps of proposed approach applied on a scan: FALKO interest point detection, segmentation, and finally, B-spline curve fitting. Note: Control points of B-splines are not shown in the figure.

The detailed explanation of the approach is presented in a later section of the paper. Firstly, the B-spline theory is presented.

4. B-Spline Theory

In this section, some basic concepts of B-spline, spline continuity, and the computation of an active B-spline curve to approximate the data points in a plane are presented.

4.1. B-Spline Curve

The B-spline curve is represented as a linear combination of basis functions that are also known as B-spline basis functions. An l-dimensional curve $S(t)$ composed of basis functions $B_{i,k}$ of degree $k(order = k+1)$ and control points $P_i \epsilon R^l (i = 0, \dots, n)$ can be expressed as a function of parameter $t \epsilon R$ such as

$$S(t) = \sum_{i=0}^{n} B_{i,k}(t) P_i \tag{1}$$

where basis functions $B_{i,k}$ are defined by the Cox–de Boor recursion formulas [29,30]

$$B_{i,0}(t) = \begin{cases} 1, & \text{if } \xi_i \leq t \leq \xi_{i+1} \\ 0, & \text{otherwise.} \end{cases} \tag{2}$$

and for all $k > 0$

$$B_{i,j}(t) = \frac{t - \xi_i}{\xi_{i+j} - \xi_i} B_{i,j-1}(t) + \frac{\xi_{i+j+1} - t}{\xi_{i+j+1} - \xi_{i+1}} B_{i+1,j-1}(t) \tag{3}$$

where knot vector $\Xi = \xi_0, \dots, \xi_{n+k}$ is any non-decreasing sequence of real numbers. The multiplicity of the extreme values of the knot vector equal to the curve's order makes it a clamped B-spline; otherwise, it is an unclamped one.

4.2. Continuity

With respect to the knot value, the continuity, which is also known as the degree of smoothness of the B-spline of order O, is at its maximal value, that is, C^{O-2} continuity of all derivatives up to $(O-2)$th [31]. Continuity at a knot multiplicity m is reduced to C^{O-m-1}. In contrast, coinciding n consecutive control points reduce the continuity of the B-spline curve by $n-1$. For example, two control points of the quadratic $(O=3)$ B-spline coincide to form a hinge (discontinuous first derivative) at the point of coincidence on the curve.

4.3. Active B-Spline Curve Fitting

Approximating a noise-contaminated or non-uniformly distributed set of data points $X_s \epsilon R^2, s = 1, \ldots, N$ by using an active B-spline curve can be formulated as a non-linear optimization problem. A data point X_s that lies on the B-spline must satisfy

$$X_s = B_{0,k}(t_s)P_0 + \ldots + B_{n,k}(t_s)P_n \tag{4}$$

The process of finding t_s is known as data parameterization. The main approach of approximation is based on the idea of starting an active B-spline with an appropriately chosen initial curve and converging via iterative optimizations towards targeted data points [32]. The knot vector and the order of the curve are assumed to be fixed throughout the fitting procedure. At first, keeping the parameter values constant, control points $P_i, i = 0, \ldots, n$ are found such that objective function

$$f = \frac{1}{2}\sum_{s=1}^{N} d^2(S(t), X_s) + \lambda f_{req} \tag{5}$$

is minimized, where $d(S(t), X_s)$ is the orthogonal distance from data point X_s to the point on the initial curve, $S(t)$. This scheme of minimization is called point distance minimization (PDM) [32]. Secondly, control points produced in the minimization step are kept fixed while the data parameterization method is used to select t_s such that St_s (also known as footpoint of X_s) is the nearest point on the updated curve from data point X_s. Regularization term f_{reg} governs the smoothness by looking at the differential Σ and curvature ρ of curve at every point, whereas positive constant λ represents its weight. Differential operations or stretching of the curve are defined as

$$\Sigma = \int ||S'(t)||^2\, dt \tag{6}$$

and the bending or curvature energy, ρ, of the curve is defined as

$$\rho = \int ||S''(t)||^2\, dt \tag{7}$$

In our implementation, only the bending energy has been considered. The approximation error of the curve can be calculated as the root mean squared error E_{curve}, which is defined as

$$E_{curve} = \left[\frac{1}{m}\sum_{q=1}^{m} ||S(t_q) - c_q||^2\right]^{1/2} \tag{8}$$

where m is the total number of data points c_q upon which curve approximation is applied.

5. B-Spline Feature Extraction

In this section, we present the various types and methods of finding the B-spline features from 2D laser scan data. In a laser scan, the state-of-the-art low-level features or corner detectors FALKO [2] and BID [3] can be used to detect stable and view-point-invariant interest points. The high-level B-spline feature extraction approaches are presented in [3,10]. The B-spline feature, which has at least one associated corner (interest point) or vertex in its representation, is defined as an auxiliary B-spline. In contrast, the B-splines that do not have any associated corner (interest point) or vertex in the representation are called bleak B-splines [3]. The B-spline feature extraction involves two important steps that are a segmentation of the scan data and B-spline curve fitting on the segmented data.

5.1. Segmentation

The data points obtained from the laser scan may not be represented using a single B-spline feature. Therefore, the first step in the extraction of high-level B-spline features is the segmentation step. The B-spline feature extraction approaches can be categorized based on the segmentation methodology they adopt.

5.1.1. Relative Position and Orientation-Based Segmentation (RPOS)

The segmentation methodology based on the relative position and orientation between two consecutive data points was presented in [10]. It can be called the relative position and orientation-based segmentation (RPOS). The RPOS is centered on the evaluation of the relative positions of two successive laser data points, as shown in Figure 3, and mathematically represented as follows.

$$p_i = d_i - d_{i-1} \tag{9}$$

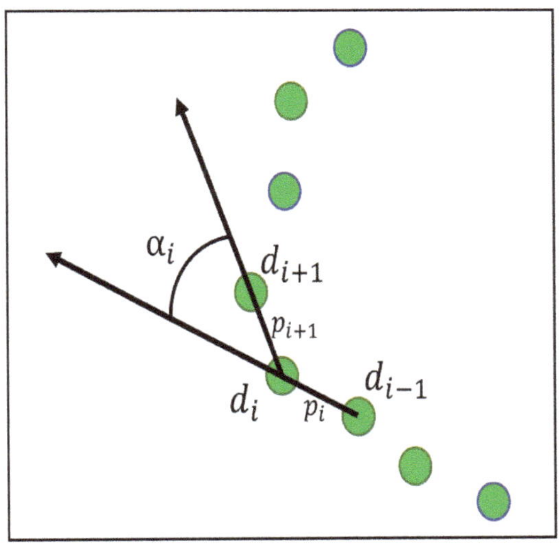

Figure 3. An RPOS-based segmentation presented in [10].

Then, the following assessments were performed.

$$|\alpha_i| \leq \alpha_{max} \leftrightarrow \cos(\alpha_{max}) \leq \cos(\alpha_i) \tag{10}$$

$$max(\|p_i\|, \|p_{i+1}\|) \leq \eta min(\|p_i\|, \|p_{i+1}\|) \tag{11}$$

When a set of successive data points fulfills both relationships described above, they are taken to be part of the same feature. In RPOS, the proposed value for angular threshold $\alpha_{max} = [0, \pi/4]$ results in the extraction of bleak segments only. The larger values of α_{max} may associate the corner points in the segment, but it results in an incorrect and excessive number of corners identified in scanned data [10]. Therefore, the values for parameters $\alpha_{max} = \pi/4$ and $\eta = 1.5$ were chosen for implementation.

5.1.2. Segmentation Using Varying Euclidean Distances (SVEDs)

Another segmentation approach presented in [3] is based on Euclidean cluster extraction [33], as shown in Figure 4. Instead of using a constant distance [33], a varying Euclidean distance threshold was used in [3]. It can be called segmentation using varying Euclidean distance SVED-based cluster extraction. For every point, the neighborhood-radius r_q is the function of the datapoint distance $\|d_q\|$ from a sensor's origin, and it is represented in [2] as

$$r_q = a\exp(b\|d_q\|) \tag{12}$$

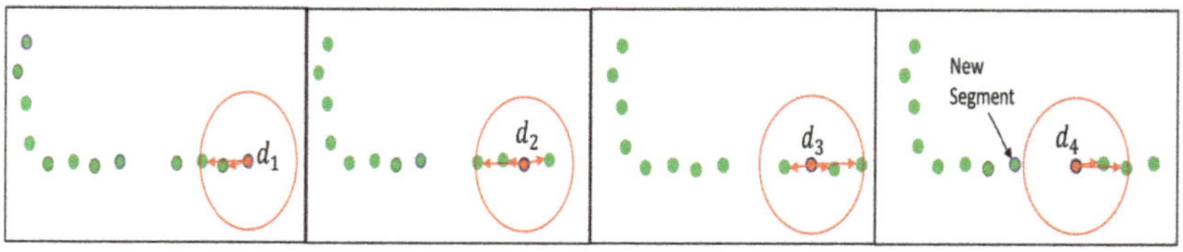

Figure 4. The Euclidean cluster extraction methodology [33] adopted in [3] with varying r_q.

Parameters $a = 0.2$ and $b = 0.07$ were selected for scan ranges between 0.5 and 30 m. Radius r_q is taken as the varying euclidean distance. The SVED approach may produce both auxiliary and bleak segments [3], as shown in Figure 5.

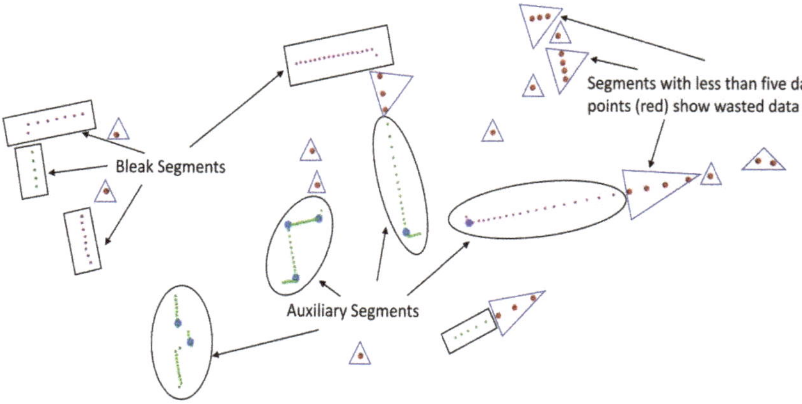

Figure 5. An SVED segmentation-based approach containing the Bleak segment (purple datapoints with no interest points in blue) and auxiliary segments (with associated Interest points in blue color) where the wasted data points are shown in red. The BID-based interest point detection approach was used to detect corners (shown in blue).

5.2. B-Spline Features

The flexibility of splines curves to approximate noise-contaminated data is one of their main attractions. Every segment in a scan is represented by the B-spline curve approximation. The two curve-fitting approaches for LIDAR data are presented. The first is a least-squares B-spline approximation LSBA in [10], and tje other is point distance minimization PDM in [3].

5.2.1. Least-Squares B-Spline Approximation (LSBA)

The B-spline curve fitting of a set of data points $X_s \epsilon R^2, s = 1, \ldots, N$ by using a least-squares solution can be devised as an approximation problem. If a data point, X_s, lies on the B-spline, then it must satisfy Equation (13) as follows:

$$X_s = B_{0,k}(t_s)P_0 + \ldots + B_{n,k}(t_s)P_n \tag{13}$$

The problem can be represented by using the equations as follows:

$$X = BP \begin{cases} X = \begin{bmatrix} X_0 & X_1 & \ldots & X_N \end{bmatrix}^T \\ P = \begin{bmatrix} P_0 & P_1 & \ldots & P_N \end{bmatrix}^T \\ B = \begin{bmatrix} B_{0,k}(t_0) & \ldots & B_{n,k}(t_0) \\ \ldots & \ddots & \ldots \\ B_{0,k}(t_m) & \ldots & B_{n,k}(t_m) \end{bmatrix} \end{cases} \tag{14}$$

where matrix B is the collocation matrix. The process of finding t_s, which is the position of datapoint along the curve, is known as data parameterization. The cumulated chord length can be used to find the position between consecutive datapoints as follows.

$$\begin{cases} t_0 = 0 \\ t_j = \sum_{s=1}^{i} \|X_s - X_{s-1}\| \end{cases} \tag{15}$$

The total length of the curve, l, which is the maximum value of the knot vector, can be calculated as follows.

$$l = \sum_{s=1}^{m} \|X_s - X_{s-1}\| \tag{16}$$

Finally, the least-squares solution of the approximation problem can be calculated by using the pseudoinverse matrix of B.

$$P = [B^T B]^{-1} B^T X \tag{17}$$

In the case of bleak segments, the order (or degree) of the B-spline curve, the number of control points, and the parameter values along the curve are predefined to approximate the segments using the least-squares solution [10]. It provides a solution for bleak segments only, because in this approximation, the corner position cannot be provided, and only the number of control points is given.

5.2.2. Point Distance Minimization (PDM)

Another method for representing the segments using B-splines is active B-spline curve fitting, which is presented in Section 4. The approach involves two steps. The first one is the initialization of a B-spline curve of a specific order and shape using a known position and the number of control points and secondly, the point distance minimization (PDM) scheme [32] is applied to the initial curve to approximate the segments. The main appeal of this approach is that not only bleak segments but the point distance minimization scheme can also be used to approximate auxiliary segments that have at least one associated corner detected in the laser scan, as shown in Figure 5.

The significant idea of the auxiliary segments is the representation of the corners. In the quadratic B-spline curve, a corner can be represented by the knot multiplicity $M_{knot} = 2$ or the control point multiplicity $M_{cp} = 2$ [31]. The knot multiplicity approach was used in [3] to minimize the number of control points used in the B-spline's representation. However, representing the corner using control point multiplicity M_{cp} results in the part of the curve on either side of the corner constrained to be linear [31]. The importance of linearity constraints on either sides of the corner is significant for the representation of corners in most of the building's structures.

The control points in the auxiliary segment are categorized into three types. The control point used in the representation of the corner is defined as a vertex control point [3]. The boundary control points are initialized at the boundary data points, whereas the precision control points are used in the representation of the fragments [3]. The fragment is the region of the B-spline curve between two vertex control points or the region between the boundary control point and the vertex control point [3]. The precision control points define the accuracy of the B-spline approximation of a segment. Therefore, the number of precision control points was proposed to be chosen as a function of the Euclidean distance between the fragment's endpoints [3]. Once the curve is initialized using the control points, the PDM scheme is applied to approximate the segment.

5.3. Extraction Methodologies

Various approaches can be used in the extraction of B-spline features from 2D laser scan data. The approaches are categorized on the basis of the phenomena they adopt.

5.3.1. RPOS-LSBA

The RPOS-LSBA [10] used a typical approach involving B-spline extraction, which involved the segmentation of the laser scan and then B-spline curve fitting on the segments. In the segmentation step, it employed the relative position and orientation-based segmentation RPOS, which is based on the relative position of d_i and orientation α_i of the two consecutive scan data points, as shown in Figure 3. As segmentation relies on the geometrical (angle and distance) property of the two very next neighboring data points, therefore, a distant point from the neighbors not only creates a new segment but also limits the upcoming neighbor in becoming a part of a segment rather than isolating it. This increases wasted data points. Segments containing less than five data points are considered wasted. The second step is the application of the least-square solution [10] to the segments obtained in RPOS to obtain a B-spline approximation. It is named least square B-spline approximation LSBA. This B-spline feature extraction approach not only increases wasted data points but also lacks a representation of stable points (corners) in the scan. The B-spline feature extraction based on RPOS-LSBA is presented in Algorithm 1.

Algorithm 1: RPOS-LSBA Features Etraction

```
SEGMENTATION
rpos_seg(r_th, α_max) // position and orientation thresholds
for each datapoint d_i in scan S do
  if position_and_orientation < (r_th, α_max) then
    φ_n ← d_i
  end if
end for
return φ // creates and return Bleak segments
B-SPLINE APPROXIMATION
lsba_curve(φ)
for each segment φ_n in scan S do
  BSplines = least_square_Approx(P_c, Order_curve, φ_n) // no. of control points P_c
end for
return BSplines
```

5.3.2. BID-SVED-PDM

Another method for B-spline feature extraction was presented in [3]. In this approach, an additional step of corner detection was proposed. Firstly, the stable and state-of-the-art B-spline-based interest point detector BID [3] was used for corner detection, as shown in Figure 6. Roughly, a set of a uniform number of neighbor data points $N(c_q)$ is acquired using radius $r_q = a\exp(b\|d_q\|)$. The clamped and uniform quadratic B-spline curve is initialized with four control points evenly spaced along the axis defined by the largest eigenvector of the covariance matrix of $N(c_q)$, as shown in red color in Figure 6 (left). Then, the point distance minimization (PDM) method is used to estimate the shape specified by the data points, as in Figure 6 (middle). Finally, the inverse of the normalized Euclidean distance, N_q, among two median control points, P_{q1} and P_{q2}, of the estimated curve, $S_q(t)$, is taken as the measure of corner occurrence as in Figure 6 (right).

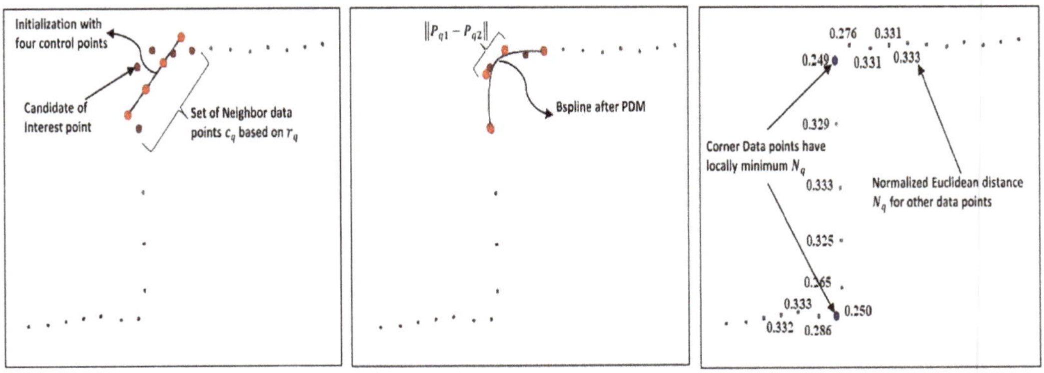

Figure 6. The B-spline-based interest point detection BID method with initialization step (**left**), PDM (**middle**) step for the corner data point, and the normalized Euclidean distance N_q values (**right**) calculated for candidates of the interest points around the actual corner points [3]. The four control points are represented in red color.

$$N_q = \frac{\|P_{q1} - P_{q2}\|}{\sum_{i=0}^{2} \|P_{qi} - P_{q(i+1)}\|} \tag{18}$$

The local maxima of N_q^{-1} beyond threshold T_{th} can be employed to detect the indices of scan points and ultimately corner point I_p by using non-maxima suppression NMS. The data points with N_q values 0.249 and 0.250 in Figure 6 (right) are the detected interest points.

In the second step, segmentation using a varying Euclidean distance-based cluster extraction (SVED) approach was used [3] and derived from [33], which produces segments with a minimum wastage of data points. Finally, the active B-spline curve approximation based on point distance minimization PDM [32] was used for the representation of the segments using B-spline features. The active B-spline approach enables the representation of detected corners in the laser scan. Algorithm 2 shows the B-spline feature extraction based on BID-SVED-PDM.

Algorithm 2: BID-SVED-PDM Feature Extraction

CORNER DETECTION
bid_$I_p(T_{th})$ // T_{th} normalized Euclidean distance threshold [3]
for each datapoint d_i with its neighbors in scan S **do**
  if $E_q^{-1} > T_{th}$ **then**
    $I_p \leftarrow d_q$
  end if
end for
$I \leftarrow Non\_Maxima\_Suppression(I_p)$
return I // returns I set of interest points
SEGMENTATION
sved_seg(r_q) // varying Euclidean radius r_q
for each datapoint d_i in scan S **do**
  if $Euclidean\_distance \leq r_q$ **then**
    $\phi_n \leftarrow d_q$
  end if
end for
return ϕ // creates and return Bleak segments
ACTIVE B-SPLINE APPROXIMATION
active_Bspline_curve(ϕ)
for each segment ϕ_n in scan S **do**
  $(Aux, Bleak)$ = find_aux_bleak(ϕ_i, I_p) // search for Auxiliary and Bleak Segments
end for
Bleak Segments B-spline Approximation
for each bleak segment ϕ_n in scan S **do**
  $B_{cp} \leftarrow$ Find_Boundary_Control_Points (ϕ_n)
  $P_{cp} \leftarrow$ Find_Precision_Control_Points (B_{cp})
  $curve_{init} \leftarrow$ Curve_Initialization ($B_{cp}, P_{cp}, order_{curve}$)
  $Bspline \leftarrow$ point_distance_minimization ($curve_{init}, \phi_i$)
end for
Auxiliary Segments B-spline Approximation
for each Auxiliary segment ϕ_n in scan S **do**
  $B_{cp}, V_{cp} \leftarrow$ Find_Boundary_Vertex_Cntrl_Pnts (ϕ_n, I_p)
  $P_{cp} \leftarrow$ Find_Precision_Control_Points (B_{cp})
  $curve_{init} \leftarrow$ Curve_Initialization ($B_{cp}, V_{cp}, P_{cp}, order_{curve}$)
  $Bspline \leftarrow$ point_distance_minimization ($curve_{init}, \phi_i$)
end for
return $BSplines$

5.3.3. FALKO-SVED-PDM

The B-spline-based interest point or corner detection is a computationally very expensive and time-consuming approach. Therefore, in this section, we proposed to use the fast adaptive laser keypoint orientation-invariant FALKO [2] for corner detection. It manipulates the simple idea of edge crossing in 2D range data, as shown in Figure 7, which makes it faster. After finding the potential corner candidates based on the minimum number of neighbors (cardinality) on each side, they are evaluated geometrically (by measuring the height of the triangle as in Figure 7) to obtain a rough approximation of the corner. Then, for each candidate point, a cornerness score is computed.

$$score(p_i) = score_L(p_i) + score_R(p_i) \tag{19}$$

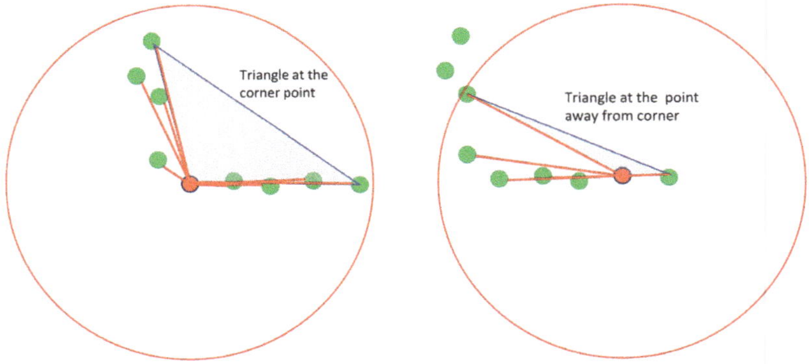

Figure 7. A FALKO-based corner detection example presented in [2].

This score function measures the alignment of the two point sets on each side of the candidate point as follows:

$$score_L(p_i) = \sum_{h=i-1}^{j_{min}} \sum_{k=h-1}^{j_{min}} |d_\theta(\phi_h), \phi_k| \tag{20}$$

$$score_L(p_i) = \sum_{h=i+1}^{j_{min}} \sum_{k=h+1}^{j_{min}} |d_\theta(\phi_h), \phi_k| \tag{21}$$

where d_θ is the distance function of the quantized orientations ϕ_1 and ϕ_2 concerning the candidate point as

$$d_\theta(\phi_1, \phi_2) = (\phi_1 + \phi_2 + \frac{s_n}{2}) mod(s_n) - \frac{s_n}{2} \tag{22}$$

where the number of circular sectors in the polar grid is represented as s_n. Secondly, segmentation SVED and, finally, the active B-spline curve approximation using PDM are employed. This approach can be named FALKO-SVED-PDM. Although, FALKO detects a few suporious interest points, it is fast; therefore, the proposed methodology considerably reduces the overall execution time of the extraction of the B-spline curve. The FALKO-based corner detection algorithm is shown in Algorithm 3.

Algorithm 3: FALKO-Based Corner Detection.

CORNER DETECTION
falko_$I_p(score_{th})$ // $score_{th}$ FALKO score threshold
for each datapoint d_i using its neighbors in scan S **do**
  if $cardinality \geq 2$ **then**
    potential_candidate $\leftarrow d_q$
  end if
end for
for each potential candidate d_c with its neighbors in scan S **do**
  compute_corner_score()
  if $corner\_score \geq score_{th}$ **then**
    $I_p \leftarrow d_q$
  end if
  $I \leftarrow Non\_Maxima\_Suppression(I_p)$
  return I // returns I set of interest points
end for

6. Experiments

In this section, the experimental setup and results are presented based on the performance evaluation of the B-spline feature extraction approach for compact and faster mapping using B-spline features extracted from the 2D range data. The data sets, three indoor (Intel, Fr079, and MIT-csail) and one outdoor (Fr-clinic), provided by [1] are tested in our experiments. The data sets given in [1] contain both the corrected ground truth and the corresponding original scans.

6.1. Experimental Setup

The method of B-spline features extraction is evaluated based on the four metrics of performance, which are retrievability, compactness, accuracy, and time. Although the first three metrics were proposed in [3], the fourth metric time is proposed in this paper as the execution time becomes very important when the robot performs real time tasks.

6.1.1. Retrievability (Γ)

It determines the extent of scan data points that are allotted to the extracted B-spline features. It is described by the percentage of the data points designated for the B-spline features over the total number of data points N in a scan. It can be represented as

$$\Gamma = \frac{1}{N}\sum_{i=1}^{m} \Xi_{\Phi_i} \tag{23}$$

where Ξ shows the number of data points in a non-discarded segment (which has more than five data points), and m represents the total number of non-discarded segments in a scan. It can be used as the measure of the wasted data points in a scan.

6.1.2. Compactness (η)

The second metric of performance is the compactness η of the depiction of a laser scan using B-spline features. It is calculated by the ratio of the total number of control points of B-splines to the total number of data points allocated to those B-splines in a scan as follows:

$$\eta = \frac{\sum_{i=1}^{m} N_{cp_i}}{\sum_{i=1}^{m} \Xi_{\Phi_i}} \tag{24}$$

where N_{cp_i} shows the number of control points in the ith B-spline feature of ϕ_i in a scan. The smaller value of η means that it is a more precise and compact representation of a scan. This means that lesser control points are used in the representation of a scan.

6.1.3. Accuracy ($\wedge$)

The accuracy $\wedge$ of the representation of the scan using B-spline features is assessed by the fitting error, E_{curve}, of the curves. The approximation error, E_{curve}, of the extracted features in the scan is calculated and then the average of the approximation error was taken as a measure of accuracy. It can be represented as follows

$$\wedge = \frac{1}{m}\sum_{i=1}^{m} E_{curve_i} \tag{25}$$

where m shows the number of B-spline features of non-discarded segments, and E_{curve_i} is the approximation error of the ith feature in the scan.

6.1.4. Execution Time (t)

The execution time t of the B-spline curve approximation is the time taken by the method in the representation of a laser scan using B-spline features. The B-spline extraction time of complete data set is averaged over the total number of scan in the set to determine the average time of one scan.

6.2. Results Evaluation

By contrasting it with the approaches described in [2,10] using the performance criteria presented, the suggested B-spline features extraction approach was evaluated. Additionally, feature extraction execution time criteria were also proposed to assess the performance of all approaches. For segmentation, the parameter values were set exactly as suggested in [10], but for feature representation, a least-square uniform quadratic B-spline approximation was employed. In the SVED approach of segmentation, the same values of the parameters were set for varying radius r_q, as shown in [3]. The same parameter values that were chosen for FALKO [2] and BID [3] were set for the assessment of both methodologies. Only the Bleak segments are produced by the RPOS method described in [10]. Therefore, four control points were used in the approximation of all B-splines in RPOS-LSBA-based approach.

Figures 8 and 9 show the segmentation results of RPOS and SVED approaches, respectivey. The noisy scan was taken from the Intel data set. For the same scan, RPOS produces a lot more wasted data points (red markers), as shown in Figure 6, whereas SVED produce few wasted data points (red markers), as shown in Figure 9. The green and purple markers are shown to represent two separate and consecutive segments. The wasted data points will not be used in the representation of B-splines; therefore, it will be taken as empty space in the environment.

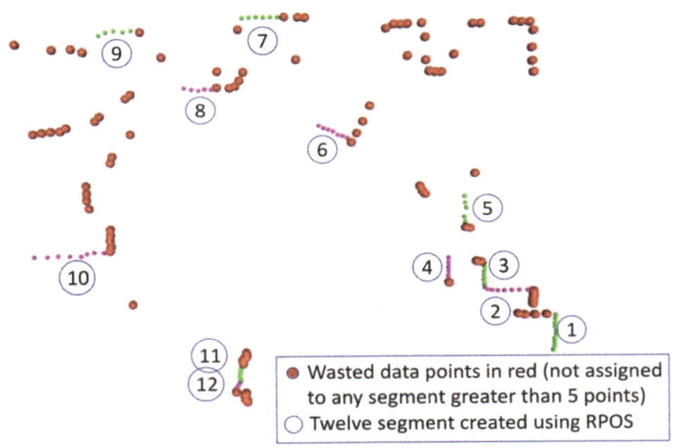

Figure 8. The relative position and orientation-based segmentation RPOS approach with red marks representing wasted datapoints, which are not part of any segment in the scan. A total of 12 green and purple segments were created by this approach. The green and purple colors are used to show the data points of two consecutive segments.

Figure 10a shows the implementation of RPOS-LSBA methodology representing the retrievability and compactness performance measures. In a noise-free scan taken as a sample, 34 out of 180 data points were wasted, whereas in SVED, only 14 data points were wasted, as shown in Figure 10b,c. The number of wasted data points increases significantly in the case of noise scan data, as shown in Figure 6. The retrievability increased significantly in BID-SVED-PDM and FALKO-SVED-PDM. Figure 10a also shows that 40 control points were used in the representation of 144 data points, whereas 43 control points were used in the representation of 166 data points, as shown in Figure 10b. In noisy scan data, a significantly larger value of η is achieved while using RPOS-LSBA, which shows poor performance.

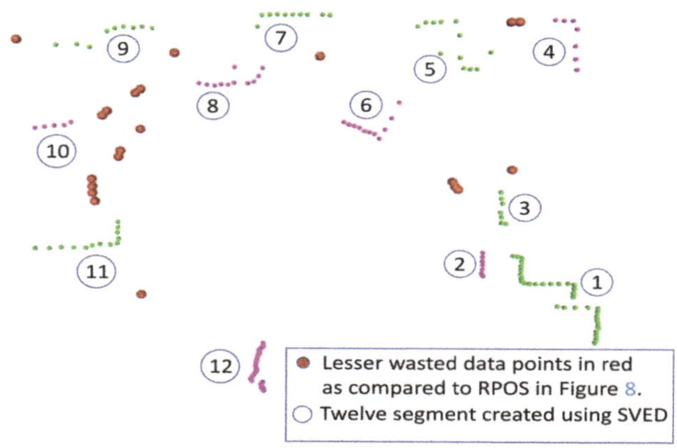

Figure 9. The segmentation using varying Euclidean-distance-based cluster extraction (SVED) approach with red marks representing lesser wasted datapoints, which are not part of any segment in the scan. The green and purple colors are used to show the data points of two consecutive segments.

Figure 11 shows the resultant map of the Intel data set created using the FALKO-SVED-PDM approach. It not only accuratly approximated the straight but also the curved features. The map contains the B-spline segments with approximation error $E_{curve} \leq 0.01$.

For all data sets, the graphical results of all performance measures are shown in Figure 9 every time the datasets were scanned and all four metrics were computed. After that, the outcomes are averaged across all scans.

In all four data sets, the state-of-the-art [3] approach results in two outstanding metrics: retrievability and compactness; however [3], they have very poor execution times. The bar graph demonstrates that our proposed approach's performance significantly improved in terms of execution times when compared with the state-of-the-art approach [3], as shown in Figure 12d. In terms of retrievability and compactness, the proposed approach's performance is comparable with the state-of-the-art method and significantly better than the RPOS-LSBA. However, the performance of FALKO-SVED-PDM and BID-SVED-PDM in the approximation error (accuracy) is slightly low but within an acceptable range.

Table 1 lists the summarized outcomes of the three methods. The results reveal our proposed approach's superior performance in contrast to the RPOS-LSBA [10] in terms of retrievability and compactness across all datasets, which is analogous to BID-SVED-PDM [3]. However, in the three data sets, the computation time of the proposed method has been reduced by more than 300% if we compare it with the execution time of the state-of-the-art approach [3]. Although the proposed method's approximation accuracy was slightly low, the overall curve-fitting results are still good and are within the acceptable range. All tests were performed on an Intel Core i7-750 CPU at 2.70 GHz 8 GB RAM with an Ubuntu 16.04 LTS and ROS Kinetic operating system.

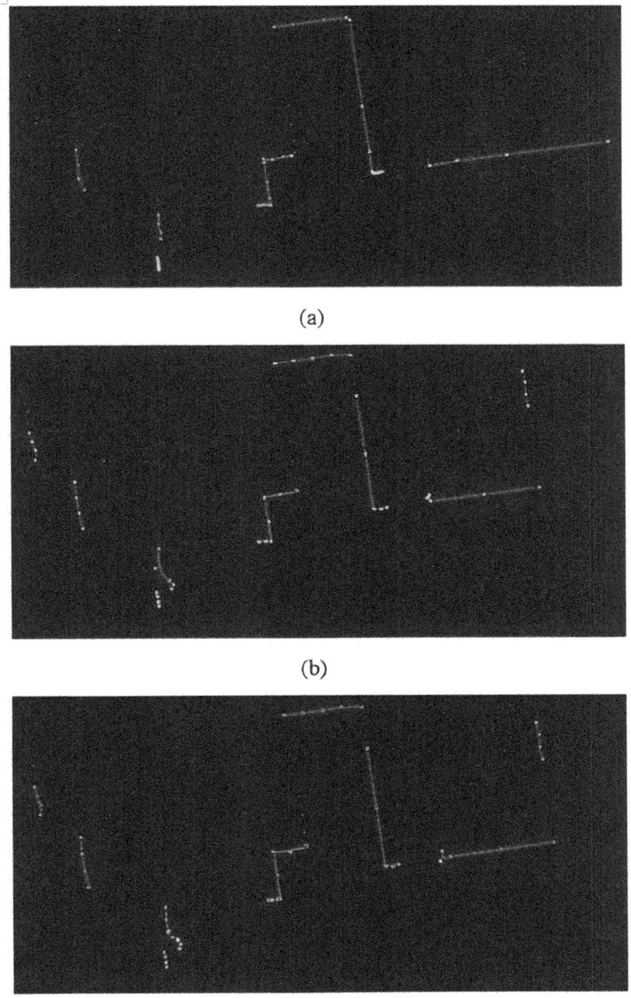

(c)

Figure 10. The retrievability and compactness results of (**a**) RPOS-LSBA, (**b**) BID-SVED-PDM, and (**c**) FALKO-SVED-PDM feature extraction approaches. White datapoints show wasted data, and green points are the control points used to represent the B-splines in the scan.

Figure 11. The map of a part of the Intel data set showing the curved and straight walls represented using FALKO-SVED-PDM. The map contains the B-spline segments with approximation error $E_{curve} \leq 0.01$.

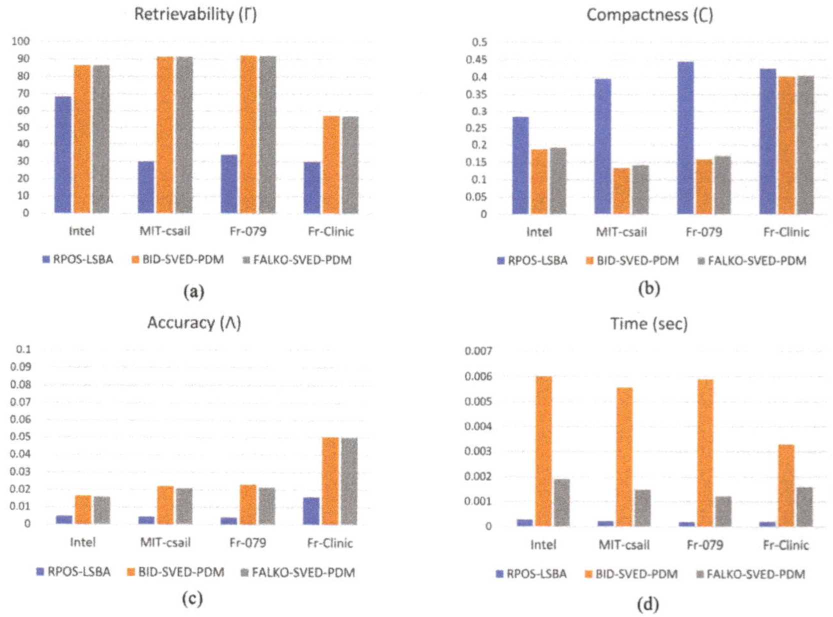

Figure 12. The retrievability (**a**), compactness (**b**), accuracy (**c**), and execution time (**d**) results of complete data sets for all three approaches using all three methods. The vertical axis shows the quantitative measure of the performance of all three methods, and the horizontal axis shows data set types.

Table 1. B-spline features assessement.

Data Set	Methodology	Γ	η	$\wedge$	t (s)
Intel	RPOS-LSBA	68.4	0.284	0.0049	0.00028
	BID-SVED-PDM	86.4	0.187	0.0163	0.00600
	FALKO-SVED-PDM	86.4	0.192	0.0158	0.00190
MIT-csail	RPOS-LSBA	30.0	0.395	0.0045	0.00022
	BID-SVED-PDM	91.4	0.133	0.0221	0.00557
	FALKO-SVED-PDM	91.4	0.142	0.0207	0.00148
Fr-079	RPOS-LSBA	33.9	0.445	0.0041	0.00018
	BID-SVED-PDM	91.9	0.158	0.0230	0.00590
	FALKO-SVED-PDM	91.9	0.169	0.0214	0.00122
Fr-Clinic (1000 Scan)	RPOS-LSBA	30.0	0.425	0.0157	0.00020
	BID-SVED-PDM	57.1	0.402	0.0501	0.00329
	FALKO-SVED-PDM	57.1	0.405	0.0501	0.00159

7. Conclusions

The properties of high-level features and low-level point features, respectively, were used in this article to address the mapping and localization issues for the 2D range data. We presented the classification of the B-spline feature extraction approaches based on their techniques. We proposed a method for the fast and accurate mapping of the environment based on high-level characteristics such as B-splines suited for both curved and straight-line geometric shapes. We also proposed a benchmark time metric for the evaluation of the proposed methodology and compared it with the state-of-the-art methods. It was found that the performance of the proposed approach was comparable in retrievability and compactness for all indoor and outdoor data sets compared with the state-of-the-art approach [3] of B-spline feature extractions; however, this was obtained at the cost of a slight reduction in the accuracy of the curves that was still satisfactory. The time taken for the approximation of B-spline features was greatly reduced in the proposed method compared with the state-of-the-art methodology. We finally presented algorithms for all methods that were very effective when implementing the procedure. In future, we will explore more efficient methods for B-spline approximations, test our method for B-spline associations, and expand the application of our work for the 3D range sensors as well as for the visual SLAM.

Author Contributions: Conceptualization, M.U. and A.A.; formal analysis, M.U., A.A., A.T., M.Z.U.R. and A.M.K.; investigation, M.U., A.A., A.T., M.Z.U.R. and A.M.K.; methodology, M.U., A.A. and A.T.; writing—original draft, M.U., A.A., A.T., M.Z.U.R. and A.M.K.; writing—review and editing, M.U., A.T. and M.Z.U.R. All authors have read and agreed to the published version of the manuscript.

Funding: This research received no external funding.

Institutional Review Board Statement: Not applicable.

Informed Consent Statement: Not applicable.

Data Availability Statement: Anlayzed datasets in this paper were also used in [1–3]. The datasets with corrected poses were taken from https://radish.sourceforge.net/index.php (accessed on 3 February 2019).

Acknowledgments: I acknowledge the support of my friends, family, and specifically the Department of Mechanical, Mechatronics and Manufacturing Engineering of the University of Engineering and Technology, Lahore, Pakistan.

Conflicts of Interest: There is no conflict of interest declared by the authors.

Abbreviations

The following abbreviations are used in this manuscript:

RPOS	Relative position- and orientation-based segmentation;
SVED	Segmentation based on varying Euclidean distance;
BID	B-spline-based interest point detector;
FALKO	Fast adaptive laser keypoint orientation-invariant;
PDM	Point distance minimization.

References

1. Tipaldi, G.D.; Arras, K.O. FLIRT—Interest regions for 2D range data. In Proceedings of the IEEE International Robotics and Automation Society (ICRA), Anchorage, AK, USA, 3–7 May 2010; pp. 3616–3622.
2. Kallasi, F.; Rizzini, D.L.; Caselli, S. Fast keypoint features from laser scanner for robot localization and mapping. *IEEE Robot. Autom. Lett.* **2016**, *1*, 176–183. [CrossRef]
3. Usman, M.; Khan, A.M.; Ali, A.; Yaqub, S.; Zuhaib, K.M.; Lee, J.Y.; Han, C.S. Active B-splines: An extensive approach to feature detection and description for 2D range data. *IEEE Robot. Autom. Lett.* **2019**, *1*, 176–183.
4. Moravec, H.; Elfes, A. High resolution maps from wide angle sonar. In Proceedings of the IEEE International Conference on Robotics and Automation, St. Louis, MO, USA, 25–28 March 1985; Volume 2, pp. 116–121.
5. Rodriguez-Losada, D.; Matia, F.; Galan, R. Building geometric feature based maps for indoor service robots. *Robot. Auton. Syst.* **2006**, *54*, 546–558. [CrossRef]
6. Veeck, M.; Veeck, W. Learning polyline maps from range scan data acquired with mobile robots. In Proceedings of the IEEE/RSJ International Conference on Intelligent Robots and Systems (IROS), New Orleans, LA, USA, 26 April–1 May 2004; Volume 2, pp. 1065–1070.
7. Zhang, S.; Adams, M.; Tang, F.; Xie, L. Geometrical feature extraction using 2D range scanner. In Proceedings of the 4th International Conference on Control and Automation Proceedings, Las Vegas, NV, USA, 12 June 2003; pp. 901–905.
8. Nunez, P.; Vazquez-Martin, R.; del Toro, J.C.; Bandera, A.; Sandoval, F. Feature extraction from laser scan data based on curvature estimation for mobile robotics. In Proceedings of the IEEE International Conference on Robotics and Automation, Orlando, FL, USA, 15–19 May 2006; pp. 1167–1172.
9. Meier, K.; Chung, S.-J.; Hutchinson, S. Visual-inertial curve simultaneous localization and mapping: Creating a sparse structured world without feature points. *J. Field Robot.* **2018**, *35*, 516–544. [CrossRef]
10. Pedraza, L.; Rodriguez-Losada, D.; Matia, F.; Dissanayake, G.; Miró, J.V. Extending the limits of feature-based SLAM with B-splines. *IEEE Trans. Robot.* **2009**, *25*, 353–366. [CrossRef]
11. Smith, R.; Self, M.; Cheeseman, P. Estimating uncertain spatial relationships in robotics. In *Autonomous Robot Vehicles*; Springer: Berlin/Heidelberg, Germany, 1990; pp. 167–193.
12. Lowry, S.; Sünderhauf, N.; Newman, P.; Leonard, J.J.; Cox, D.; Corke, P.; Milford, M.J. Visual place recognition: A survey. *IEEE Trans. Robot.* **2015**, *32*, 1–19. [CrossRef]
13. Ok, K.; Liu, K.; Frey, K.; How, J.P.; Roy, N. Robust object-based slam for high-speed autonomous navigation. In Proceedings of the International Conference on Robotics and Automation (ICRA), Montreal, QC, Canada, 20–24 May 2019; pp. 669–675.
14. Taubner, F.; Tschopp, F.; Novkovic, T.; Siegwart, R.; Furrer, F. Lcd–line clustering and description for place recognition. In Proceedings of the International Conference on 3D Vision (3DV), Fukuoka, Japan, 25–28 November 2020; pp. 908–917.
15. Besl, P.J.; McKay, N.D. Method for registration of 3-D shapes. In *Sensor Fusion IV: Control Paradigms and Data Structures, Proceedings of the ROBOTICS '91, Boston, MA, USA, 14–15 November 1991*; SPIE: Bellingham, WA USA 1992; Volume 1611, pp. 586–606.
16. Bernreiter, L.; Ott, L.; Nieto, J.; Siegwart, R.; Cadena, C. PHASER: A robust and correspondence-free global pointcloud registration. *IEEE Robot. Autom. Lett.* **2021**, *6*, 855–862. [CrossRef]
17. Le, H.M.; Do, T.-T.; Hoang, T.; Cheung, N.-M. SDRSAC: Semidefinite-based randomized approach for robust point cloud registration without correspondences. In Proceedings of the IEEE/CVF Conference on Computer Vision and Pattern Recognition, Long Beach, CA, USA, 15–20 June 2019; pp. 124–133.
18. Bosse, M.; Zlot, R. Keypoint design and evaluation for place recognition in 2D lidar maps. *Robot. Auton. Syst.* **2009**, *57*, 1211–1224. [CrossRef]
19. Li, Y.; Olson, E.B. Extracting general-purpose features from LIDAR data. In Proceedings of the IEEE International Conference on Robotics and Automation, Anchorage, AK, USA, 3–7 May 2010; pp. 1388–1393.
20. Li, Y.; Olson, E.B. Structure tensors for general purpose LIDAR feature extraction. In Proceedings of the IEEE International Conference on Robotics and Automation, Shanghai, China, 9–13 May 2011; pp. 1869–1874.
21. Zhang, G.; Lee, J.H.; Lim, J.; Suh, I.H. Building a 3-D line-based map using stereo SLAM. *IEEE Trans. Robot.* **2015**, *31*, 1364–1377. [CrossRef]
22. Lu, Y.; Song, D. Visual navigation using heterogeneous landmarks and unsupervised geometric constraints. *IEEE Trans. Robot.* **2015**, *31*, 736–749. [CrossRef]
23. Smith, P.; Reid, I.; Davison, A.J. Real-Time Monocular SLAM with Straight Lines. *BMVC* **2006**, *6*, 17–26.

24. Nuske, S.; Choudhury, S.; Jain, S.; Chambers, A.; Yoder, L.; Scherer, S.; Chamberlain, L.; Cover, H.; Singh, S. Autonomous exploration and motion planning for an unmanned aerial vehicle navigating rivers. *J. Field Robot.* **2015**, *32*, 1141–1162. [CrossRef]
25. Yang, J.; Dani, A.; Chung, S.-J.; Hutchinson, S. Vision-based localization and robot-centric mapping in riverine environments. *J. Field Robot.* **2017**, *34*, 429–450. [CrossRef]
26. Vandorpe, J.; Brussel, H.V.; Xu, H. Exact dynamic map building for a mobile robot using geometrical primitives produced by a 2D range finder. In Proceedings of the IEEE International Conference on Robotics and Automation, Minneapolis, MN, USA, 22–28 April 1996; Volume 1, pp. 901–908.
27. Pfister, S.T.; Roumeliotis, S.I.; Burdick, J.W. Weighted line fitting algorithms for mobile robot map building and efficient data representation. In Proceedings of the IEEE International Conference on Robotics and Automation, Taipei, Taiwan, 14–19 September 2003; Volume 1, pp. 1304–1311.
28. Grisetti, G.; Stachniss, C.; Burgard, W. Improved techniques for grid mapping with rao-blackwellized particle filters. *IEEE Trans. Robot.* **2007**, *23*, 34–46. [CrossRef]
29. Rogers, D.F. *An Introduction to NURBS: With Historical Perspective*; Morgan Kaufmann: Burlington, MA, USA, 2001.
30. De Boor, C. A Practical Guide to Spline. *Math Comput.* **1978**, *27*, 325. [CrossRef]
31. Blake, A.; Isard, M. *Active Contours: The Application of Techniques from Graphics, Vision, Control Theory and Statistics to Visual Tracking of Shapes in Motion*; Springer Science & Business Media: Berlin/Heidelberg, Germany, 2012.
32. Wang, W.; Pottmann, H.; Liu, Y. Fitting B-spline curves to point clouds by curvature-based squared distance minimization. *ACM Trans. Graph. ToG* **2006**, *25*, 214–238. [CrossRef]
33. Rusu, R.B.; Cousins, S. 3D is here: Point cloud library (pcl). In Proceedings of the IEEE International Conference on Robotics and Automation, Shanghai, China, 9–13 May 2011; pp. 1–4.

sensors

MDPI

Article

Deep-Learning-Based ADHD Classification Using Children's Skeleton Data Acquired through the ADHD Screening Game

Wonjun Lee, Deokwon Lee, Sanghyub Lee, Kooksung Jun and Mun Sang Kim *

School of Integrated Technology, Gwangju Institute of Science and Technology, Gwangju 61005, Republic of Korea
* Correspondence: munsang@gist.ac.kr

Abstract: The identification of attention deficit hyperactivity disorder (ADHD) in children, which is increasing every year worldwide, is very important for early diagnosis and treatment. However, since ADHD is not a simple disease that can be diagnosed with a simple test, doctors require a large period of time and substantial effort for accurate diagnosis and treatment. Currently, ADHD classification studies using various datasets and machine learning or deep learning algorithms are actively being conducted for the screening diagnosis of ADHD. However, there has been no study of ADHD classification using only skeleton data. It was hypothesized that the main symptoms of ADHD, such as distraction, hyperactivity, and impulsivity, could be differentiated through skeleton data. Thus, we devised a game system for the screening and diagnosis of children's ADHD and acquired children's skeleton data using five Azure Kinect units equipped with depth sensors, while the game was being played. The game for screening diagnosis involves a robot first travelling on a specific path, after which the child must remember the path the robot took and then follow it. The skeleton data used in this study were divided into two categories: standby data, obtained when a child waits while the robot demonstrates the path; and game data, obtained when a child plays the game. The acquired data were classified using the RNN series of GRU, RNN, and LSTM algorithms; a bidirectional layer; and a weighted cross-entropy loss function. Among these, an LSTM algorithm using a bidirectional layer and a weighted cross-entropy loss function obtained a classification accuracy of 97.82%.

Keywords: ADHD; deep learning; screening; skeleton

Citation: Lee, W.; Lee, D.; Lee, S.; Jun, K.; Kim, M.S. Deep-Learning-Based ADHD Classification Using Children's Skeleton Data Acquired through the ADHD Screening Game. *Sensors* **2023**, *23*, 246. https://doi.org/10.3390/s23010246

Academic Editors: Sašo Blažič, Gregor Klancar and Marija Seder

Received: 25 November 2022
Revised: 19 December 2022
Accepted: 21 December 2022
Published: 26 December 2022

1. Introduction

Attention deficit hyperactivity disorder (ADHD) is a disorder that occurs frequently in childhood and refers to a condition in which attention is continuously insufficient, resulting in distraction, hyperactivity, and impulsivity [1]. ADHD patients range from children to adults, but the proportion of adolescents and children among ADHD patients accounts for more than 80% of the total number of patients. In addition, the number of children and adolescents with ADHD is increasing every year [2]. If ADHD is left untreated, its symptoms can lead to difficulties throughout childhood [3], and children with ADHD symptoms have a 70% chance of developing ADHD into adolescence. Furthermore, if the condition remains untreated, it can persist into adulthood for more than 50% of children. For these reasons, early diagnosis and treatment of ADHD are important [4,5].

ADHD is a psychiatric disease, not a fracture or cancer that can be diagnosed relatively simply using MRI or CT scan results. Psychiatric diseases are not simple diseases that can be diagnosed with simple tests or symptoms, and therefore, doctors require a large amount of time and effort for diagnosis. Currently, the most commonly used method for diagnosis of ADHD by doctors in hospitals involves aggregating the results of consultations with child patients, consultations with their parents or teachers, and survey results such as the child behavior checklist (CBCL) [6,7]. In addition, the results of audio–visual attention tests such as the continuous performance test (CPT) are used as auxiliary data for the diagnosis of ADHD [8,9]. However, it is not easy to make an accurate diagnosis only on the basis of the above results because surveys and counseling, which have the greatest influence on diagnosis, involve a large number of subjective opinions of parents and the children's teachers [10]. In addition, since it takes a large amount of time to undertake counseling and surveying, doctors require a great amount of effort to conduct ADHD diagnostic tests. There is also a global shortage of trained specialists able to diagnose ADHD, which often delays diagnosis.

As mentioned above, the existing method of diagnosing ADHD by synthesizing the results of counseling and questionnaire results is based on the activity level of children observed by parents or teachers, and thus subjective errors may occur depending on the observer. In order to reduce errors due to subjective judgment, studies are underway to reduce subjective errors by using accelerometers or simulators to objectively measure and judge the activity level of children and to make more accurate diagnoses with objective results.

While various studies as above are being conducted, research on machine learning and deep learning has accelerated, and we have started to apply machine learning and deep learning and develop algorithms for classification in various fields in various medical fields [11,12]. In particular, with the development of machine learning and deep learning technologies, studies on the diagnosis of ADHD through artificial intelligence using a variety of objective data have been actively conducted since 2010. Research is underway for faster and more efficient ADHD diagnosis through machine learning and deep learning using data acquired through accelerometers, simulators, and games, as well as biometric data such as MRI, EEG, and ECG as input data. These research results are expected to help doctors improve the accuracy of ADHD diagnosis and shorten the time taken for this process. In particular, in the case of MRI, the Neuro Bureau ADHD-200 Dataset [13] was released, and various research institutes are now actively conducting research to improve the accuracy of ADHD diagnosis using this dataset by developing machine learning and deep learning algorithms.

In this study, a game was developed for children to screen for ADHD in a child-friendly environment such as a school without direct intervention from experts. While the children were playing the game, the children's skeleton data were acquired using five depth sensors. Using the acquired skeleton data and deep learning algorithm, ADHD, ADHD-RISK, and normal were screened with 97% accuracy. ADHD screening evaluation was conducted using several simple models, and among them, it was verified that the LSTM-bidirectional model had the highest accuracy. In particular, ADHD-RISK is a class that has not been seen in other studies and is a taxon that is difficult for doctors to discriminate in actual clinical practice. However, it is important to screen ADHD-RISK that has the potential to develop into ADHD with high accuracy in actual ADHD clinical practice, and this system is expected to be helpful in the future.

The detailed structure of this paper is as follows. The Introduction section introduces the existing methods for screening ADHD. The Materials and Methods section introduces the materials used in this study and the methods proposed. The Results section describes the experimental results. The Discussion section describes the discussion in this study. Finally, the Conclusion section presents the conclusion.

2. Related Work

Various data acquisition systems and artificial intelligence algorithms for the diagnosis of ADHD are being developed. As a first example, various research teams have developed deep learning or machine learning algorithms using the Neuro Bureau ADHD-200 Dataset and have conducted studies on the diagnosis of ADHD. The ADHD-200 Dataset consists of 776 resting-state fMRI and structural MRI data [14–21].

The team of Peng et al. developed a CNN-based deep learning algorithm to obtain an ADHD diagnosis accuracy of 72.9% [16]. In addition, the research team of Chen et al. developed an SVM-based machine learning algorithm and acquired an ADHD diagnosis accuracy of 88.1% [21]. Since a very good public dataset is publicly available for MRI-data-based ADHD diagnosis research, it is expected that many research teams will be able to achieve higher accuracy research results through continuous algorithm development.

The second example comprises studies on the diagnosis of ADHD using EEG data [22–28]. The team of Tosun et al. obtained 92.2% ADHD classification accuracy using an LSTM-based deep learning algorithm for 1088 ADHD patients and 1088 normal groups [25]. In addition, the research team of Altinkaynak et al. obtained an accuracy of 91.3% using an MLP-based machine learning algorithm using EEG data of 23 ADHD patients and 23 normal subjects [27].

A third related type of study involves the use of continuous performance test (CPT) test results; this test is widely used for the diagnosis of ADHD in hospitals. CPT test results are used as input data for ADHD classification research [29,30]. Using the CPT results of 213 ADHD patients and 245 normal subjects, the research team of Slobodin et al. obtained an ADHD classification accuracy of 87% using a random-forest-based machine learning algorithm [29].

The last related study was conducted by the research team of O'Mahony et al., who classified ADHD using data measured during the TOVA test by having the children under study wear two IMU sensors on their waist and ankle. A classification accuracy of 95.1% was obtained using the SVM-based machine learning algorithm [31].

As mentioned above, various datasets and algorithms are being used and researched for the classification of ADHD. However, studies on the classification of ADHD using skeleton data have not yet been conducted. Skeleton data comprise the subject's joint movements. After giving children a specific task, we acquired skeleton data and RGB images while the children were performing the task. After all data were acquired, four psychiatrists divided the children into three categories on the basis of the RGB image analysis results and the CBCL and K-ARS results: ADHD, ADHD-RISK, or normal. In this study, it was assumed that there would be a significant difference in the behaviors of children with ADHD, children with ADHD-RISK, and normal children while performing tasks. Using the skeleton data measured on the basis of the above hypothesis as input data, a deep-learning-based algorithm was used to classify the ADHD, ADHD-RISK, and normal groups.

3. Materials and Methods

3.1. Description of the Game for the Screening Diagnosis of ADHD

In this study, an ADHD diagnosis game was used to acquire skeleton data for children's ADHD screening diagnosis.

The game consists of a total of five stages, comprising two practice games and three main games. In the game, a robot first moves randomly on the nine numbered boards marked on the floor, as shown in Figure 1a. At this time, the child memorizes the path the robot takes while waiting in the wait zone, as shown in Figure 1b. After the robot has completed moving on the path, the child has to follow the path the robot has taken after the start signal given by the robot. While following the path, the child performs one more task.

As the child follows the path, the characters of the witch or the Wizard of Oz appear on the screen. As shown in Figure 2a, when a witch appears, the children sit down, and when a character appears, the children wave their hands, as shown in Figure 2b. The higher the level, the more complex the path. For more information about the game, please see Ref. [32].

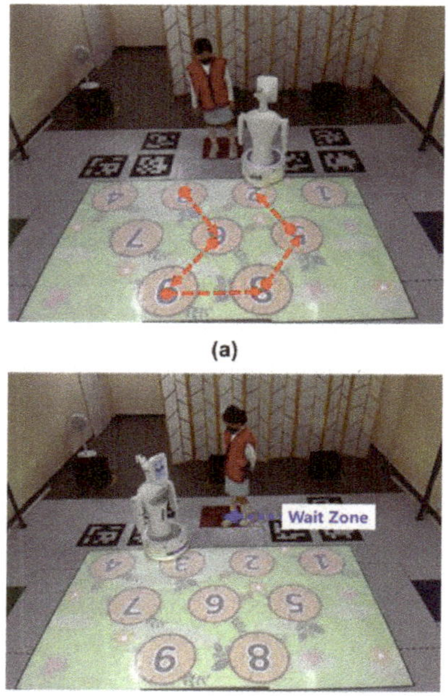

Figure 1. A game for the diagnosis of ADHD. (**a**) The robot moves on the numbered board first. (**b**) A child waiting in the waiting zone for the robot to take a certain path remembers the path taken by the robot and follows it.

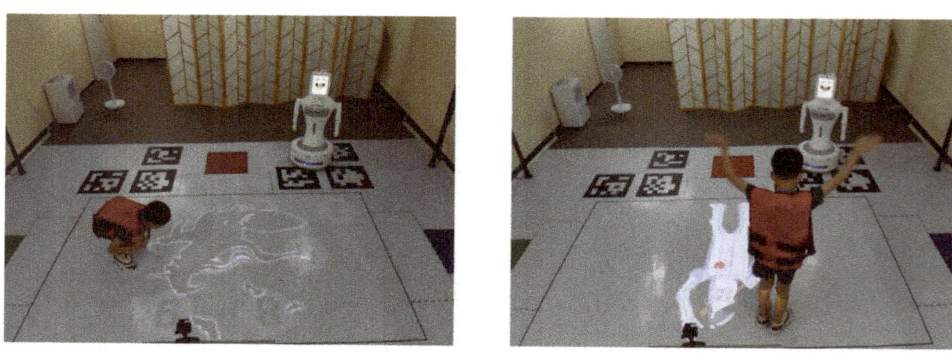

(a) (b)

Figure 2. Children perform additional tasks while following the robot's path. (**a**) When the witch appears on the floor, the child sits down. (**b**) When the character appears, the child waves both hands over their head.

3.2. Skeleton Capture System Using Azure Kinect

As shown in Figure 3, five Azure Kinect (Microsoft Corp., Redmond, WA, USA) units from Microsoft and a beam projector (HU85LA) from LG were used to acquire the children's skeleton data while the children were playing the game. Moreover, Robocare's robot (Silbot) was used to progress the game and follow the path. In this study, the human tracking and skeleton merge algorithm using Azure Kinect was adopted as in a previous study by this research team (please refer to [33] for details).

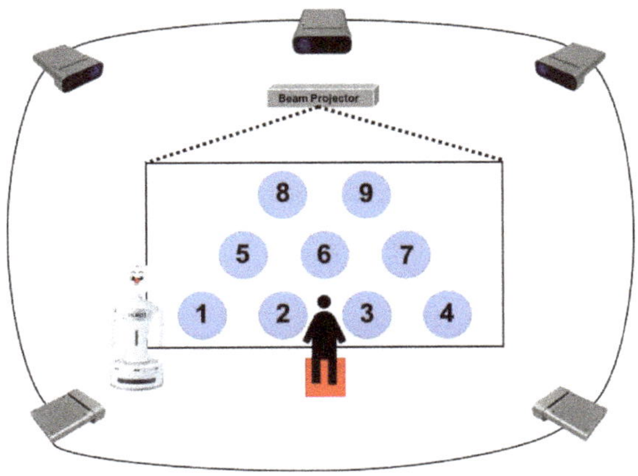

Figure 3. Game setup for children, consisting of five Azure Kinect units to acquire the children's skeleton data, a beam projector to project the game screen, and a robot to help the game progress.

In the sensor system, a calibration procedure was performed to match the coordinate system of each sensor. For this, trajectory data consisting of 3D centroid coordinates of a sphere object with a specific color were used. The trajectory data recognized by each subordinate sensor were used to calculate the rotation matrix using the trajectory data extracted by the master sensor. In addition, the coordinate system of the entire system calibrated with the coordinate system of the master sensor was converted to the world coordinate system set by the user using the Aruco marker. After that, the skeleton data of the target in the capturing area can be extracted from the depth image using Kinect Azure Body Tracking SDK. The skeleton data extracted from each sensor are merged with the skeleton data extracted from the other sensors that recognize objects from different angles to overcome the occlusion problem that occurs in a single sensor. The merging procedure is performed for each joint and includes a method of filtering the noise candidate group using DBSCAN based on the location information of each joint. As a result, the system was able to extract the skeleton data of the object in the capturing area of 3 m × 5 m from the world coordinate system set by the user by using five sensors installed at a height of 2 m. Finally, through the above system, the movement of each waiting stage and game stage acquired continuous movement at 15 frames per second.

The beam projector was used to show the path for the main game and the witch or character corresponding to the additional task. Finally, the robot was responsible for explaining and demonstrating the game to the children.

3.3. Data Acquisition Methods and Target Candidates

For this study, elementary school students from first to sixth grade living in Seoul, South Korea, were recruited. Data were acquired from October 2019 to December 2021 from a total of three elementary schools and one research center. All children participating in the experiment were given parental consent and went through an ethical review process.

The children's skeleton data and RGB image data were saved while the children played the game. The data were largely divided into standby data, which were acquired while the robot explained the game and demonstrated the path the child should follow, and game data, which were acquired while the child played the game. All the children played the game in five stages. Therefore, for each stage, two pieces of data were generated during standby and during the game, resulting in a total of 10 pieces of data. Additionally, the standby data acquired while the robot explained the whole game before the first game was included. In conclusion, each child had six standby data points and five game data points, totaling 11 data points. A total of 596 children participated in this study. By synthesizing the CBCL and K-ARS results performed before the game for this study and the RGB image viewing results among the obtained children's data, four clinicians divided them into the ADHD, ADHD-RISK, and normal groups.

As shown in Table 1, as a result of classification, 66 ADHD group children, 181 ADHD-RISK group children, and 349 normal group children were classified.

Table 1. Number of children by ADHD class of children participating in this study.

Classes	Number of Participants
ADHD	66
ADHD-RISK	181
Normal	349
Total	596

3.4. Deep Learning Model Using GRU, RNN, and LSTM for ADHD Classification

The RNN-based deep learning algorithm was used to classify the ADHD, ADHD-RISK, and normal groups. The size of the data for each level was different because the standby time and the completion time of the game were different for each level. Therefore, the longest frame among all children's step-by-step data was defined as the reference size. If each data length was smaller than the standard size, the remaining frames were set to zero. The Azure Kinect devices used in this study can provide data for a total of 32 joints. However, in this study, a total of 18 joints were used, except for the low-accuracy and unstable end point joints, as shown in Figure 4 [19].

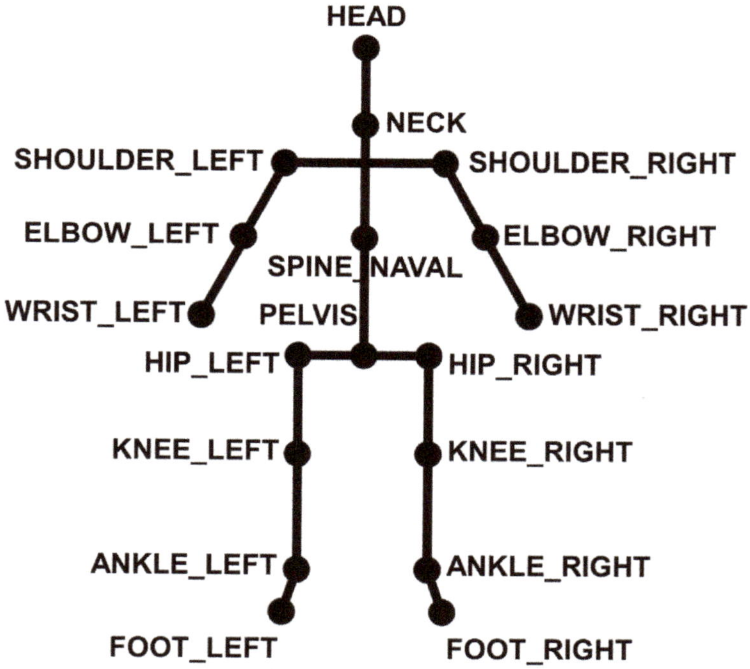

Figure 4. Joints used to classify ADHD classes in children.

A total of six standby skeleton data points and a total of five game data points were used as input data. As shown in Figure 5, each sequential data point passed through the RNN layer. Each feature extracted after passing through the RNN was concatenated. Finally, they passed through the classification layer and children were finally classified as ADHD, ADHD-RISK, or normal. The RNN models used were GRU, RNN, and LSTM.

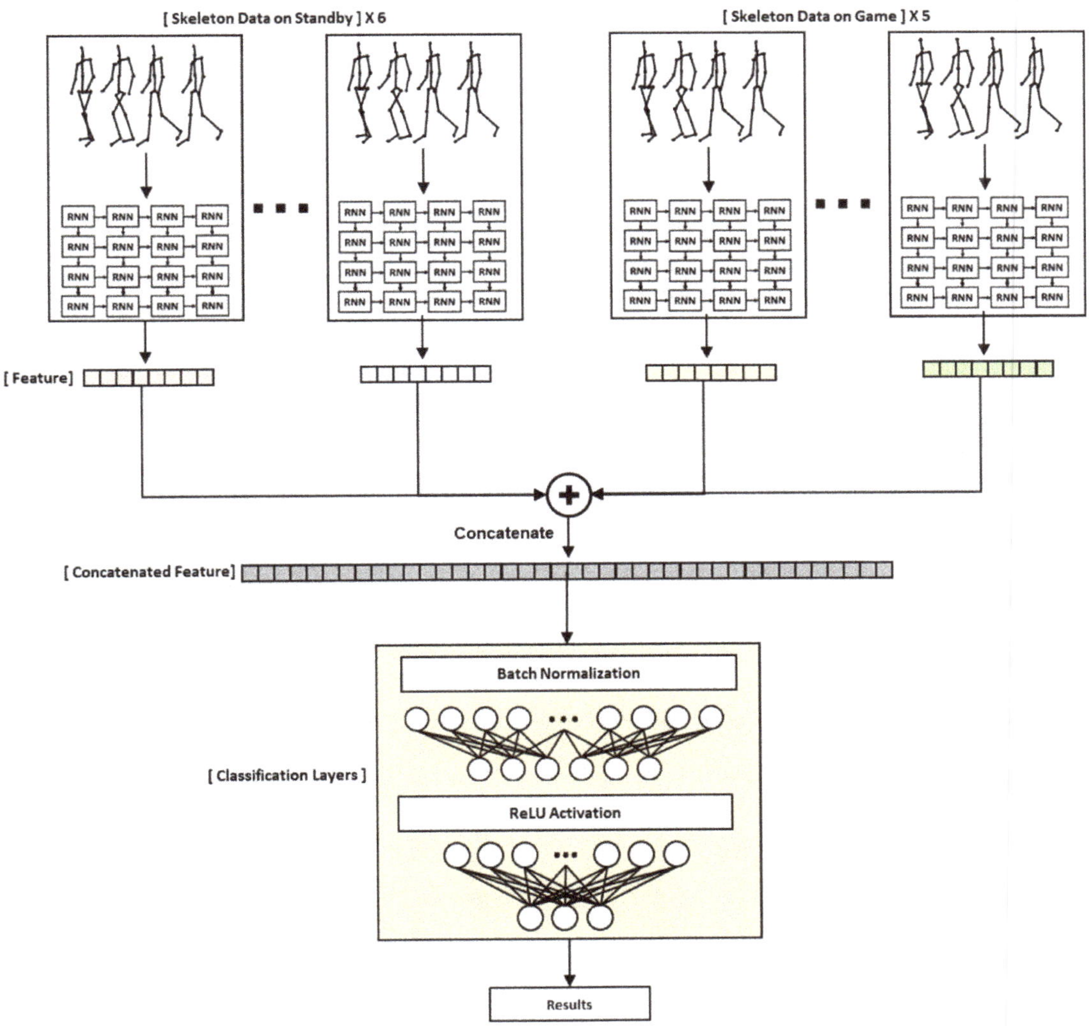

Figure 5. Description of an RNN-based model designed to classify children's ADHD class.

Moreover, as shown in Figure 6, we experimented by adding a bidirectional layer to each model to improve performance. The final output $\boldsymbol{y}$ was obtained by concatenating the hidden states $\overrightarrow{h}$ and $\overleftarrow{h}$ of the t cell of the forward RNN-based model and the $t-1$ cell of the backward RNN-based model. ReLU was used as the activation function. In addition, as shown in Equation (1), a weighted cross-entropy loss function was used among the loss functions to prevent overfitting and to improve performance because the number of data points for each class was different.

$$Weighted\ Cross\ Entropy\ Loss = -\sum_{i=1}^{C} w_i t_i \log(p_i) \quad (1)$$

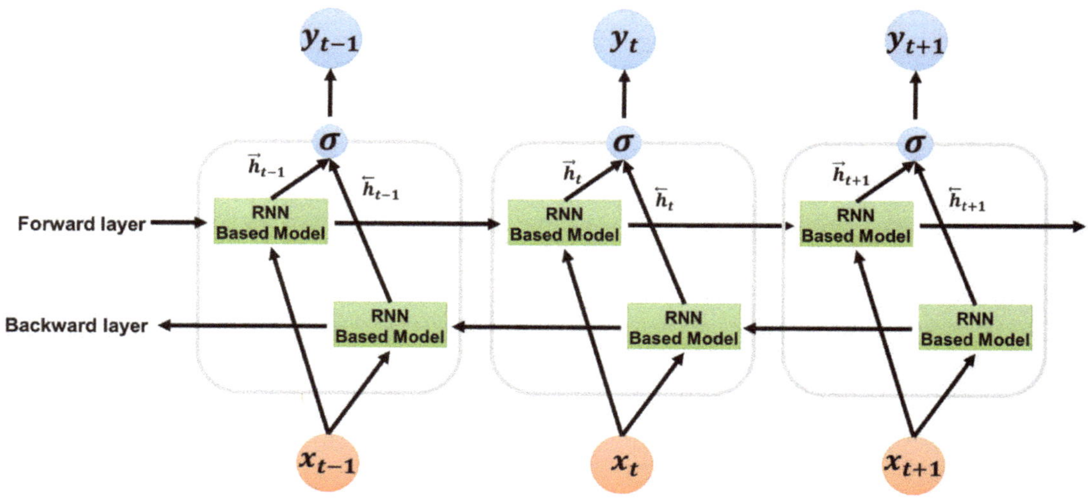

Figure 6. Bidirectional layer description for improving the performance of the RNN-based ADHD classification model.

In formula (1), t_i is the truth label, p_i is the Softmax probability for the ith class, and w_i represents the weight of the loss function. The weight is given as the inverse of the rate of input data. In addition, this experiment was verified by leave-one-person-out cross-validation.

4. Results

In this study, classification of ADHD, ADHD-RISK, and normal groups was performed using the RNN model using skeleton data as input data. Three types of RNN models were used: GRU, RNN, and LSTM, and each model used a bidirectional layer and a weighted loss function to improve performance and prevent overfitting. Moreover, in this paper, accuracy, precision, recall, and the F1-score were used to evaluate the model. The formula for each parameter is as shown in the following Formulas (2)–(5):

$$Accuracy = \frac{TP + TN}{TP + TN + FP + FN} \quad (2)$$

$$Precision = \frac{TP}{TP + FP} \quad (3)$$

$$Recall = \frac{TP}{TP + FN} \quad (4)$$

$$F1 = \frac{2 \times Precision \times Recall}{Precision + Recall} \quad (5)$$

In the above formulas, TP (true positive) is the result of predicting an answer that is actually true as true; FP (false positive) is the result of predicting an answer that is actually false as true; FN (false negative) is the result of predicting an answer that is actually true as false; and finally, TN (true negative) is the result of predicting an answer that is actually false as false.

As can be seen in Table 2, GRU, RNN, and LSTM showed 94.04%, 88.35%, and 88.35% accuracy, respectively. However, it was confirmed that the F1-score of the ADHD and ADHD-RISK classes did not exceed 90 and was low.

Table 2. Comparison of results of three RNN-based ADHD classification models using skeleton data as input data.

GRU				
Label	**Accuracy**	Precision	Recall	F1-Score
Normal	94.04	0.93	1	0.96
ADHD-RISK		1	0.70	0.82
ADHD		0.96	0.82	0.89
RNN				
Label	Accuracy	Precision	Recall	F1-Score
Normal	88.35	0.87	1	0.93
ADHD-RISK		1	0.54	0.70
ADHD		0.89	0.48	0.62
LSTM				
Label	Accuracy	Precision	Recall	F1-Score
Normal	88.35	0.86	1	0.92
ADHD-RISK		1	0.63	0.77
ADHD		1	0.34	0.51

Thus, in order to improve the performance, we used a bidirectional layer and a weighted loss function in the above model.

As a result, as shown in Table 3, the accuracy of each model slightly increased to 96.81%, 96.81%, and 97.82% for GRU, RNN, and LSTM, respectively. The accuracy increased slightly, but the F1-score increased significantly. Among the F1-scores of each model, the ADHD and ADHD-RISK classes scored 51~89% before using the bidirectional layer and weighted loss function but these scores increased significantly to 87~100% after using the bidirectional layer and weighted loss function. In particular, the ADHD class of the GRU-bidirectional weighted loss function recorded an F1-score of 100%.

Table 3. Using skeleton data as input data and comparing the results of adding a bidirectional layer and a weighted loss function to three RNN-based ADHD classification models.

GRU-Bidirectional with Weighted Loss				
Label	Accuracy	Precision	Recall	F1-Score
Normal	96.81	0.95	1	0.97
ADHD-RISK		1	0.89	0.94
ADHD		1	1	1
RNN-Bidirectional with Weighted Loss				
Label	Accuracy	Precision	Recall	F1-Score
Normal	96.81	0.98	0.99	0.99
ADHD-RISK		0.99	0.90	0.94
ADHD		0.83	1	0.91
LSTM-Bidirectional with Weighted Loss				
Label	Accuracy	Precision	Recall	F1-Score
Normal	97.82	0.97	1	0.98
ADHD-RISK		1	0.93	0.96
ADHD		0.92	0.98	0.95

After confirming the above results, we conducted one additional experiment. The skeleton data used in this study were divided into two main categories: the skeleton data of children waiting while the robot explained and demonstrated the game before the game started, and skeleton data obtained while the children directly played the game. We verified which of the above two large categories of data were helpful for the classification of ADHD. This test used models that applied a bidirectional layer and a weighted loss function to each RNN model on the basis of the previous experiment. Table 4 below shows the results of using only the skeleton data during standby and the game.

Table 4. ADHD class classification results when using the skeleton data acquired during waiting and skeleton data acquired during the game as input data.

Standby Skeleton					**Game Skeleton**				
GRU-Bidirectional with Weighted Loss					GRU-Bidirectional with Weighted Loss				
Label	Accuracy	Precision	Recall	F1-Score	Label	Accuracy	Precision	Recall	F1-Score
Normal		0.96	0.98	0.97	Normal		0.95	1	0.97
ADHD-RISK	95.57	0.97	0.78	0.86	ADHD-RISK	95.39	1	0.69	0.81
ADHD		0.87	0.97	0.91	ADHD		0.89	1	0.94
RNN-Bidirectional with Weighted Loss					RNN-Bidirectional with Weighted Loss				
Label	Accuracy	Precision	Recall	F1-Score	Label	Accuracy	Precision	Recall	F1-Score
Normal		0.95	0.99	0.97	Normal		0.98	1	0.99
ADHD-RISK	96.14	1	0.90	0.94	ADHD-RISK	95.12	1	0.67	0.80
ADHD		0.89	0.93	0.91	ADHD		0.81	1	0.89
LSTM-Bidirectional with Weighted Loss					LSTM-Bidirectional with Weighted Loss				
Label	Accuracy	Precision	Recall	F1-Score	Label	Accuracy	Precision	Recall	F1-Score
Normal		0.94	1	0.97	Normal		0.95	1	0.97
ADHD-RISK	94.85	0.95	0.70	0.81	ADHD-RISK	94.85	1	0.65	0.79
ADHD		0.94	0.91	0.92	ADHD		0.83	1	0.90

When only the skeleton data were used during standby, the RNN model obtained the highest accuracy with 96.14%, and when only the skeleton data were used during the game, the GRU model obtained the highest accuracy with 95.39%. The accuracy was similar for standby data and game data, but there was a significant difference in the F1-scores of ADHD and ADHD-RISK. We think that it is more important to screen ADHD-RISK or ADHD than to screen children in the normal group in the ADHD screening diagnosis. This is because, if a child with ADHD is screened as normal, an opportunity for early treatment may be missed [18]. When only waiting data were used, the F1-Score of ADHD-RISK was higher by at least 5% to maximum 14% for each model. In addition, in the case of ADHD, the F1-Score of the waiting data was 2% higher, except for the GRU model. On the basis of the above results, it was judged that the skeleton data during waiting were more helpful in classifying ADHD and ADHD-RISK. Since the validation method of the model used in this study is leave-one-out cross-validation, in order to obtain the standard deviation of the model result, it is necessary to check the results of learning several times for each model. The standard deviation of accuracy after learning five times for each model is 0.17~0.67%.

5. Discussion

We classified ADHD, ADHD-RISK, and normal groups using only the skeleton data of children acquired through games. Currently, there are no studies using skeleton data among studies for the screening diagnosis of children's ADHD. All children who participated in the experiment for this study performed the same task in the same limited environment. The advantage of the children's data collected under these conditions was that they were objective and that high-quality data could be acquired in a much shorter time than the method of acquiring data by wearing an IMU device for a long period of time. This study is expected to serve as a cornerstone for future research on the screening and diagnosis of ADHD in children using skeleton data. Since the data used in this study were time series data, they were compared using an RNN-type model, and as a result, an accuracy of 97.82% was obtained. High accuracy was verified in this study, even when compared with the results of other studies that classified ADHD using various data. The model with the highest accuracy was an LSTM model with a bidirectional layer added, and the confusion matrix obtained through the model is shown in Figure 7 below.

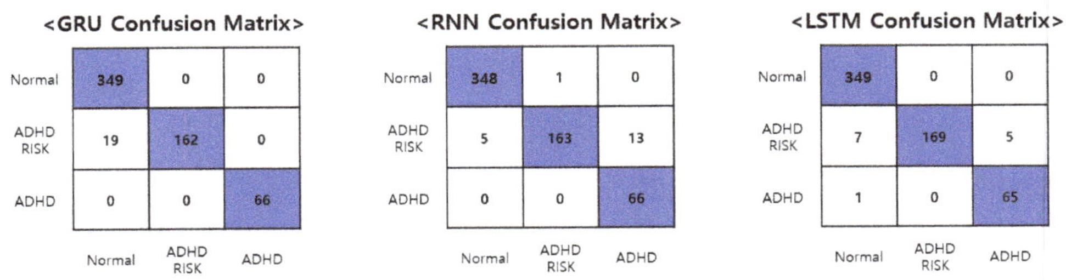

Figure 7. The confusion matrix of the GRU, RNN, and LSTM model with the best classification results.

Figure 7 is the confusion matrix of the best results obtained using the GRU, RNN, and LSTM models with bidirectional layers added and the entire skeleton data, respectively. All models correctly selected the normal group and the ADHD group. However, differences in overall accuracy occurred due to differences in ADHD-RISK screening results. In particular, the LSTM model classified all 349 out of 349 people into the normal group, and 65 out of 66 people into the ADHD group. However, in the case of ADHD-RISK, 7 out of a total of 181 ADHD-RISK patients were classified into the normal group, and 5 people were classified into the ADHD group. Classifying the ADHD-RISK group into the ADHD group is not a major problem but classifying the ADHD-RISK group as normal is to be overcome through future research on the performance improvement of the model. The diagnosis of ADHD is actually conducted in hospitals with various indicators. ADHD is not a disease that can be objectively diagnosed through the results of MRI or CT scans such as with cancer, but instead requires a mental evaluation. For this reason, this study was established, and the deep learning algorithm used in this study and the skeleton data obtained through the game comprised a system created to selectively diagnose ADHD in order to help doctors prior to diagnosis. The goal of this study was to enable children to easily perform screening tests through games, not only in schools but also in educational centers, before receiving an accurate diagnosis by a doctor using this system. Through this system, it is possible to judge the situation of children on the basis of the children's ADHD screening results, to recommend hospital treatment to children with ADHD, and to guide parents and teachers in charge of ADHD-RISK children to monitor those children.

This study used a restricted game environment to observe children's behavior. During the game, the children's skeleton data were acquired. ADHD, ADHD-RISK, and normal were classified using the RNN-based deep learning algorithm using the acquired children's skeleton data. ADHD-RISK was used as a new class in this paper. When doctors diagnose ADHD and normal, the difference between the two taxa is clear, but ADHD-RISK, which lies on the border between ADHD and normal, is difficult to classify. In fact, by classifying ADHD-RISK, which is more needed by doctors in clinical practice, we have developed a system that can help doctors screen for ADHD in future clinical practice.

6. Conclusions

In this study, we classified ADHD, ADHD-RISK, and normality by using children's skeleton data acquired through a game for the screening and diagnosis of children's ADHD.

The performance of various RNN models was compared, and among them, the LSTM-bidirectional model showed the best results. In the future, we plan to conduct research on performance improvement using various models such as the GCN model optimized for skeleton data. We also intend to analyze which joints affect the model at any point in time using the attention layer. On the basis of the above results, we would like to simplify the skeleton data acquisition system used in this study by utilizing a different number of existing sensors and proposing a new game system that can focus on the major joints in ADHD classification.

Author Contributions: W.L. designed the algorithm, performed the experimental work, wrote the manuscript. D.L. organized the experiment setup. S.L. performed the experiment and organized the manuscript contents. K.J. edited the manuscript and analyzed the results of the experiment. Corresponding author: M.S.K. All authors have read and agreed to the published version of the manuscript.

Funding: This work was supported by the Ministry of Trade, Industry and Energy of Korea under-grant (20003762) and The Open AI Dataset Project (AI-Hub, S.Korea) in 2022 and GIST Research Project grant funded by the GIST in 2022.

Institutional Review Board Statement: The study was approved by the Institutional Review Board of Gwangju Institute of Science and Technology (20220419-HR-66-04-02).

Informed Consent Statement: Informed consent was obtained from all subjects involved in the study.

Conflicts of Interest: The authors declare no conflict of interest.

References

1. Furman, L. What is attention-deficit hyperactivity disorder (ADHD)? *J. Child Neurol.* **2005**, *20*, 994–1002. [CrossRef] [PubMed]
2. Zikra Faiz, D.; Iqbal, T.; Azeem, A.; Siddique, M.; Warraich, W.Y. A Comparative Study between Online and Traditional Counseling for Students with Attention Deficit Hyperactivity Disorder (ADHD): School Psychologists Perspective in the Obsequies of Pandemic COVID-19. *Linguist. Antverp.* **2021**, *3*, 5763–5777.
3. Antoniou, E.; Rigas, N.; Orovou, E.; Papatrechas, A. ADHD and the Importance of Comorbid Disorders in the Psychosocial Development of Children and Adolescents. *J. Biosci. Med.* **2021**, *9*, 1. [CrossRef]
4. Rocco, I.; Corso, B.; Bonati, M.; Minicuci, N. Time of onset and/or diagnosis of ADHD in European children: A systematic review. *BMC Psychiatry* **2021**, *21*, 1–24. [CrossRef]
5. Öztekin, I.; Finlayson, M.A.; Graziano, P.A.; Dick, A.S. Is there any incremental benefit to conducting neuroimaging and neurocognitive assessments in the diagnosis of ADHD in young children? A machine learning investigation. *Dev. Cogn. Neurosci.* **2021**, *49*, 100966. [CrossRef]
6. Biederman, J.; DiSalvo, M.; Vaudreuil, C.; Wozniak, J.; Uchida, M.; Woodworth, K.Y.; Green, A.; Farrell, A.; Faraone, S.V. The child behavior checklist can aid in characterizing suspected comorbid psychopathology in clinically referred youth with ADHD. *J. Psychiatr. Res.* **2021**, *138*, 477–484. [CrossRef]
7. Bernanke, J.; Luna, A.; Chang, L.; Bruno, E.; Dworkin, J.; Posner, J. Structural brain measures among children with and without ADHD in the Adolescent Brain and Cognitive Development Study cohort: A cross-sectional US population-based study. *Lancet Psychiatry* **2022**, *9*, 222–231. [CrossRef]
8. Chen, I.-C.; Chang, C.-H.; Chang, Y.; Lin, D.-S.; Lin, C.-H.; Ko, L.-W. Neural dynamics for facilitating ADHD diagnosis in preschoolers: Central and Parietal Delta synchronization in the kiddie continuous performance test. *IEEE Trans. Neural Syst. Rehabil. Eng.* **2021**, *29*, 1524–1533. [CrossRef]
9. Scimeca, L.M.; Holbrook, L.; Rhoads, T.; Cerny, B.M.; Jennette, K.J.; Resch, Z.J.; Obolsky, M.A.; Ovsiew, G.P.; Soble, J.R. Examining conners continuous performance test-3 (CPT-3) embedded performance validity indicators in an adult clinical sample referred for ADHD evaluation. *Dev. Neuropsychol.* **2021**, *46*, 347–359. [CrossRef]
10. Sempere-Tortosa, M.; Fernández-Carrasco, F.; Navarro-Soria, I.; Rizo-Maestre, C. Movement patterns in students diagnosed with adhd, objective measurement in a natural learning environment. *Int. J. Environ. Res. Public Health* **2021**, *18*, 3870. [CrossRef]
11. Gasmi, K. Improving Bert-Based Model for Medical Text Classification with an Optimization Algorithm. 2022, pp. 101–111. Available online: https://link.springer.com/chapter/10.1007/978-3-031-16210-7_8 (accessed on 18 December 2022).
12. Gasmi, K. Medical Text Classification based on an Optimized Machine Learning and External Semantic Resource. *J. Circuits Syst. Comput.* **2022**. Available online: https://www.worldscientific.com/doi/10.1142/S0218126622502917 (accessed on 18 December 2022). [CrossRef]
13. Bellec, P.; Chu, C.; Chouinard-Decorte, F.; Benhajali, Y.; Margulies, D.S.; Craddock, R.C. The neuro bureau ADHD-200 preprocessed repository. *Neuroimage* **2017**, *144*, 275–286. [CrossRef] [PubMed]
14. Zhang, C.; Ma, X.; Qin, P. LiDAR-IMU-UWB-Based Collaborative Localization. *World Electr. Veh. J.* **2022**, *13*, 32. [CrossRef]

15. Zhao, K.; Duka, B.; Xie, H.; Oathes, D.J.; Calhoun, V.; Zhang, Y. A dynamic graph convolutional neural network framework reveals new insights into connectome dysfunctions in ADHD. *Neuroimage* **2022**, *246*, 118774. [CrossRef]
16. Peng, J.; Debnath, M.; Biswas, A.K. Efficacy of novel Summation-based Synergetic Artificial Neural Network in ADHD diagnosis. *Mach. Learn. Appl.* **2021**, *6*, 100120. [CrossRef]
17. Riaz, A.; Asad, M.; Alonso, E.; Slabaugh, G. DeepFMRI: End-to-end deep learning for functional connectivity and classification of ADHD using fMRI. *J. Neurosci. Methods* **2020**, *335*, 108506. [CrossRef] [PubMed]
18. Liu, S.; Zhao, L.; Wang, X.; Xin, Q.; Zhao, J.; Guttery, D.S.; Zhang, Y.-D. Deep spatio-temporal representation and ensemble classification for attention deficit/hyperactivity disorder. *IEEE Trans. Neural Syst. Rehabil. Eng.* **2020**, *29*, 1–10. [CrossRef]
19. Luo, Y.; Alvarez, T.L.; Halperin, J.M.; Li, X. Multimodal neuroimaging-based prediction of adult outcomes in childhood-onset ADHD using ensemble learning techniques. *NeuroImage Clin.* **2020**, *26*, 102238. [CrossRef] [PubMed]
20. Sun, Y.; Zhao, L.; Lan, Z.; Jia, X.-Z.; Xue, S.-W. Differentiating boys with ADHD from those with typical development based on whole-brain functional connections using a machine learning approach. *Neuropsychiatr. Dis. Treat.* **2020**, *16*, 691. [CrossRef]
21. Chen, Y.; Tang, Y.; Wang, C.; Liu, X.; Zhao, L.; Wang, Z. ADHD classification by dual subspace learning using resting-state functional connectivity. *Artif. Intell. Med.* **2020**, *103*, 101786. [CrossRef]
22. Vahid, A.; Bluschke, A.; Roessner, V.; Stober, S.; Beste, C. Deep learning based on event-related EEG differentiates children with ADHD from healthy controls. *J. Clin. Med.* **2019**, *8*, 1055. [CrossRef] [PubMed]
23. Dubreuil-Vall, L.; Ruffini, G.; Camprodon, J.A. Deep learning convolutional neural networks discriminate adult ADHD from healthy individuals on the basis of event-related spectral EEG. *Front. Neurosci.* **2020**, *14*, 251. [CrossRef] [PubMed]
24. Chen, H.; Song, Y.; Li, X. Use of deep learning to detect personalized spatial-frequency abnormalities in EEGs of children with ADHD. *J. Neural Eng.* **2019**, *16*, 066046. [CrossRef] [PubMed]
25. Tosun, M. Effects of spectral features of EEG signals recorded with different channels and recording statuses on ADHD classification with deep learning. *Phys. Eng. Sci. Med.* **2021**, *44*, 693–702. [CrossRef] [PubMed]
26. Müller, A.; Vetsch, S.; Pershin, I.; Candrian, G.; Baschera, G.-M.; Kropotov, J.D.; Kasper, J.; Rehim, H.A.; Eich, D. EEG/ERP-based biomarker/neuroalgorithms in adults with ADHD: Development, reliability, and application in clinical practice. *World J. Biol. Psychiatry* **2019**, *21*, 172–182. [CrossRef] [PubMed]
27. Altınkaynak, M.; Dolu, N.; Güven, A.; Pektaş, F.; Özmen, S.; Demirci, E.; İzzetoğlu, M. Diagnosis of Attention Deficit Hyperactivity Disorder with combined time and frequency features. *Biocybern. Biomed. Eng.* **2020**, *40*, 927–937. [CrossRef]
28. Koh, J.E.; Ooi, C.P.; Lim-Ashworth, N.S.; Vicnesh, J.; Tor, H.T.; Lih, O.S.; Tan, R.-S.; Acharya, U.R.; Fung, D.S.S. Automated classification of attention deficit hyperactivity disorder and conduct disorder using entropy features with ECG signals. *Comput. Biol. Med.* **2022**, *140*, 105120. [CrossRef]
29. Slobodin, O.; Yahav, I.; Berger, I. A machine-based prediction model of ADHD using CPT data. *Front. Hum. Neurosci.* **2020**, *14*, 560021. [CrossRef]
30. Yasumura, A.; Omori, M.; Fukuda, A.; Takahashi, J.; Yasumura, Y.; Nakagawa, E.; Koike, T.; Yamashita, Y.; Miyajima, T.; Koeda, T. Applied machine learning method to predict children with ADHD using prefrontal cortex activity: A multicenter study in Japan. *J. Atten. Disord.* **2020**, *24*, 2012–2020. [CrossRef]
31. O'Mahony, N.; Florentino-Liano, B.; Carballo, J.J.; Baca-García, E.; Rodríguez, A.A. Objective diagnosis of ADHD using IMUs. *Med. Eng. Phys.* **2014**, *36*, 922–926. [CrossRef]
32. Choi, M.-T.; Yeom, J.; Shin, Y.; Park, I. Robot-assisted ADHD screening in diagnostic process. *J. Intell. Robot. Syst.* **2019**, *95*, 351–363. [CrossRef]
33. Lee, S.-h.; Lee, D.-W.; Jun, K.; Lee, W.; Kim, M.S. Markerless 3D Skeleton Tracking Algorithm by Merging Multiple Inaccurate Skeleton Data from Multiple RGB-D Sensors. *Sensors* **2022**, *22*, 3155. [CrossRef] [PubMed]

sensors

MDPI

Article

A Novel Path Planning Strategy for a Cleaning Audit Robot Using Geometrical Features and Swarm Algorithms

Thejus Pathmakumar, M. A. Viraj J. Muthugala *, S. M. Bhagya P. Samarakoon, Braulio Félix Gómez and Mohan Rajesh Elara

Engineering Product Development, Singapore University of Technology and Design, Singapore 487372, Singapore; pathmakumar_thejus@mymail.sutd.edu.sg (T.P.); bhagya_samarakoon@mymail.sutd.edu.sg (S.M.B.P.S.); braulio_felix@mymail.sutd.edu.sg (B.F.G.); rajeshelara@sutd.edu.sg (M.R.E.)

* Correspondence: viraj_jagathpriya@sutd.edu.sg

Abstract: Robot-aided cleaning auditing is pioneering research that uses autonomous robots to assess a region's cleanliness level by analyzing the dirt samples collected from various locations. Since the dirt sample gathering process is more challenging, adapting a coverage planning strategy from a similar domain for cleaning is non-viable. Alternatively, a path planning approach to gathering dirt samples selectively at locations with a high likelihood of dirt accumulation is more feasible. This work presents a first-of-its-kind dirt sample gathering strategy for the cleaning auditing robots by combining the geometrical feature extraction and swarm algorithms. This combined approach generates an efficient optimal path covering all the identified dirt locations for efficient cleaning auditing. Besides being the foundational effort for cleaning audit, a path planning approach considering the geometric signatures that contribute to the dirt accumulation of a region has not been device so far. The proposed approach is validated systematically through experiment trials. The geometrical feature extraction-based dirt location identification method successfully identified dirt accumulated locations in our post-cleaning analysis as part of the experiment trials. The path generation strategies are validated in a real-world environment using an in-house developed cleaning auditing robot BELUGA. From the experiments conducted, the ant colony optimization algorithm generated the best cleaning auditing path with less travel distance, exploration time, and energy usage.

Keywords: audit robot; geometrical feature; cleaning auditing; swarm algorithms

Citation: Pathmakumar, T.; Muthugala, M.A.V.J.; Samarakoon, S.M.B.P.; Gómez, B.F.; Elara, M.E. A Novel Path Planning Strategy for a Cleaning Audit Robot Using Geometrical Features and Swarm Algorithms. *Sensors* **2022**, *22*, 5317. https://doi.org/10.3390/s22145317

Academic Editors: Gregor Klancar, Marija Seder and Sašo Blažič

Received: 6 June 2022
Accepted: 14 July 2022
Published: 16 July 2022

1. Introduction

Cleanliness is one of the inevitable factors that span from an individual's living space to the growth index of developing and developed nations. The professional cleaning industry is a steeply growing field, valued at over $46 Billion U.S Dollars in 2020 forecast to grow 10% by the end of 2026 [1]. Amidst coping with pandemics like COVID-19, the demand for cleaning services has increased steeply in recent years [2]. For the past decade, we can observe a successful usage of leading-edge technologies, including robots for efficient cleaning in both domestic and professional cleaning domains [3–6]. The necessity for high-quality performance, efficiency, and favorable factors from the industry has paved the way for the successful application of robotics technology in the field of automated cleaning.

A vast volume of research has been reported on enhancing the quality of robot-aided cleaning for the past five years. Most reported research focuses on complete coverage planning, energy-aware cleaning, multi-robot cooperation, etc. For instance, a scalable approach for full coverage planning for cleaning robots has been reported [7]. Furthermore, a sector-based online complete coverage planning to bridge the shortcomings of cleaning robots with limited sensing and computation resource is proposed by Lee et al. [8]. The problem of ensuring cleaning efficiency is addressed from a different perspective by introducing optimal footprint for robots alongside the conventional complete coverage path planning approaches [9]. The work mentioned above is validated on a vertical surface cleaning robot by performing an effective hydro-blasting for ship-hull cleaning.

Noh et al. [10] presented an energy-aware cleaning path to enhance the cleaning efficiency using deep reinforcement learning-based approach for energy-aware cleaning. Besides the coverage planning, multi-robot collaboration is another area widely utilized for improving the cleaning quality. The work mentioned in [11] proposed an online heterogeneous multi-robot collaboration system for cleaning robots. The work discussed above highlights a scalable approach for robots with limited sensing capabilities to maximize cleaning performance. The concept of a multi-robot system for cleaning is also explored in the case of oil spill cleaning [12]. The system mentioned above proposes an aerial multi-robot system for an optimal strategy for the contaminated area with minimal wastage of dispersants. The efforts towards improving automated cleaning are centered on improving the efficiency of cleaning robots or machines rather than analyzing the cleaning quality it delivers. Even though the technical advancements in the field of automated cleaning are significant, the analysis of cleaning quality provided by the cleaning robots remains as naive as manual inspection.

Auditing the cleanliness of a region is an important factor to be considered in every cleaning and maintenance task. The attempts to audit the cleanliness are reported in the field of food processing industries and hospitals. For example, the work presented by Giske et al. [13] explores a comparative study between the quality of cleaning delivered by robots and manual cleaning methods. As mentioned earlier, the research discusses validating the effectiveness of cleaning using micro-biological analysis. An *Adenosine Triphosphate* (ATP) bioluminescence technique is another method used to assess the quality of cleaning by estimating the microbial colony presence on the surface of interest. Lewis et al. presented the usage of ATP bioluminescence to benchmark the quality of cleanliness [14]. In the work mentioned above, the authors benchmarked the quality of cleanliness for hospitals in relative light units (RLU). Similarly, Asgharian et al. presented the systematic procedures and guidelines for cleaning quality analysis in the pharmaceutical industry [15].

Similarly, a cleaning assessment report generation based on surface swabbing followed by a laboratory analysis as detailed in [16]. Even though the cleaning quality assessment is regarded as an essential practice in every domain, effort towards establishing a quality assessment is reported in a few areas and is limited to hospitals, pharmaceuticals, and food processing industries [17,18]. Besides, the current methods for micro-biological analysis are laborious and not scalable for cleaning auditing for built infrastructure and domestic cleaning. The primary effort to establish a scalable cleaning auditing is reported in [19] which is the robot-aided cleaning auditing framework using a sample auditing sensor. The sample audit sensor performs adhesive dust-lifting followed by visual analysis to provide a sample audit score for a sample region of 2 cm × 2 cm. The framework uses repeated sample auditing for a vast area using an audit robot that carries the sample auditing sensor as the functional payload. Using the approach mentioned above, the levels of cleanliness for a region are estimated by combining the result of all the sample audits performed in the area of operation. Besides computing the overall auditing score for a region, the work mentioned above provides an empirical estimation of the quality of cleaning. The autonomous robot-aided cleaning auditing can be a potential solution for performing cleanliness inspection effectively without human interventions and laboratory-oriented procedures compared to the existing cleaning quality analysis. A significant challenge in the development of robot-aided cleaning auditing is the formulation of an efficient exploration strategy for inspection. The work mentioned in [19] used a frontier exploration strategy for the robot to explore its area of operation and periodic pattern to determine the region for auditing. However, in work mentioned above, frontier exploration does not guarantee the robot to explore dirt-accumulated areas. The shortcomings of the frontier-exploration-based auditing strategy and sampling decision are improved using reinforcement learning-based exploration [20]. In the research mentioned above, the audit robot uses its experience learned from the modeled environment for exploration and making sampling decisions. This work presents a first-of-its-kind path planning approach dedicated to cleaning auditing robots. The autonomous mobile robots that perform similar tasks to the auditing, especially floor cleaning and ground patrolling robots, use the complete coverage planning strategy using an optimization technique to yield the best path [21,22]. However, adopting a similar complete coverage planning for audit robots is not effective because:

- for a large environment, collecting samples uniformly across the complete region are not time-efficient
- the dirt accumulation pattern is chaotic and often clustered in regions left unattended by cleaning robots.
- lack of an effective method that determines from where to gather dirt samples for an effective cleaning auditing.

This research work bridges the challenges mentioned above by proposing a novel path planning strategy that is driven by geometric signatures of the environment for cleaning auditing robots. To the best of the author's knowledge, a path-planning approach considering geometrical signatures of the environment has not been explored in the domain of cleaning robots or cleaning audit robots so far. Due to less accessibility and the inherent navigation behavior of cleaning robots to avoid obstacles, the walls and corners often become the site for possible dirt accumulation compared to the other regions for a regularly cleaned area. Therefore, a geometrical feature extraction framework is devised to identify the probable dirt accumulation region from a 2D map. Swarm algorithms are exploited to plan an efficient way to cover the sample locations.

The general objective of this research work is subdivided into three:

1. formulate the dirt location identification method by extracting the geometrical signatures from the environment
2. devise an efficient path that connects all the identified dirt location (i.e., sample locations) by minimizing the energy cost
3. validate the geometry-based dirt location identification in a real environment after repeated cleaning trials
4. experimentally validate the optimal path generated by the proposed method in a real environment using an in-house developed cleaning audit robot

The rest of the article is structured as the Section 2 provides the background study conducted for devising the proposed approach. This is followed by Section 3 that presents a bird's eye view of the proposed approach. The detailed description of path generation strategy is provided in Sections 4 and 5. The Section 6 provides a detailed description of the conducted validation trials, followed by the Section 7.

2. Related Works

Two significant aspects of the proposed path planning strategy are (1) identification of probable dirt locations in a 2D map using geometrical features and (2) generating an optimal path through probable dirt locations using swarm algorithms.

2.1. Geometrical Feature Extraction

Feature extraction and description is a key element in perception for autonomous robots. Moreover, a geometrical feature like line segments, curvature, and corners is used over the conventional feature extraction techniques like SIFT [23], ORB [24], AKAZE, [25] etc. A SLAM algorithm that uses descriptors for line segments is reported in [26]. Similarly, the curvature property of road lane is used as a key feature for pose estimation for autonomous vehicles is reported in [27]. Yan et al. used corner features for two-dimensional SLAM [28]. Besides, the geometrical feature extraction methods are also widely used for hand gesture recognition [29], finger-knuckle-print verification [30], etc. Among the applications of geometrical feature extraction, the classical method for the extraction of boundaries in a 2D matrix is canny edge detection [31]. Another popular approach is estimating the boundaries based on Hough transformation for the line geometry detection [32–34]. The popular approach for corner detection is the Harris Corner detector [35]. Similarly, a robust Anisotropic Directional Derivative (ANDD) filter-based corner detection is another classic method for corner extraction [36]. Chord-to-Point Distance Accumulation (CPDA) is another method reported to have low localization error and high repeatability [37]. Similarly, machine learning-based detectors have known for accuracy and repeatability in detecting corners from a given 2D image [38,39].

2.2. Optimal Path Generation

Swarm algorithms and evolutionary algorithms are widely adopted for solving the optimal path planning algorithms with multiple constraints [40–44]. Whale optimization (WO) based path planning for an underwater robot is mentioned in [45], where the optimization techniques are used for generating a path with safe and minimal fuel consumption. An improved sparrow search algorithm for path-planning for a mobile robot is reported in [46]. Furthermore, hybrid Quantum-behaved Particle Swarm Optimization (HQPSO), a variant of classical PSO, has been used for path generation for an unmanned underwater vehicle (UUV) [47]. Modified ant-colony optimization (ACO) is used for path planning for AGV-based parking system is detailed in [48].

3. System Overview

3.1. Cleaning Auditing Overview

The robot-aided cleaning auditing is a three-step procedure detailed in [19]. For completeness, we briefly explain the process of the robot-aided cleaning auditing framework. Figure 1d shows the overview of robot-aided cleaning auditing. The first step in robot-aided cleaning auditing is called sample auditing, where the cleanliness of a sample area is determined. The second step is space auditing, where repeated sample auditing for a larger area is performed. The space auditing is facilitated by an in-house developed audit robot called BELUGA, facilitated by exploration algorithms to achieve efficient sample auditing in different locations. The BELUGA robot is a differential drive mobile robot equipped with sensors for navigation and perception (shown in Figure 1c). The robot maps a given area and does localization within the map using Adaptive Monte Carlo Localization (AMCL) method. The perception, localization, and control algorithm are executed using an onboard embedded computer. The key payload carried by the BELUGA robot is the sample auditing sensor (shown in Figure 1a). The sensor consists of adhesive tape for dust-lifting and a digital microscope for analyzing the adhesive tape surface. The field of view of the digital microscope has an active light source to maintain constant ambient light throughout the operation such that variation in light intensity does not affect the sensor operation. Figure 1b shows the dust extracted by the sample audit sensor viewed by its built-in digital microscope. The sensor uses the structural similarity index(SSIM) of the tape surface images before and after dust-lifting to estimate the magnitude of dirtiness of the surface [19]. The sensor reports the magnitude of cleanliness as a sample audit score to evaluate the overall cleanliness of the sample region (sample auditing). The comprehensive cleanliness estimation is done by conducting repeated sample auditing for a larger area with the help of the BELUGA robot (space auditing). With the exploration strategies, the robot goes to specific locations to perform sample auditing.

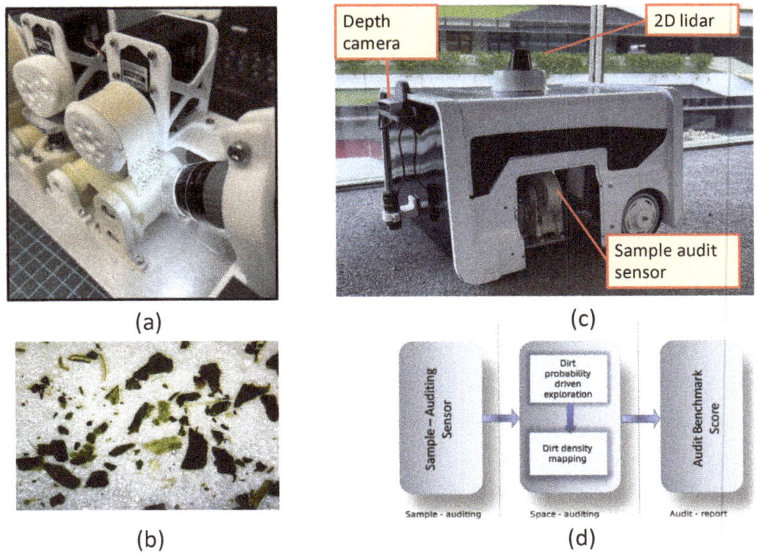

Figure 1. The dust extracted by the sample audit sensor (**a**), the dust particles viewed by the microscope of the sample audit sensor (**b**), the BELUGA audit robot equipped with sample audit sensor (**c**), and the overview of robot-aided cleaning auditing framework [19] (**d**).

3.2. Path Generation Overview

A bird's eye view of the proposed path planning framework is detailed in this section. Figure 2 shows the overall architecture of the proposed path generation framework. The process pipeline has three components—geometry extraction, sample selection, and path generation. The input to the system is the 2D occupancy grid of the environment, which is generated by the simultaneous localization and mapping (SLAM), a.k.a mapping method. From the 2D occupancy grid, the locations for sampling are selected based on the environment's geometrical signatures that contribute to the dirt accumulation. The geometry extraction and sample selection procedures achieve the process mentioned above for dirt location identification. The geometry extraction procedure comprises boundary extraction, corner extraction, and free-space extraction from the given occupancy grid. The sample selection procedure selects the sample locations in the nearest proximity to boundary maps and corner maps. The general sample points are a set of locations in the occupancy grid obtained by uniform grid sampling. The sample locations are identified by combining the boundary samples, corner samples, and random samples. The random samples are a few random locations in the free space. A detailed information regarding the dirt location identification strategy mentioned above is presented in Section 4.

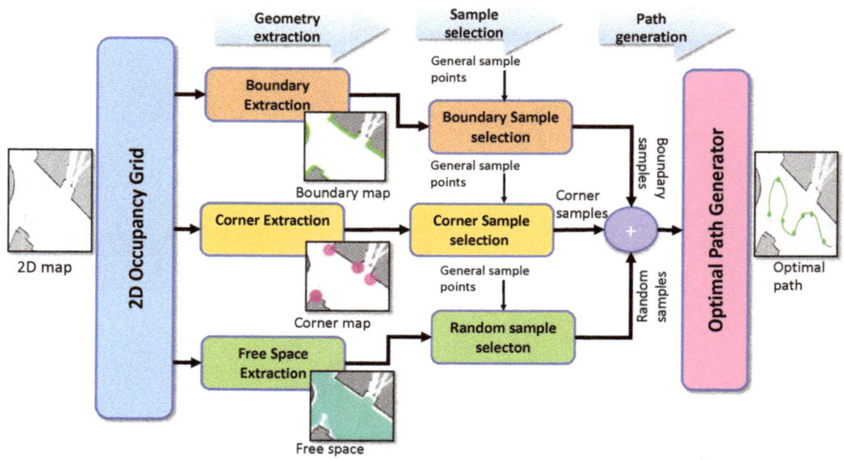

Figure 2. The overview of the proposed path generation strategy for dirt sample gathering using geometrical signatures extracted from the 2D map of the environment combined with an optimal path generator.

Once the sample locations are identified, the robot can visit the locations in any order to do sampling and auditing. However, choosing the optimal way to visit the locations and perform the sample auditing is important. The optimal path generator in the proposed framework takes in the locations identified after the sample selection procedure and generates an efficient path with minimum time and energy. With the help of navigation algorithms, [49], the robot follows the path and performs auditing for a given area.

4. Dirt Location Identification

This section presents the identification of probable dirt locations (sample audit locations) from a given occupancy grid. Considering the natural tendency of dirt accumulation nearer to the corners and wall, the probable dirt locations are identified by selecting sample points in close proximity to walls and corners. To identify the dirt locations for the robot to perform sample auditing, the entire occupancy grid is evenly sampled with a resolution proportional to the robot footprint and locations corresponding to the samples are regarded as general sample locations. We defined the general sample locations as the set of locations that are equally spaced in the free-space region in an occupancy grid Figure 3c. While generating the general sample locations, the occupied cells and unknown cells in the grid are discarded. The sample audit locations are a subset of the general sample location that lies closer to the corners and boundaries in the occupancy grid. Hence, the main task involved in sample location identification or the probable dirt location is to extract the corners and boundaries from the occupancy grid. Since the occupancy grid a.k.a map is represented as a two-dimensional 8-bit matrix, where each element represent the three states of occupancy (occupied, free and unknown). Hence the conventional image processing algorithms can be applied over an occupancy grid. Figure 3a shows the representation of an occupancy grid.

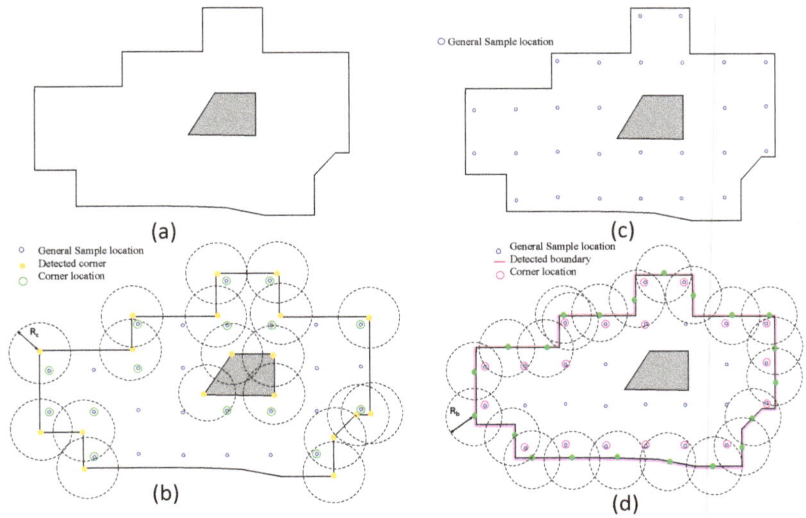

Figure 3. The representation of occupancy grid (**a**), corner location extraction (**b**), general sample locations after uniform sampling (**c**), and boundary location extraction (**d**).

4.1. Boundary Sample Extraction

The Figure 3d illustrates the critical operations involved in the boundary extraction procedure. For boundary extraction, we performed the following methods:

1. General sample location identification: The general sample locations are obtained by sampling the free space in a uniform fashion. The general sample location $P^{general}$ is obtained by grid sampling the occupancy grid $S_{i,j}$ in intervals Δu and Δv along rows and columns.
2. Occupancy grid thresholding: Each cell in an occupancy grid shows the probability of occupancy. A two state binary map B has been generated by applying a thresolding such that $B(i,j) = 1$ if $S(i,j) < T_{max}$ else $B(i,j) = 0$, where $S(i,j)$ is the occupancy grid and T_{max} is the maximum occupancy threshold.
3. Boundary extraction: On the binary map, we used Zhang-Suen thinning algorithm [50] for extracting the contours. However, the contour extraction results in detecting all closed contours in the occupancy grid, including the undesirable contours that form over the obstacles. The largest contour C_{large} from all the set of contours C_i is regarded as the boundary region.
4. Selection of boundary location: The contour corresponding to the boundary region C_{large} is sampled in a regular interval to obtain a set of points $B_{i,j}$ that lies on the contour. The boundary locations $L^{boundary}$ are selected from the general sample $S(i,j)$ points that lie in the distance less than R_b.

The location $L^{boundary}$ is the sample location that lies closer to the walls such that performing sample auditing at $L^{boundary}$ will result in analyzing dirt accumulations contributed by the walls in a region. The distance R_b decides how many samples closer to the walls have to be considered for sample auditing.

4.2. Corner Sample Extraction

The corner Sample extraction follows a similar approach to boundary extraction.

1. The general sample location $P^{general}$ is identified by performing uniform sampling of occupancy grid $S_{i,j}$.

2. Corner extraction: The machine learning-based fast corner extraction algorithm [39] is used for identifying the corner locations.
3. Selection of corner location: Similar to the boundary location identification, the corner locations L^{corner} are selected from the general sample $S(i,j)$ points that lie in the distance less than R_c.

The distance R_c decides how many samples closer to the sharp corners have to be considered for sample auditing.

Random Samples

Few random locations L^{random} are selected from the general sample location $P^{general}$ for spanning auditing to the complete area. However, the random locations are smaller in number, and it is selected based on the size of the occupancy grid. The locations for auditing are a combination of corner locations, boundary locations, and random locations. The Equation (1) represented the set of locations for auditing.

$$\mathcal{S} = L^{corner} \cup L^{boundary} \cup L^{random} \tag{1}$$

where L^{corner}, $L^{boundary}$, and L^{random} represents corner locations, boundary locations and random locations respectively.

5. Optimal Path Planing

After determining the probable dirt locations, the robot has to visit each location once and perform the auditing. Here, the determined probable dirt locations (P_i) are considered as the waypoints of the robot's navigation path. The robot is assumed to be initialized at a designated starting location in the workspace (denoted as P_1). Two example scenarios are depicted in Figure 4, where there is N number of locations to be visited, including the starting point. Similarly, there can be many sequences of waypoints where the robot can visit all the locations at one time.

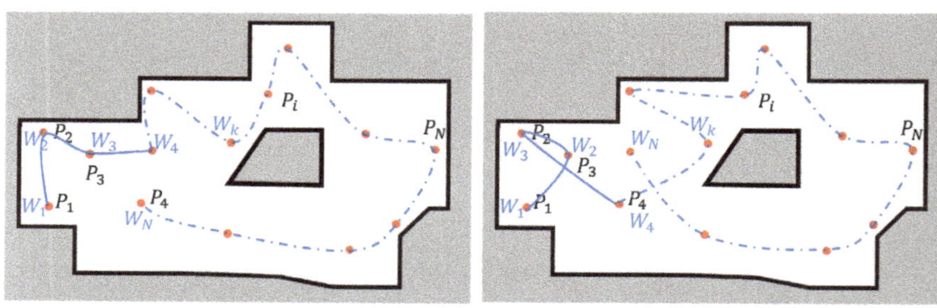

Figure 4. Examples for waypoint sequences that can be followed by the robot for auditing process.

The robot must determine an optimum sequence of these waypoints that minimizes the energy usage for an efficient auditing process, as the energy-efficient coverage is vital for a robot deployed in maintenance and inspection domains [51,52]. The optimum sequence of waypoints is defined as $\{W_k\}$ such that $k = 1$ to N, where N is the number of waypoints, including the starting position. The energy used by the robot for navigation is proportional to the distance traveled by the robot in the event of a level workspace. The energy usage of navigating from kth waypoint to lth waypoint, $E_{k,l}$ can be estimated as (2) where $D(W_i, W_j)$ represents the navigation distance between kth and lth waypoints, and M is a proportional constant. The navigation distance $D(W_i, W_j)$ is estimated considering the $A*$ algorithm for a collision-free path between W_i to W_j.

$$E_{k,l} = MD(W_k, W_l) \;\; \text{where } k \neq l \tag{2}$$

The total energy usage of the robot, E, for accomplishing the navigation for auditing can be formulated as in (3).

$$E = \sum_{k=1}^{N-1} E_{k,k+1} \quad (3)$$

The energy usage of the robot for navigation between all the pairs of the waypoints could be estimated, and the total energy requirement could be estimated by considering all the possible sequences to find the sequence that results in the lowest energy usage. However, the number of possible pairs of waypoints becomes $N(N-1)/2$. There exist $(N-1)!$ combinations for joining these waypoint pairs. Thus, determining the path of lowest energy usage through evaluating the energy usage for all the possible paths becomes inefficient and complex when N is high. Moreover, this problem is a no polynomial-time known solution problem (NP hard problem).

Swarm optimization algorithms are effective techniques for finding a proper solution for this kind of problem [53,54]. Two versatile swarm algorithms, Ant Colony Optimization (ACO) [55] and Particle Swarm Optimization (PSO) [56] are used here to find the optimal sequence of waypoints considering the minimization of the cost function given in (3). These two optimization techniques are selected since they are well known for the convergence to the global optima in similar problems.

The path generation problem considered here is analogous to the classical Travelling Salesman Problem (TSP). However, in most of the off the shelf tools for classical TSP, it is assumed that there are no obstacles in between the nodes and hence Euclidean distances between the nodes are considered for the optimization. In contrast, for the cleaning auditing, the robot has to operate in obstacle-cluttered environments where the robot must find a collision-free path between two nodes. The $A*$ algorithm is used to find the collision-free path between two nodes. The distance of the $A*$ path is used as the distance between two nodes.

5.1. Particle Swarm Optimization (PSO)

Particle Swarm Optimization (PSO) is inspired by the social behavior of birds flock or fish school. The cooperation of individuals in a swarm based on individual and group knowledge toward finding a goal is utilized in this technique. The flow of the PSO algorithm is given in Figure 5. Here, each individual is considered as a particle with a position and a moving velocity. Each particle is randomly initialized with a velocity and a position at the start. Then, the algorithm iteratively attempts to find the optimal solution. The fitness of each particle is evaluated for the current solution, and the global and the local best positions are updated as per the evaluated finesses. Then, the new velocity and position of each particle in the swarm are calculated. This process is repeated until a stopping criterion is met. The global best at the time of stopping the iteration is the final optimum solution.

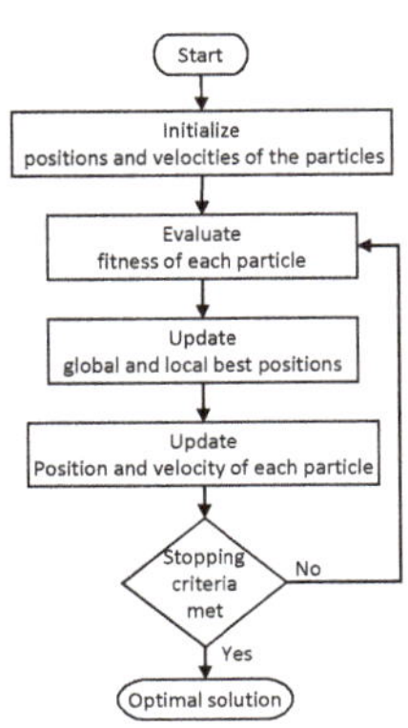

Figure 5. Flow of the PSO algorithm.

The parameters of the PSO algorithm have been set as follows in this work by observing the performance variation. The population size was chosen as 100. The inertial weight and the inertial weight damping ratio were configured to 0.9 and 0.95, respectively. Global and local learning coefficients were set to 0.85. Reaching 1000 iterations was defined as the stopping criterion.

5.2. Ant Colony Optimization (ACO)

Ant Colony Optimization (ACO) is inspired by the foraging behavior of some ant species. The ants lay pheromone on the ground to mark direct peer ants toward resources such as food sources while exploring the habitat. The flow of the ACO algorithm is depicted in Figure 6. At the start, ants and the pheromone trails are initialized. Each ant represents a solution. Then, the paths found by ants are compared. In other words, the fitness value of each ant is evaluated. Subsequently, the pheromones are updated based on their fitness levels. This process is iterated until a stopping criterion is met. The best solution at the termination of the algorithm is the optimal solution.

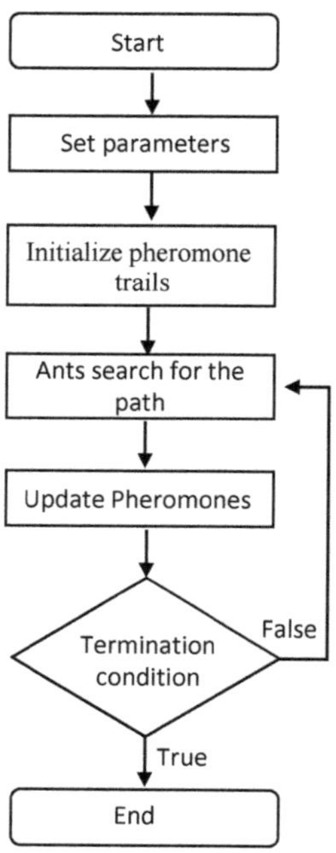

Figure 6. Flow of the ACO algorithm.

The parameters of the ACO algorithm have been configured as follows based on performance observations. The number of ants was set to 100. The evaporation coefficient was set to 0.15. The effect of ant sight and traces were chosen as 1 and 4, respectively. Reaching 1000 iterations is defined as the stopping criterion.

6. Results and Discussion

We have carried out multiple experiment trials to validate the proposed approach. We conducted two sets of experiments to analyze the performance of the proposed approach. The first set of experiments quantifies the performance of dirt sampling with proposed geometrical feature-based dirt location identification in real-time. The second set of experiments analyzes the behavior of the path generated by the proposed framework in different environments.

6.1. Trial-1: Sample Location Identification

The performance of sampling with geometrical features based on dirt location is analyzed by defining the dirt gathering efficiency, which is the ratio of the number of dirt samples collected to the total samples collected as given in Equation (4),

$$\eta_{dirt} = \frac{N_d}{N_d + N_c} \tag{4}$$

where N_d and N_c represent the number of dirt samples collected and the number of clean samples collected. The factors N_d and N_c are determined by counting the number of dirt particles gathered using dirt lifting followed by computer vision-based dirt counting. The dirt specks on the adhesive tape are treated as the connected pixels on the images (also known as blobs) captured by the microscope. Steps involved in computer vision-based dirt counting include:

1. Apply thresholds on the source image and convert the image to binary
2. Using contour extraction, identify the connected pixels from the binary image and estimate blob centroid
3. The blob centroid is regarded as the location of the dirt particle
4. The number of centroids is regarded as the dirt count

The blob detection algorithm is implemented using OpenCV libraries given in [57,58].

The Figure 7 shows the dirt speck identified and counted using blob detection. Our first set of the experiment consists of two trials, trial-1.a and trial-1.b, in which the trials replicates a cleaning routine carried out using a commercial cleaning robot that does a zig-zag path planning. For validating the probable dirt location identification, we created a sample environment of dimensions 4.5 m × 4.5 m. The domestic floor cleaning robot with autonomous navigation capabilities is operated for 15 min to replicate a cleaning routine. To analyze the dirt accumulation, dirt particles are uniformly distributed (fine ground tea powder with a particle size 0.5 mm–1 mm) before the operation of the cleaning robot. The cleaning routine is repeated for five rounds. After five rounds of cleaning operation, the dirt samples are gathered based on uniform sampling and at the location candidates obtained from the proposed geometrical features.

The comparison of dirt particle counts at locations provided by the algorithm with the locations selected by uniform sampling is recorded. The trial-1.a and trial-1.b differ in terms of the obstacle density. Figure 8a shows the sample locations identified for the trial-1.a. Figure 8b shows the locations identified by the proposed geometrical feature-based dirt location identification.

Figure 8c shows the dirt counts recorded for uniform sample gathering and Figure 8d shows dirt count obtained from the proposed approach. Similarly, Figure 9a shows the sample locations identified for the trial-1.b and Figure 9b shows the locations identified by the proposed method. Figure 9c shows the dirt counts recorded for uniform sample gathering and Figure 8d shows the dirt count obtained from the proposed approach. In both trials, a dirt count equal to 10 is regarded as the threshold for classifying the gathered sample as dirty or clean. This implies that all gathered samples having a count above 10 is regarded as dirt sample. The dirt gathering efficiency (Equation (4)) is computed for trial-1.a and trial-1.b. The result of dirt gathering efficiency calculation for trial-1.a and trial-1.b is tabulated in Tables 1 and 2, respectively.

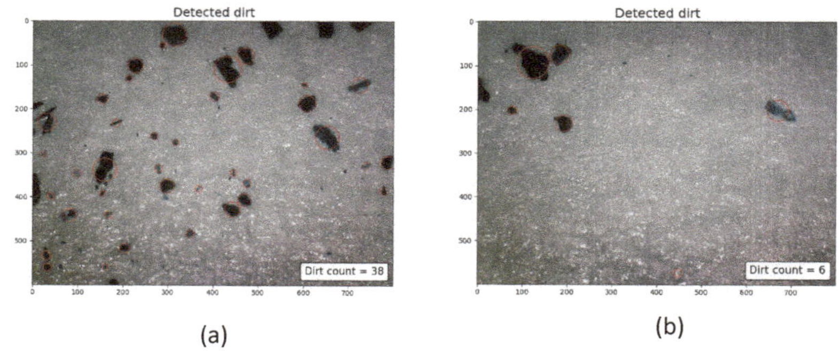

Figure 7. The dirt particles gathered from a dense dirt accumulated region (**a**), the dirt particles gathered from a sparse dirt accumulated region (**b**).

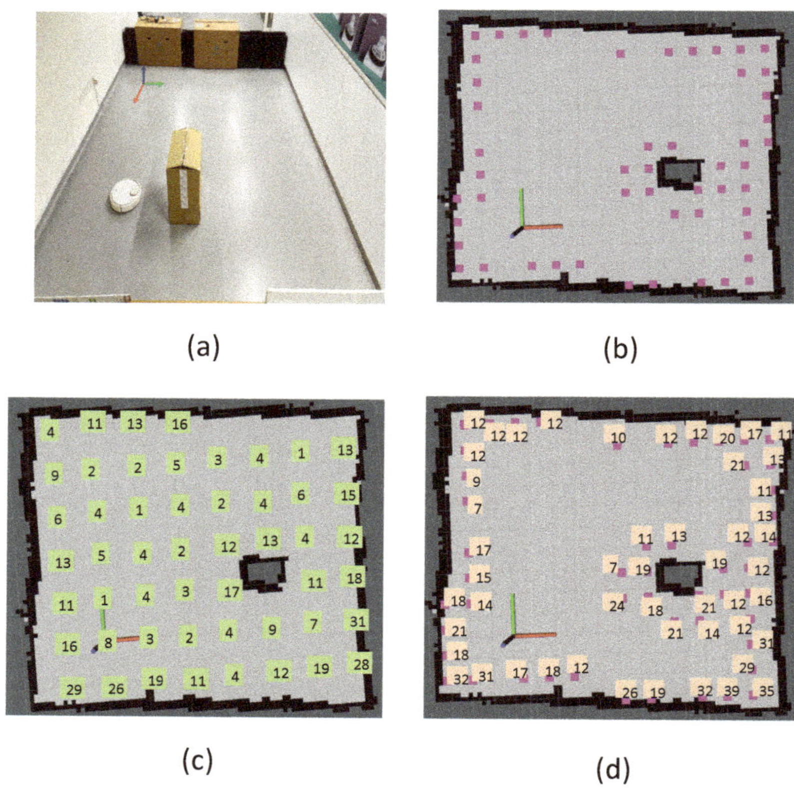

Figure 8. The environment chosen for trial-1.a (**a**), identified sample locations based on the proposed approach (**b**), dirt counts recorded corresponding to locations in a uniform grid sampling (**c**), dirt counts recorded corresponding to identified sample locations based on the proposed approach (**d**).

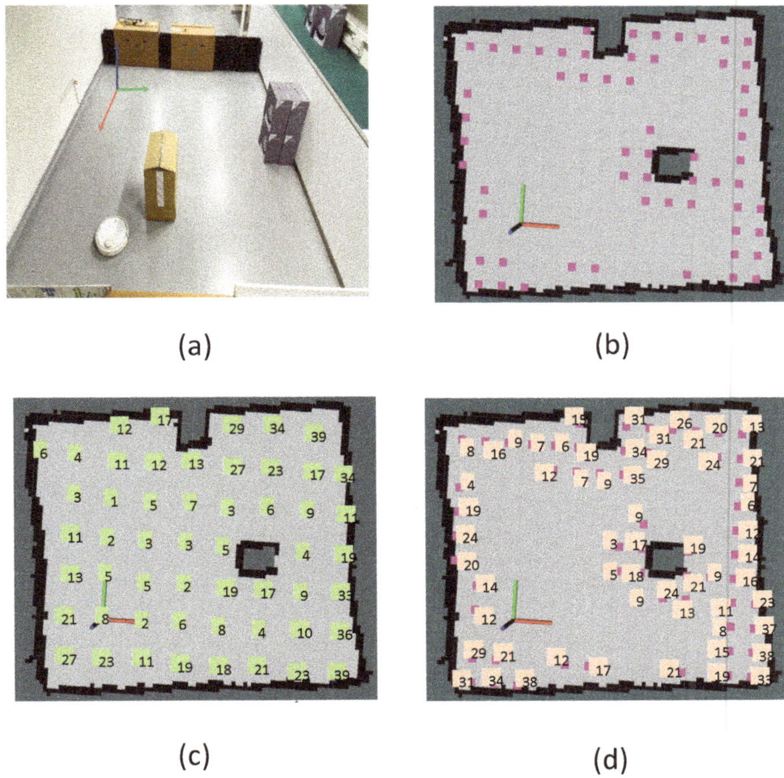

Figure 9. The environment chosen for trial-1.b (**a**), identified sample locations based on the proposed approach (**b**), dirt counts recorded corresponding to locations in a uniform grid sampling (**c**), dirt counts recorded corresponding to identified sample locations based on the proposed approach (**d**).

Table 1. The sampling efficiency for a dirt count threshold 10 in trial-1a.

	Trial-1a	
	Uniform Sampling	**Proposed Approach**
Number of dirt samples N_d	22	47
Number of clean samples N_s	29	4
Total sampling $(N_d + N_s)$	51	51
Dirt gathering efficiency $N_d/(N_d + N_s)$	0.43	0.92

In trial-1.a, the sample locations identified using the geometrical signatures extracted from the maps were more concentrated toward the walls and corners. In trial-1.a, 51 locations are identified for using the proposed approach. The locations closer to the wall and corners had more relatively dirt counts than other locations. It is also observed that few locations near the walls, the robot cleaned well and left fewer dirt counts. However, the general pattern observed is the dense accumulation of dirt nearer to the location identified by the proposed approach. The dirt gathering efficiency of 0.92 for the proposed approach confirms the above-mentioned observations. In trial-1.a and trial-1.b, we have observed sample locations are not identified by the proposed approach in certain regions around the walls. This is because of the imperfections in the LiDAR scan while generating the map. In trial-1.b, the sample locations identified using the geometrical signatures extracted from the maps were also more concentrated towards the walls and corners and sparse dirt accumulation in the central locations. In trial-1.b, 59 locations are identified using the proposed approach. The proposed method showed more number of locations for trial-1.b because of more number of walls and corners introduced by additional obstacles in the environment. Similar to the previous trial, with few locations near the walls, the robot cleaned well, leaving fewer dirt counts. However, the dirt accumulation was dense near the location identified by the proposed approach. A dirt gathering efficiency of 0.74 is recorded for the proposed approach, and 0.54 is recorded for the uniform sampling method. We observed the robot took more turns near the region bounded between two obstacles and walls, resulting in multiple passes through the same location. This resulted in a drop in dirt gathering efficiency for the proposed approach. However, there is a significant improvement in dirt gathering efficiency in both experiment trials.

Table 2. The sampling efficiency for a dirt count threshold 10 in trial-1b.

	Trial-1b	
	Uniform Sampling	**Proposed Approach**
Number of dirt samples N_d	29	44
Number of clean samples N_s	24	15
Total sampling $(N_d + N_s)$	53	59
Dirt gathering efficiency $N_d/(N_d + N_s)$	0.54	0.74

6.2. Trial-2: Path Generation

Our second set of experiment trials validates our proposed path generation framework. Trial-2 consists of experiments conducted in three real-world environments with the BELUGA robot. Each environment is different regarding the area of operation for cleaning auditing. The first environment (environment-1) is an indoor lab space with approximately 58 m^2 of the total area accessible. The obstacle density in environment-1 is higher; however, the obstacles are placed in a well-ordered manner. The environment-2 is a semi-indoor pantry area. The obstacle density in environment-2 is slightly more than in environment-1. The third environment, environment-3 is a ramp entrance where the environment is more complex in terms of shape and orientation of obstacles. The Figure 10 shows the operation of the BELUGA robot in all three environments. Four sets of experiment trials are conducted in each environment considering PSO, ACO, Zig-Zag, and random path planning. The test environments considered for the experiments have a moderate number of points to visit. However, we are targeting the cleaning audit in the large environment such as shopping malls. In that case, the number of points will be much higher, and swarm-based optimization methods would be more suitable than the other approaches. Zig-zag path planning is one of the common path planning methods used in the domain of cleaning robotics. Hence, it would be worthy of considering zig-zag path planning as a baseline for comparison in cleaning auditing applications along with random path selection. Here, Zig-zag path planning is considered along the Y-axis where the robot starts the selection of the point which it has the least coordinate in Y-axis. Then, the point with the second least Y coordinate is selected as the second point. This ordering pattern is continued until all the points are selected.

Figure 10. BELUGA robot operating in different environments (map given in inset). Environment-1 (**a**), environment-2 (**b**), and environment-3 (**c**).

In every experiment trial, the total path covered by the robot, the total time taken for completing the sampling, and the current consumption from the robot are recorded. The total energy taken for the exploration is computed using the Equation (5):

$$E = \int_0^T v(t)i(t)\,dt \tag{5}$$

where $v(t)$ is the terminal voltage of the battery, T is the total exploration time, and $i(t)$ is the instantaneous current reading from the battery management system of the robot. The overall observations recorded in trial-2 are tabulated in Table 3. The convergence results of the PSO and ACO algorithms are given in Figure 11.

Table 3. The overall observations from trial-2.

Algorithm	Parameter	Validation Trials		
		Environment-1 N = 29	Environment-2 N = 39	Environment-3 N = 59
PSO	Path length (D)	12.27 m	37.66 m	58.21 m
	Time taken (T)	470 s	889 s	1320 s
	Total energy (E)	23.72 kJ	44.82 kJ	66.84 KJ
ACO	Path length (D)	9.5 m	17.04 m	23.8 m
	Time taken (T)	427 s	642 s	951 s
	Total energy (E)	22.55 kJ	32.35 kJ	47.89 KJ
Zig-Zag	Path length (D)	16.33 m	17.85 m	53.42 m
	Time taken (T)	511 s	665 s	1248 s
	Total energy (E)	27.98 kJ	35.13 kJ	65.88 KJ
Random	Path length (D)	25.15 m	73.31 m	106.75 m
	Time taken (T)	602 s	1220 s	1785 s
	Total energy (E)	31.78 kJ	64.44 kJ	94.39 KJ

In environment-1, the algorithm has identified 29 sample points. The experiment trials in environment-1 are represented as trial-2.a, trial-2.b, trial-2.c and trial-2.d. In trial-2.a, the robot did sample auditing in the identified locations by selecting the points randomly and following an $A*$ path connecting the selected points (shown in Figure 12a). In trial-2.b, the robot did sample auditing in the identified locations by selecting the points in a zig-zag fashion along the Y-axis. Similar to the previous trial, the robot followed an $A*$ path connecting the selected points (shown in Figure 12b). In trial-2.c the point selection is made based on the PSO algorithm; in trial-2.d, the point selection is made based on the ACO algorithm. From trial-2.a to trial-2.d, we could observe that the optimal path generated by the ACO algorithm has a shorter path length than all other methods. The most sub-optimal strategy was a random selection of points. The robot took 427 s to complete the path generation in the case of the ACO algorithm with a total energy consumption of 22.55 kJ (Kilo-Joule). From the observations made from environment-1, it is evident that ACO shows the best convergence in terms of optimizing the path length and energy consumption.

The convergence results shown in Figure 11a,b also verify the proper convergence of the ACO and PSO algorithms in this case.

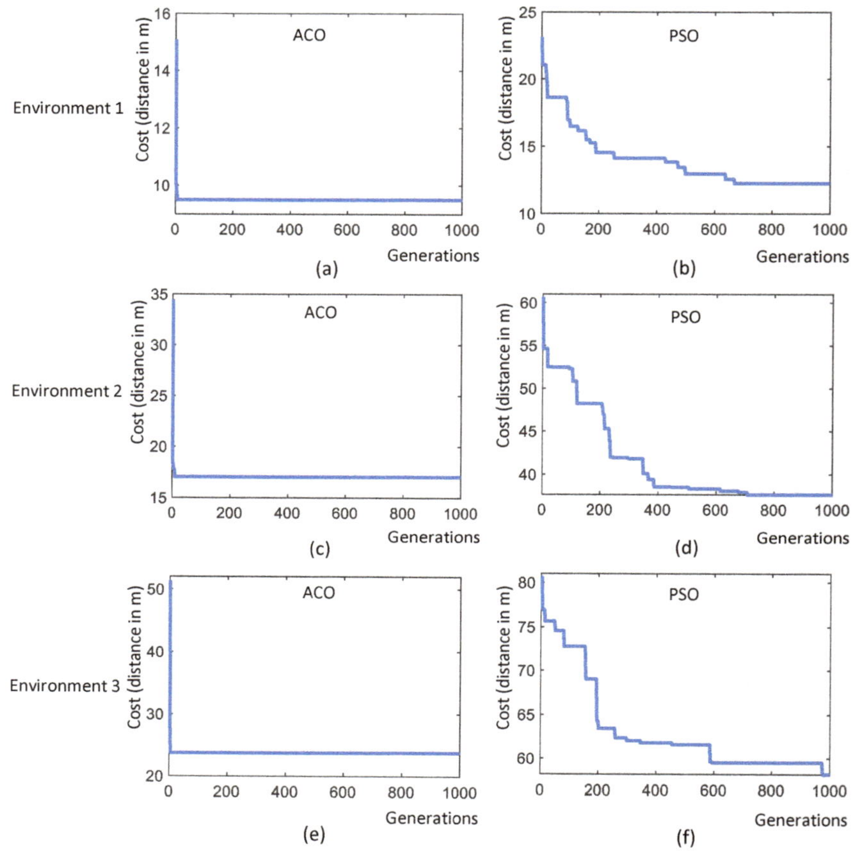

Figure 11. Convergence results of ACO and PSO. (**a**): Environment 1 using ACO, (**b**): Environment 1 using PSO, (**c**): Environment 2 using ACO, (**d**): Environment 2 using PSO, (**e**): Environment 3 using ACO, and (**f**): Environment 3 using PSO.

In environment-2, the algorithm has identified 41 sample points from 93 m^2 of area. However, the robot could access only 39 points and 2 points were too close to the obstacle and omitted during navigation.

Four sets of experiment trials are conducted in environment-2, represented as trial-2.e, trial-2.f, trial-2.g, and trial-2.h, respectively. Similar to environment-1, the robot did sample auditing by selecting the sample points randomly and following an $A*$ path connecting the selected points in trial-2.e (as shown in Figure 13a). Similarly, the robot did sample auditing in the identified locations by selecting the points in a zig-zag fashion along the y-axis in trial-2.f. The path followed by the robot is generated using the $A*$ path connecting the selected points (as shown in Figure 13b). In trial-2.f, the point selection is done based on the PSO algorithm and in trial-2.d (Figure 13b), the point selection is done based on the ACO algorithm (Figure 13b). From trial-2.e to trial-2.h, we could observe that the optimal path generated by the ACO algorithm has a shorter path length than all other methods. The most sub-optimal strategy was a random selection of points. In the case of the ACO algorithm, the robot took 642 s to complete the exploration with a total energy consumption of 32.35 kJ. In environment-2, ACO shows the best convergence in optimizing the path length and energy consumption. Another important observation is zig-zag selection (trial-2.f) has outperformed PSO (trial-2.g) in yielding a shorter path length and less energy consumption. Similar to environment-1, the random selection of points recorded the most energy-expensive path generation strategy.

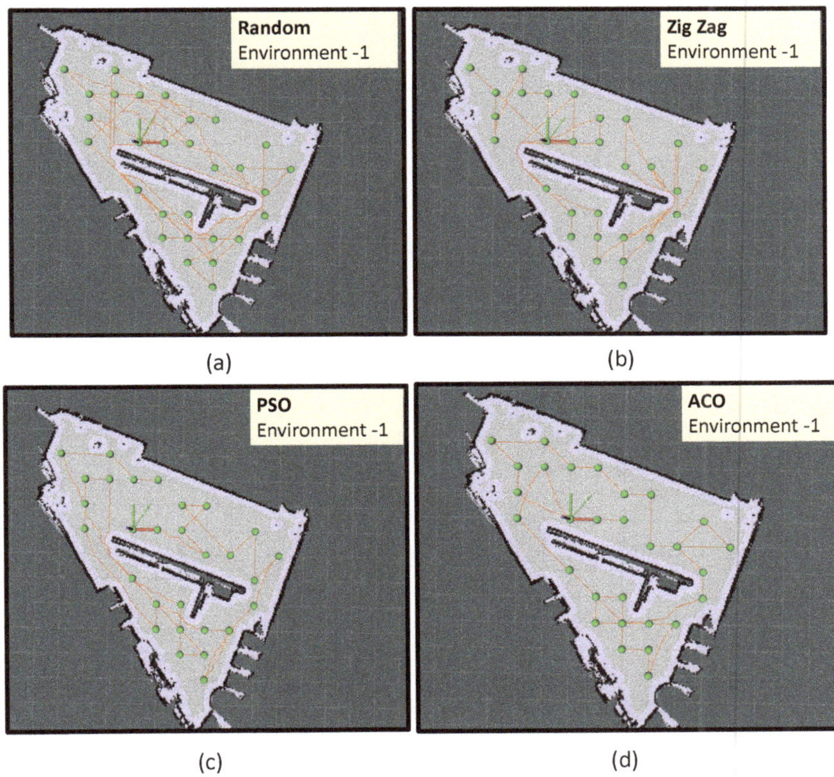

Figure 12. The path followed by the robot in environment-1 using random selection of points (**a**), using zig-zag selection of points (**b**), PSO algorithm (**c**), and ACO algorithm (**d**).

Environment-3 is the biggest area among all other environments where the algorithm has identified 59 sample points from 124 m^2 of area. The trials conducted in environment-3 is represented as trial-2.i, trial-2.j, trial-2.k and trial-2.l respectively. In trial-2.i, the robot did sample auditing in the identified locations by selecting the points randomly and following an $A*$ path connecting the selected points (shown in Figure 14a). In trial-2.j, the robot did sample auditing in the identified locations by selecting the points in a zig-zag fashion along the y-axis. Similar to the previous trial, the robot followed an $A*$ path connecting the selected points (shown in Figure 14b). In trial-2.k, the point selection is made based on the PSO algorithm and in trial-2.d, the point selection is made based on the ACO algorithm. From trial-2.i to trial-2.l, we could observe that the optimal path generated by the ACO algorithm has a shorter path length compared to all other methods. The most sub-optimal strategy was the random selection of points. The robot took 951 s to complete the path generation in the case of the ACO algorithm with a total energy consumption of 47.89 kJ. From the observations made from environment-3, it is evident that ACO shows the best convergence in optimizing the path length and energy consumption.

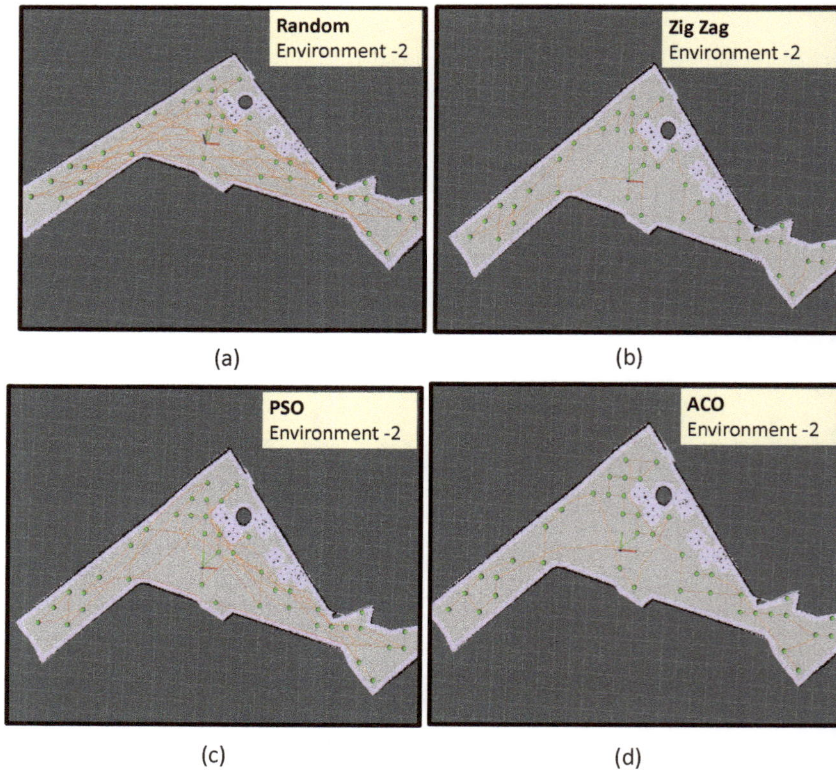

Figure 13. The path followed by the robot in environment-2 using random selection of points (**a**), using zig-zag selection of points (**b**), PSO algorithm (**c**) and ACO algorithm (**d**).

From the experiments conducted in environment-1, environment-2 and environment-3, ACO showed a better performance in yielding shorter and energy-optimized paths for sample auditing. After ACO, the second-best path generation in terms of shorter path length and less energy consumption is given by PSO for environment-1 and Zig-Zag for environment-2 and environment-3. In larger environments, the PSO algorithm was showing a sub-optimal performance. The robot skipped few sampling points during the audit process since navigation algorithms in BELUGA robot was not allowing the robot to visit the narrow location. However, skipping a few samples (2 out of 39) while auditing has a negligible effect on the overall auditing process. Besides, the compact dimension of the sample-audit sensor allows using compact robot platforms to perform auditing in narrow space by trading-off requirements for large power sources and computation modules.

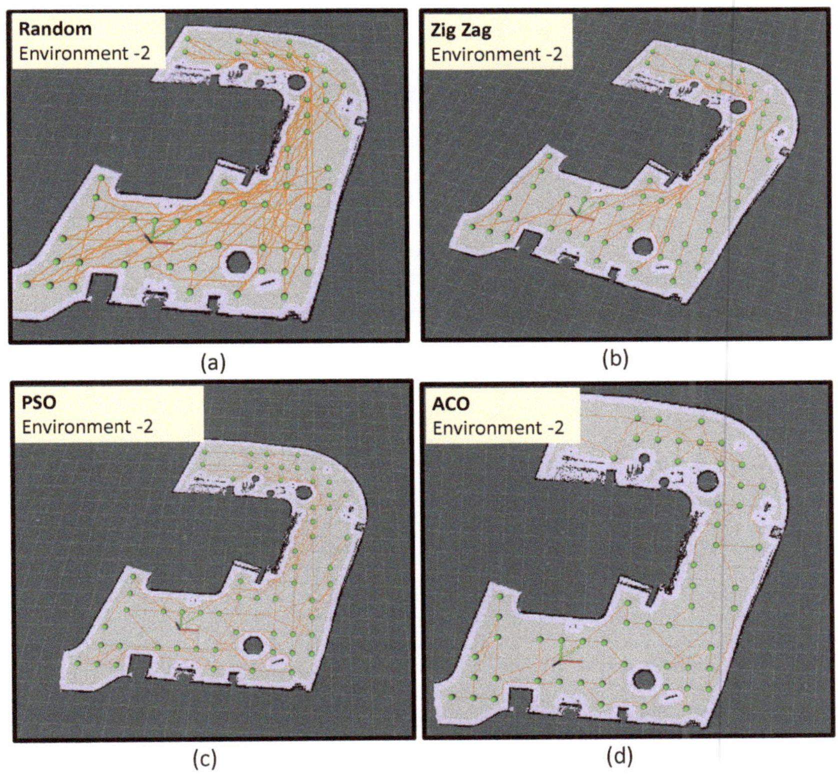

Figure 14. The path followed by the robot in environment-3 using random selection of points (**a**), using zig-zag selection of points (**b**), PSO algorithm (**c**) and ACO algorithm (**d**).

7. Conclusions and Future Works

A novel path planning strategy for robot-aided cleaning auditing has been devised by extracting the geometrical features from the map. Considering the boundaries and corners as the geometrical signatures that contributes to the dirt accumulation in an indoor environment, the locations for performing the auditing process are identified as part of the path planning strategy. To generate an optimal path that covers the identified sample locations, swarm algorithms like ACO and PSO are utilized. The optimization algorithm identified an efficient path covering all the sampling locations by minimizing the energy consumption by the robot. The dirt gathering efficiency of formulated geometry-based sampling locations and the behavior of the paths generated by the proposed approach are evaluated in real-time. Experiment results show that the geometry feature-based sample location identification aligned with dirt accumulation spots after multiple cleaning iterations in the same environment. The ACO-based path generation showed better performance by yielding the shortest exploration path with the smallest energy footprint compared to PSO and other path generation strategies like zig-zag and random point selection in our in-house developed BELUGA robot.

The future works of this research will be focusing on:

- Study the effect of variation of dirt patterns in auditing algorithms and consider dirt pattern distribution for audit path planning
- A comprehensive dirt dataset generation for machine learning-based sample auditing

- Inclusion of more geometrical signatures that contribute to the dirt accumulation in a region
- Extending the present cleaning auditing framework by including olfactory sensing techniques

Author Contributions: Conceptualization, T.P.; methodology, M.A.V.J.M.; software, T.P., M.A.V.J.M., and B.F.G.; validation, S.M.B.P.S.; analysis, T.P., M.A.V.J.M., and S.M.B.P.S.; funding acquisition, M.R.E.; Original draft, T.P. All authors have read and agreed to the published version of the manuscript.

Funding: This research is supported by the National Robotics Program under its Robotics Enabling Capabilities and Technologies (Funding Agency Project No. 192 25 00051), National Robotics Program under its Robot Domain Specific (Funding Agency Project No. 192 22 00058), National Robotics Program under its Robotics Domain Specific (Funding Agency Project No. 192 22 00108), and administered by the Agency for Science, Technology and Research .

Institutional Review Board Statement: Not applicable.

Informed Consent Statement: Not applicable.

Data Availability Statement: Not applicable.

Conflicts of Interest: The authors declare no conflict of interest.

References

1. 27 Janitorial Services Industry Statistics and Trends. Available online: https://brandongaille.com/27-janitorial-services-industry-statistics-and-trends (accessed on 12 July 2022).
2. Tay, S. How can Singapore's cleaning industry prepare for an 'endemic'? Available online: https://govinsider.asia/future-of-work/how-can-singapores-cleaning-industry-prepare-for-an-endemic-nea-dalson-chung/ (accessed on 12 July 2022).
3. Seo, T.; Jeon, Y.; Park, C.; Kim, J. Survey on glass and façade-cleaning robots: Climbing mechanisms, cleaning methods, and applications. *Int. J. Precis. Eng. Manuf.-Green Technol.* **2019**, *6*, 367–376. [CrossRef]
4. Perminov, S.; Mikhailovskiy, N.; Sedunin, A.; Okunevich, I.; Kalinov, I.; Kurenkov, M.; Tsetserukou, D. Ultrabot: Autonomous mobile robot for indoor uv-c disinfection. In Proceedings of the 2021 IEEE 17th International Conference on Automation Science and Engineering (CASE), Lyon, France, 23–27 August 2021; pp. 2147–2152.
5. Grando, M.T.; Maletz, E.R.; Martins, D.; Simas, H.; Simoni, R. Robots for cleaning photovoltaic panels: State of the art and future prospects. *Rev. Tecnol. Cienc.* **2019**, *35*, 137–150. [CrossRef]
6. Wang, X.V.; Wang, L. A literature survey of the robotic technologies during the COVID-19 pandemic. *J. Manuf. Syst.* **2021**, *60*, 823–836. [CrossRef]
7. Miao, X.; Lee, J.; Kang, B.Y. Scalable coverage path planning for cleaning robots using rectangular map decomposition on large environments. *IEEE Access* **2018**, *6*, 38200–38215. [CrossRef]
8. Lee, T.K.; Baek, S.; Oh, S.Y. Sector-based maximal online coverage of unknown environments for cleaning robots with limited sensing. *Rob. Auton. Syst.* **2011**, *59*, 698–710. [CrossRef]
9. Pathmakumar, T.; Sivanantham, V.; Anantha Padmanabha, S.G.; Elara, M.R.; Tun, T.T. Towards an Optimal Footprint Based Area Coverage Strategy for a False-Ceiling Inspection Robot. *Sensors* **2021**, *21*, 5168. [CrossRef]
10. Noh, D.; Lee, W.; Kim, H.R.; Cho, I.S.; Shim, I.B.; Baek, S. Adaptive Coverage Path Planning Policy for a Cleaning Robot with Deep Reinforcement Learning. In Proceedings of the 2022 IEEE International Conference on Consumer Electronics (ICCE), Las Vegas, NV, USA, 7–9 January 2022; pp. 1–6.
11. Nemoto, T.; Mohan, R.E. Heterogeneous multirobot cleaning system: State and parameter estimation. *Autom. Constr.* **2020**, *109*, 102968. [CrossRef]
12. Kaviri, S.; Tahsiri, A.; Taghirad, H.D. Coverage Control of Multi-Robot System for Dynamic Cleaning of Oil Spills. In Proceedings of the 2019 7th International Conference on Robotics and Mechatronics (ICRoM), Tehran, Iran, 20–21 November 2019; pp. 17–22.
13. Giske, L.A.L.; Bjørlykhaug, E.; Løvdal, T.; Mork, O.J. Experimental study of effectiveness of robotic cleaning for fish-processing plants. *Food Control* **2019**, *100*, 269–277. [CrossRef]
14. Lewis, T.; Griffith, C.; Gallo, M.; Weinbren, M. A modified ATP benchmark for evaluating the cleaning of some hospital environmental surfaces. *J. Hosp. Infect.* **2008**, *69*, 156–163. [CrossRef]
15. Asgharian, R.; Hamedani, F.M.; Heydari, A. Step by step how to do cleaning validation. *Int. J. Pharm. Life Sci.* **2014**, *5*, 3345–3365.
16. Malav, S.; Saxena, N. Assessment of disinfection and cleaning validation in central laboratory, MBS hospital, Kota. *J. Evol. Med Dent. Sci.* **2018**, *7*, 1259–1263. [CrossRef]
17. Bjørlykhaug, E.; Egeland, O. Vision system for quality assessment of robotic cleaning of fish processing plants using CNN. *IEEE Access* **2019**, *7*, 71675–71685. [CrossRef]
18. Cloutman-Green, E.; D'Arcy, N.; Spratt, D.A.; Hartley, J.C.; Klein, N. How clean is clean—Is a new microbiology standard required? *Am. J. Infect. Control* **2014**, *42*, 1002–1003. [CrossRef] [PubMed]
19. Pathmakumar, T.; Kalimuthu, M.; Elara, M.R.; Ramalingam, B. An autonomous robot-aided auditing scheme for floor cleaning. *Sensors* **2021**, *21*, 4332. [CrossRef]

20. Pathmakumar, T.; Elara, M.R.; Gómez, B.F.; Ramalingam, B. A Reinforcement Learning Based Dirt-Exploration for Cleaning-Auditing Robot. *Sensors* **2021**, *21*, 8331. [CrossRef]
21. Shang, Z.; Bradley, J.; Shen, Z. A co-optimal coverage path planning method for aerial scanning of complex structures. *Expert Syst. Appl.* **2020**, *158*, 113535. [CrossRef]
22. Galceran, E.; Carreras, M. A survey on coverage path planning for robotics. *Robot. Auton. Syst.* **2013**, *61*, 1258–1276. [CrossRef]
23. Bay, H.; Tuytelaars, T.; Gool, L.V. Surf: Speeded up robust features. In Proceedings of the European Conference on Computer Vision, Graz, Austria, 7–13 May 2006; Springer: Cham, Switzerland, 2006; pp. 404–417.
24. Rublee, E.; Rabaud, V.; Konolige, K.; Bradski, G. ORB: An efficient alternative to SIFT or SURF. In Proceedings of the 2011 International Conference on Computer Vision, Barcelona, Spain, 6–13 November 2011; pp. 2564–2571.
25. Sharma, S.K.; Jain, K. Image stitching using AKAZE features. *J. Indian Soc. Remote Sens.* **2020**, *48*, 1389–1401. [CrossRef]
26. Hirose, K.; Saito, H. Fast Line Description for Line-based SLAM. In *BMVC*; Keio University: Tokyo, Japan, 2012; pp. 1–11.
27. Suh, J.; Choi, E.Y.; Borrelli, F. Vision-based race track slam based only on lane curvature. *IEEE Trans. Veh. Technol.* **2019**, *69*, 1495–1504. [CrossRef]
28. Yan, T.; Zhang, Y. Mobile Robots Location and Mapping Based on Corner Features. In Proceedings of the 2018 5th International Conference on Information Science and Control Engineering (ICISCE), Zhengzhou, China, 20–22 July 2018; pp. 833–838.
29. Mahmood, M.R.; Abdulazeez, A.M. Different model for hand gesture recognition with a novel line feature extraction. In Proceedings of the 2019 International Conference on Advanced Science and Engineering (ICOASE), Zakho-Duhok, Iraq, 2–4 April 2019; pp. 52–57.
30. Kim, J.; Oh, K.; Oh, B.S.; Lin, Z.; Toh, K.A. A line feature extraction method for finger-knuckle-print verification. *Cogn. Comput.* **2019**, *11*, 50–70. [CrossRef]
31. Canny, J. A Computational Approach to Edge Detection. *IEEE Trans. Pattern Anal. Mach. Intell.* **1986**, *6*, 679–698. [CrossRef]
32. Fernandes, L.A.; Oliveira, M.M. Real-time line detection through an improved Hough transform voting scheme. *Pattern Recognit.* **2008**, *41*, 299–314. [CrossRef]
33. Mochizuki, Y.; Torii, A.; Imiya, A. N-Point Hough transform for line detection. *J. Vis. Commun. Image Represent.* **2009**, *20*, 242–253. [CrossRef]
34. Herout, A.; Dubská, M.; Havel, J. Review of Hough transform for line detection. In *Real-Time Detection of Lines and Grids*; Springer: Berlin/Heidelberg, Germany, 2013; pp. 3–16.
35. Derpanis, K.G. *The Harris Corner Detector*; York University: York, UK, 2004; 2p.
36. Shui, P.L.; Zhang, W.C. Corner detection and classification using anisotropic directional derivative representations. *IEEE Trans. Image Process.* **2013**, *22*, 3204–3218. [CrossRef]
37. Awrangjeb, M.; Lu, G.; Fraser, C.S.; Ravanbakhsh, M. A fast corner detector based on the chord-to-point distance accumulation technique. In Proceedings of the 2009 Digital Image Computing: Techniques and Applications, Melbourne, VIC, Australia, 1–3 December 2009; pp. 519–525.
38. Rosten, E.; Drummond, T. Machine learning for high-speed corner detection. In Proceedings of the European Conference on Computer Vision, Graz, Austria, 7–13 May 2006; pp. 430–443.
39. Rosten, E.; Porter, R.; Drummond, T. Faster and better: A machine learning approach to corner detection. *IEEE Trans. Pattern Anal. Mach. Intell.* **2008**, *32*, 105–119. [CrossRef]
40. Xin, J.; Zhong, J.; Li, S.; Sheng, J.; Cui, Y. Greedy mechanism based particle swarm optimization for path planning problem of an unmanned surface vehicle. *Sensors* **2019**, *19*, 4620. [CrossRef]
41. Huo, L.; Zhu, J.; Li, Z.; Ma, M. A hybrid differential symbiotic organisms search algorithm for UAV path planning. *Sensors* **2021**, *21*, 3037. [CrossRef]
42. Samarakoon, S.B.P.; Muthugala, M.V.J.; Elara, M.R. Metaheuristic based navigation of a reconfigurable robot through narrow spaces with shape changing ability. *Expert Syst. Appl.* **2022**, *201*, 117060. [CrossRef]
43. Fu, J.; Lv, T.; Li, B. Underwater Submarine Path Planning Based on Artificial Potential Field Ant Colony Algorithm and Velocity Obstacle Method. *Sensors* **2022**, *22*, 3652. [CrossRef]
44. Yuan, X.; Yuan, X.; Wang, X. Path Planning for Mobile Robot Based on Improved Bat Algorithm. *Sensors* **2021**, *21*, 4389. [CrossRef]
45. Yan, Z.; Zhang, J.; Yang, Z.; Tang, J. Two-dimensional optimal path planning for autonomous underwater vehicle using a whale optimization algorithm. *Concurr. Comput. Pract. Exp.* **2021**, *33*, e6140. [CrossRef]
46. Zhang, Z.; He, R.; Yang, K. A bioinspired path planning approach for mobile robots based on improved sparrow search algorithm. *Adv. Manuf.* **2022**, *10*, 114–130. [CrossRef]
47. Ma, Y.; Feng, W.; Mao, Z.; Li, H.; Meng, X. Path planning of UUV based on HQPSO algorithm with considering the navigation error. *Ocean Eng.* **2022**, *244*, 110048. [CrossRef]
48. Wang, X.; Shi, H.; Zhang, C. Path planning for intelligent parking system based on improved ant colony optimization. *IEEE Access* **2020**, *8*, 65267–65273. [CrossRef]
49. Zheng, K. Ros navigation tuning guide. In *Robot Operating System (ROS)*; Springer: Berlin/Heidelberg, Germany, 2021; pp. 197–226.
50. Chen, W.; Sui, L.; Xu, Z.; Lang, Y. Improved Zhang-Suen thinning algorithm in binary line drawing applications. In Proceedings of the 2012 International Conference on Systems and Informatics (ICSAI2012), Yantai, China, 19–20 May 2012; pp. 1947–1950.

51. Muthugala, M.V.J.; Samarakoon, S.B.P.; Elara, M.R. Tradeoff between area coverage and energy usage of a self-reconfigurable floor cleaning robot based on user preference. *IEEE Access* **2020**, *8*, 76267–76275. [CrossRef]
52. Muthugala, M.V.J.; Samarakoon, S.B.P.; Elara, M.R. Toward energy-efficient online Complete Coverage Path Planning of a ship hull maintenance robot based on Glasius Bio-inspired Neural Network. *Expert Syst. Appl.* **2022**, *187*, 115940. [CrossRef]
53. Zhang, H.Y.; Lin, W.M.; Chen, A.X. Path planning for the mobile robot: A review. *Symmetry* **2018**, *10*, 450. [CrossRef]
54. Muthugala, M.A.V.J.; Samarakoon, S.M.B.P.; Mohan, R.E. Design by Robot: A Human-Robot Collaborative Framework for Improving Productivity of a Floor Cleaning Robot. In Proceedings of the 2022 IEEE International Conference on Robotics and Automation (ICRA), Philadelphia, PA, USA, 23–27 May 2022.
55. Dorigo, M.; Birattari, M.; Stutzle, T. Ant colony optimization. *IEEE Comput. Intell. Mag.* **2006**, *1*, 28–39. [CrossRef]
56. Poli, R.; Kennedy, J.; Blackwell, T. Particle swarm optimization. *Swarm Intell.* **2007**, *1*, 33–57. [CrossRef]
57. Bradski, G.; Kaehler, A. *Learning OpenCV: Computer Vision with the OpenCV Library*; O'Reilly Media, Inc.: Sebastopol, CA, USA, 2008.
58. Raihan, F.; Ce, W. PCB defect detection USING OPENCV with image subtraction method. In Proceedings of the 2017 International Conference on Information Management and Technology (ICIMTech), Special Region of Yogyakarta, Indonesia, 15–17 November 2017; pp. 204–209.

sensors

MDPI

Article

Toward a Comprehensive Domestic Dirt Dataset Curation for Cleaning Auditing Applications

Thejus Pathmakumar, Mohan Rajesh Elara, Shreenhithy V Soundararajan and Balakrishnan Ramalingam *

Engineering Product Development, Singapore University of Technology and Design, Singapore 487372, Singapore; pathmakumar_thejus@mymail.sutd.edu.sg (T.P.); rajeshelara@sutd.edu.sg (M.R.E.); shreenhithy@sutd.edu.sg (S.V.S.)

* Correspondence: balakrishnan@sutd.edu.sg

Abstract: Cleaning is an important task that is practiced in every domain and has prime importance. The significance of cleaning has led to several newfangled technologies in the domestic and professional cleaning domain. However, strategies for auditing the cleanliness delivered by the various cleaning methods remain manual and often ignored. This work presents a novel domestic dirt image dataset for cleaning auditing application including AI-based dirt analysis and robot-assisted cleaning inspection. One of the significant challenges in an AI-based robot-aided cleaning auditing is the absence of a comprehensive dataset for dirt analysis. We bridge this gap by identifying nine classes of commonly occurring domestic dirt and a labeled dataset consisting of 3000 microscope dirt images curated from a semi-indoor environment. The dirt dataset gathered using the adhesive dirt lifting method can enhance the current dirt sensing and dirt composition estimation for cleaning auditing. The dataset's quality is analyzed by AI-based dirt analysis and a robot-aided cleaning auditing task using six standard classification models. The models trained with the dirt dataset were capable of yielding a classification accuracy above 90% in the offline dirt analysis experiment and 82% in real-time test results.

Keywords: domestic dirt; dirt dataset; audit robot; cleaning benchmark; dirt classification; robot-aided cleaning auditing

Citation: Pathmakumar, T.; Elara, M.R.; Soundararajan, S.V.; Ramalingam, B. Toward a Comprehensive Domestic Dirt Dataset Curation for Cleaning Auditing Applications. *Sensors* **2022**, *22*, 5201. https://doi.org/10.3390/s22145201

Academic Editors: Sašo Blažič, Gregor Klancar and Marija Seder

Received: 18 June 2022
Accepted: 8 July 2022
Published: 12 July 2022

1. Introduction

Cleaning is an inevitable routine associated with every domain. According to the recent market studies, the professional cleaning industry is steeply growing and expected to reach a market size of USD 88.9 billion by 2025 [1,2]. The growth of the cleaning industry is further boosted up by the increasing demand during the pandemic outbreak. A plethora of leading edge technologies have been introduced to the field of domestic and professional cleaning to enhance the performance of cleaning and maximize the productivity for the past decade [3–5]. This includes the usage of novel cleaning strategies using floor cleaning robots [3], Ultra-Violet-C (UVC) disinfection robots [4], cable-driven wall cleaning robots [5], etc. Currently, the reported studies about the cleaning performance enhancement are centered on the development of novel classes of cleaning robots and its associated components including robot efficient navigation, control, multi-robot cooperation, etc. For example, a morphology switching strategy for maximizing the area coverage in reconfigurable cleaning robots is reported [6]. Fuzzy inference systems used for enhanced adhesion awareness in vertical glass wall cleaning robots are reported [7]. An adaptive floor cleaning strategy by analyzing the human density is detailed in [8]. The research work mentioned in [9] presents an efficient charging mechanism for cleaning robots using infrared spots and neural network-based location estimators. A novel functional footprint-based efficient ship hull cleaning method using evolutionary algorithms is reported in [10]. The preceding analysis shows that a significant amount of research has focused on adapting the latest technologies for enhancing the performance of cleaning. However, a systematic method for assessing the quality of cleaning delivered by various cleaning methods has not been studied predominantly.

Dirt analysis is one of the critical elements for cleanliness auditing. The major challenges in the field of dirt analysis targeting cleaning auditing include the following:

- Even though there are proven AI models that could yield superior accuracy, there is no comprehensive dataset that the experts can use to train the AI models for dirt analysis.
- The dirt particles are often detected by typical computer vision-based analysis. However, capturing the finer features of the dirt is essential for AI-based analysis.
- The domestic dirt has identical visual features, which makes classification challenging for the AI models, which demands a high-quality dataset highlighting the visually distinct features of the dirt particles.

This research work presents a novel domestic dirt image dataset for cleaning auditing application including AI-based dirt analysis and robot-assisted cleaning inspection. Unlike the conventional vision-based dirt detection algorithms, whose application is limited to cleaning robots, the proposed method presents a dirt detection and classification strategy for cleaning auditing, where an audit sensor gathers the dirt samples via adhesive dirt lifting. As part of this work, a comprehensive dataset consisting of microscope images of commonly occurring domestic dirt is acquired using the sample audit sensor. To distinguish the features, each dirt image is analyzed in under 10× magnification. In addition, the selected dataset's usability is further analyzed using training and validation accuracy in different deep-learning architectures that enable deep-learning-based dirt analysis. To the best of the author's knowledge, a dataset of microscopic domestic dirt images has not been reported so far. The proposed research work opens the door toward new fronts for AI-driven dirt analysis targeting the domain of cleaning auditing. The The general objective of this research work is subdivided into the following:

1. Gathering of magnified images of domestic dirt particles using adhesive dirt lifting;
2. Analyze the usability of an acquired dataset in training and classifying the domestic dirt in standard classification models, using a cross-validation technique;
3. Analyze the performance of the proposed scheme in a real-time scenario by rolling out the trained classification model for real-time dirt composition estimation for an in-house developed audit robot.

The rest of the article is structured as follows. Section 2 provides a detailed study on the related works, Section 3 provides a birds-eye view of the adopted methods and methodology; Section 4 reports a detailed description of our analysis and experiments conducted in this research effort followed by Section 6, which concludes our findings.

2. Related Works

Despite the importance, cleaning auditing is analyzed in very few domains. For example, the formation of tests for cleaning quality analysis in the fish processing industry is reported in [11]. Lewis et al. proposed a modified adenosine triphosphate (ATP) benchmark for estimating the quality of cleaning for hospital environments [12]. Efforts toward establishing cleaning standards for the hospital environment are reported [13]. Similarly, Aziz et al. proposed microbiological monitoring for cleaning analysis targeting hospital contamination estimation [14]. The state-of-the-art cleaning auditing methods are centered around microbiological analysis and the ATP bioluminescence method [15]. However, the analysis method mentioned above is not scalable to most domains, including professional and domestic floor cleaning. The pioneering effort in cleaning auditing is proposed by Pathmakumar et al., where cleaning auditing is done with the help of a autonomous mobile robot and a sample audit sensor [16]. The research work mentioned above uses a cleaning auditing sensor that extracts the dirt from the floor by adhesive dust lifting followed by analysis using computer vision-based techniques. The auditing of a larger area is achieved with the audit robot using dirt exploration methods [17]. The robot-aided cleaning auditing is a viable approach for post-cleaning analysis compared to the laborious microbiological methods, which are limited only to a specific domain. The robot-aided cleaning auditing executes the cleaning auditing in a fully automated fashion, which bridges the lack of proper post-cleaning analysis in the domain of automated cleaning. One of the main challenges in robot-aided cleaning auditing is the development of an effective method for dirt analysis. Conventionally, visual detection methods are often used for dirt identification. For instance, Grunauer et al. proposed an unsupervised dirt spot detection where the problem is addressed as a binary classification problem [18]. Similarly, Bormann et al. proposed a training-free dirt detection framework in an office environment [19]. The above-mentioned method is further improved by a multi-class machine learning-based dirt detection method using a modified YOLOv3 framework [20]. A similar approach is reported where a cascaded neural network is used for multi-class debris detection for floor cleaning robots [21]. The current dirt detection algorithms are reported in [18–21]; here, the dirt detection is targeted at a cleaning robot to perform a selective cleaning. However, for the concerned scenario of cleaning auditing, a microscopic analysis of domestic dirt sampled after a dust-lifting process is inevitable. However, the present dirt detection frameworks are not designed for the above-mentioned application.

3. Methodology

This work adopts our in-house developed cleaning auditing robot (BELUGA) for dirt dataset collection. In order to classify the domestic dirt, it is essential to build a comprehensive dataset of domestic dirt. The domestic dirt is referred to as particulate contaminates (usually measured in microns) that are carried by the airflow and settled down in undisturbed air. The samples are collected from indoor and semi-indoor regions by attaching the sensors to the BELUGA robot. The overview of the sample collection procedures and method for dirt inference is depicted in Figure 1. The sample collection procedure involves the gathering of different sets of dirt samples from floor surfaces. The samples can be collected either by using the sample audit sensor in a standalone way or by using an autonomous robot carrying a auditing sensor payload. The sample audit sensor gather images using the adhesive dirt lifting principle, which is an established principle used for forensic trace collection [22,23]. For the analysis part, the images of the adhesive tape are taken using a microscopic camera with 10× magnification. The images gathered using the sample audit sensor are stored in a remote database on a cloud server. In case of limited connectivity to the remote server, the captured images are enhanced and stored locally onboard the robot. The sample collection is controlled and monitored using a web application hosted on the local device that serves as an interface for the user to control the process. In the remote server, the images are stored in a labeled manner.

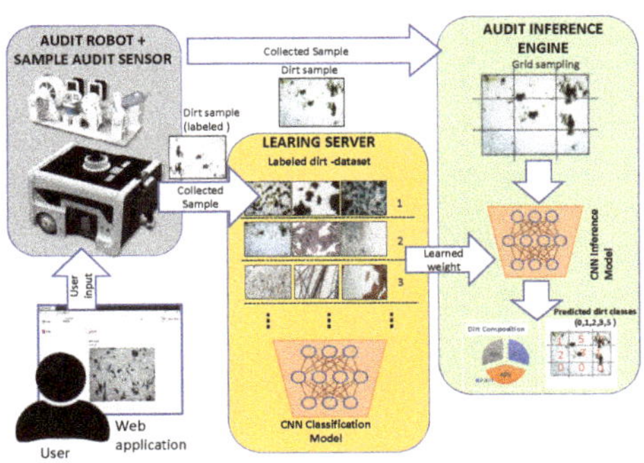

Figure 1. Proposed overview for domestic dirt collection using autonomous robot.

3.1. Robot Architecture

The audit robot is the carrier of a sample audit sensor, and it helps perform the dirt sample gathering from a vast area of space where the manual sample gathering is not feasible. The audit robot called BELUGA is an in-house developed autonomous cleaning audit robot integrated with the sample audit sensor. The Figure 2 shows BELUGA robot and mounting of sample audit sensor onto the robot. The BELUGA robot is comprised of a locomotion module, power distribution module, navigation module, audit sensor and processing module (shown in Figure 3).

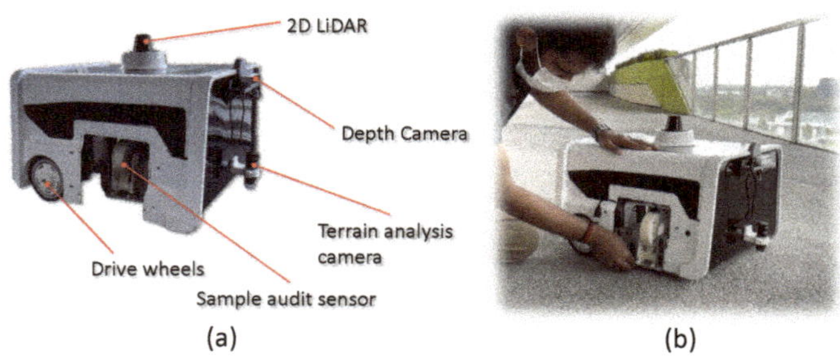

Figure 2. Cleaning audit robot—BELUGA (**a**); Attaching of sample audit sensor with BELUGA (**b**).

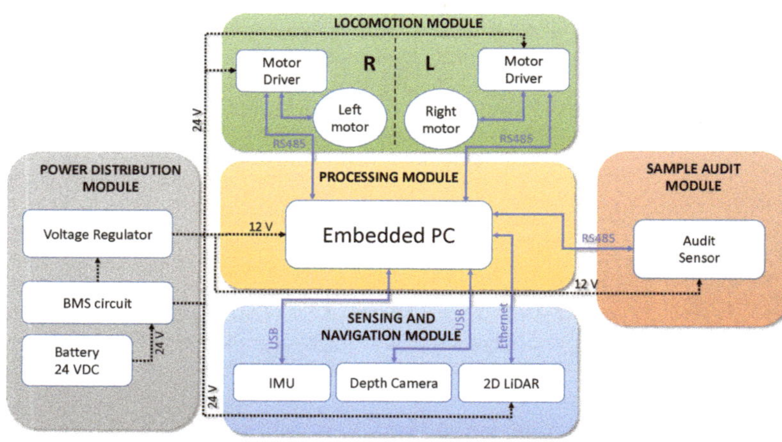

Figure 3. The system architecture of the BELUGA robot with the main subsystems.

The locomotion unit of the robot is composed of a pair of brushless DC (BLDC) motors that form a differential drive wheel configuration. The third point of contact of the robot is made using a free rotating castor wheel. The BLDC motor drivers established a closed-loop velocity control for the driving wheels. The velocity feedback is obtained from the BLDC motor drivers using the incremental encoders associated with the motor. The velocity feedback from the motors is used for computing the odometry information. The communication with the drive motors is established using MODBUS-RTU communication. The robot uses a 24 V DC lithium-ion phosphate battery to power all subsystems. The main supply from the battery is regulated and distributed further to power-sensitive components. A 2D LiDAR is the primary perception sensor associated with the robot, which is supported by a depth camera to add 3D perception capabilities and detect obstacles below the level of LiDAR. The LiDAR used in the robot is SICK TIM 581 with a range of 20 m in semi-outdoor conditions. The Intel Real-sense D435i is the depth camera with a resolution of 640 × 480 and 87° × 58° angular field of view. Using Kalman filter-based state-estimation techniques, the wheel odometry information is generated by fusing the wheel velocity feedback from the motor drivers and the inertial measurement unit (IMU). The IMU used in the BELUGA robot is VectorNav VN100, which is a nine-axis IMU with built-in noise filtration.

The robot performs autonomous navigation in real time using the input from the perception sensors and odometry information. The robot possesses an embedded computer with an Intel core i7 processor and runs with an operating system of Ubuntu 20.04. The perception and navigation algorithms run alongside the ROS middleware in the embedded computer. Apart from the navigation sensors, the robot is integrated with the sample audit sensor. The communication between the sample audit sensor and the embedded PC is established through USB and RS485 communication. The USB established a communication link with the digital microscope, and RS485 is used to control the servo motors' actuation inside the sensor. On the BELUGA robot, the sensor is attached to a removable sensor bay, which allows detaching the sensor from the robot.

3.2. Sample Audit Sensor

The major components associated with the sample audit sensor are the adhesive tape, winding motor, pressing motor, pressing ramp, and a digital microscope. The gathering of the dirt samples is done by the synchronized motion of winding motors and pressing motors. The winding motors and pressing motors are digital servos that can be position-controlled according to the data passed over the RS485 communication protocol. The winding motor of the sensor rotates an adhesive tape, and the pressing motor actuates the pressing ram that exerts downward force to press and release the sticking surface of the tape against the surface. The winding motor displaces the tape surface toward the field of view of the digital microscope to capture the magnified image of the surface of the tape. The pressing of the adhesive tape is completed for a 2 cm × 2 cm area; hence, analysis of the dirt is localized to the same surface area. Every sampling completed by the BELUGA robot takes 13 to 16 s, which includes 4 s to lower the sensor bay to the floor, 2 s to conduct a pre-sampling winding, 4 s for stamping action, and 3 to 6 s to perform a post-sampling winding. The post-sampling winding varies from 3 to 6 s since the length to displace varies as the adhesive tape radius decreases linearly as more and more samples are taken. The movement of the adhesive tape is guided through silicon idlers so that the adhesive tape will not become stuck while displacing the stamped area to the field of view of the microscope. Figure 4 shows the sample collection sensor module. The sensor consists of the (a) collection mechanism to gather dirt from the floor and pass it to the (b) microscope camera to capture, transfer and process the sample images.

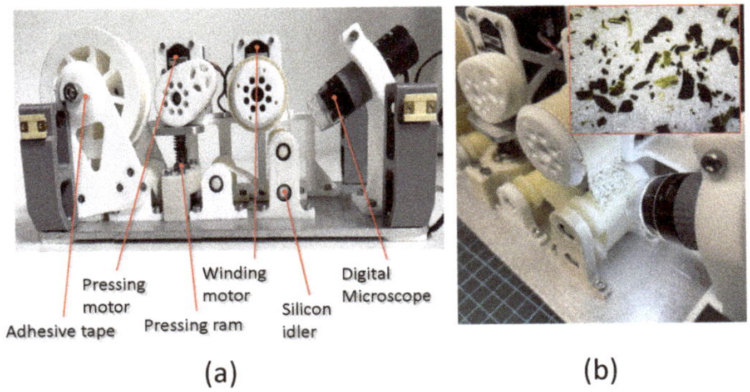

Figure 4. The sample audit sensor with major components (**a**) and the extracted dirt particles under the view of the microscope (the 10× magnified image captured in the inset) (**b**).

4. Experiments and Analysis

This section describes the experimental and analysis procedure of dirt dataset collection and evaluation methods. The experiment has been performed in four phases. The first phase involves dataset curation. The second phase validates the collected dirt dataset through a k-fold cross-validation scheme and converging time of the various image classification algorithm. The third phase involves validation of the dirt dataset through the AI-based dirt classifier algorithm. The final phase involves validating the trained dirt classification models with the BELUGA robot on a real-time field trail.

4.1. Dataset Curation

The BELUGA robot is set to exploration mode, where it explores the region using the frontier exploration method, and the sample gathering is completed every 10 s. Figure 5 shows sample collection performed at different locations using the BELUGA robot. We selected a food court (Figure 5a), semi-indoor walkway (Figure 5b), office pantry (Figure 5c), long corridor (Figure 5d), office space (Figure 5e) and warehouse (Figure 5f) as the sites for data collection. Once the robot explores 98% of the deployed area, the robot resets its map and completes the frontier exploration again.

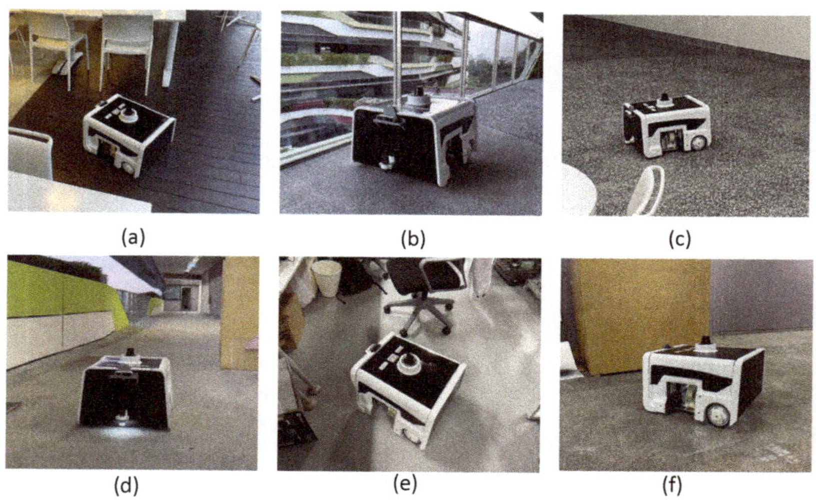

Figure 5. The dirt sample collection using the BELUGA robot in different locations such as a food court (**a**), semi-indoor walkway (**b**), office pantry (**c**), long corridor (**d**), office space (**e**), and warehouse (**f**).

4.2. Dirt Class Identification

The sample collection procedure is repeated for 2 days in each location, and the composition of dirt is analyzed. The main dust particles lifted by the adhesive tape include ashes, hair (mostly from walkways and corridors), tiny bits of paper, sand, soil, etc. The bits of paper are collected from all the six locations taken for the data collection. The traces of sand and soil particles were also captured from all locations—however, the sand and soil particle concentration was higher in the walkways and corridors. The traces of food particles were also identified from every location; however, the most frequent occurrence of food particles in the collected samples was from the food court and pantry. Paint particles are captured in the sample collection completed in warehouses often, and it is observed that more traces are captured when the robot completes sampling near the corners and walls of the warehouse. Traces of seeds, lint, etc. were also identified from the collected samples across all locations. From the observations made from the dirt sample collection, the specified classes for the dirt data includes ash, hair, sand, soil, paper, paint, food, and fiber. These classes are finalized for the dirt dataset considering the following:

- Frequent occurrence during sample collection;
- Presence in every location;
- The health and environmental factors associated with identified dirt particles [24–26].

The identified dirt classes for the dataset are provided in the Table 1.

Table 1. The classes of identified domestic dirt.

Class (Number)	Sample Images
Ash	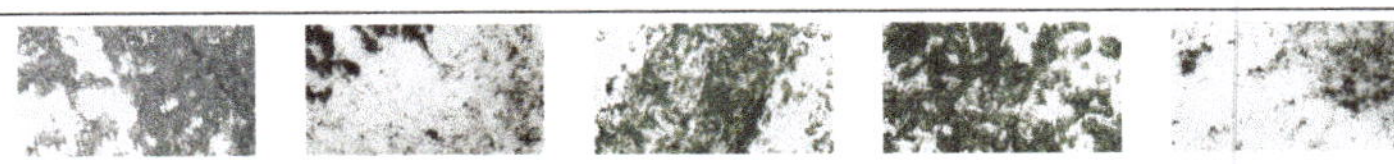
Hair	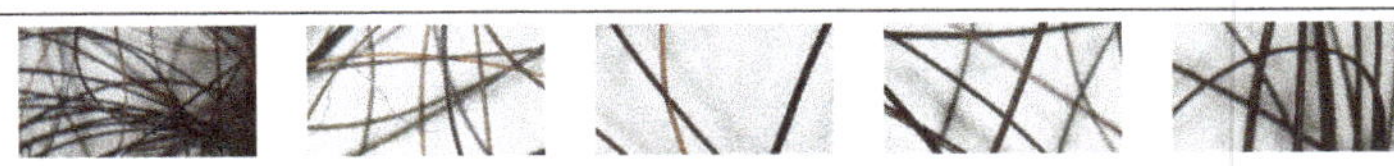
Sand	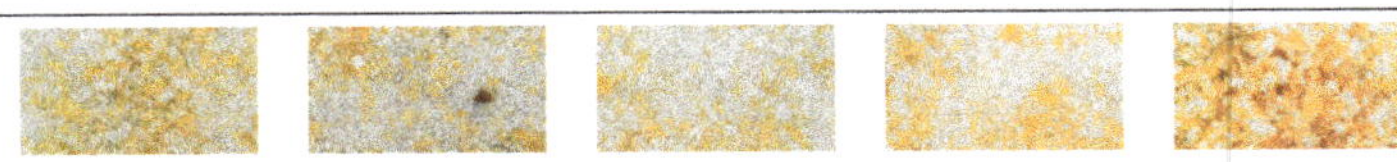
Soil	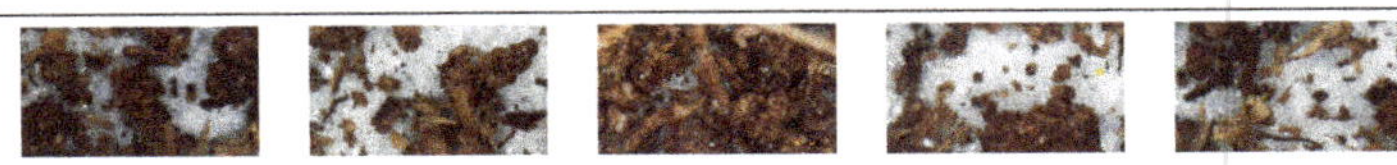
Paper	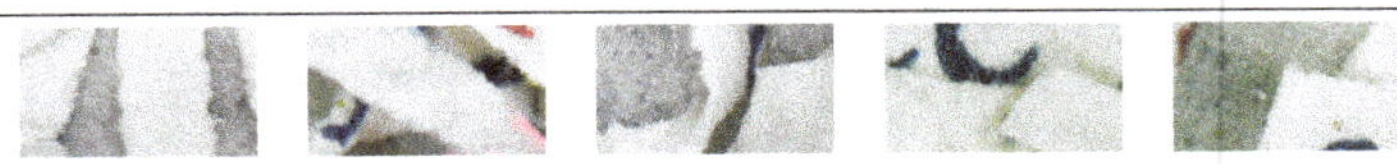
Paint	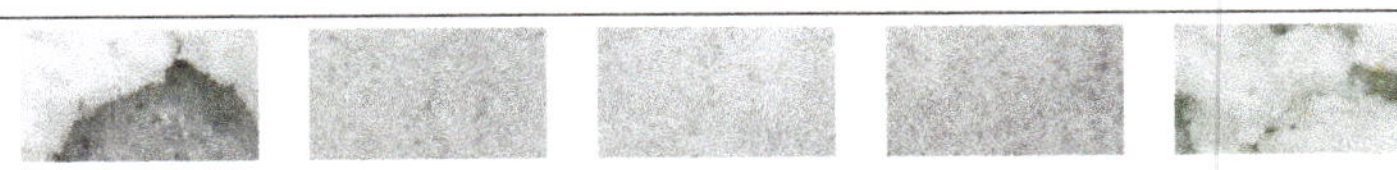
Food	
Fibre	
No-dirt	

4.3. Dataset Preparation and Training

Upon identifying the distinct dirt classes, the image captured by the microscope at a resolution of 1280 × 720 was decimated to 16 image samples, each of size 320 × 120 images. The dirt samples collected by the robot as well as the samples manually collected were labeled and formed the dataset for domestic dirt. The samples where the dirt is spread less than 60% in the image area are discarded. Images overlapping with different classes are also discarded. After discarding the invalid images, a dataset of 3000 samples from each class is curated. From every class of labels images, 2500 images are taken for training and cross-validation, and the remaining 500 are used for offline testing.

4.4. Dirt Dataset Validation

The quality of the dataset is analyzed by inspecting the accuracy of the state-of-the-art AI-based image classification models trained using it. The quality of the collected dirt dataset was evaluated through training accuracy with a k-fold cross-validation scheme and statistical measure. For analysis, we choose two sets of classification models. The first set is composed of three less dense neural network architectures, and the second set is composed of three dense-layer neural networks. For the less dense models, we took VGG-11 [27,28], VGG-16 [27] and MobileNetV2 [29]. For the second set, for the dense-layer models, we choose ResNet50 [30,31], ResNet101 [30] and Darknet53 [32].

4.4.1. K-Fold Cross-Validation Method

The k-fold cross-validation is completed by following the procedures given below:

1. Select the fold $k = 5$.
2. Split the dataset to k groups, which are also known as folds.
3. Select $k - 1$ folds for training the model and one fold for testing.
4. For every iteration, a new model is trained independent of the previous iteration.
5. Repeat the training and cross-validation k times; in every iteration, the remaining fold will serve as the test set.
6. The accuracy is determined on the kth iteration as the average of all iterations.

The six-image classifier model was trained with an early stopping condition to avoid the over-fitting of the model. The models are trained in NVIDIA GeForce 3080 GPU with a batch size of 32 and a learning rate of 4×10^{-3}. Cross-entropy loss is used to estimate the model's prediction performance in every forward pass [33]. Comparing the accuracy of every iteration provides insight into the curated dataset's reliability and trustworthiness. However, a k-fold gives a more stable and trustworthy result since training and testing are performed on multiple combinations of test–train set decimation from the dataset.

Figure 6 reports the convergence profile for the models for the first 30 epochs of training. All models converged above 90% accuracy in every fold for all the six models. The average accuracy of every fold of training in VGG-11, VGG-16, and MobleNetV2 is observed in the first 10 epochs of training. A slightly delayed convergence is observed for dense-layer models such as ResNet50, ResNet101, and Darknet53. The dense-layer models took 15–20 epochs to see a consistency in the convergence trend. This trend is attributed to the relatively complex nature of ResNet50, ResNet101, and Darknet53 compared to the other models. One of the key indicators that is attributed to the quality of the dataset is the deviation of accuracy in every fold of training. The average training accuracy for k-fold and standard deviation in accuracy is tabulated in Table 2. All models show a small standard deviation in the k-fold training, which indicates a non-biased and balanced dataset. The variation in accuracy in different folds of training is comparatively less when it comes to dense-layer models. The minimal deviation is reported by ResNet101 (1.89), and maximum accuracy in training is reported by ResNet50 (96.58%).

Table 2. Statistical measures for dirt classification.

Model	Average Accuracy (%)	K-Fold Standard Deviation
VGG-11	94.77	5.82
VGG-16	92.77	3.90
MobileNetV2	94.65	3.44
ResNet50	96.58	2.05
ResNet101	96.54	1.89
Darknet53	95.801	3.06

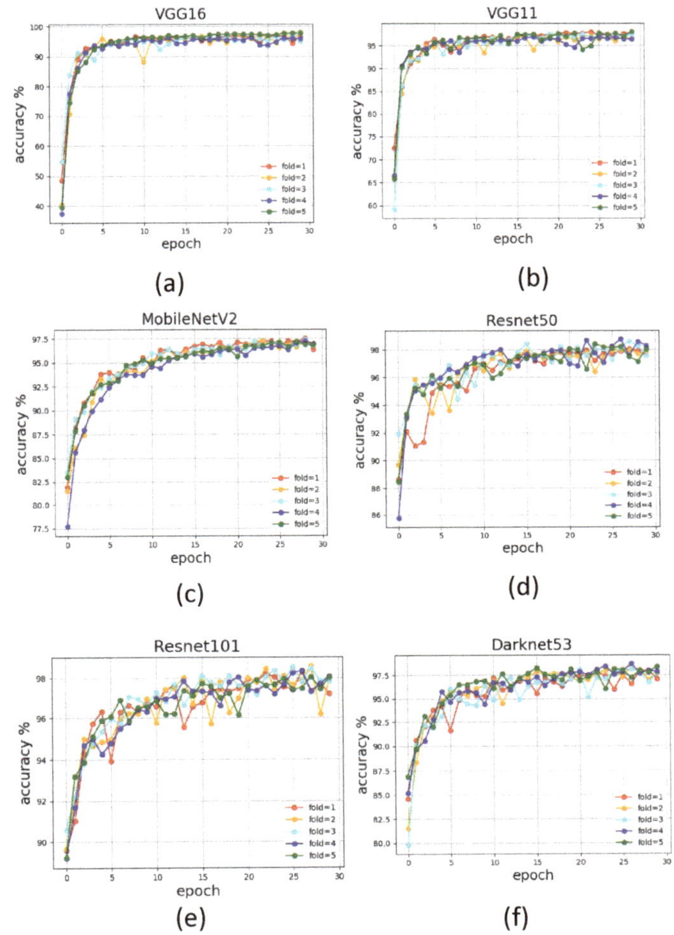

Figure 6. The accuracy profile for different classification models trained using curated dirt dataset; VGG-16 (**a**), VGG-11 (**b**), MobileNetV2 (**c**), Resnet50 (**d**), Resnet101 (**e**), Darknet53 (**f**).

4.4.2. Dirt Dataset Validation Through Statistical Measure

The dirt dataset's efficiency was validated through a statistical measure function. Here, the models trained using our dirt dataset classification accuracy were chosen as the evaluation matrix to assess the dirt dataset quality. Accuracy Equation (1), precision Equation (2), recall Equation (3) and $F_{measure}$ Equation (4) were used to evaluate the trained model's classification accuracy. The confusion matrix function was used to find the variables tp (true positives), fp (false positives), tn (true negatives) and fn (false negatives) through which accuracy, precision, recall and $F_{measure}$ were calculated. The evaluation metrics includes:

$$Accuracy(Acc) = \frac{tp + tn}{tp + fp + tn + fn} \tag{1}$$

$$Precision(Prec) = \frac{tp}{tp + fp} \tag{2}$$

$$Recall(Rec) = \frac{tp}{tp + fn} \tag{3}$$

$$F_{measure}(F_1) = \frac{2 \times precision \times recall}{precision + recall} \tag{4}$$

A set of 500 images from each class is used to compute the confusion matrix parameter (tp (true positives), fp (false positives), tn (true negatives) and fn (false negatives)), and these images were not used for training the image classifier. Table 3 gives the statistical measures results computed through the confusion matrix parameter.

Table 3. Statistical measures for dirt classification.

Model	Class	Measurements			
		Precision	Recall	F1	Accuracy %
VGG-16	Ash	0.9803	0.9970	0.9886	99.70
	Food	0.9918	0.9675	0.9795	96.65
	Fibre	0.9559	0.9644	0.9601	96.44
	Paper	0.8621	0.9336	0.8964	89.64
	Paint	0.9689	0.9291	0.9485	92.91
	Soil	0.9993	0.9970	0.9981	99.70
	Sand	0.9878	0.9943	0.9910	99.43
	Hair	0.9897	0.9998	0.9948	99.99
	No-dirt	0.9971	0.9941	0.9956	99.41
VGG-11	Ash	0.9900	0.9960	0.9930	99.60
	Food	0.9709	0.9787	0.9748	97.87
	Fibre	0.9847	0.9555	0.9699	95.55
	Paper	0.8482	0.9458	0.8944	94.58
	Paint	0.9914	0.9110	0.9495	91.10
	Soil	0.9983	0.9985	0.9982	99.85
	Sand	0.9906	0.9820	0.9863	98.20
	Hair	0.9622	0.9977	0.9796	99.77
	No-dirt	0.9798	0.9985	0.9891	99.85
MobileNet V2	Ash	0.9950	0.9920	0.9935	99.20
	Food	0.9907	0.9638	0.9770	96.38
	Fibre	0.9700	0.9585	0.9642	95.85
	Paper	0.9108	0.9038	0.9073	90.38
	Paint	0.9707	0.9516	0.9610	95.16
	Soil	0.9963	0.9993	0.9978	99.93
	Sand	0.9843	0.9970	0.9906	99.70
	Hair	0.9589	0.9954	0.9768	99.54
	No-dirt	0.9812	0.9956	0.9883	99.56

Table 3. *Cont.*

Model	Class	Measurements			
		Precision	Recall	F1	Accuracy %
ResNet50	Ash	0.9930	0.9880	0.9905	99.30
	Food	0.9956	0.9606	0.9778	96.06
	Fibre	0.9761	0.9703	0.9732	97.03
	Paper	0.9454	0.9392	0.9423	93.92
	Paint	0.9660	0.9758	0.9709	97.58
	Soil	0.9991	0.9985	0.9993	99.85
	Sand	0.9825	0.9973	0.9899	99.73
	Hair	0.9852	0.9965	0.9908	99.65
	No-dirt	0.9963	0.9985	0.9974	99.85
ResNet101	Ash	0.9990	0.9970	0.9980	99.70
	Food	0.9919	0.9755	0.9836	97.55
	Fibre	0.9910	0.9822	0.9866	98.22
	Paper	0.9016	0.9624	0.9310	96.24
	Paint	0.9889	0.9499	0.9690	94.99
	Soil	0.9997	0.9985	0.9993	99.85
	Sand	0.9899	0.9970	0.9934	99.70
	Hair	0.9730	0.9965	0.9846	99.65
	No-dirt	0.9971	0.9985	0.9978	99.85
Darknet53	Ash	0.9979	0.9699	0.9837	96.99
	Food	0.9382	0.9307	0.9344	93.07
	Fibre	0.9939	0.9614	0.9774	93.07
	Paper	0.9001	0.6681	0.7570	66.81
	Paint	0.9861	0.9583	0.9720	95.83
	Soil	0.9901	0.9993	0.9996	99.93
	Sand	0.9568	0.9111	0.9334	91.11
	Hair	0.6506	0.9965	0.7872	99.65
	No-dirt	0.9728	0.9985	0.9855	99.65

The offline test results show that all the six classification frameworks show an accuracy above 90%. Among the classes, ash, soil, sand, and hair showed the best classification accuracy, since the images were visually distinct in color and texture. The lowest classification accuracy was reported for the class paper and paint, since a scrap of paint and paper bit possesses almost the same visual features under 10× magnification by the camera. The darknet53 model showed the lowest classification accuracy for the class paper with 66.81%. On the other hand, the no-dirt class representing the samples devoid of dirt, which is a critical factor in determining the surface's cleanliness, showed high accuracy in all trained models. Even though the evaluation is performed on GPU, the ResNet101, Darknet53, and ResNet50 reported comparatively lower inference time than VGG-16, VGG-11, and MobileNetV2. The difference in inference time is attributed to the number of operations within the network. Regarding ResNet50, ResNet101, and Darknet53, there are 23 M, 44.5 M, and 40.5 M parameters, respectively. Whereas in the case of VGG-11, VGG-16, and MobilenetV2, there were 133 M, 138 M, and 3.4 M parameters trained, respectively. The statistical measurements reported in the offline test results show that the dataset gathered for dirt classification is un-biased between the classes and it can be used alongside the standard deep-learning models.

4.5. Real-Time Robot-Aided Cleaning Inspection

In addition to the offline test, the usability of the curated dataset is analyzed in a real-time cleaning inspection use case with the BELUGA robot. We have chosen an environment for testing which is similar to the chosen regions for data collection. DarkNet53 was used for rolling out the real-time inference considering the best performance in k-fold cross-validation and offline testing. Since the BELUGA robot's embedded computer is devoid of GPU, the inference is completed by establishing a communication with a remote server running with GPU. Using the BELUGA robot, the five dirt samples are collected each from the food court, walkway and indoor office space. For every sample image captured by the digital microscope after dust lifting, the sample image is divided into 16 images, matching the training dataset. The models loaded with weights were trained, and the classification of dirt is completed for every 16 images from the collected sample. Figure 7 shows the images classified from the dirt sample and the estimated dirt composition.

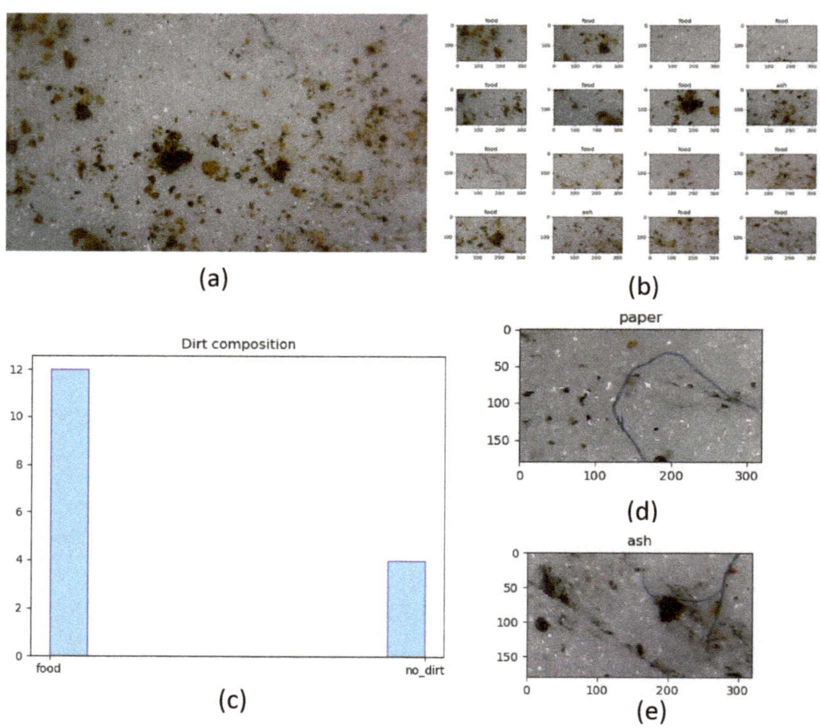

Figure 7. The test images collected in real time using the BELUGA robot (**a**), classified images (**b**), histogram of classified dirts from single sample (**c**), an exmple for wrongly classified image (**d**), and overlapping of dirt specks (**e**).

Comparison with Offline Test Results

Out of the 15 sample images, which were divided into 240 test image samples, 203 samples reported the right classification with an admissible accuracy of 84.58%. The model's real-time accuracy was less than the offline test results. Despite the above-mentioned shortcomings, the model trained with the curated dataset showed a good accuracy, which is acceptable for the dirt composition estimation in cleaning auditing.

5. Discussion

The experiments results conducted showed the prepared dataset's usability on popular deep learning models such as VGG-16, VGG-11, MobileNet V2, ResNet50, ResNet101, and Darknet53. Upon rolling out the trained with the prepared dataset for real-time inference, an admissible accuracy was observed. During the course of our dataset curation, certain limitations identified include the narrow field of view of the camera and the overlapping of multiple dirt classes in the sample image. Moreover, certain dirt particles share similar textures, making them difficult to be distinguished. Although the data inference is substantially faster, the dirt data collection is found to be slow, since it involves adhesive dirt lifting. Unlike the offline test results, false-positive occurrences are reported in the real-time test results (shown in Figure 7d), which are contributed to the following factors:

- Overlapping of multiple specks of dirt classes in a sample image (shown in Figure 7e);
- The shaking of adhesive tape during the actuation of the sensor may result in blur images that eventually lead to a wrong classification;
- Encountered dirt specks with very close visual resemblance make it indistinguishable for the model to classify.

6. Conclusions and Future Works

A dirt image dataset was proposed for AI-based dirt classification for automated cleaning auditing. Our in-house developed cleaning inspection robot BELUGA was used to gather the dirt sample images from a semi-indoor environment. We identified nine visually distinct dirt classes, and 3000 10× magnified microscope images for each class are gathered for the dataset. The usability of the collected dirt dataset was evaluated by analyzing the training and evaluation accuracy in six state-of-the-art image classifier models. The k-fold cross-validation method with a cross-entropy loss function was used to compute the model's training accuracy, and the statistical measure function was used to assess the classification accuracy of models trained using our dirt image dataset. During the training, a minimal standard deviation for training accuracy for every k-fold cross-validation iteration is observed, which indicates the unbiased nature of the collected dataset. The offline test results indicate that all the trained models scored above 90% accuracy for all classes. The quality of the dataset is further validated by rolling out the trained dataset for real-time cleaning auditing using an in-house developed BELUGA robot. The accuracy of real-time testing was comparatively less compared to the offline test results, which is mainly attributed to the overlapping of multiple specks in the same region of the sample image. The motion blur was also introduced in the dirt lifting process, which diminished the accuracy of real-time dirt analysis. In addition, the time taken for overall dirt sample collection was slower because the adhesive dirt-lifting process was time-consuming. Our future research will be focused on the following areas:

- Combining microbial and chemical analysis in the process of sample auditing;
- Incorporating novel autonomous algorithms toward dirt exploration;
- A comprehensive study comparing the different algorithms with respect to cleaning auditing;
- Exploring the usability of the current dataset for instance segmentation of dirt particles;
- Improving the current dataset by expanding the number of dirt classes and open-sourcing the dataset.

Author Contributions: Conceptualization, T.P.; methodology, B.R.; software, T.P. and S.V.S.; validation, B.R. and M.R.E.; analysis, T.P., M.R.E. and B.R.; Original draft, T.P. All authors have read and agreed to the published version of the manuscript.

Funding: This research is supported by the National Robotics Program under its Robotics Enabling Capabilities and Technologies (Funding Agency Project No. 1922500051), National Robotics Program under its Robot Domain Specific (Funding Agency Project No. 192 22 00058), National Robotics Program under its Robotics Domain Specific (Funding Agency Project No. 1922200108), and administered by the Agency for Science, Technology and Research.

Institutional Review Board Statement: Not applicable.

Informed Consent Statement: Not applicable.

Data Availability Statement: Not applicable.

Conflicts of Interest: The authors declare no conflict of interest.

References

1. Global Cleaning Services Market Size to be $88.9 Billion by 2025, Says Beroe Inc. Available online: https://www.prnewswire.com/news-releases/global-cleaning-services-market-size-to-be-88-9-billion-by-2025--says-beroe-inc-301272305.html (accessed on 5 June 2022).
2. How Can Singapore's Cleaning Industry Prepare for An "Andemic"? | GovInsider. Available online: https://govinsider.asia/future-of-work/how-can-singapores-cleaning-industry-prepare-for-an-endemic-nea-dalson-chung/ (accessed on 5 June 2022).
3. Toprak, J.; Kamiloglu, A.; KIRCI, P. Design of a Smart Vacuum Cleaner with Indoor Localization. In Proceedings of the Conference of Open Innovations Association, FRUCT, Helsinki, Finland, 12–14 May 2021; pp. 495–499.
4. Guettari, M.; Gharbi, I.; Hamza, S. UVC disinfection robot. *Environ. Sci. Pollut. Res.* **2021**, *28*, 40394–40399. [CrossRef]
5. Shao, Z.; Xie, G.; Zhang, Z.; Wang, L. Design and analysis of the cable-driven parallel robot for cleaning exterior wall of buildings. *Int. J. Adv. Robot. Syst.* **2021**, *18*, 1729881421990313. [CrossRef]
6. Prabakaran, V.; Mohan, R.E.; Sivanantham, V.; Pathmakumar, T.; Kumar, S.S. Tackling area coverage problems in a reconfigurable floor cleaning robot based on polyomino tiling theory. *Appl. Sci.* **2018**, *8*, 342. [CrossRef]
7. Muthugala, M.; Vega-Heredia, M.; Mohan, R.E.; Vishaal, S.R. Design and control of a wall cleaning robot with adhesion-awareness. *Symmetry* **2020**, *12*, 122. [CrossRef]
8. Sivanantham, V.; Le, A.V.; Shi, Y.; Elara, M.R.; Sheu, B.J. Adaptive Floor Cleaning Strategy by Human Density Surveillance Mapping with a Reconfigurable Multi-Purpose Service Robot. *Sensors* **2021**, *21*, 2965. [CrossRef]
9. Chang, C.L.; Chang, C.Y.; Tang, Z.Y.; Chen, S.T. High-efficiency automatic recharging mechanism for cleaning robot using multi-sensor. *Sensors* **2018**, *18*, 3911. [CrossRef]
10. Pathmakumar, T.; Sivanantham, V.; Anantha Padmanabha, S.G.; Elara, M.R.; Tun, T.T. Towards an Optimal Footprint Based Area Coverage Strategy for a False-Ceiling Inspection Robot. *Sensors* **2021**, *21*, 5168. [CrossRef]
11. Giske, L.A.L.; Bjørlykhaug, E.; Løvdal, T.; Mork, O.J. Experimental study of effectiveness of robotic cleaning for fish-processing plants. *Food Control* **2019**, *100*, 269–277. [CrossRef]
12. Lewis, T.; Griffith, C.; Gallo, M.; Weinbren, M. A modified ATP benchmark for evaluating the cleaning of some hospital environmental surfaces. *J. Hosp. Infect.* **2008**, *69*, 156–163. [CrossRef]
13. Al-Hamad, A.; Maxwell, S. How clean is clean? Proposed methods for hospital cleaning assessment. *J. Hosp. Infect.* **2008**, *70*, 328–334. [CrossRef]
14. Aziz, E.; Aouane, M.; Abdeljabbar, R.; Nabyl, B. Microbiological Study of Surfaces in the Hospital Environment Case of the Provincial Hospital of Sidi Kacem, Morocco. *Indian J. Forensic Med. Toxicol.* **2022**, *16*, 419–426.
15. Aycicek, H.; Oguz, U.; Karci, K. Comparison of results of ATP bioluminescence and traditional hygiene swabbing methods for the determination of surface cleanliness at a hospital kitchen. *Int. J. Hyg. Environ. Health* **2006**, *209*, 203–206. [CrossRef] [PubMed]
16. Pathmakumar, T.; Kalimuthu, M.; Elara, M.R.; Ramalingam, B. An autonomous robot-aided auditing scheme for floor cleaning. *Sensors* **2021**, *21*, 4332. [CrossRef] [PubMed]
17. Pathmakumar, T.; Elara, M.R.; Gómez, B.F.; Ramalingam, B. A Reinforcement Learning Based Dirt-Exploration for Cleaning-Auditing Robot. *Sensors* **2021**, *21*, 8331. [CrossRef]
18. Grünauer, A.; Halmetschlager-Funek, G.; Prankl, J.; Vincze, M. The power of GMMs: Unsupervised dirt spot detection for industrial floor cleaning robots. In Proceedings of the Annual Conference Towards Autonomous Robotic Systems, Lincoln, UK, 8–10 September 2017; pp. 436–449.
19. Bormann, R.; Weisshardt, F.; Arbeiter, G.; Fischer, J. Autonomous dirt detection for cleaning in office environments. In Proceedings of the 2013 IEEE International Conference on Robotics and Automation, Karlsruhe, Germany, 6–10 May 2013; pp. 1260–1267.
20. Bormann, R.; Wang, X.; Xu, J.; Schmidt, J. DirtNet: Visual Dirt Detection for Autonomous Cleaning Robots. In Proceedings of the 2020 IEEE International Conference on Robotics and Automation (ICRA), Online, 31 May–31 August 2020; pp. 1977–1983.
21. Ramalingam, B.; Lakshmanan, A.K.; Ilyas, M.; Le, A.V.; Elara, M.R. Cascaded machine-learning technique for debris classification in floor-cleaning robot application. *Appl. Sci.* **2018**, *8*, 2649. [CrossRef]
22. Damsteeg-van Berkel, S.; Beemster, F.; Dankelman, J.; Loeve, A.J. The influence of contact force on forensic trace collection efficiency when sampling textiles with adhesive tape. *Forensic Sci. Int.* **2019**, *298*, 278–283. [CrossRef]

23. Wiesner, S.; Tsach, T.; Belser, C.; Shor, Y. A comparative research of two lifting methods: Electrostatic lifter and gelatin lifter. *J. Forensic Sci.* **2011**, *56*, S58–S62. [CrossRef]
24. Dris, R.; Gasperi, J.; Mirande, C.; Mandin, C.; Guerrouache, M.; Langlois, V.; Tassin, B. A first overview of textile fibers, including microplastics, in indoor and outdoor environments. *Environ. Pollut.* **2017**, *221*, 453–458. [CrossRef]
25. Iskander, F. Cigarette ash as a possible source of environmental contamination. *Environ. Pollut. Ser. Chem. Phys.* **1986**, *11*, 291–301. [CrossRef]
26. Luczynska, C.; Sterne, J.; Bond, J.; Azima, H.; Burney, P. Indoor factors associated with concentrations of house dust mite allergen, Der p 1, in a random sample of houses in Norwich, UK. *Clin. Exp. Allergy J. Br. Soc. Allergy Clin. Immunol.* **1998**, *28*, 1201–1209. [CrossRef]
27. Simonyan, K.; Zisserman, A. Very deep convolutional networks for large-scale image recognition. *arXiv* **2014**, arXiv:1409.1556.
28. Gan, Y.; Yang, J.; Lai, W. Video object forgery detection algorithm based on VGG-11 convolutional neural network. In Proceedings of the 2019 International Conference on Intelligent Computing, Automation and Systems (ICICAS), Chongqing, China, 6–8 December 2019; pp. 575–580.
29. Sandler, M.; Howard, A.; Zhu, M.; Zhmoginov, A.; Chen, L.C. Mobilenetv2: Inverted residuals and linear bottlenecks. In Proceedings of the IEEE Conference on Computer Vision and Pattern Recognition, Salt Lake City, UT, USA, 18–23 June 2018; pp. 4510–4520.
30. He, K.; Zhang, X.; Ren, S.; Sun, J. Deep residual learning for image recognition. In Proceedings of the IEEE Conference on Computer Vision and Pattern Recognition, Las Vegas, NV, USA, 27–30 June 2016; pp. 770–778.
31. Targ, S.; Almeida, D.; Lyman, K. Resnet in resnet: Generalizing residual architectures. *arXiv* **2016**, arXiv:1603.08029.
32. Redmon, J.; Farhadi, A. Yolov3: An incremental improvement. *arXiv* **2018**, arXiv:1804.02767.
33. Zhang, Z.; Sabuncu, M. Generalized cross entropy loss for training deep neural networks with noisy labels. *Adv. Neural Inf. Process. Syst.* **2018**, *31*, 8792–8802.

sensors

MDPI

Article

AI-Enabled Predictive Maintenance Framework for Autonomous Mobile Cleaning Robots

Sathian Pookkuttath, Mohan Rajesh Elara, Vinu Sivanantham and Balakrishnan Ramalingam *

Engineering Product Development Pillar, Singapore University of Technology and Design (SUTD), Singapore 487372, Singapore; sathian_pookkuttath@mymail.sutd.edu.sg (S.P.); rajeshelara@sutd.edu.sg (M.R.E.); vinu_sivanantham@sutd.edu.sg (V.S.)
* Correspondence: balakrishnan@sutd.edu.sg

Abstract: Vibration is an indicator of performance degradation or operational safety issues of mobile cleaning robots. Therefore, predicting the source of vibration at an early stage will help to avoid functional losses and hazardous operational environments. This work presents an artificial intelligence (AI)-enabled predictive maintenance framework for mobile cleaning robots to identify performance degradation and operational safety issues through vibration signals. A four-layer 1D CNN framework was developed and trained with a vibration signals dataset generated from the in-house developed autonomous steam mopping robot 'Snail' with different health conditions and hazardous operational environments. The vibration signals were collected using an IMU sensor and categorized into five classes: normal operational vibration, hazardous terrain induced vibration, collision-induced vibration, loose assembly induced vibration, and structure imbalanced vibration signals. The performance of the trained predictive maintenance framework was evaluated with various real-time field trials with statistical measurement metrics. The experiment results indicate that our proposed predictive maintenance framework has accurately predicted the performance degradation and operational safety issues by analyzing the vibration signal patterns raised from the cleaning robot on different test scenarios. Finally, a predictive maintenance map was generated by fusing the vibration signal class on the cartographer SLAM algorithm-generated 2D environment map.

Keywords: artificial intelligence; mobile cleaning robot; vibration source classification; predictive maintenance; deep learning; 1D CNN

Citation: Pookkuttath, S.; Rajesh Elara, M.; Sivanantham, V.; Ramalingam, B. AI-Enabled Predictive Maintenance Framework for Autonomous Mobile Cleaning Robots. *Sensors* **2022**, *22*, 13. https://doi.org/10.3390/s22010013

Academic Editors: Gregor Klancar, Marija Seder and Sašo Blažič

Received: 12 November 2021
Accepted: 19 December 2021
Published: 21 December 2021

1. Introduction

Mobile cleaning robots with various capacities are ubiquitous today, for instance in food courts, hypermarkets, hospitals, industries, airports, and homes, and are used for vacuuming, mopping, and sanitizing the environment. Market studies show that the personal and professional mobile cleaning robot growth is expected to reach 24 billion USD by 2026 [1]. However, proper maintenance and deployment in a robot-friendly workspace are crucial for autonomous mobile cleaning robots to avoid malfunction, catastrophic failure, and environmental-related safety issues, including customer dissatisfaction. Currently, manual supervision is widely used to monitor professional cleaning robots' performance degradation and safety-related issues. However, it is time-consuming, labor and skill-set-dependent, and challenging to deploy due to the lack of historical failure data, especially for the newly developed advanced cleaning robots. Moreover, this periodical manual approach may trigger other issues such as extended downtime, the under-utilization of components, safety issues due to abrupt failure, and high operational and maintenance costs.

Automated predictive maintenance strategies overcome these pitfalls. They are widely used in industrial robots and autonomous vehicles for continuous health monitoring, performance degradation prediction, hazardous operational environment identification, and safety system failure indication. Various methods and techniques were proposed in the literature to implement automated predictive maintenance. A fuzzy inference approach is used in [2] to predict battery power status of robotics systems, and a non-intrusive methodology using torque sensor data for monitoring industrial robot joints in [3]. Similarly, a programmable motion-fault detection method for collaborative robots in [4], a framework to assess the future dynamic behavior and the remaining useful life of industrial robots in [5], and a data-driven predictive maintenance methodology using time-series electrical power data is used to detect manipulator errors in [6].

In recent years, artificial intelligence (AI) powered predictive maintenance (PdM) has been widely studied for automated PdM design. It adopts Machine Learning (ML) and Deep Learning (DL) algorithms for fault detection and classification. These works include a K-means clustering algorithm-based PdM for wafer transfer robot to avoid unplanned downtime proposed in [7], an automatic ML tool based health monitoring system to predict safe stops in a collaborative robot in [8], an Artificial Neural Network (ANN) model to predict the system failure of a packaging robot in [9], an ML-based PdM to detect drive belt looseness in a Cartesian robot in [10], and a DL-based fault diagnosis of industrial robots in [11] using multi-sensor fusion technology. Similarly, an ML-based fault diagnosis for the vehicle brake system is studied in [12] using wavelet applications, a terrain classification study for the autonomous ground vehicles is conducted in [13] adopting a probabilistic neural network, a hierarchical component-based diagnosis and prognosis system proposed for autonomous vehicles in [14] using a Dynamic Bayesian Network (DBN) model, and a DL-model developed to forecast the health of multi-sensor autonomous vehicles by training health index networks in [15].

Though several works are available for industrial robots and autonomous vehicle applications, predictive maintenance of autonomous mobile cleaning robots is not widely studied yet. The PdM system is a mandate function to autonomous mobile cleaning robots to deliver a safe and efficient service when operating in a complex and dynamic change environment, identify any performance degradation, and avoid operational safety issues. Generally, vibration is a key indicator for industrial robots and autonomous vehicles to predict performance degradation and hazardous operating environment identification, which is applicable for autonomous mobile cleaning robot platforms.

This work proposes an AI-powered predictive maintenance framework using vibration as a source for autonomous mobile cleaning robots. A one-dimensional Convolutional Neural Network- (1D CNN) based vibration source classification model was developed and trained to classify the vibration signals arising from mobile cleaning robots. An in-house developed autonomous steam mopping robot–'Snail' is used to test and validate the proposed framework with different health conditions and hazardous operational environments.

The rest of this paper is organized as follows: Section 2 explains the literature survey on various vibration signal-based PdM works. Section 3 states the problem definition and motivation of this work. Section 4 gives an overview of the proposed system. Section 5 presents various experiments conducted and the results. Finally, Section 6 concludes the summary of the works.

2. Literature Survey

Currently, advanced industries use vibration-based health monitoring systems to detect early signs of the failure of machines and industrial robots. The vibration signals, measured using various vibration measuring sensors like piezoelectric or micro-electro-mechanical systems (MEMS) accelerometers, contain the health information of the equipment. In the literature, many reviews and research works have been conducted on vibration signal-based PdM using DL techniques, primarily for machine components, and industrial robots. The 1D CNN based vibration signal analysis studies for PdM includs a survey on 1D CNN benefits and applications in [16], fault diagnosis of machine components in [17–19], structural health monitoring in [20–22], and real-time fault diagnosis for power assets in [23]. Toh and Park reported the impact of vibration responses for early structural health monitoring in [24]. The study uses different DL architectures for vibration analysis. Pham et al. applied CNN for fault diagnosis of bearings operated under different shaft speeds in [25]. They represented vibration signals as spectrograms to classify faults with high accuracy. Kolar et al. in [26] proposed a vibration signal-based fault diagnosis framework using the CNN algorithm. The authors used a three-axis accelerometer-generated vibration signal to detect and classify the faults. In [27], Chen et al. evaluated three deep neural networks models, including Deep Boltzmann Machines (DBM), Deep Belief Networks (DBN), and Stacked Auto-Encoders (SAE), to detect rolling bearing faults using vibration signals. Chen and Lee discussed in [28] a DL approach for vibration signal analysis for machining applications. Their study covers the optimisation method for CNN, 1D CNN, and 2D CNN structures with different types of inputs such as raw signal data and time-frequency spectra images. A DL model was developed by Luo et al. to detect early faults of a CNC machine in [29]. The authors used impulse responses from the vibration signals to detect the early mechanical fault under time-varying conditions. A fault diagnosis for the industrial robot was proposed by Wu et al. in [30]. Here, the authors combined three algorithms, including Manifold learning, Treelet Transform, and Naive Bayes, to detect the fault in industrial robots.

The studies mentioned above show that vibration signals are the decisive analyzing element, and DL-based techniques are suitable for the feature extraction from vibration signals for predicting a system's performance degradation. However, most of the works available are focused on finding a specific fault, severity, or its remaining useful life, and do not consider the external factors for degradation. Moreover, the works are intended mainly for various machine components and industrial robots only. Hence, there is a research gap for monitoring autonomous cleaning robots' health status and identifying hazardous environmental factors.

3. Problem Definition

A properly designed and developed cleaning robot works as required in its planned work environment without abnormal vibration. However, due to continuous operation or the impact of various internal and external environmental factors, the robot performance degrades and generates abnormal vibration signals. In most indoor cleaning robots, external terrain factors such as rough pebble pathways or tactile pavers produce high amplitude vibrations that cause performance degradation issues, such as assembly looseness, sensor misalignment, and faster component deterioration, for example. Vibration due to collisions with walls and other obstacles is an indicator of hazardous operation. It may arise due to the failure or malfunction of obstacle avoidance sensors or the misalignment of safety sensors, or the absence of hazardous object registration in the robot navigation map, such as tiny objects and glass walls (LiDAR sensor is sometimes not accurate in detecting tiny objects below object detection range or glass walls in [31–33]). A structural imbalance vibration signal is another indicator for both performance degradation and a hazardous operational environment. It may arise due to wheel damage, wear, a loose assembly of heavier components (battery, water tank), or a robot operating in a poor ground clearance area. Hence, identifying the vibration signal source is mandated to predict the cause of failure or performance degradation and identify the operational safety issues. It will enhance the predictive maintenance actions, for instance, isolating the potential hazardous region quickly, determining the severity and fixing the issues, redirecting to its intended workspace, securing a robot-friendly environment, and includes design improvement planning.

4. Overview of the Proposed System

Figure 1 shows an overview of the AI-enabled PdM framework for an autonomous cleaning robots platform. It uses vibration signals to predict cleaning robot performance degradation and any hazardous operational environment. The overview of the framework is explained as follows. This involves the robot platform details used for tests, the vibration data acquisition unit, the four-layer 1D CNN algorithm, the vibration source mapping module, and the remote monitoring unit.

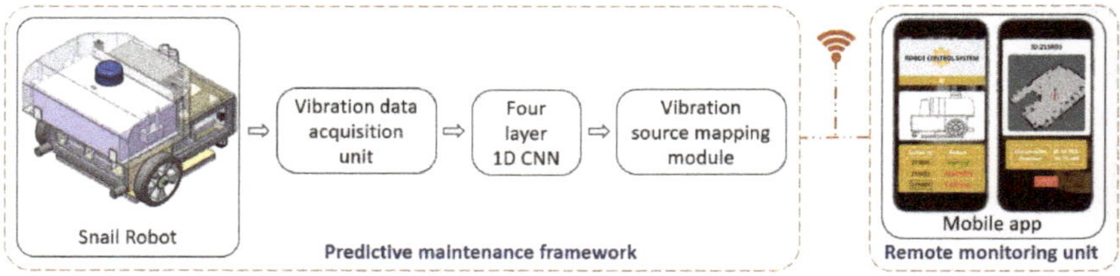

Figure 1. Overview of the proposed DL-based PdM framework.

4.1. Autonomous Steam Mopping Robot 'Snail'

An in-house autonomous steam mopping robot (Snail) designed for cleaning and disinfection of indoor floors is used in this case study, as shown in Figure 2. The overall size of the robot is 40 × 40 × 38 cm, and the total weight is 18 Kg, including inverter, battery, steam boiler, and mop head assembly. An NVIDIA Jetson AGX single-board computer is used to steer the entire operation of the robot, including autonomous navigation, executing predictive maintenance framework, and controlling all other sensors. In addition, the RPLIDAR A2 scanner is used for localization and mapping of the environment, and the Inertial Measurement Unit (IMU) Vectornav VN-100 sensor is used to estimate the motion, orientation, and heading angles of the robot. The robot locomotion was accomplished by a differential wheel drive mechanism with two supporting caster wheels. A D-Link 4G/LTE mobile router is used to control the robot remotely and monitor the robot's health.

Figure 2. Autonomous steam mopping robot 'Snail'.

4.2. Vibration Data Acquisition Unit

In our study, five vibration classes are used as critical indicators for mobile cleaning robots' performance degradation and operational safety issues. It is classified under three categories: normal, external factors, and internal factors, as shown in Figure 3. Here, terrain and collision-induced vibration belong to external factors, while assembly and structure-induced vibrations are due to internal factors.

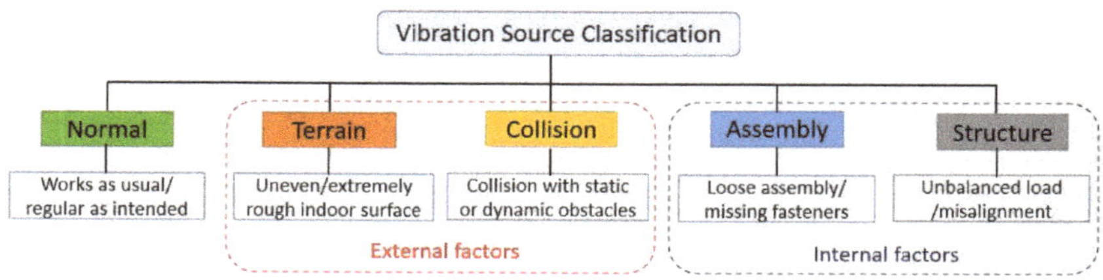

Figure 3. Vibration source classification—Normal and Potential source of failure.

The linear and angular motion of the robot will be affected due to internal or external causes of vibration. Hence, we measured the linear acceleration (X, Y, and Z-axis) and angular velocity (roll, pitch, and yaw) of the robot using the onboard IMU sensor (Vectornav VN-100), which reflects the vibration level of the robot. Our IMU data subscription includes angular acceleration calculated from each instance's current and previous angular velocity. Hence, the robot measures three signal data (linear acceleration, angular velocity, and angular acceleration) in three axes. This total of nine-sensor data collected during the exposure of five vibration source classes was used as the vibration signal data (feature values). The IMU subscription rate was set at 40 Hz, and each sample is grouped into 128 (time steps) data elements, which is every 3.2 s. The collected data is converted into a three-dimensional input array comprising samples, time steps, and features values. Figure 4 shows the data acquisition unit details, mainly the sensor position on the chassis at the center of the differential wheel drive axis.

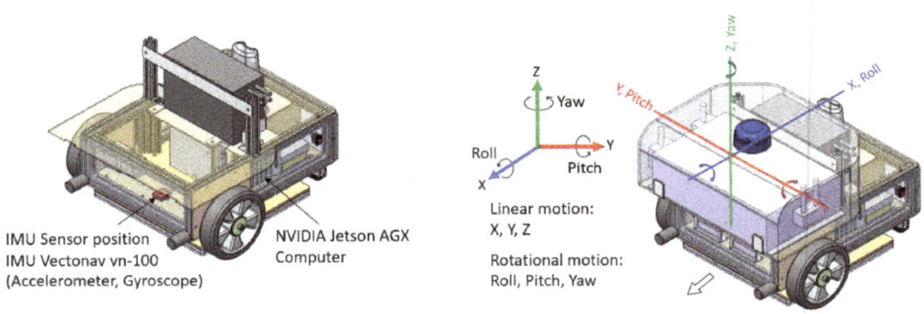

Figure 4. Data acquisition system and Linear-rotational motion of the Snail robot.

4.3. 1D Convolutional Neural Network (1D CNN)

A 1D CNN model is adopted to build the proposed vibration source classification framework. The 1D CNN is formulated by convolution operations on data vectors as explained in [23]. It consists of signal input data vector x (length N), a filter vector ω (length L), a bias term b to best fit for given data, and a nonlinear activation function. The input vector is convolved with the filter vector, and its output layer is represented as in Equation (1). This representation of output layer $\mathbf{c}$, of length $(N-L+1)$ is without zero padding. To reduce the number of parameters and highlight the key feature, a max pooling output vector $\mathbf{d}$ is defined, after each convolution layer, with a kernel size $m \times 1$, window function u, and filter moving stride s as in Equation (2).

$$\mathbf{c}(j) = f\left(\sum_{i=0}^{L-1} \omega(i)x(j-i) + b\right), j = 0, 1, \ldots, N-1 \quad (1)$$

$$\mathbf{d} = max(u(m \times 1, s)c) \quad (2)$$

Figure 5 shows the structure of the predictive maintenance 1D CNN framework. It involves mainly an input layer, four convolutional layers, a Fully Connected (FC) layer, and an output layer. The raw sensor data of the 3D array [n × 128 × 9] gets normalised first. Then each sample is flattened to a 1D array [1 × 1152] for feeding into the CNN. The first two CNN layers use 64 filters and follow kernel size (convolution window) 3. The following two CNN layers use 32 filters and apply the same kernel size. A Rectified Linear Activation Unit (ReLU) is applied to each convolutional layer to learn complex nonlinear patterns in the signal data. After each convolutional layer, max pooling with a stride size of 2 and a dropout layer (dropout rate 0.2) is applied to reduce the feature map dimension and avoid over-fitting. Finally, a flattening function is used to convert the pooled feature map into a 1D array and pass it into the FC layer. A softmax layer is added as the final activation function in the output layer that predicts the multinomial probability. The output layer contains five neurons for the five vibration classes. Table 1 shows the details of the 1D CNN architecture.

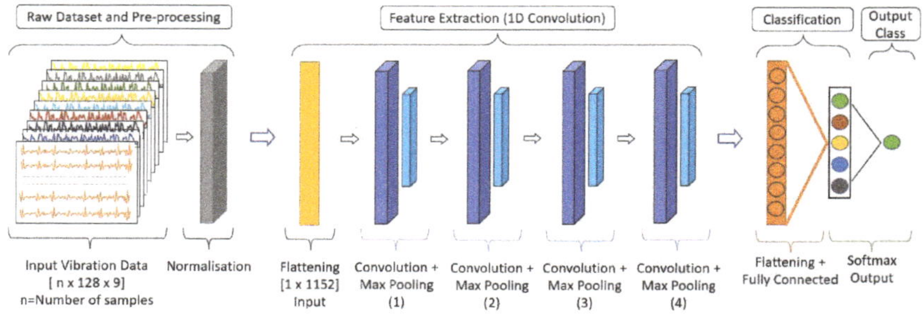

Figure 5. 1D CNN Structure.

Table 1. 1D CNN Architecture.

Layer	Kernel Size	Stride	Filters	Data Shape
Input				(1152, 1)
Conv. 1D-1	3 × 1	1	64	(1150, 64)
Max Pool 1D-1	2 × 1	2 × 1		(575, 64)
Conv. 1D-2	3 × 1	1	64	(573, 64)
Max Pool 1D-2	2 × 1	2 × 1		(287, 64)
Conv. 1D-3	3 × 1	1	32	(285, 32)
Max Pool 1D-3	2 × 1	2 × 1		(143, 32)
Conv. 1D-4	3 × 1	1	32	(141, 32)
Max Pool 1D-4	2 × 1	2 × 1		(71, 32)
Fully Connected				(100)
Output (Softmax)				(5)

4.4. Vibration Source Mapping Module

The mapping module performs 2D mapping of the environment using the onboard RPLidar and fuses the CNN predicted vibration source class on the 2D map generated. The Cartographer SLAM (Simultaneous Localisation and Mapping) algorithm is used to generate the 2D environment. It builds a grid-based map for the given environment. The prediction algorithm-generated vibration source classes are fused continuously into this grid map to generate a predictive maintenance map (PdM map). The user or maintenance team can visualize the type of performance degradation and safety-related issues on deployment space through the PdM map.

4.5. Remote Monitoring Unit

A smartphone app is developed to visualize the Snail robot's real-time prediction results for remote health monitoring and control the robot in teleoperation mode, as shown in the overview layout Figure 1. The app is connected through the robot using the MQTT messaging protocol and collects the predicted information in request-based or continuous mode.

5. Experiments and Results

This section describes the experimental methods and results. The experiments were performed in four phases: dataset preparation, training the predictive maintenance CNN framework, validating the trained model with test dataset, and real-time field trials.

5.1. Data-Set Preparation and Pre-Processing

The dataset preparation involves collecting the vibration signals from the robot with different health states deployed on varying surface conditions, operational speeds, and cleaning patterns. Figure 6 shows the robot test set up for collecting the five classes of vibration including normal operational vibration, rough terrain induced vibration, collision-induced vibration, loose assembly induced vibration, and structure imbalance vibration.

Here, the normal operational and rough surface-induced vibrations were collected by deploying the robot over smooth indoor floors and rough terrain, respectively. Collision-induced vibrations were collected by hitting the robot on different obstacles, including walls and other static and dynamic (human) objects in the environment. The loose assembly-induced vibrations were collected by loosening the robot's components, such as wheel coupling and mounting brackets. Finally, the unbalanced load-induced (structure) vibration data has been collected by using damaged/worn out wheels, asymmetrically placing heavier components, and operating the robot in a poor ground clearance area. The above mentioned five vibration classes were collected under different surface conditions (tile, concrete, carpet, wooden, vinyl, small and medium-size pebble pathways, and tactile pavers), operational speeds (linear 0.02–0.4 m/s and angular 0.3–1.3 rad/s), and cleaning patterns (straight, zig-zag, and spiral).

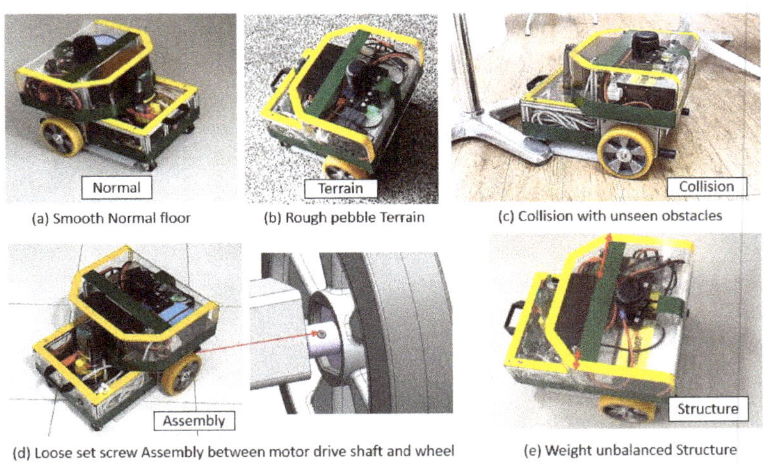

Figure 6. Robot test set up for vibration data collection of five classes.

The Figures 7–11 shows the time-amplitude graph for the vibration signals raw data collected for all the five classes across each signal type (linear acceleration, angular velocity, angular acceleration) and its three-axis. The graphs provide a visual representation of how the signals vary through different vibration source classes. It is plotted for one sample (128 data), i.e., captured in 3.2 s.

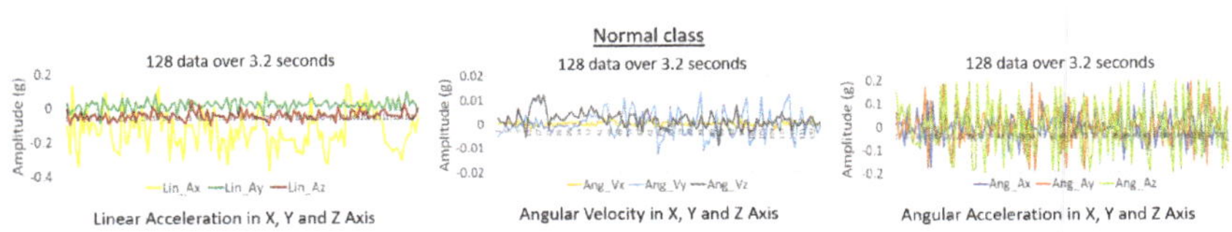

Figure 7. Vibration signals—Normal class.

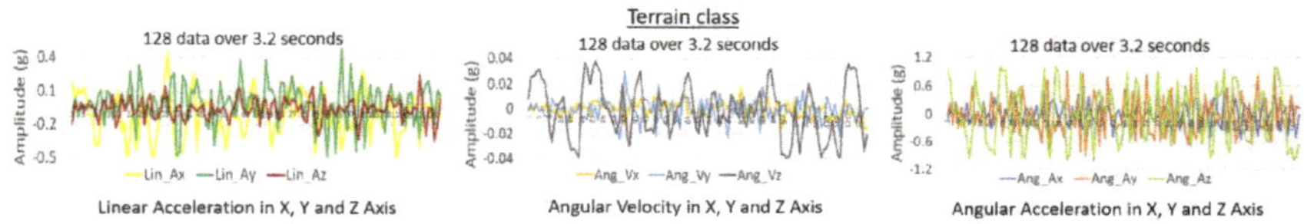

Figure 8. Vibration signals—Terrain class.

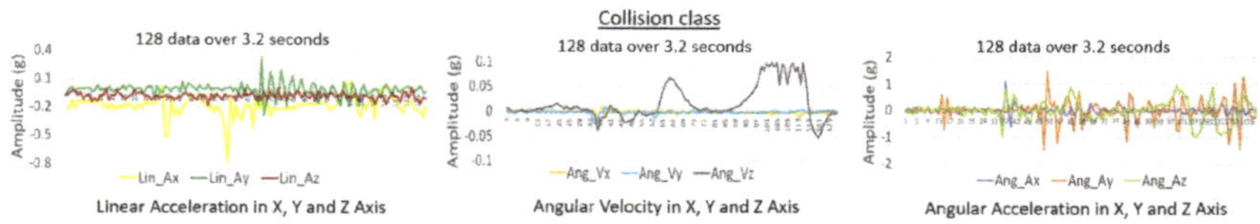

Figure 9. Vibration signals—Collision class.

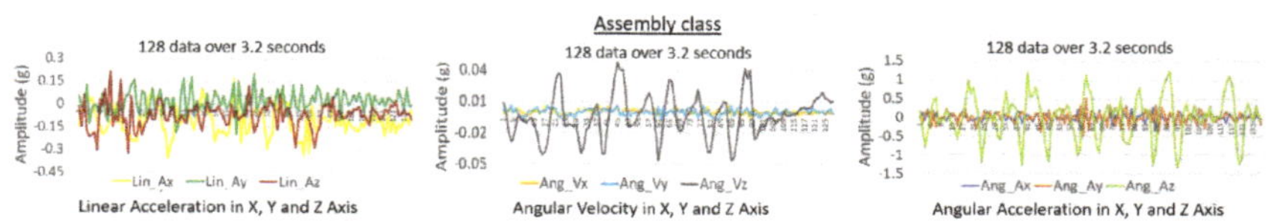

Figure 10. Vibration signals—Assembly class

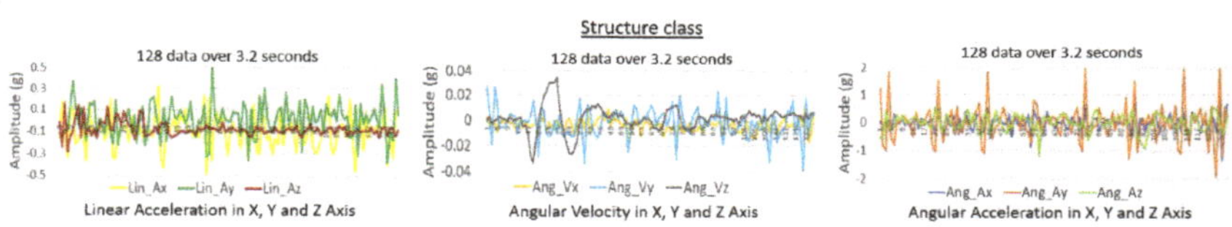

Figure 11. Vibration signals—Structure class.

Data normalisation is applied in the pre-processing stage. The normalisation process involves bringing the raw data into a standard scale without losing information. In our case, the collected data x of all the nine features of each class is normalised into -1 to $+1$ using the Equation (3). Then, the pre-processed dataset is split into training, validation, and test data sets. The training and validation datasets are used to train the model, and the test dataset is used to evaluate the model after training. A total of 2500 samples for each class were recorded and split into 80% for training and 20% for validation. Furthermore, for evaluating the model, a total of 500 samples were collected for each class as test data sets.

$$x_{Normalised} = 2\frac{x - min(x)}{max(x) - min(x)} - 1 \tag{3}$$

5.2. Training and Validation

A supervised learning strategy was used in this PdM framework using our unique dataset. Tensorflow [34] deep learning library was used to develop the predictive maintenance CNN framework, and Nvidia GeForce GTX 1080 Ti-powered workstation was used to train the model with the collected dataset. Table 2 shows the hyperparameter settings for training the model. Momentum with gradient descent was used as the optimising strategy to speed up learning and not get stuck with local minima. Adam optimiser (adaptive moment optimization) [35] with three parameters showed better training results: a learning rate of 0.001, the exponential decay rate for the first moment of 0.9, and for the second moment of 0.999. Different epochs were used for testing, and better accuracy was found with 100 epochs and a batch size of 32. The categorical cross-entropy loss function is used while compiling the model to reduce the loss during training and to improve prediction probability.

Table 2. HyperParameters setting.

Parameter	Values/Function
Optimizer	Adam
Learning rate	0.001
Batch Size	32
Epochs	100

In the training phase, a K-fold (in our study, K = 5) cross-validation technique is used to evaluate the dataset's quality, improve generalization, avoid over-learning, and choose the best model for this application. In k-fold cross-validation, the datasets are split into k subsets and k-1 subsets to train the model. The remaining one is for evaluating the model's performance.

5.3. Prediction with Test Dataset

The vibration class prediction efficiency of the trained model is evaluated with the test dataset. A total of 500 test samples were used for each class to evaluate the model. These test datasets have not been used in the training and cross-validation of the model. Accuracy, precision, recall, and F1 Score (Equations (4)–(7)) statistical measure metrics [36] were used to assess the model performance. Here, TP, FP, TN, FN represents the true positives, false positives, true negatives, and false negatives, respectively, as per the standard confusion matrix.

$$Accuracy = \frac{TP + TN}{TP + FP + TN + FN} \tag{4}$$

$$Precision = \frac{TP}{TP + FP} \tag{5}$$

$$Recall = \frac{TP}{TP + FN} \tag{6}$$

$$F1Score = \frac{2 \times Precision \times Recall}{Precision + Recall} \tag{7}$$

Table 3 shows the statistical measure result for the test dataset. Accordingly, the model classifies the normal operational vibration class with 89% accuracy, and the hazardous terrain-induced vibration and collision-induced vibration accuracy were predicted with 95% and 92% accuracy, respectively. Furthermore, loose assembly-induced vibration and structure imbalance vibration classes were predicted with 94% and 91% accuracy. The model's overall classification accuracy (average of five classes) is 92.2%. The above analysis shows that the trained model has accurately classified the five vibration classes collected from the mobile cleaning robot run on different surfaces. Hence, this proposed model is suitable for real-time deployment in mobile cleaning robots for performance degradation and hazardous operational region prediction.

Table 3. Offline test result.

Vibration Source	Precision	Recall	F1 Score	Accuracy
Normal	0.86	0.90	0.89	0.89
Terrain	0.97	0.95	0.96	0.95
Collision	0.92	0.92	0.92	0.92
Assembly	0.93	0.94	0.94	0.94
Structure	0.86	0.92	0.93	0.91

5.4. Comparison with Other Algorithms

The performance of the proposed predictive maintenance 1D CNN framework was compared with other commonly used ML/DL classifier models, such as Support Vector Machine (SVM) [37], Multilayer Perceptron (MLP) [38], Long Short-Term Memory (LSTM) [39], and CNN-LSTM [40]. The same training and test dataset, processing resources, and conditions used for the 1D CNN model have been applied for a fair comparison. The CNN-LSTM, LSTM, and MLP models were trained using the TensorFlow library and SVM with Scikit-learn [41] package. The key parameter settings such as optimiser (Adam), learning rate (0.001), and loss function (categorical cross-entropy) for the CNN-LSTM, LSTM, and MLP comparison models were used the same as 1D CNN. For the SVM comparison model, the key parameters 'C' and 'gamma' values used 100 and 0.01, respectively, and the Radial Basis Function (RBF) kernel was applied. The overall accuracy of each model over five classes and the inference time (millisecond) to process one sample data are given in Table 4.

Table 4. Accuracy comparison with other models.

Model	Accuracy (%)	Inference Time (ms)
1D CNN	92.2	0.162
CNN-LSTM	88.1	0.258
LSTM	85.4	0.276
MLP	79.8	0.193
SVM	77.5	1.675

The comparison results show that our proposed predictive maintenance 1D CNN framework has scored better classification accuracy and took less inference time than the other four algorithms. Hence, it is evident that our proposed system is an optimal algorithm to predict the performance degradation and hazardous operational environment.

5.5. Real-Time Prediction

The real-time prediction experiment was tested with the Snail robot in four different environments at the SUTD campus include the lobby, food court, corridor, and lab workspace. These environments were not used for the training data-set collection process. Before the real-time prediction experiment, all the environments were mapped by the cartographer SLAM algorithm using the on-board RPLIDAR A2 for autonomous navigation and predictive maintenance map generation. In our experiment, the continuous field trial was conducted on multiple days to observe the robot performance degradation and hazardous operational prediction. The trained model was configured in robot on-board computer NVIDIA Jetson AGX. Its prediction results (terrain, collision, assembly, or structure-induced vibrations signals) were fused into the cartographer SLAM generated environment map with different colors to identify performance degradation and hazardous operational prediction.

The first case study was trialed in a lobby environment consisting of a glass sidewall and carpet floor. Here, collision-induced vibrations were registered in the cartographer SLAM algorithm-generated map due to incorrect mapping of the glass wall. The glass wall was covered with a raising curtain in the original map during the mapping time, whereas it was removed while testing. As a result, the robot lost the previously mapped navigation awareness, and RPLidar could not locate the glass as an obstacle. The robot hit on the glass randomly, and the repeated collision-induced vibration marks were captured in the lobby environment 2D map as shown in Figure 12a.

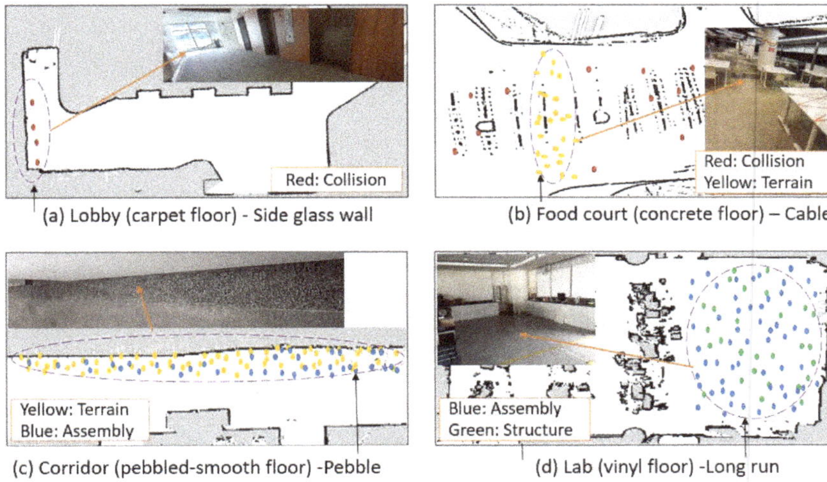

Figure 12. Real time field test case studies.

The SUTD food court was our second case study testbed. The food court environment has smooth concrete flooring with dining furniture. The robot trial was conducted during different operational times, including peak hours. Figure 12b shows the SUTD food court set up and its cartographer SLAM map registered with predicted vibration signals. Here, the collision-induced vibration and hazardous surface-induced (terrain) vibrations were recorded in the environment map. Specifically, collision-induced vibration was registered primarily during peak business hours, which arises due to accidental collision with humans, undetected furniture legs, and changes in the position of the dining furniture. Similarly, tactile pavers and cables set on the floor caused hazardous terrain-induced vibrations.

Case study three was conducted in a corridor environment with mixed style flooring with a smooth concrete floor, and a pebble paved rough surface. During our experiment, the robot was first deployed to clean the smooth concrete floor. Here, no abnormal vibrations were reported in the 2D environment map. Later, when the robot started cleaning rough pebbled surfaces, hazardous terrain-induced vibrations marks were seen on the map. When the robot was deployed for a ten day trial on the pebbled surface, a mix of terrain and loose assembly-induced vibrations were registered, as shown in Figure 12c. This is mainly due to the loose assembly of mechanical systems such as loosening of mounting brackets screws and deterioration of mop cloth.

The fourth case study was performed in our SUTD ROAR lab workspace, where the environment filled dynamic objects and the terrain consisted of vinyl flooring. The experiment lasted for four weeks, and its cumulative results in the lab environment were observed. In the first three weeks, the map showed regular drive without capturing any abnormal vibration signature. However, the fourth-week map showed a loose assembly pattern observed from the wobbled wheels due to the loosening of the wheel coupling set screws. The test continued in this wheel-wobbled state and noticeable imbalanced structural vibrations arose due to the battery becoming detached from its bracket and creating an unbalanced weight distribution. Figure 12d shows the four-week test results for case study four, where the environment map depicts loose assembly and unbalanced weight structural patterns.

As per the experimental results, we observed that one source of failure might lead to another if no action is taken. This way, if several sources of failure are present, the model will predict the predominant failure of the robot in that particular instance. Hence, through the mobile app, the maintenance person can remotely stop the robot as soon as the initial abnormal class is registered, avoiding chances for multiple failures or hazards to the environment.

Table 5 shows the statistical measure result for 100 test samples collected from the real-time case study. Here, the sample data were collected from the Snail robot through the mobile app on request-based sample mode and performed the statistical measure using the confusion matrix function. The table results indicate that the algorithm classified the five different vibration sources with an average accuracy of 88.9%, 93.5%, 91.8%, 92.1%, and 88.7% for normal, terrain, collision, assembly, and structure, respectively. In contrast with offline results, the model average prediction accuracy of five classes in real-time tests is slightly less (91%) due to external noise and various sampling periods. However, it can be reduced by fussing multiple IMU sensors or adding a noise cancellation function in the preprocessing stage.

Table 5. Real-time prediction accuracy of five classes.

Vibration Source	Normal	Terrain	Collision	Assembly	Structure
Prediction (%)	88.9	93.5	91.8	92.1	88.7

6. Conclusions

An AI-enabled predictive maintenance framework was proposed for mobile cleaning robots to monitor performance degradation and identify operational safety issues. The proposed framework was tested and validated with our in-house developed autonomous steam mopping robot 'Snail'. A four-layer 1D CNN model was developed using TensorFlow API and trained with five vibration signal datasets generated from the Snail robot with different health conditions. The efficiency of the proposed predictive maintenance framework was evaluated with offline and real-time field tests. The experimental results show that the model scored 92.2% accuracy for classifying the performance degraded and hazardous operational vibration signals in offline tests and took 0.162 ms to process one test sample. In the real-time field test, the algorithm accurately predicted robot performance degradation and operational safety issues with an accuracy of 91%. The predicated vibration signal class was fused into the cartographer SLAM-generated environment map to track the performance degradation and identify the operational safety issues. This will help manufacturers and cleaning maintenance companies to choose the right maintenance strategy, rental policy, or improve the robot design and assembly.

Author Contributions: Conceptualization, S.P.; Methodology, S.P. and B.R.; Software, V.S. and S.P.; Validation, S.P.; Formal analysis, B.R. and S.P.; Investigation, B.R. and S.P.; Resources, M.R.E.; Data curation, S.P. and V.S.; Writing—original draft, S.P. and B.R.; Writing—review & editing, B.R. and S.P.; Visualization, S.P.; Supervision, M.R.E.; Project administration, M.R.E.; Funding acquisition, M.R.E. All authors have read and agreed to the published version of the manuscript.

Funding: This research is supported by the National Robotics Programme under its Robotics Enabling Capabilities and Technologies (Funding Agency Project No. 192 25 00051), National Robotics Programme under its Robotics domain Specific (Funding Agency Project No. 192 22 00058, 192 22 00108) and administered by the Agency for Science, Technology and Research.

Acknowledgments: The authors would like to thank, the National Robotics Programme, the Agency for Science, Technology and Research and SUTD for their support.

Conflicts of Interest: There are no conflict of interest.

References

1. Research and Markets. Worldwide Cleaning Robot Industry to 2026-Key Market Drivers and Restraints. Available online: https://www.prnewswire.com/news-releases/worldwide-cleaning-robot-industry-to-2026---key-market-drivers-and-restraints-301293632.html (accessed on 10 June 2021).
2. Huang, H.P.; Wu, S.H. Diagnostic and predictive maintenance systems for abnormal behavior of power scheduling loading and its application to robotics systems. In Proceedings of the 2011 9th World Congress on Intelligent Control and Automation, Taipei, Taiwan, 21–25 June 2011; pp. 908–913. [CrossRef]
3. Izagirre, U.; Andonegui, I.; Egea, A.; Zurutuza, U. A methodology and experimental implementation for industrial robot health assessment via torque signature analysis. *Appl. Sci.* **2020**, *10*, 7883. [CrossRef]
4. Park, Y.S.; Yoo, D.Y.; Lee, J.W. Programmable Motion-Fault Detection for a Collaborative Robot. *IEEE Access* **2021**, *9*, 133123–133142. [CrossRef]
5. Aivaliotis, P.; Arkouli, Z.; Georgoulias, K.; Makris, S. Degradation curves integration in physics-based models: Towards the predictive maintenance of industrial robots. *Robot. Comput.-Integr. Manuf.* **2021**, *71*, 102177. [CrossRef]
6. Borgi, T.; Hidri, A.; Neef, B.; Naceur, M.S. Data analytics for predictive maintenance of industrial robots. In Proceedings of the 2017 International Conference on Advanced Systems and Electric Technologies (IC_ASET), Hammamet, Tunisia, 14–17 January 2017; pp. 412–417.
7. Kim, H.G.; Yoon, H.S.; Yoo, J.H.; Yoon, H.I.; Han, S.S. Development of Predictive Maintenance Technology for Wafer Transfer Robot using Clustering Algorithm. In Proceedings of the 2019 International Conference on Electronics, Information, and Communication (ICEIC), Auckland, New Zealand, 22–25 January 2019; pp. 1–4.
8. Aliev, K.; Antonelli, D. Proposal of a Monitoring System for Collaborative Robots to Predict Outages and to Assess Reliability Factors Exploiting Machine Learning. *Appl. Sci.* **2021**, *11*, 1621. [CrossRef]
9. Onur, K.; Kaymakci, O.T.; Mercimek, M. Advanced Predictive Maintenance with Machine Learning Failure Estimation in Industrial Packaging Robots. In Proceedings of the 2020 International Conference on Development and Application Systems (DAS), Suceava, Romania, 21–23 May 2020; pp. 1–6.
10. Pierleoni, P.; Belli, A.; Palma, L.; Sabbatini, L. Diagnosis and Prognosis of a Cartesian Robot's Drive Belt Looseness. In Proceedings of the 2020 IEEE International Conference on Internet of Things and Intelligence System (IoTaIS), BALI, Indonesia, 27–28 January 2021; pp. 172–176.
11. Wang, J.; Wang, D.; Wang, X. Fault diagnosis of industrial robots based on multi-sensor information fusion and 1D convolutional neural network. In Proceedings of the 2020 39th Chinese Control Conference (CCC), Shenyang, China, 27–29 July 2020; pp. 3087–3091.
12. Alamelu Manghai, T.; Jegadeeshwaran, R. Vibration based brake health monitoring using wavelet features: A machine learning approach. *J. Vib. Control* **2019**, *25*, 2534–2550. [CrossRef]
13. Dupont, E.M.; Moore, C.A.; Collins, E.G.; Coyle, E. Frequency response method for terrain classification in autonomous ground vehicles. *Auton. Robot.* **2008**, *24*, 337–347. [CrossRef]
14. Gomes, I.P.; Wolf, D.F. Health Monitoring System for Autonomous Vehicles using Dynamic Bayesian Networks for Diagnosis and Prognosis. *J. Intell. Robot. Syst.* **2021**, *101*, 1–21. [CrossRef]
15. Safavi, S.; Safavi, M.A.; Hamid, H.; Fallah, S. Multi-Sensor Fault Detection, Identification, Isolation and Health Forecasting for Autonomous Vehicles. *Sensors* **2021**, *21*, 2547. [CrossRef]
16. Kiranyaz, S.; Avci, O.; Abdeljaber, O.; Ince, T.; Gabbouj, M.; Inman, D.J. 1D convolutional neural networks and applications: A survey. *Mech. Syst. Signal Process.* **2021**, *151*, 107398. [CrossRef]
17. Eren, L.; Ince, T.; Kiranyaz, S. A generic intelligent bearing fault diagnosis system using compact adaptive 1D CNN classifier. *J. Signal Process. Syst.* **2019**, *91*, 179–189. [CrossRef]
18. Abdeljaber, O.; Sassi, S.; Avci, O.; Kiranyaz, S.; Ibrahim, A.A.; Gabbouj, M. Fault detection and severity identification of ball bearings by online condition monitoring. *IEEE Trans. Ind. Electron.* **2018**, *66*, 8136–8147. [CrossRef]
19. Ince, T.; Kiranyaz, S.; Eren, L.; Askar, M.; Gabbouj, M. Real-time motor fault detection by 1-D convolutional neural networks. *IEEE Trans. Ind. Electron.* **2016**, *63*, 7067–7075. [CrossRef]
20. Abdeljaber, O.; Avci, O.; Kiranyaz, S.; Gabbouj, M.; Inman, D.J. Real-time vibration-based structural damage detection using one-dimensional convolutional neural networks. *J. Sound Vib.* **2017**, *388*, 154–170. [CrossRef]
21. Abdeljaber, O.; Avci, O.; Kiranyaz, M.S.; Boashash, B.; Sodano, H.; Inman, D.J. 1-D CNNs for structural damage detection: Verification on a structural health monitoring benchmark data. *Neurocomputing* **2018**, *275*, 1308–1317. [CrossRef]
22. Avci, O.; Abdeljaber, O.; Kiranyaz, S.; Hussein, M.; Inman, D.J. Wireless and real-time structural damage detection: A novel decentralized method for wireless sensor networks. *J. Sound Vib.* **2018**, *424*, 158–172. [CrossRef]
23. Mitiche, I.; Nesbitt, A.; Conner, S.; Boreham, P.; Morison, G. 1D-CNN based real-time fault detection system for power asset diagnostics. *IET Gener. Transm. Distrib.* **2020**, *14*, 5766–5773. [CrossRef]
24. Toh, G.; Park, J. Review of vibration-based structural health monitoring using deep learning. *Appl. Sci.* **2020**, *10*, 1680. [CrossRef]
25. Pham, M.T.; Kim, J.M.; Kim, C.H. Accurate bearing fault diagnosis under variable shaft speed using convolutional neural networks and vibration spectrogram. *Appl. Sci.* **2020**, *10*, 6385. [CrossRef]
26. Kolar, D.; Lisjak, D.; Pająk, M.; Pavković, D. Fault diagnosis of rotary machines using deep convolutional neural network with wide three axis vibration signal input. *Sensors* **2020**, *20*, 4017. [CrossRef]

27. Chen, Z.; Deng, S.; Chen, X.; Li, C.; Sanchez, R.V.; Qin, H. Deep neural networks-based rolling bearing fault diagnosis. *Microelectron. Reliab.* **2017**, *75*, 327–333. [CrossRef]
28. Chen, H.Y.; Lee, C.H. Deep Learning Approach for Vibration Signals Applications. *Sensors* **2021**, *21*, 3929. [CrossRef] [PubMed]
29. Luo, B.; Wang, H.; Liu, H.; Li, B.; Peng, F. Early fault detection of machine tools based on deep learning and dynamic identification. *IEEE Trans. Ind. Electron.* **2018**, *66*, 509–518. [CrossRef]
30. Wu, Y.; Fu, Z.; Fei, J. Fault diagnosis for industrial robots based on a combined approach of manifold learning, treelet transform and Naive Bayes. *Rev. Sci. Instrum.* **2020**, *91*, 015116. [CrossRef]
31. Tibebu, H.; Roche, J.; De Silva, V.; Kondoz, A. LiDAR-Based Glass Detection for Improved Occupancy Grid Mapping. *Sensors* **2021**, *21*, 2263. [CrossRef] [PubMed]
32. Foster, P.; Sun, Z.; Park, J.J.; Kuipers, B. Visagge: Visible angle grid for glass environments. In Proceedings of the 2013 IEEE International Conference on Robotics and Automation, Karlsruhe, Germany, 6–10 May 2013; pp. 2213–2220.
33. Schwartz, M.; Zarzycki, A. The effect of building materials on LIDAR measurements. In Proceedings of the 35th eCAADe Conference-Volume 2, Rome, Italy, 20–22 September 2017.
34. Abadi, M.; Barham, P.; Chen, J.; Chen, Z.; Davis, A.; Dean, J.; Devin, M.; Ghemawat, S.; Irving, G.; Isard, M.; et al. Tensorflow: A system for large-scale machine learning. In Proceedings of the 12th USENIX Symposium on Operating Systems Design and Implementation (OSDI 16), Savannah, GA, USA, 2–4 November 2016; pp. 265–283.
35. Kingma, D.P.; Ba, J. Adam: A method for stochastic optimization. *arXiv* **2014**, arXiv:1412.6980.
36. Grandini, M.; Bagli, E.; Visani, G. Metrics for multi-class classification: An overview. *arXiv* **2020**, arXiv:2008.05756.
37. Marius, H. Multiclass Classificaton with Support Vector Machines (SVM) Dual Problem and Kernel Function. *Towards Data Sci.* **2020**. Available online: https://towardsdatascience.com/multiclass-classification-with-support-vector-machines-svm-kernel-trick-kernel-functions-f9d5377d6f02 (accessed on 8 October 2021).
38. Taud, H.; Mas, J. Multilayer perceptron (MLP). In *Geomatic Approaches for Modeling Land Change Scenarios*; Springer: Berlin/Heidelberg, Germany, 2018; pp. 451–455.
39. Hochreiter, S.; Schmidhuber, J. Long short-term memory. *Neural Comput.* **1997**, *9*, 1735–1780. [CrossRef]
40. Mohamad, T.H.; Abbasi, A.; Kim, E.; Nataraj, C. Application of Deep CNN-LSTM Network to Gear Fault Diagnostics. In Proceedings of the 2021 IEEE International Conference on Prognostics and Health Management (ICPHM), Detroit (Romulus), MI, USA, 7–9 June 2021; pp. 1–6.
41. Pedregosa, F.; Varoquaux, G.; Gramfort, A.; Michel, V.; Thirion, B.; Grisel, O.; Blondel, M.; Prettenhofer, P.; Weiss, R.; Dubourg, V.; et al. Scikit-learn: Machine learning in Python. *J. Mach. Learn. Res.* **2011**, *12*, 2825–2830.

sensors

MDPI

Article

Autonomous UAV System for Cleaning Insulators in Power Line Inspection and Maintenance

Ricardo Lopez Lopez, Manuel Jesus Batista Sanchez, Manuel Perez Jimenez and Begoña C. Arrue * and Anibal Ollero

GRVC Robotics Laboratory, University of Seville, Avenida de los Descubrimientos, S/N, 41092 Seville, Spain; ricloplop2@us.es (R.L.L.); manbatsan@alum.us.es (M.J.B.S.); mpjimenez@us.es (M.P.J.); aollero@us.es (A.O.)

* Correspondence: barrue@us.es

Abstract: The inspection and maintenance tasks of electrical installations are very demanding. Nowadays, insulator cleaning is carried out manually by operators using scaffolds, ropes, or even helicopters. However, these operations involve potential risks for humans and the electrical structure. The use of Unmanned Aerial Vehicles (UAV) to reduce the risk of these tasks is rising. This paper presents an UAV to autonomously clean insulators on power lines. First, an insulator detection and tracking algorithm has been implemented to control the UAV in operation. Second, a cleaning tool has been designed consisting of a pump, a tank, and an arm to direct the flow of cleaning liquid. Third, a vision system has been developed that is capable of detecting soiled areas using a semantic segmentation neuronal network, calculating the trajectory for cleaning in the image plane, and generating arm trajectories to efficiently clean the insulator. Fourth, an autonomous system has been developed to land on a charging pad to charge the batteries and potentially fill the tank with cleaning liquid. Finally, the autonomous system has been validated in a controlled outdoor environment.

Keywords: UAVs; inspection and maintenance; mobile robots; insulators

Citation: Lopez Lopez, R.; Batista Sanchez, M.J.; Perez Jimenez, M.; Arrue, B.C.; Ollero, A. Autonomous UAV System for Cleaning Insulators in Power Line Inspection and Maintenance. *Sensors* **2021**, *21*, 8488. https://doi.org/10.3390/s21248488

Academic Editors: Gregor Klancar, Marija Seder and Sašo Blažič

Received: 21 November 2021
Accepted: 16 December 2021
Published: 20 December 2021

1. Introduction

The inspection and maintenance of power lines represent a significant economic cost for electricity supply companies. These tasks need to be performed periodically and can result in failures, leading to economic and material losses.

The power transmission system often has to cover long distances, and it has routes that are difficult to access by land.

Moreover, maintenance tasks are performed at high altitudes and require the use of helicopters, ropes, and elevating platforms for tasks such as the installation of bird flight diverters or cleaning power line devices. Particularly, power line insulators are cleaned while the line is energized.

An electrical flash-over may occur through the air between the tower and the conductor, causing blackouts, brownouts, and damage to the installation if the insulators are contaminated. The aim of cleaning it is to remove oxidation and other deposits using a water jet.

Research in aerial robotics for industrial inspection is providing autonomous solutions to traditional methods. The performance of aerial robotics has been proven in fields such as civil engineering [1,2], agriculture [3], the mining industry [4], or conveying systems [5].

In addition, the use of UAVs for tasks that are not purely inspection has increased. Particularly, UAVs are already being used for agriculture monitoring and the spraying of crops [6].

Moreover, in [7], a low-cost fumigation system which has been redesigned and implemented in a UAV is shown. The results show that there is a big difference in application costs between using the system on UAVs or using a conventional hydro-pneumatic sprayer.

Visual inspection operations of power lines are of relevance due to the cost, complexity, and risk of human inspection caused by the need to cover large areas using manned helicopters.

Traditional monocular [8,9] and stereo [10] cameras have been implemented in aerial platforms to detect power line defects. For instance, landing algorithms on power lines have been developed and validated in real environments with an aerial system composed of a LiDAR and a monocular camera for inspection and maintenance [11]. Thermal cameras have been widely used to detect damaged components with an over-heating problem [12] including insulators [13,14]. The recent advances in the computational power of onboard computers have led to real-time applications on UAVs. It is common to find applications in which neural networks are used to detect or recognize elements to interact with them [15]. The detection of insulators has been addressed by different approaches. Due to the different aspect ratios and scales, a faster region-based convolutional neuronal network (R-CNN) model has been implemented [16]. Moreover, the detection of composite and porcelain insulators has been developed on cluttered backgrounds with a single shot multibox detector (SSD) [17].

Despite the considerable potential of UAVs for inspection and maintenance, flight autonomy has been a major constraint. A wide range of wireless charging systems has been developed [18]. A platform for UAVs consisting of two coils transmitting energy using an oscillating magnetic field [19] has been designed. In addition, a self-leveling platform for small UAVs that can be installed on any type of surface [20] has been implemented. Usually, vision-based control is needed to ensure the necessary precision to land on the charging pad [21].

The use of control algorithms for landing UAVs on platforms is well studied. Image-based visual servoing algorithms have been developed to control the UAV while tracking the platform [22]. Moreover, a model of predictive control for autonomous landing under wind disturbances on a moving platform [23] has been developed.

The main contribution of this paper is the design and experimental validation of a fully autonomous application for cleaning insulators on power lines. First, a lightweight cleaning tool that can be installed on a UAV has been designed. Second, the soiled areas of the insulator have been segmented, and an algorithm has been developed to obtain a sequence of optimal points to clean the insulator. Finally, an algorithm has been developed that estimates the fluid trajectory to hit the soiled areas.

The remainder of the paper is organized as follows. Section 2 describes the application and how the cleaning tool was designed and developed, the insulator detection and soil segmentation, and the descent algorithm to land on the charging pad. Section 3 shows the experimental validation of the system in an outdoor environment. Finally, Section 4 outlines the conclusions of this work.

2. System Description

The main goal of this work is to develop an autonomous UAV system for insulator cleaning in power lines as shown in Figure 1. First, the system uses a global positioning system (GPS) point to approach the target area. Then, a vision algorithm using a convolutional neuronal network (CNN) provides the location of the insulator to control the UAV. The cleaning tool aims and shoots a water jet on a point sequence given by an algorithm with a semantic segmentation neuronal network that detects the soiled areas of the insulator. The maximum payload that the UAV can carry will be mostly used for storing the cleaning liquid. For that reason, the tool developed for cleaning the insulators has been designed to minimize weight. When the battery or liquid tank level is low, the system returns to the charging pad. Finally, a vision-based algorithm for autonomous landing on a charging pad has been developed. The overall system structure can be seen in Figure 2.

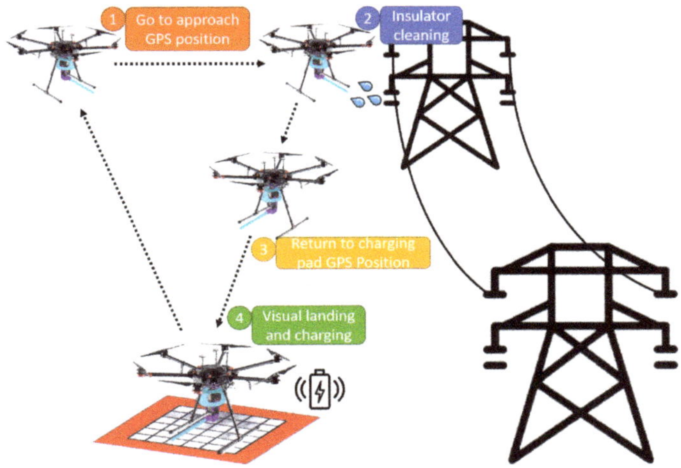

Figure 1. Conceptual design of the operation. (**1**) The UAV is sent to the GPS position of an insulator. (**2**) Visual local control is performed while locating and cleaning the areas that need the maintenance of the insulator. (**3**) When the operation is finished or batteries need to be recharged, the UAV returns to the GPS position of a charging station. (**4**) When it reaches the position, an autonomous vision-based descent is performed.

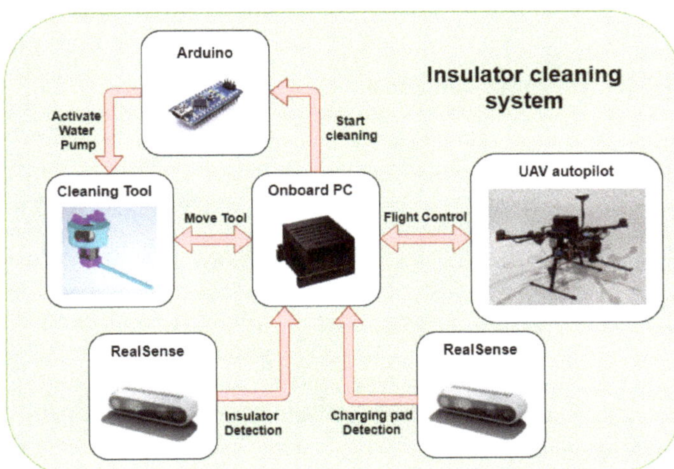

Figure 2. Conceptual scheme of the systems involved in the application.

2.1. Insulator Cleaning Tool

This section describes the custom tool used to clean the insulators. It has been designed to be small and lightweight, so it can be embedded even in small aerial robots.

The cleaning tool has been built with two Dynamixel AX-12A servomotors that provide the necessary degrees of freedom (DOFs) and a good power-to-weight ratio. The purpose of these DOFs is to allow the tool to aim at the insulator while flying, compensating for the oscillations and disturbances that the UAV might suffer.

Meanwhile, the cleaning liquid is driven by a water pump through a flexible rubber tube to the nozzle.

A micro board and a relay have been implemented to activate the water pump to control the flow of the liquid. Given the capacity of the tank, 0.44 L, and the flow rate of the water pump, 240 L/H, the algorithm estimates when the tank is going to be empty and activates a signal to decide whether to return to the charging pad or not.

In addition, a custom nozzle has been designed to increase the range and horizontal jet dispersion as shown in Figure 3.

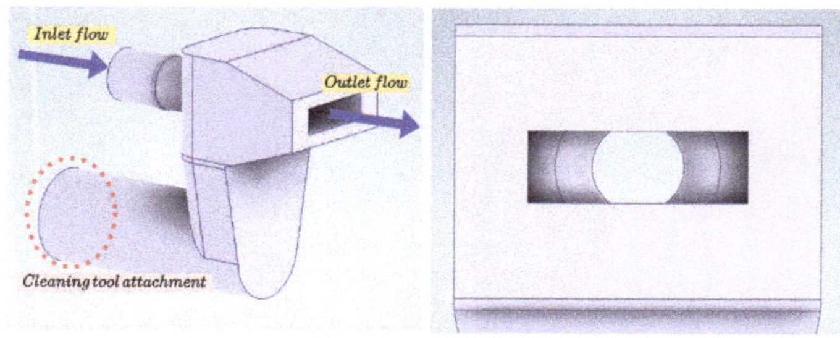

Figure 3. Nozzle with variable cross-section design to increase range and dispersion.

The tool has been printed in polylactic acid (PLA). It is designed to aim downwards to prevent droplets of water from reaching critical systems such as propellers or the electronics.

One of the main elements to prevent a breakage of the yaw servomotor and maintain its performance is the red part in Figure 4, which holds the servomotor. The movement is transmitted to the purple piece, and as can be seen, if there were torsional moments, the blue piece that surrounds the previous ones would prevent the possible breakage of the servomotor. The purple piece at the bottom is the one that holds the cleaning tool to the UAV. This system has four silicone dampers, which connect the UAV body with the pointing tool, to avoid excessive vibrations between the UAV and the tool.

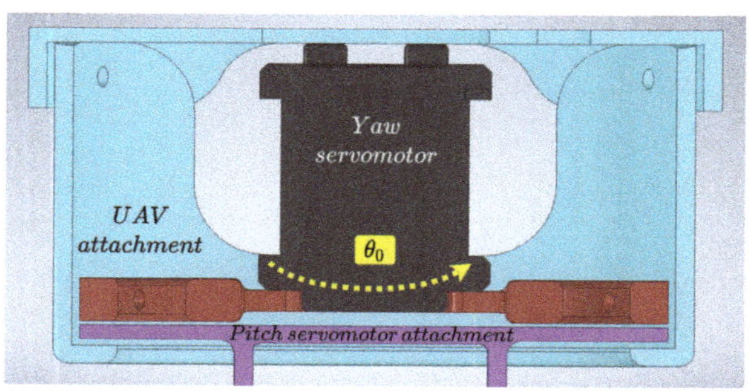

Figure 4. Design of the two-DOF cleaning tool.

To move the described system, an analysis of its kinematics was carried out. The direct (2) and inverse (2) kinematics were obtained using the Denavit–Hartenberg method [24,25]. The scheme used can be seen in Figure 5, and the parameters are shown in Table 1. These parameters were applied to obtain the direct kinematics of the tool, which is used to know the transformation from the camera to the end-effector. In the equations, θ_0, and θ_1 are the joints that control the yaw and pitch, and $P = (X_0^e, Y_0^e, Z_0^e)$ is the position of the end-effector from the tool coordinate frame.

$$\begin{aligned} X_0^e &= L_1 \cos\theta_0 \cos\theta_1 \\ Y_0^e &= L_1 \cos\theta_1 \sin\theta_0 \\ Z_0^e &= L_0 + L_1 \sin\theta_1 \end{aligned} \quad (1)$$

$$\begin{aligned} \theta_0 &= \arctan \frac{Y_0^e}{X_0^e} \\ \theta_1 &= \arcsin \frac{Z_0^e - L_0}{L_1} \end{aligned} \tag{2}$$

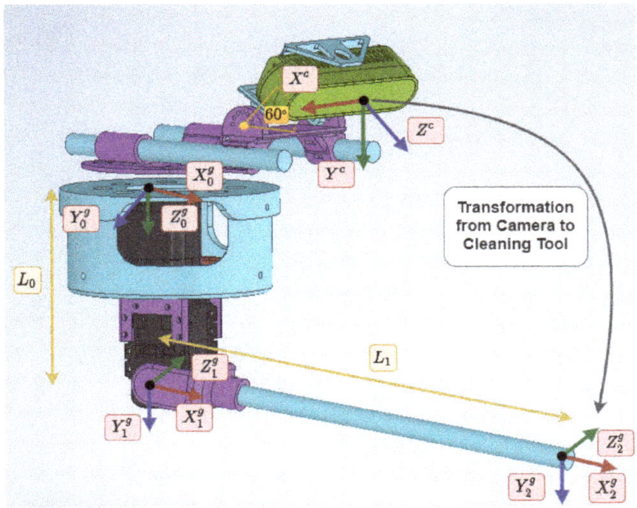

Figure 5. Schematic used to calculate the kinematics of the system along with the reference axes employed.

Table 1. Denavit–Hartenberg parameters of the manipulator.

	θ_i	d_i	α_i	a_i
Link 1	θ_0	L_0	90°	0
Link 2	θ_1	0	0	L_1

2.2. Targeting System

The targeting system has two goals: to evaluate whether the target point for cleaning can be reached and to minimize the distance between the fluid trajectory and the target point.

The flow trajectory is determined by the velocity that the pump is capable of transmitting to the fluid at the nozzle outlet. Therefore, the fluid outlet parameters have been estimated experimentally to choose a mathematical model. Due to the low velocity of the fluid, the air resistance can be neglected. Therefore, it has been determined that the parabolic trajectory can be accurate for the range where the UAV is at least 1.5 m away from the insulator [26,27]. The deviation from the estimated trajectory is mostly caused by the wind.

The algorithm starts when the system detects a point to clean that is less than 1.5 m from the UAV. First, the target point is transformed to the reference frame of the cleaning tool. Second, the joints θ_0 and θ_1 are defined to evaluate the best configuration. Third, using direct kinematics, the point of the end-effector is obtained using direct kinematics. Then, the parabolic trajectory of the fluid is calculated for that configuration. A point is obtained from the intersection of the trajectory in the horizontal plane (Π_h) with the target point. This algorithm attempts to minimize the distance $d = \sqrt{\delta_x^2 + \delta_y^2}$, as shown in Figure 6. Finally, when the entire workspace has been covered, the joints are sent to the cleaning tool. Algorithm 1 summarizes the process.

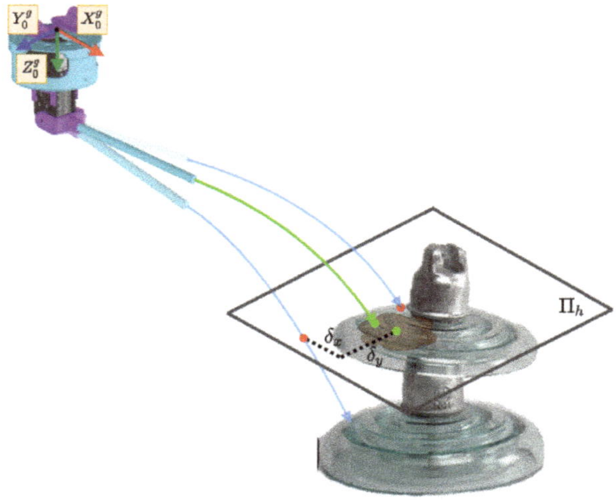

Figure 6. Diagram of the targeting system for choosing the optimal joint variables to hit the target point of the insulator.

Algorithm 1 Algorithm to obtain the optimal joints to hit the target.

$TargetPoint^{camera} \leftarrow CleaningZoneDetection(RGB, Deph)$
$TargetPoint^{tool} \leftarrow TransformToTool(TargetPoint^{camera})$
$MinDistance \leftarrow 0.5m$
for $\theta_0^i = Min(\theta_0)$ to $Max(\theta_0)$ **do**
  for $\theta_1^j = Min(\theta_1)$ to $Max(\theta_1)$ **do**
    $Joints \leftarrow (\theta_0^i, \theta_1^j)$
    $WorkSpacePoint \leftarrow DirectKinematic(Joints)$
    $Point^{tool} \leftarrow FluidTrajectory(WorkSpacePoint)$
    $Distance \leftarrow (TargetPoint^{tool}, Point^{tool})$
    if $Distance < MinDistance$ **then**
      $MinDistance \leftarrow Distance$
      $FinalJoints \leftarrow Joints$

2.3. Insulator Detection

The cleaning task requires the UAV to be positioned close enough to the insulator for the cleaning liquid jet to reach the desired point. For this purpose, a vision algorithm to detect and track the insulator has been developed. The detection has been implemented by a CNN. Then, a NVIDIA Jetson Xavier AGX was chosen, which allows the detection to be performed with a low inference time.

To locate the insulator, a positioning system has been developed. First, the YOLO v4 Tiny neural network [28] is used to detect the device in the image. The CNN has been trained using open datasets and around 4000 additional images of insulators that have been collected in several environments with the aerial platform.

This detection system is executed in real-time on the UAV's onboard computer. To improve the detection rate, an implementation with TensorRT has been used, providing a higher frequency of the bounding box, which improves the control loop. Table 2 shows a comparison between the different implementations on a validation dataset.

Table 2. Inference time between different object detector models.

	Inference Time (ms)
YOLO v4 Tiny	29.4
YOLO v4 Tiny + TensorRT	20.4

Once the device is detected, the resulting bounding box is used to feed a lightweight Kalman tracker [29].

This tracker ensures continuity when the detection is lost between a few frames of real detection, and it reduces the effect of outliers. Figure 7 shows the bounding box of the implemented Kalman tracker in blue and the detection provided by the YOLO v4 Tiny network in green.

Figure 7. Insulator detection using YOLO v4 Tiny with TensorRT implementation (green bounding box) and tracker result (blue bounding box).

Then, a foreground extraction algorithm has been applied on the bounding box to segment the insulator and to obtain its centroid in the depth image. RGB and depth images have been aligned using CUDA, which reduces the processing time.

Finally, to improve the continuity of the detected position, another Kalman filter is implemented. This filter provides a smoother estimation of the 3D position of the device and reduces the noise effect introduced by the depth estimation of the camera.

2.4. Cleaning Zone Detection

Once the UAV maintains its position close enough to the insulator, a segmentation network is used to differentiate the areas that need to be cleaned from the ones that are already clean.

The semantic segmentation network used is based on a fully convolutional network FCN-ResNet101 and has been implemented using the Nvidia inference library for Jetson that allows the use of TensorRT and FP16 precision to decrease the inference time. A dataset has been created with about 2000 images of the insulator in various states of soiling that have been manually labeled. The network has been trained with 80% of the images and validated with the remaining 20%.

Using the segmentation mask obtained from the network, a system that defines the path to follow by the cleaning tool was developed. To generate this path, the image resulting from the segmentation of the soiled areas has been divided into equal horizontal segments as can be seen in Figure 8. The centroids of these divisions are used alongside the aligned depth image to obtain the points to be followed by the tool. This algorithm is performed online at a rate of 40 ms, so these points and the path calculated dynamically change during the cleaning operation. The points delivered to the cleaning tool are fed from the bottom up, as the cleaning of electrical equipment is carried out in this way.

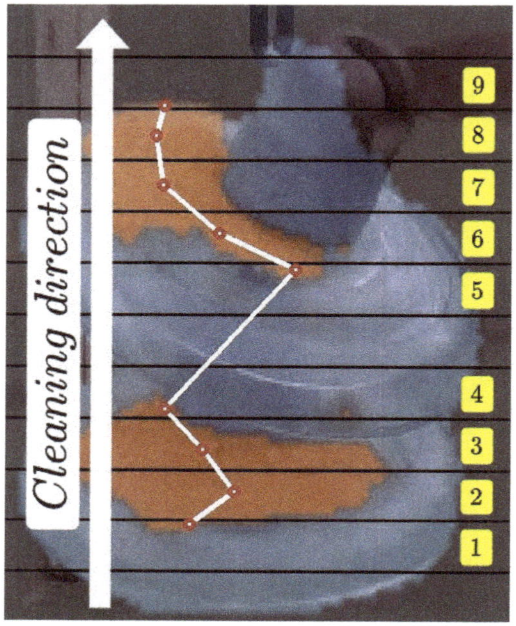

Figure 8. Soiled area segmentation and cleaning trajectory generated by the algorithm.

2.5. Autonomous Landing on Charging Pad

The application relies on the ability of the UAV to charge itself, since cleaning multiple insulators would drain the battery. Therefore, a platform that integrates a commercial charging pad [30] has been designed and built. The platform is designed to be clearly visible from a high altitude to facilitate landing, as shown in Figure 9a.

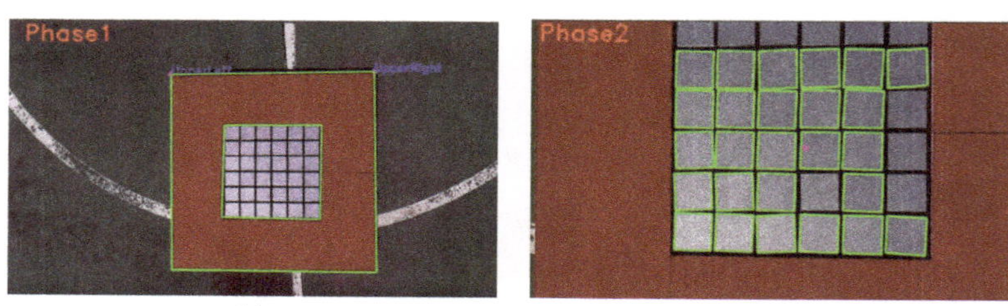

(**a**) (**b**)

Figure 9. Two-phase detection algorithm. (**a**) Phase 1; (**b**) Phase 2.

The descent maneuver is performed using a position-based visual servoing (PBVS) algorithm [31]. Due to the low velocity during the descent maneuver, it has been assumed that the image plane is parallel to the ground and the depth has been estimated using the depth image provided by the camera.

The detection and tracking of the platform are essential to make a stable and reliable landing maneuver. In many cases, the platform will be located at a distance of more than

7 m from the UAV. A lightweight algorithm is required to make it feasible to maneuver the vehicle even in adverse wind conditions that require responsive control. The detection of the charging pad was performed using an algorithm with two phases.

During the first phase, the camera sees the target partially or completely, at a minimum distance of two meters. Color segmentation [32] and shape matching were performed using HSV (hue, saturation, value) thresholding to detect the outside contour. To be able to start charging, the landing gear must be within the inner boundary of the platform. It is necessary to estimate the relative yaw difference between the platform and the UAV. Due to the small available area, a safe landing can only be achieved if the yaw difference is less than 15° and the distance to the center is less than 15 cm.

In the second phase, the algorithm tracks the platform center and yaw misalignment in close range. The system detects multiple inner squares using the HSV threshold and the Hough transform. Since the cells of the charging platform are highly reflective, an online algorithm has been developed to maximize the number of squares it detects by preprocessing the image by applying convolutional filters with different kernel sizes and changing HSV thresholds. The two phases can be seen in Figure 9.

A landing state machine has been developed to control the UAV through the descent. When the system needs to return to the charging pad, the state machine is started along with the first detection phase. The algorithm consists of three stages as can be seen in Figure 10.

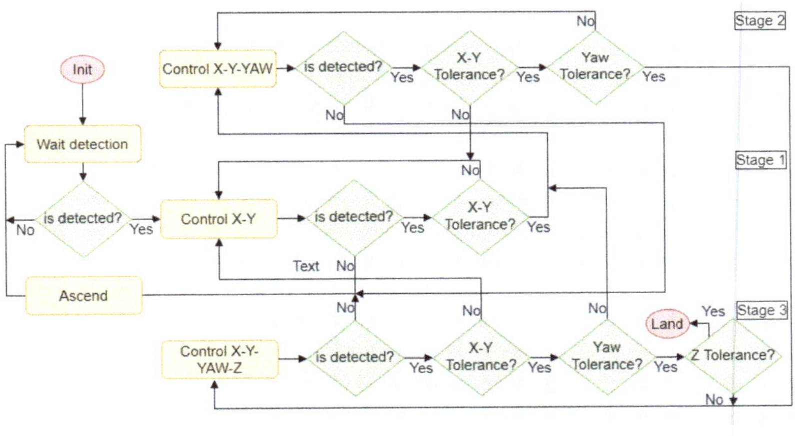

Figure 10. State machine used for the descent and landing maneuver.

The first stage detects the landing station for the first time and controls the UAV to center the target in the X–Y plane. When the UAV has succeeded in reducing the distance to the center of the platform by a threshold, it continues to the next stage. The second is used to align the UAV with the platform by adding to the control a yaw rotation. The third stage initiates when the X–Y distance and yaw rotation thresholds are reached. Then, the descent maneuver commences while keeping the platform centered and aligned, controlling four DOFs. If any of the conditions are not fulfilled, the system will return to the previous stage until they are satisfied. When the distance to the platform is less than 50 cm on the Z-axis, the landing is performed. The limits set for switching between stages are dynamically calculated depending on the height of the UAV.

3. Experimental Validation

This section shows the experimental results and analyzes each subsystem that composes the application. The autonomous insulator cleaning system has been validated in a controlled outdoor environment. This work does not attempt to analyze the interaction of flight systems with electromagnetic interference generated in proximity [33] or in contact [34] with high-voltage power lines. Thus, an insulator has been placed in a controlled outdoor environment.

The multirotor used in this work has been developed specifically for this application. A T-motor Navigator MN4014 of 330 KV motor has been integrated into the Tarot XS690 frame. The autopilot implemented is the Pixhawk 2. The power supply consists of a LiPo 6 s 7000 mAh battery for the motors and a LiPo 4 s 3000 mAh battery that supplies the onboard computer, the arm, the Arduino, and the pump. The total weight of the platform is 4.5 kg plus 0.44 kg of cleaning liquid that the tank holds. This setup delivers a nominal thrust of 1.5–1.6 kg per axis with a maximum flight time of 7 min. Figure 11 shows the hardware setup of the UAV.

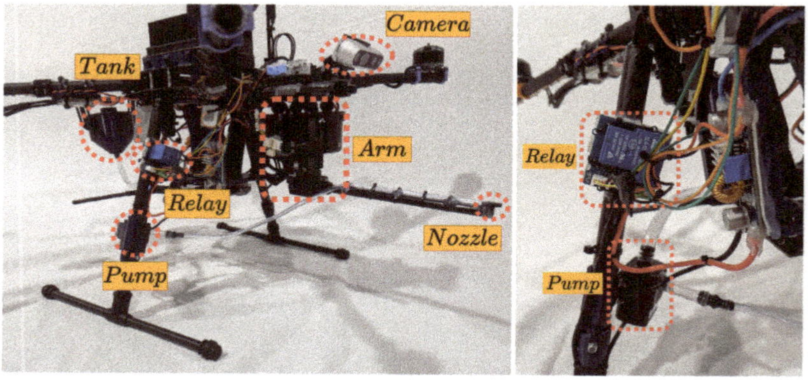

Figure 11. UAV hardware setup.

3.1. Visual Control

The experiments have been performed in an outdoor environment using the Intel RealSense D435i camera as described in Section 2.3 at a resolution of 640 × 480 px. A mission has been conducted in which the UAV takes off from the landing platform. Then, the system performs an approaching maneuver to the insulator. At a distance where the insulator can be reached, the local control algorithm begins, as can be seen in Figure 12. The targeting system is reliant on the visual control performed by the multirotor; thus, the local control has to be precise and stable to achieve the proper cleaning of the insulator.

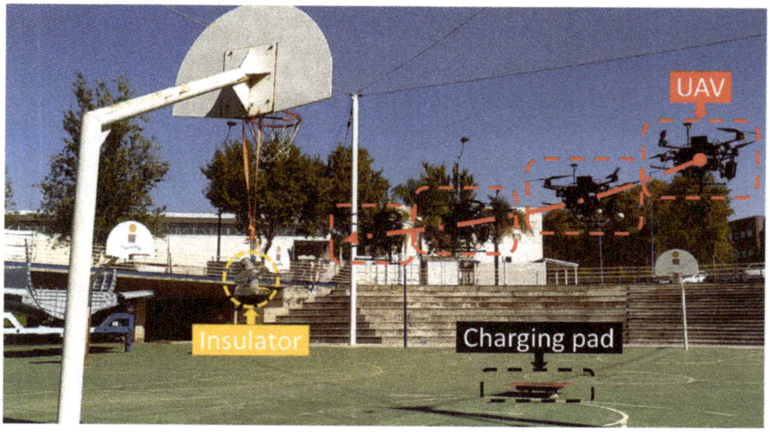

Figure 12. Path followed by the UAV.

The visual guidance has been performed using a cascade control that applies a simple PID algorithm [35] for controlling the multirotor. The position controller is fed with the position of the insulator. To have a wider range of vision, the camera has been positioned with a 30° pitch rotation. Then, it has been estimated that the optimum distance for cleaning is 1.5 m on the Y-axis and −0.5 m on the Z-axis in an ENU (east–north–up) coordinate system.

The control signal and the position of the insulator can be seen in Figure 13. First, the insulator is detected at a distance of 4.5 m on the Y-axis and the UAV starts the approach maneuver, saturating the forward velocity. On the X-axis, an overshoot of at most 20 cm of error, which is within the reach of the cleaning tool, occurs. When the visual control manages to stabilize the UAV between an error of 20 cm on the X-axis, 30 cm on the Y-axis, and 10 cm on Z-axis for 3 s, the cleaning phase starts.

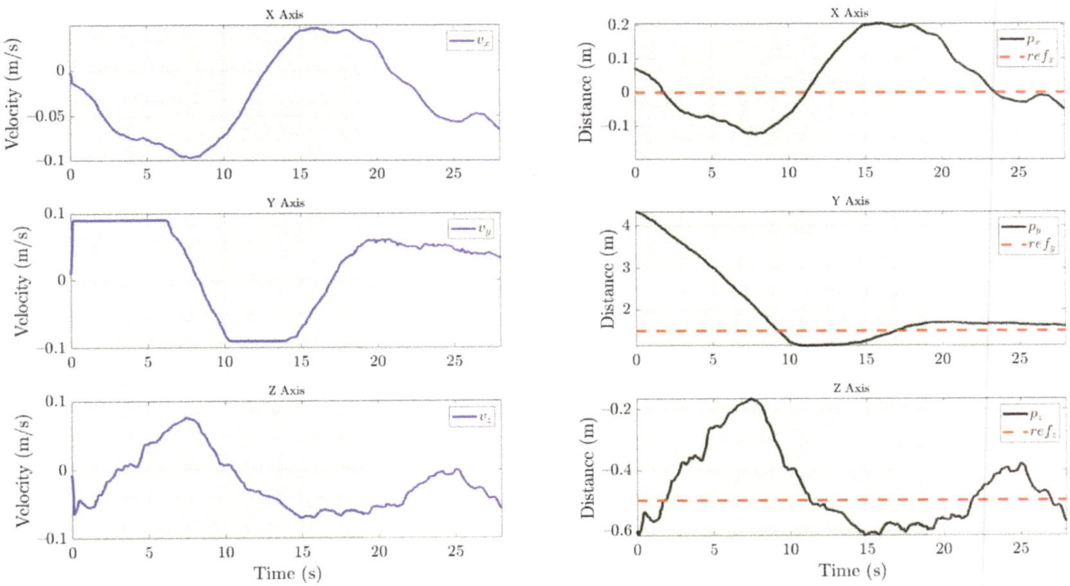

Figure 13. Visual control of the UAV with the control signal (**blue line**), position estimation of the insulator (**black line**), and reference (**dotted red line**). Once the insulator is detected, the UAV performs the approach phase indicated by the green area. When the UAV is stabilized on the three axes below a threshold for three seconds, the cleaning phase indicated by the yellow area begins.

Then, during the cleaning phase, the targeting and cleaning system is capable of performing the task while the UAV compensates for the weight dumped from the cleaning liquid tank. The drift in the X-axis and Y-axis is less than 30 cm. Moreover, the insulator is always in sight, and the tracking that feeds the position control is stable.

Finally, the results show that the UAV can be controlled with the payload during the approach and cleaning phase.

3.2. *Targeting System and Cleaning Insulator Results*

The targeting and cleaning system starts when the UAV comes within range of the insulator. First, the relay is activated. Then, the arm is driven in pitch to prune the air bubbles in the liquid transmission system. Therefore, a sequence of calibration points is performed at the start-up. This makes the cleaning liquid jet start at maximum pressure so that it can reach the estimated point from the beginning. Second, the system described in Section 2.4 estimates the points to be reached by the water jet to clean the insulator. Third, the arm guided by the points of the vision system sequentially reaches the point and clears the soiled area. Finally, the cleaning task ends when it has been detected that there is no soil in the insulator or the tank cleaning liquid has been completely emptied.

Figure 14 shows the onboard camera view of the insulator during the cleaning process, where it can be better appreciated how the cleaning tool works.

The cleaning tool is able to sequentially reach the target point and decrease the soil. The graph shows the path followed by the tool and the path generated from the coordinates of the camera transformed to the reference system of the tool. The UAV is in constant motion during cleaning, so the joints are compensating such motion to impact the target. However, the inertia of the lines means that when it varies in X, Y, or Z directions, there is the same variation in the tool to compensate for this new movement towards the achievable position.

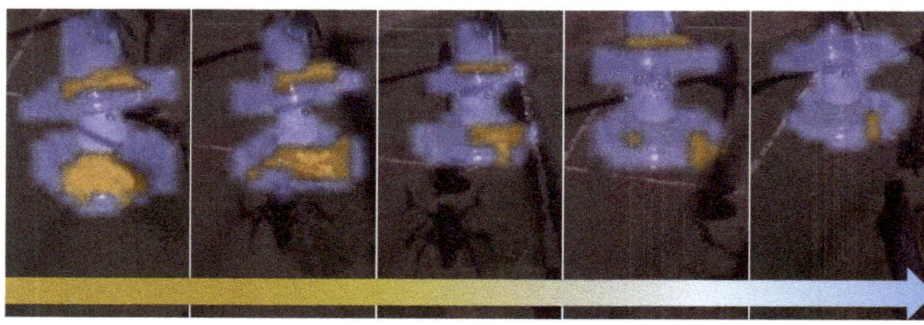

Figure 14. Evolution of soiled areas in the insulator.

The percentage of the area that is cleaned has been measured using the area in pixels from the segmentation of the clean and soiled areas. As can be seen in Figure 15, about 90% of the soil is removed from the insulator. However, during the first three estimated points, the soil appears to be spread out. Although, once the next points are reached, this is significantly reduced as more cleaning liquid is applied to the surface.

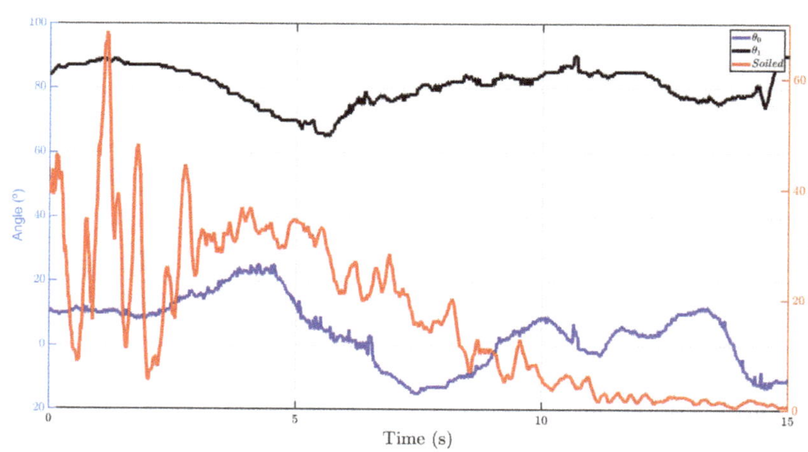

Figure 15. Joint variables and percentage of soiled area cleaned over time by the cleaning tool.

It should be noted that the parabolic trajectory estimation has been essential to obtain the best configuration of the joints, since the drop of the water jet is significant due to the low pressure and velocity of the fluid at the nozzle outlet. Nearly 30 cleaning experiments have been carried out under different soiling states, of which an average of about 70% of the dirty area of the insulator has been reduced.

3.3. Landing System

After cleaning the insulator, the UAV moves to a safe position and returns near the charging station via GPS waypoints. The algorithm described in Section 2.5 is then activated, and the descent maneuver begins once the platform is detected. We chose to use a resolution of 640 × 480 px, since the Intel RealSense D435i camera can stream images at 60 Hz with this resolution. This configuration is optimal because the detection and tracking algorithm can estimate the position of the center of the platform in around 25 ms and therefore send the control signal at 40 Hz.

During the descent maneuver, the arm is placed in a symmetrical position to minimize the moments that can be generated by the weight distribution. In addition, the cleaning liquid tank is empty or nearly empty. Therefore, the descent control system is more precise due to inertia reduction. This is a key factor since the landing maneuver must be precise (around 20 cm in the horizontal plane) due to the dimensions of the platform, so control in the horizontal plane of the UAV becomes crucial.

Figure 16 shows the trajectory followed by the UAV during the descent maneuver. The state machine and the safety cone that is applied ensure that the UAV cannot land if the safety conditions are not satisfied.

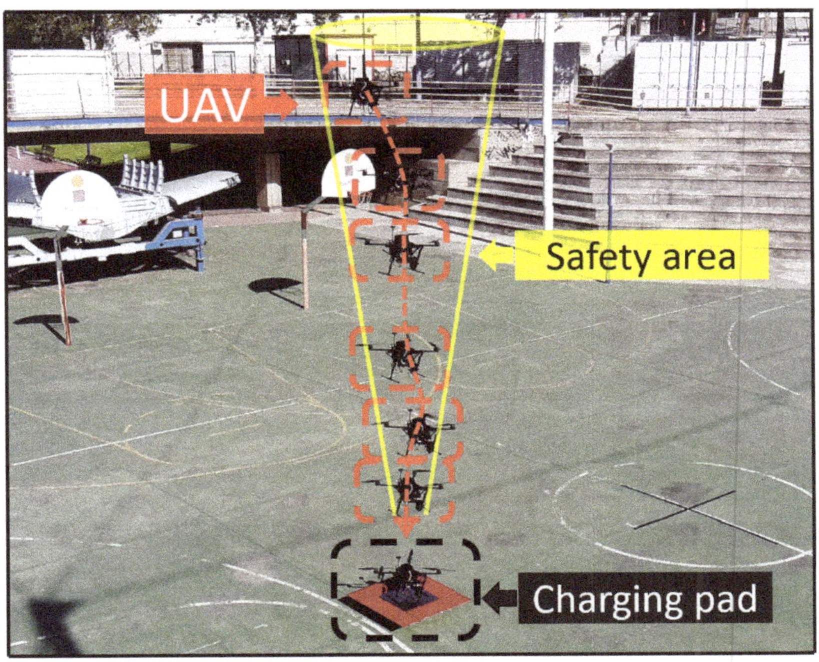

Figure 16. Trajectory followed by the UAV during landing—example of the cone made for a safer descent.

Nearly 30 descent maneuvers have been made. The visual descent control can be seen in Figure 17. The mean error obtained during the experiments in the X and Y axes is 12 cm from the center of the platform. This range is acceptable for the platform to charge since at least one leg must be within the cell. The results show that the landing is safe and robust. However, the UAV manages to land at an altitude of 7 m on average in 25 s. The autonomous cleaning operation video of an insulator can be seen in Supplementary Materials.

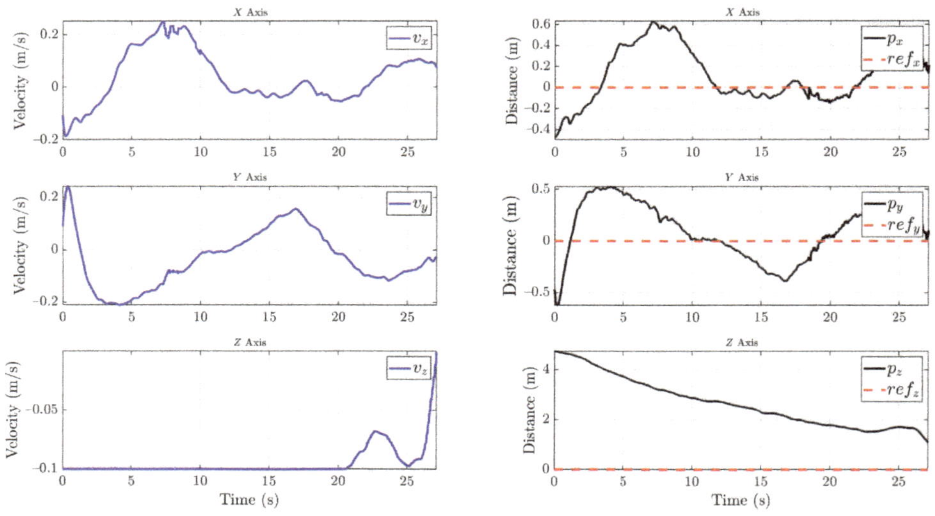

Figure 17. Control signals sent to the autopilot during landing (**blue line**) and charging pad position (**black line**) with the reference (**dotted red line**). The UAV aligns with the platform in the horizontal plane while descending at a constant speed. The second phase of detection begins at 1.8 m, slowing the descent speed and waiting for the optimum horizontal plane alignment to achieve a safe landing.

4. Conclusions and Future Work

The work presented in this article shows a completely autonomous cleaning system for power line insulators. This system offers an alternative to reduce risk exposure during power line maintenance. Additionally, the system can detect when it needs to recharge its batteries or cleaning liquid, return to the charging station, and land autonomously.

Due to the interference caused by power lines, this system offers more precise local positioning than GPS in the proposed environment. Given the ability of UAVs to move nimbly, a good position for cleaning the insulators can be obtained. The system is adaptable from small to larger aerial robots. Furthermore, any controller for UAVs that allows the vehicle to be moved using speed commands can be used.

Nevertheless, the application has to be validated with insulators on an energized power line. The system has been developed not to be dependent on GPS signals near the power line due to electromagnetic interference; therefore, it is necessary to measure the optimal range to start local control for cleaning. If a longer range is required, the UAV must be equipped with a higher resolution camera. In addition, the segmentation network has to be trained with more images of soiled insulators. Hence, the creation of a larger dataset is necessary.

Future work should focus on increasing the pump pressure to improve the cleaning. However, this would change the estimation of the fluid trajectory. By increasing the velocity at the nozzle outlet, the trajectory would no longer look like a parabola, becoming an ellipsoid that increasingly depends on air resistance. Therefore, it would be necessary to replace the method developed in this work with a more detailed study of the jet. It should be noted that the UAV must compensate this force. This disturbance can be predicted, which would increase the need for a model-based control that would combine the UAV model and the model of the force caused by the fluid outflow. In addition, fluid loss in the tank can be taken into consideration and added to the model. Moreover, it is necessary to carry out experiments with insulator cleaning liquids to validate the flow transmission and pump system.

The system has a margin for improvement. The system will refill the liquid autonomously by targeting an onshore reservoir, reversing the flow in the transmission system. The landing camera can be replaced by a higher resolution camera with a higher or equal streaming rate to increase the distance at which the platform is detected. The servo motors used in the targeting system will be replaced by more precise and smaller encoder motors. Finally, an application with a simple user interface will be developed to tune the cleaning, control, and charging parameters.

Supplementary Materials: The following are available online at https://www.mdpi.com/article/10.3390/s21248488/s1, Video S1: Autonomous UAV System for Cleaning Insulators in Power Line Inspection and Maintenance.

Author Contributions: Conceptualization, R.L.L., M.J.B.S., M.P.J. and B.C.A.; methodology, R.L.L., M.J.B.S., M.P.J. and B.C.A.; software, R.L.L., M.J.B.S. and M.P.J.; validation, R.L.L., M.J.B.S. and M.P.J.; formal analysis, R.L.L., M.J.B.S. and M.P.J.; investigation, R.L.L., M.J.B.S., M.P.J. and B.C.A.; resources, R.L.L., M.J.B.S., M.P.J. and B.C.A.; data curation, R.L.L., M.J.B.S. and M.P.J.; writing—original draft preparation, R.L.L., M.J.B.S. and M.P.J.; rewriting—review and editing, R.L.L., M.J.B.S., M.P.J. and B.C.A.; visualization, R.L.L., M.J.B.S., M.P.J. and B.C.A.; supervision, B.C.A.; project administration and supervision, A.O.; funding acquisition, A.O. All authors have read and agreed to the published version of the manuscript.

Funding: This work is funded by the AERIAL-CORE project (Horizon 2020 Grant Agreement No. 871479) and the DREAM project under CDTI – FEDER INTERCONNECTA 2018 Programme

Institutional Review Board Statement: Not applicable.

Informed Consent Statement: Not applicable.

Data Availability Statement: Not applicable.

Acknowledgments: We thank Robotics, Vision, and Control Group for supporting us during the development of this paper. This work has been developed in the framework of the project AERIAL-CORE funded under H2020 ICT-10-2019-2020 and Spanish project DREAM funded under CDTI–FEDER INTERCONNECTA 2018 Programme. The authors appreciate the support given by Robotics, Vision, and Control Group. The collaboration of F.J. García-Rubiales in carrying out the experiments was decisive.

Conflicts of Interest: The authors declare no conflict of interest.

Abbreviations

The following abbreviations are used in this manuscript:

UAV	Unmanned aerial vehicle
DOF	Degrees of freedom
R-CNN	Region-based convolutional neuronal network
SSD	Single shot multibox detector
GPS	Global positioning system
CNN	Convolutional neuronal network
PLA	Polylactic acid
CAD	Computer-aided design
CNN	Convolutional neuronal network
PBVS	Position-based visual servoing
HSV	Hue, saturation, value
ENU	East–north–up

References

1. Hallermann, N.; Morgenthal, G. Visual inspection strategies for large bridges using Unmanned Aerial Vehicles (UAV). In Proceedings of the 7th IABMAS, International Conference on Bridge Maintenance, Safety and Management, Shanghai, China, 7–11 July 2014; pp. 661–667.
2. Sanchez-Cuevas, P.; Ramon-Soria, P.; Arrue, B.; Ollero, A.; Heredia, G. Robotic System for Inspection by Contact of Bridge Beams Using UAVs. *Sensors* **2019**, *19*, 305. [CrossRef]

3. Tsouros, D.C.; Bibi, S.; Sarigiannidis, P.G. A Review on UAV-Based Applications for Precision Agriculture. *Information* **2019**, *10*, 349. [CrossRef]
4. Ren, H.; Zhao, Y.; Xiao, W.; Hu, Z. A review of UAV monitoring in mining areas: Current status and future perspectives. *Int. J. Coal Sci. Technol.* **2019**, *6*, 320–333. [CrossRef]
5. Ghassoun, Y.; Gerke, M.; Khedar, Y.; Backhaus, J.; Bobbe, M.; Meissner, H.; Tiwary, P.K.; Heyen, R. Implementation and Validation of a High Accuracy UAV-Photogrammetry Based Rail Track Inspection System. *Remote Sens.* **2021**, *13*, 384. [CrossRef]
6. Giles, D.; Billing, R. Deployment and Performance of a UAV for Crop Spraying. *Chem. Eng. Trans.* **2015**, *44*, 307–312. [CrossRef]
7. Martinez-Guanter, J.; Agüera, P.; Agüera, J.; Pérez-Ruiz, M. Spray and economics assessment of a UAV-based ultra-low-volume application in olive and citrus orchards. *Precis. Agric.* **2020**, *21*, 226–243. [CrossRef]
8. Ramesh, K.; Murthy, A.S.; Senthilnath, J.; Omkar, S. Automatic detection of powerlines in UAV remote sensed images. In Proceedings of the 2015 International Conference on Condition Assessment Techniques in Electrical Systems (CATCON), Bangalore, India, 10–12 December 2015; pp. 17–21.
9. He, T.; Zeng, Y.; Hu, Z. Research of Multi-Rotor UAVs Detailed Autonomous Inspection Technology of Transmission Lines Based on Route Planning. *IEEE Access* **2019**, *7*, 114955–114965. [CrossRef]
10. Hrabar, S.; Merz, T.; Frousheger, D. Development of an autonomous helicopter for aerial powerline inspections. In Proceedings of the 2010 1st International Conference on Applied Robotics for the Power Industry, Montreal, QC, Canada, 5–7 October 2010; pp. 1–6.
11. Hamelin, P.; Mirallès, F.; Lambert, G.; Lavoie, S.; Pouliot, N.; Montfrond, M.; Montambault, S. Discrete-time control of LineDrone: An assisted tracking and landing UAV for live power line inspection and maintenance. In Proceedings of the 2019 International Conference on Unmanned Aircraft Systems (ICUAS), Atlanta, Georgia, 11–14 June 2019; pp. 292–298. [CrossRef]
12. Luo, X.; Zhang, J.; Cao, X.; Yan, P.; Li, X. Object-aware power line detection using color and near-infrared images. *IEEE Trans. Aerosp. Electron. Syst.* **2014**, *50*, 1374–1389. [CrossRef]
13. Zhao, Z.; Xu, G.; Qi, Y. Representation of binary feature pooling for detection of insulator strings in infrared images. *IEEE Trans. Dielectr. Electr. Insul.* **2016**, *23*, 2858–2866. [CrossRef]
14. Yin, J.; Lu, Y.; Gong, Z.; Jiang, Y.; Yao, J. Edge Detection of High-Voltage Porcelain Insulators in Infrared Image Using Dual Parity Morphological Gradients. *IEEE Access* **2019**, *7*, 32728–32734. [CrossRef]
15. Vemula, S.; Frye, M. Real-Time Powerline Detection System for an Unmanned Aircraft System. In Proceedings of the 2020 IEEE International Conference on Systems, Man, and Cybernetics (SMC), Toronto, ON, Canada, 11–14 October 2020; pp. 4493–4497.
16. Zhao, Z.; Zhen, Z.; Zhang, L.; Qi, Y.; Kong, Y.; Zhang, K. Insulator Detection Method in Inspection Image Based on Improved Faster R-CNN. *Energies* **2019**, *12*, 1204. [CrossRef]
17. Miao, X.; Liu, X.; Chen, J.; Zhuang, S.; Fan, J.; Jiang, H. Insulator Detection in Aerial Images for Transmission Line Inspection Using Single Shot Multibox Detector. *IEEE Access* **2019**, *7*, 9945–9956. [CrossRef]
18. Chittoor, P.K.; Chokkalingam, B.; Mihet-Popa, L. A Review on UAV Wireless Charging: Fundamentals, Applications, Charging Techniques and Standards. *IEEE Access* **2021**, *9*, 69235–69266. [CrossRef]
19. Junaid, A.B.; Konoiko, A.; Zweiri, Y.; Sahinkaya, M.N.; Seneviratne, L. Autonomous Wireless Self-Charging for Multi-Rotor Unmanned Aerial Vehicles. *Energies* **2017**, *10*, 803. [CrossRef]
20. Godzdanker, R.; Rutherford, M.J.; Valavanis, K.P. ISLANDS: A Self-Leveling landing platform for autonomous miniature UAVs. In Proceedings of the 2011 IEEE/ASME International Conference on Advanced Intelligent Mechatronics (AIM), Budapest, Hungary, 3–7 July 2011; pp. 170–175. [CrossRef]
21. Lange, S.; Sunderhauf, N.; Protzel, P. A vision based onboard approach for landing and position control of an autonomous multirotor UAV in GPS-denied environments. In Proceedings of the 2009 International Conference on Advanced Robotics, Munich, Germany, 22–26 June 2009; pp. 1–6.
22. Lee, D.; Ryan, T.; Kim, H.J. Autonomous landing of a VTOL UAV on a moving platform using image-based visual servoing. In Proceedings of the 2012 IEEE International Conference on Robotics and Automation, St Paul, MN, USA, 14–19 May 2012; pp. 971–976. [CrossRef]
23. Feng, Y.; Zhang, C.; Baek, S.; Rawashdeh, S.; Mohammadi, A. Autonomous Landing of a UAV on a Moving Platform Using Model Predictive Control. *Drones* **2018**, *2*, 34. [CrossRef]
24. Cristoiu, C.; Nicolescu, A. New approach for forward kinematic modeling of industrial robots. *Res. Sci. Today* **2017**, *13*, 136.
25. Damic, V.; Cohodar, M.; Tvrtkovic, M. Inverse dynamic analysis of hobby robot uarm by matlab/simulink. In Proceedings of the 27th DAAAM International Symposium, Vienna, Austria, 26–29 October 2016; Volume 27.
26. McCoy, R. *Modern Exterior Ballistics: The Launch and Flight Dynamics of Symmetric Projectiles*; Schiffer Pub.: Atglen, PA, USA, 1999.
27. Peszynski, K.; Perczynski, D.; Piwecka, L. Mathematical model of a fountain with a water picture in the shape of an hourglass. In Proceedings of the EPJ Web of Conferences, Stará Lesná, Slovakia, 1–5 July 2019; EDP Sciences: Les Ulis, France, 2019; Volume 213, p. 02064.
28. Bochkovskiy, A.; Wang, C.Y.; Liao, H.Y.M. Yolov4: Optimal speed and accuracy of object detection. *arXiv* **2020**, arXiv:2004.10934.
29. Bewley, A.; Ge, Z.; Ott, L.; Ramos, F.; Upcroft, B. Simple online and realtime tracking. In Proceedings of the 2016 IEEE International Conference on Image Processing (ICIP), Phoenix, AZ, USA, 25–28 September 2016; pp. 3464–3468. [CrossRef]
30. Rugged and Automatic Battery Charging for The Outdoor Environment. Available online: https://skycharge.de/charging-pad-outdoor (accessed on 12 February 2021).

31. Hutchinson, S.; Hager, G.D.; Corke, P.I. A tutorial on visual servo control. *IEEE Trans. Robot. Autom.* **1996**, *12*, 651–670. [CrossRef]
32. Cheng, H.D.; Jiang, X.H.; Sun, Y.; Wang, J. Color image segmentation: Advances and prospects. *Pattern Recognit.* **2001**, *34*, 2259–2281. [CrossRef]
33. Li, Y.; Ding, Q.; Li, K.; Valtchev, S.; Li, S.; Yin, L. A Survey of Electromagnetic Influence on UAVs from an EHV Power Converter Stations and Possible Countermeasures. *Electronics* **2021**, *10*, 701. [CrossRef]
34. Suarez, A.; Salmoral, R.; Zarco-Periñan, P.J.; Ollero, A. Experimental Evaluation of Aerial Manipulation Robot in Contact With 15 kV Power Line: Shielded and Long Reach Configurations. *IEEE Access* **2021**, *9*, 94573–94585. [CrossRef]
35. Spinka, O.; Kroupa, S.; Hanzalek, Z. Control System for Unmanned Aerial Vehicles. In Proceedings of the 2007 5th IEEE International Conference on Industrial Informatics, Vienna, Austria, 23–27 June 2007; Volume 1, pp. 455–460. [CrossRef]

sensors

MDPI

Article

False Ceiling Deterioration Detection and Mapping Using a Deep Learning Framework and the Teleoperated Reconfigurable 'Falcon' Robot

Archana Semwal [1], Rajesh Elara Mohan [1], Lee Ming Jun Melvin [1], Povendhan Palanisamy [1], Chanthini Baskar [2], Lim Yi [1], Sathian Pookkuttath [1] and Balakrishnan Ramalingam [1,*]

[1] Engineering Product Development Pillar, Singapore University of Technology and Design (SUTD), Singapore 487372, Singapore; archana_semwal@sutd.edu.sg (A.S.); rajeshelara@sutd.edu.sg (R.E.M.); melvin_lee@sutd.edu.sg (L.M.J.M.); povendhan_palanisamy@mymail.sutd.edu.sg (P.P.); yi_lim@mymail.sutd.edu.sg (L.Y.); sathian_pookkuttath@sutd.edu.sg (S.P.)

[2] School of Electronics Engineering, Vellore Institute of Technology, Chennai 600127, India; chanthini.baskar@vit.ac.in

* Correspondence: balakrishnan@sutd.edu.sg

Abstract: Periodic inspection of false ceilings is mandatory to ensure building and human safety. Generally, false ceiling inspection includes identifying structural defects, degradation in Heating, Ventilation, and Air Conditioning (HVAC) systems, electrical wire damage, and pest infestation. Human-assisted false ceiling inspection is a laborious and risky task. This work presents a false ceiling deterioration detection and mapping framework using a deep-neural-network-based object detection algorithm and the teleoperated 'Falcon' robot. The object detection algorithm was trained with our custom false ceiling deterioration image dataset composed of four classes: structural defects (spalling, cracks, pitted surfaces, and water damage), degradation in HVAC systems (corrosion, molding, and pipe damage), electrical damage (frayed wires), and infestation (termites and rodents). The efficiency of the trained CNN algorithm and deterioration mapping was evaluated through various experiments and real-time field trials. The experimental results indicate that the deterioration detection and mapping results were accurate in a real false-ceiling environment and achieved an 89.53% detection accuracy.

Keywords: defect detection; Faster R-CNN; deep learning; object detection; IoRT; inspection robot

Citation: Semwal, A.; Mohan, R.E.; Melvin, L.M.J.; Palanisamy, P.; Baskar, C.; Yi, L.; Pookkuttath, S.; Ramalingam, B. False Ceiling Deterioration Detection and Mapping Using a Deep Learning Framework and the Teleoperated Reconfigurable 'Falcon' Robot. *Sensors* **2022**, *22*, 262. https://doi.org/10.3390/s22010262

Academic Editors: Gregor Klancar, Marija Seder and Sašo Blažič

Received: 28 November 2021
Accepted: 27 December 2021
Published: 30 December 2021

1. Introduction

False ceiling inspection is one of the most required inspections for essential maintenance and repair tasks in commercial buildings. Generally, a false ceiling is built with material such as Gypsum board, Plaster of Paris, and Poly Vinyl Chloride (PVC) and used to hide ducting, messy wires, and Heating, Ventilation, and Air Conditioning (HVAC) systems. However, poor construction and the use of substandard material in false ceilings require periodic inspection to avoid deterioration. Structural defects, degradation in HVAC systems, electrical damage, and infestation are common potential building and human safety hazards. Human visual inspection is a common technique used by building maintenance companies, where trained safety inspectors will audit the environment of a false ceiling. However, deploying human visual inspection for a false ceiling environment has many practical challenges. It requires a highly skilled inspector to access a complex false ceiling environment. Workforce shortage due to safety issues and low wages is another challenge faced by false ceiling maintenance companies. These facts highlight the need for an automated, cost-effective, and exhaustive inspection of false ceilings to prevent such risks.

Hence, the aim of this research is to automate the inspection process to detect and map various deterioration factors in false ceiling environments. Further, the literature survey

(Section 2) confirms a research gap between robot-assisted inspection and deep learning frameworks for false ceiling inspection and maintenance. Thus, this work presents an automated false ceiling inspection framework using a convolutional neural network trained with our false ceiling deterioration image dataset composed of four classes, structural defects (spalling, cracks, pitted surfaces, and water damage), degradation in HVAC systems (corrosion, molding, and pipe damage), electrical damage (frayed wires), and infestations (termites and rodents). Further, the inspection task is performed with the help of our in-house-developed crawl class robot, known as the 'Falcon', with a deterioration mapping function using Ultra-Wideband (UWB) modules. The deterioration mapping function marks the class of deteriorations with locations on a map for the inspection and maintenance of false ceilings.

This manuscript is organized as follows; after explaining the importance and contributions of the study in Section 1, Section 2 presents a literature review, and Section 3 presents an overview of the proposed system. Section 4 discusses the experimental setup and the results. Section 5 includes a discussion. Section 6 concludes.

2. Related Work

In recent years, various semi- or fully automated techniques have been reported in the literature for narrow and enclosed space inspections for building maintenance tasks. Here, computer vision algorithms were used for automatically detecting defects from images collected by inspection tools such as borescope cameras [1,2] and drones [3,4]. However, borescope cameras and drone-based methods have many practical difficulties when used as inspection tools in false ceiling environments. Because false ceiling environments have many protruding elements such as electrical wire networks, gas pipes, and ducts, it is also difficult to fly drones to inspect the complex environment of a false ceiling.

Robot-based inspection is a better solution than borescope cameras and drone-based inspection. It has been widely used for various narrow and enclosed space inspection applications, such as crawl space inspection [5,6], tunnel inspection [7,8], drain inspection [9,10], defect detection in glass facade buildings [11,12], and power transmission line fault detection [13]. Gary et al. proposed a q-bot inspection robot for autonomously surveying underfloor voids (floorboards, joists, vents, and pipes). It uses a mask Region Convolutional Neural Network (mask-RCNN) approach with a two-stage transferring learning method. It was able to detect with an accuracy of 80% [5]. Self-reconfigurable robot 'Mantis' was used for crack detection, and glass facade cleaning in high-rise buildings [11,14], where a CNN-based deep learning framework with 15 layers is used for detecting cracks on glass panels. Similarly, a steel climbing robot was developed for steel infrastructure monitoring. The authors developed a steel crack detection algorithm using steel surface image-stitching and a 3D map building technique. The steel crack detection algorithm was able to achieve a success rate of 93.1% [15]. In [16], Gui et al. automated a defect detection and visualization task for airport runway inspection. The proposed novel robotic system employed a camera, Ground Penetrating Radar (GPR), and a crack detection algorithm based on images and GPR data. An F1-measure of 70% and 67% was achieved for crack detection and subsurface defect detection, respectively. In [17], Perez et al. aimed at detecting building defects (mold, deterioration, and stains) using convolutional neural networks (CNNs). The authors presented a deep-learning-based detection and localization model employing VGG-16 to extract and classify features. The tests demonstrated an overall detection accuracy of 87.50%. Xing et al., in [18], proposed a CNN-based method for workpiece surface defect detection. The authors designed a CNN model with symmetric modules for feature extraction and optimized the IoU to compute the loss function of the detection method. The average detection accuracy of the CNN on the Northeastern University-Surface Defect Database (NEU-CLS) and on self-made datasets was 99.61% and 95.84%, respectively. Similary, Xian et al. (in [19]) presented automatic metallic surface defect detection and recognition using a CNN. The authors designed a novel Cascaded Autoencoder (CASAE) architecture for segmenting and localizing defects. The segmentation results demonstrated an IoU score of 89.60%. Cheon et al. presented an Automatic Defect Classification (ADC) system for wafer surface defect classification and the detection of unknown defect class [20]. The proposed model adopted a single CNN model and achieved a classification accuracy of 96.2%. Finally, Civera et al. proposed video processing techniques for the contactless investigation of large oscillations to deal with geometric nonlinearities and light structures.

Though several works are available for narrow and enclosed space inspection applications using robot and computer vision algorithms, the defect detection and mapping of false ceilings are not yet widely studied. In the literature, very few works have reported robot-assisted ceiling inspection. Robert et al. in [21] introduced a fully autonomous industrial aerial robot using a top-mounted omni wheel drive system and an AR marker system. The proposed system can perform high precision localization and positioning to perform an ink-marker placement task for measuring and maintaining the ceiling. In [22], a flexible wall and ceiling climbing robot with six permanent magnetic wheels is proposed by Yuanming et al. to climb vertical walls and reach overhead ceilings. In [23], Ozgur et al. developed a 16-legged palm-sized climbing robot using flat bulk tacky elastomer adhesives. The proposed robot has a passive peeling mechanism for energy-efficient and vibration-free detachment to climb in any direction in 3D space. In [24], a self-reconfigurable false ceiling inspection robot is presented using an induction approach [25,26] and a rodent activity detection task [6]. A Perimeter-Following Controller (PFC) based on fuzzy logic was integrated into the robot to follow the perimeter of the false ceiling autonomously, and an AI-enabled remote monitoring system was proposed for rodent activity detection in false ceilings. All of these robots used for various purposes are summarized in Table 1. However, this research mainly focused on the robot design for various crawl spaces and does not involve the deterioration detection and mapping of false ceilings.

Table 1. Summary of research.

Reference	Typology of the Platform	Aim
Gary et al. [5]	Wheeled Robot	Detect common features of underfloor void
Balakrishnan et al. [6]	Tracked Robot	Rodent Activity Monitoring
Protopapadakis et al. [7]	Wheeled Robot	Crack detection for tunnel inspection
Menendez et al. [8]	Wheeled Robot	Tunnel structural inspection
Palanisamy et al. [9]	Wheeled Robot	Drain Structural Defect Detection
Melvin et al. [10]	Wheeled Robot	Drain blockage inspection
Kouzehgar et al. [11]	Wheeled Robot	Automatic glass crack detection
Hung et al. [15]	Wheeled Robot	Inspection of steel structures and bridges
Gui et al. [16]	Wheeled Robot	Airport runway inspection
Ladig et al. [21]	Aerial Robot	Platform to do high precision localization and positioning
Ozgur et al. [23]	Sixteen-legged climbing robot	To climb in any direction in 3D space
Hayat et al. [24]	Reconfigurable wheeled robot	False-ceiling inspection

The literature survey indicates that there is a research gap in the robot-assisted false ceiling inspection field. Therefore, this work proposes a false ceiling inspection and deterioration mapping framework using a Deep-Learning (DL)-based deterioration detection algorithm and our in-house-developed teleoperated reconfigurable false ceiling inspection robot, known as the 'Falcon'.

3. Overview of the Proposed System

Figure 1 shows an overview of the false ceiling inspection and deterioration mapping framework. Our in-house-developed crawl class Falcon robot was used for false ceiling inspection, and a deep-learning-based object detection algorithm was trained for deterioration detection from robot captured images. Further, a UWB localization module was used to localize the deterioration location and generate a deterioration map of a false ceiling. The detail of each module and functional integration is given as follows.

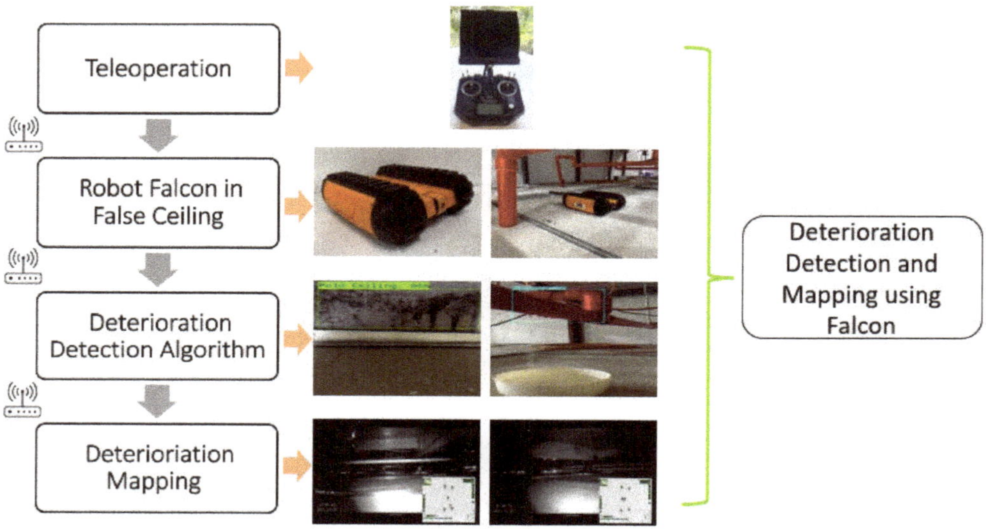

Figure 1. Overview of the proposed system.

3.1. The Falcon Robot

A false ceiling panel is built using a fragile material such as Gypsum board or Plaster of Paris. Moreover, a false ceiling environment is crowded with components such as piping, electrical wiring, suspended cables, and protruding elements. Therefore, the Falcon was designed as a lightweight robot that can easily traverse obstacles. Furthermore, the camera used for image capturing or recording videos of the false ceiling environment is able to tilt the angle from 0 to 90 degrees for better accessibility in the crawled spaces. During the development stage, three versions of the Falcon robot were built due to changing requirements and design considerations, shown in Figure 2. In Version 2, the track mechanism is reinforced with a fork structure to avoid slippage while crossing obstacles in a false ceiling environment. Furthermore, a closed-form design approach was applied due to excessive dust-settling on electronic components. In Version 3 (as shown in Figure 3), a more precise IMU and more powerful motor is used. The robot height was further reduced to travel in spaces with an 80 mm height. All of the specifications of the Falcon robots are detailed in Table 2. The Falcon robot was powered with a 3 × 3.7 V, 3400 mAh battery that operates between 0.5 to 1.5 h with full autonomy functionalities. The operating range of the Falcon robot is directly determined by energy consumed by sensors and actuators, such as cameras, IMUs, cliffs, and motors. During autonomous operations, the motor operating at 1.6 A and 12 V consumes 33.2 W. Considering the battery power of 65.3 W, motors consume the highest fraction of the energy used for locomotion and to overcome tall obstacles. Further, the exploration tasks during false ceiling inspection also drain the energy, affecting the range of autonomy.

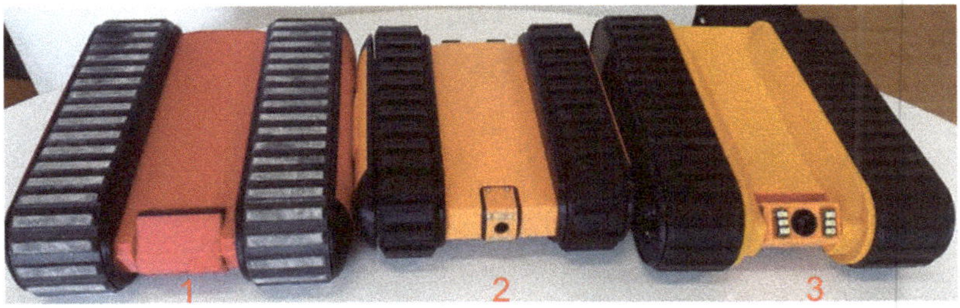

Figure 2. Different Version of Falcon Robot (Version 1, 2 and 3).

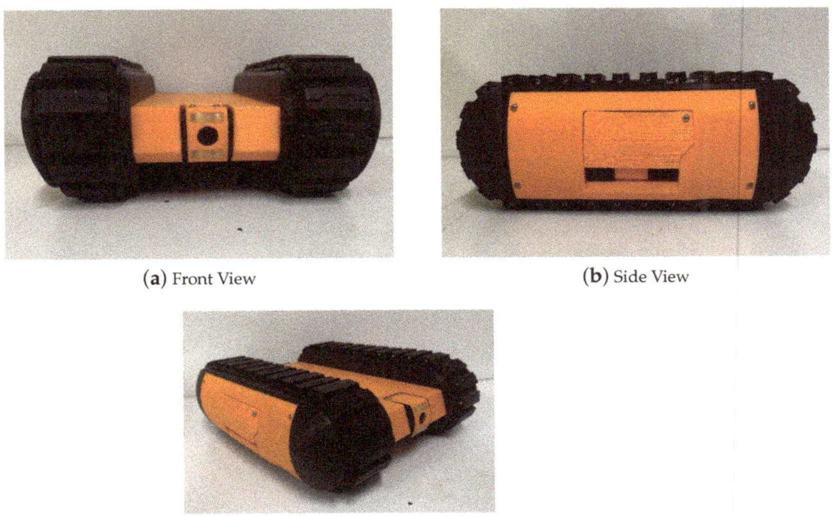

(**a**) Front View

(**b**) Side View

(**c**) Isometric View

Figure 3. Different view of the Falcon Robot (Version 3).

Locomotion Module: An important design consideration for a false ceiling robot is the form factor to overcome obstacles with a height of 55 mm and to traverse through low hanging spaces under 80 mm. In order to overcome these narrow spaces and tall obstacles, a locomotion module in the form of tracks that has maximized the contact area was used. The tracks extended along the dimension of the vehicle and were configured to be 236 (L) $\times$ 156 (W) $\times$ 72 (H) [mm $\times$ mm $\times$ mm]. The Falcon can operate regardless of the direction it flips over, as both sides of the locomotion modules are consolidated with hemispherical attachments to avoid stabilizing laterally. However, the operational terrain of the false ceiling may impose uncertainty on the Falcon robot. Therefore, the motors with higher specifications were chosen; e.g., a safety factor of 2 on a maximum inclined slope of 12 degrees.

Control system: A small-footprint, low-power ARM, Cortex-M7-powered, Teensy-embedded computing system was used as the onboard processor for the Falcon robot. It processes the velocity commands from the user and computes motor speeds using an inverse kinematic model. The MQTT server was employed to send the velocity command from the control station. In addition, the control unit is responsible for vital safety layer functionality to prevent the free-fall of the robots. Thus, the processor calibrates the IMU and cliff sensors to differentiate openings in the ceilings from the false noises while overcoming obstacles.

Table 2. Technical specifications of the Falcon robot.

Description	Specification
Dimension [L × W × H]	0.236 m × 0.156 m × 0.072 m
Weight (including battery)	1.3 kg
Type of Locomotion Drive	Track
Top & Bottom Ground Clearance	0.011 m, 0.011 m
Operating Speed	0.1 m/s
Maximum Obstacle Height	0.055 m
Operational Duration	0.5 h–0.75 h
Battery	3-cell Lithium Ion
Operation Mode	Teleoperation (with integrated sensors to detect falls and stops autonomously)
Communication Mode	Wi-Fi through a local MQTT server
Camera Specifications (onboard light source)	VGA 640 × 480, up to 30 fps, 60 degree view angle, 20 cm-infinity focusing range

System Architecture

Figure 4 illustrates the system architecture of the Falcon robot. It consists of the following units: (1) a locomotion module, (2) a control unit, (3) a power distribution module, (4) a wireless communication module, and (5) a perception sensor.

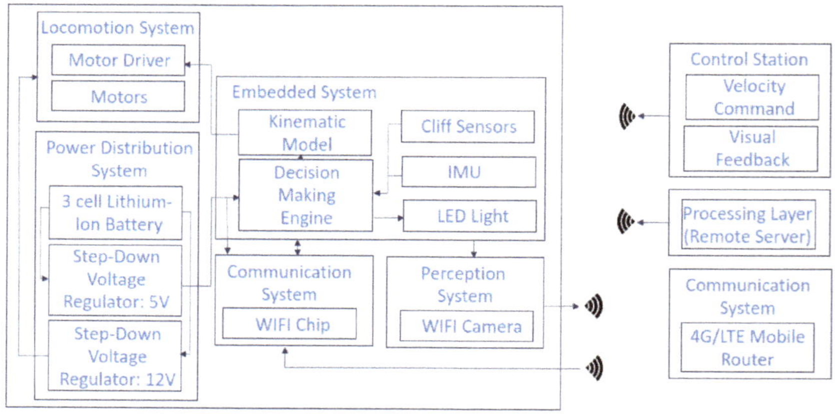

Figure 4. System architecture of the Falcon.

Perception Module The WiFi camera module operates with a 5 V power rating and a 640 × 480 pixel density at 30 fps. The encoded video feed is a recorder and is additionally used to process the data through computer vision and machine learning algorithms to identify defects. Since the perception system relies heavily on lighting conditions, a NeoPixel stick with an 8∼50 RGB LED strip is used as the robot's light source. Furthermore, a dedicated router is used to avoid data loss and for improved data security. Finally, a titling camera (up to 90 degrees) was incorporated considering broader field of view requirements using a servo motor controlled by A Teensy-embedded computing system.

3.2. Deterioration Detection Algorithm

Generally, deterioration factors in false ceiling environments are tiny and cover only a small number of the pixels of an image. Therefore, there is a requirement of a detection algorithm able to detect small objects to mitigate overlap or pixelated issues. Furthermore, the information extracted from images is lost due to multiple layers of the convolution neural network. The inspection algorithm needs an extensive, accurate, and apt framework with a small object detection capability. A Faster R-CNN model is an optimal framework when compared with similar CNN architectures and was used to detect small deterioration factors of the false ceiling environment in our case study [9,10]. Figure 5 shows an overview of the Faster R-CNN framework. Its architecture comprises three main components: the feature extractor network, the Region Proposal Network (RPN), and the detection network. All three components are briefly described in the following section.

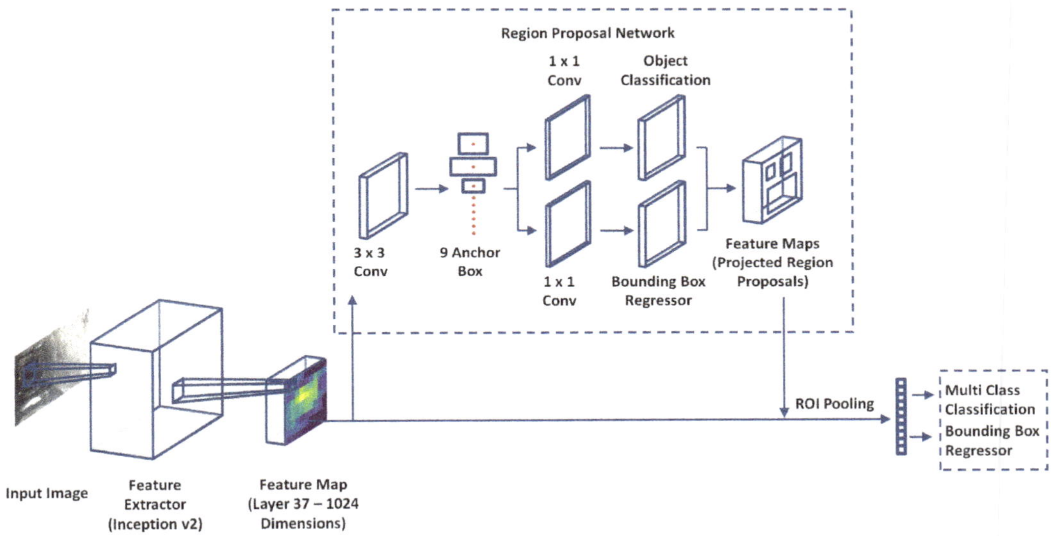

Figure 5. Functional block diagram of the deterioration detection algorithm.

3.2.1. Feature Extractor Network

In our case study, Inception v2 performed the feature extraction task. It is an upgraded version of Inception v1, providing better accuracy and reducing computational complexity. Here, the input image size was 768 × 1024, and a total of 42 deep convolutional layers were used to build the feature extractor network. The number of feature maps directly controlled the task complexity, so an optimal 1024-size feature map (extracted from Layer 37 via a transfer learning scheme on a pre-trained dataset of COCO [27]) was fed into the Faster R-CNN. Further, in Inception v2, filter banks were expanded to reduce the loss of information, known as a 'representational bottleneck.' Finally, the convolution 5 × 5 and 3 × 3 was factorized into two 3 × 3 convolutions and a combination of 1 × 3 and 3 × 1 convolutions, respectively, to boost the performance and reduce the computational cost. Further, Table 3 summarizes the layer details and input dimensions.

Table 3. Inception v2 backbone.

Layer Details	Input Dimensions
Conv	299 × 299 × 3
Conv	149 × 149 × 32
Conv	147 × 147 × 32
Pool	147 × 147 × 64
Conv	73 × 73 × 64
Conv	71 × 71 × 80
Conv	35 × 35 × 192
3× Inception module A	35 × 35 × 288
5× Inception module B	17 × 17 × 768
2× Inception module C	8 × 8 × 1280
Pool	8 × 8 × 2048
Linear	1 × 1 × 2048
Softmax	1 × 1 × 1000

3.2.2. Region Proposal Network

The Region Proposal Network (RPN) shares the output of the feature extractor network to the object detection and classification network. The RPN takes the feature map as an input (the output of the feature extractor network) and generates a bounding box with an objectness score using the anchor box technique first proposed by Shaoqing Ren et al. [28]. The anchor boxes are predefined, fixed-size boxes and detect objects of varying sizes and overlapping objects. It performs a 3 × 3 sliding window operation to generate anchor boxes in a 256-size feature map. Nine anchor boxes can be created from the combinations of sizes and ratios. Further, a stride of 8 (each kernel is offset by eight pixels from its predecessor) is used to determine the actual position of the anchor box in the original image. The output of the above convolution is fed into two parallel convolution layers, one for classification and the other for the boundary box regression. Finally, Non-Max Suppression (NMS) is applied to filter out the overlapping bounding boxes based on their objectness scores.

3.2.3. Detection Network

The detection network consists of the Region of Interest (RoI) pooling layer and a fully connected layer. The shared feature map from the feature network and the object proposals generated by the RPN are fed into the RoI pooling layer to extract fixed-sized feature maps for each object proposal generated by the RPN. The fixed-sized feature maps are then fed two different fully connected layers with a softmax function. The first fully connected layer seeks to classify the object proposals into one of the object classes, plus a background class for removing bad proposals (N + 1 units, where N is the total number of object classes). The second fully connected layers seeks to better adjust the bounding box for the object proposal according to the predicted object class (4N units for a regression prediction of the x_{center}, y_{center}, $width_{center}$, and $height_{center}$ of each of the N possible object classes). Similar to the RPN, NMS is applied to filter out redundant bounding boxes and retain a final list of objects using a probability threshold and a limit on the number of objects for each class.

3.3. *Deterioration Mapping*

In our case study, the deterioration mapping function was accomplished using the UWB module. Explicitly, the UWB module was employed to track the mobile robot and localize the deterioration position. This location estimation feature was utilized and combined with the object detection module to identify, locate, and mark the deteriorations on the map of the false ceiling. At least three beacons must be installed where the actual number of beacons required is dependent on the complexity of the false ceiling infrastructure. In addition, sensor fusion was used to reduce localization errors and to calculate the exact position. It combines wheel odometry, IMU data, and UWB localization data to offer a more accurate location estimate. The beacon map was generated with the origin (0,0) as the location of the first beacon initiated and the relative position of other stationary beacons as landmarks. The mobile beacons within this relative map reflect the real-world location of the Falcon robot. As the robot explores and identifies deteriorations using the deterioration detection module, the location of the detected deterioration's class is marked on the beacon map with their corresponding color code. It marks the deterioration's class with an accuracy of a 30 cm radius on the map and is useful for the efficient inspection and maintenance of false ceilings.

3.4. Remote Console

The remote console is used to monitor and control the mobile Falcon robot for performing experiments. The primary mode of interaction happens via a transmitter and receiver system and directly in nature. The user controls the machine by sending signals that are transmitted through a remote. In our case study, the Taranis Q X7 from FrSky was used considering full telemetry capabilities as well as the RSSI signal strength feedback. The battery compartment uses two 18650 Li-Ion batteries and can be balance-charged via the Mini USB interface.

4. Experiments and Results

This section elaborates on the experimental setup and results of the proposed false ceiling deterioration detection and mapping framework. The experiments were carried out in five steps: dataset preparation, training and validation, prediction with a test dataset, a real-time field trial, and a comparison with other models.

4.1. Data-Set Preparation

The false ceiling deterioration training dataset was prepared by collecting images from various online sources and defect image dataset libraries (a surface defect database [29] and a crack image dataset [30,31]). In our dataset collection process, the common false ceiling deteriorations are categorized into four classes, namely, structural defects (spalling, cracks, and pitted surfaces), infestation (termites and rodents), electrical damage (frayed wires), and degradation in HVAC systems (molding, corrosion, and water leakage). Five thousand images were collected from an online source, and around 800 images were collected from a real false ceiling environment to train the deep learning algorithm. The CNN model was trained and tested using images with a 768 × 1024 pixel resolution. The "LabelImg" GUI was used for bounding boxes and class annotations. Annotations were recorded as XML files in the PASCAL Visual Object Classes (VOC) format.

Further, the data augmentation process was applied on labeled images to help control the over-fitting and class imbalance issues in the model training stage. Data augmentation processes such as horizontal flips, scaling, cropping, rotations of the image, blurring, grayscale colors, and color enhancing were applied to the collected images. Figure 6 shows a sample of the data augmentation of one image. Further, Table 4 elaborates the settings of the various types of augmentations applied.

4.2. Training and Validation

The object detection model, the Faster R-CNN, was built using the TensorFlow (v1.15) API and the Keras wrapper library. The pre-trained Inception V2 model was used as a feature extraction module. It was trained on the COCO dataset. A Stochastic Gradient Descent (SGD) optimizer was used for the training of the Faster R-CNN module. The hyper-parameters used were 0.9 for momentum, an initial learning rate of 0.0002, which decays over time, and a batch size of 1.

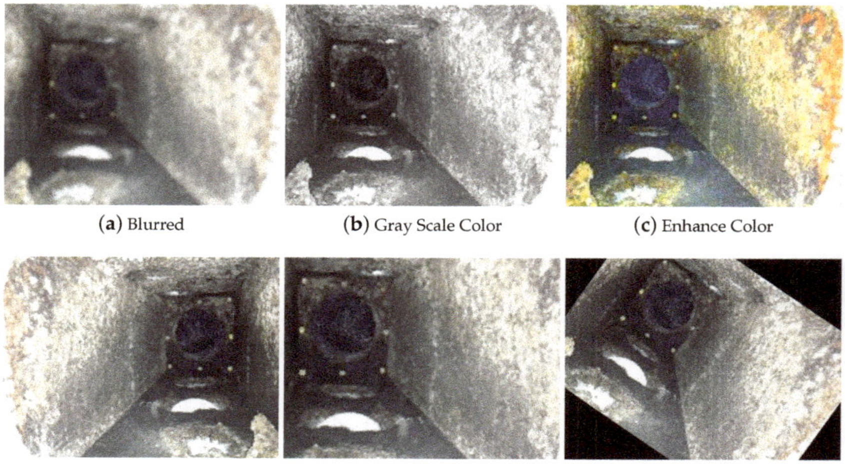

(a) Blurred **(b)** Gray Scale Color **(c)** Enhance Color **(d)** Horizontal Flip **(e)** Scale **(f)** Rotated

Figure 6. Sample of data augmentation of one image.

Table 4. Augmentation types and settings.

Augmentation Type	Augmentation Setting
Blurring	gaussianblur (from sigma 1.0× to 3.0×)
Grayscale	individual rgb spectrum (from factor 0 to 1.5×)
Color Enhancing	contrast (from 0.5× to 1.5×)
Horizontal Flip	flip the image horizontally
Scaling	0.5× to 1.5×
Rotation	from −45 degree to +45 degree
Translation	X-axis (−0.3× to 0.3×) Y-axis (−0.3× to 0.3×)

The model was trained and tested on the Lenovo ThinkStation P510. It consists of an Intel Xeon E5-1630V4 CPU running at 3.7 GHz, 64 GB of Random Access Memory (RAM), and a Nvidia Quadro P4000 GPU (1792 Nvidia CUDA Cores and 8 GB GDDR5 memory size running at a 192.3 GBps bandwidth). The same hardware is used to run as a local server to allow the Falcon robot to carry out inference during real-time testing.

The K-fold (here K = 10) cross-validation technique was used for validating the dataset and model training accuracy. In this evaluation, the dataset was divided into K subsets, and K−1 subsets were used for training. The remaining subset was used for evaluating the performance. This process was run K times to obtain the mean accuracy and other quality metrics of the detection model. K-fold cross-validation was performed to verify that the images reported were accurate and not biased towards a specific dataset split. The images shown were attained from the model with good precision. In this analysis, the model scored a 91.5% mean accuracy for k = 10. This indicates that the model is not biased towards a specific dataset split.

4.3. Prediction with the Test Dataset

The trained model's deterioration detection and classification accuracy were evaluated using the test dataset. In this evaluation process, 100 images were tested from each class. These test datasets were not used in the training and cross-validation of the model. Figure 7 shows the detection results of the given test dataset.

The experimental results show that the deterioration detection algorithm accurately detected and classified the deterioration in the given test images with a high confidence level average of 88%. Further, the model classification accuracy was evaluated using standard statistical metrics such as accuracy (Equation (1)), precision (Equation (2)), recall (Equation (3)), and $F_measure$ (Equation (4)).

$$Accuracy(Acc) = \frac{tp + tn}{tp + fp + tn + fn} \tag{1}$$

$$Precision(Prec) = \frac{tp}{tp + fp} \tag{2}$$

$$Recall(Rec) = \frac{tp}{tp + fn} \tag{3}$$

$$F_{measure}(F_1) = \frac{2 \times precision \times recall}{precision + recall} \tag{4}$$

Here, tp, fp, tn, fn represent the true positives, false positives, true negatives, and false negatives, respectively, as per the standard confusion matrix. Table 5 provides the statistical measure results of the offline test. Figure 8 demonstrates the graphical representation of Table 5 for improved visualization.

Table 5. Statistical measures for the deterioration detection framework (the proposed framework).

Category	Class	Precision	Recall	F1	Accuracy
Structural Defect	Spalling	93.60	92.77	93.16	88.70
	Crack	94.77	95.32	95.04	91.50
	Pitted Surface	95.70	93.13	94.39	89.70
	Water Damage	92.11	89.74	90.91	86.00
Degradation in HVAC	Molding	91.95	94.74	93.32	90.20
	Corrosion	91.88	93.63	92.74	88.50
	Pipe Damage	91.89	90.67	91.28	87.00
Infestation	Termite	96.59	92.39	94.44	90.10
	Rodent Activity	95.01	93.63	94.31	91.52
Electrical Damage	Fray wires	96.13	93.29	94.69	92.23

The statistical measures experimental result indicate that the proposed framework detected structural defects with an average accuracy of 88.9%, degradation in the HVAC system at an 88.56% accuracy, infestation at a 90.75% accuracy, and electrical damage at a 92.2 % accuracy.

4.4. Real-Time Field Trial

The real-time field trial experiments were performed in two different false ceiling environments, including the Oceania Robotics prototype false ceiling testbed and the SUTD ROAR laboratory real false ceiling. The false ceiling testbed consists of frames, dividers, pipes, and other common false ceiling elements. For experimental purposes, various deteriorations in false ceilings such as frayed wire, damaged pipes, and termite damage were manually created and placed in the prototype environment. Some of the defects, such as pitted surfaces and spalling, were fabricated using printed images of these defects. These printed images were glued at various locations in a false ceiling testbed for experimental purposes. Further, to track the robot position and identify the false ceiling deterioration location, a mobile beacon was placed on the top of the Falcon, and stationary beacons were mounted on projecting beams or sidewalls. The mobile beacons were the transmitters operating in unique frequencies, while all of the stationary beacons operated in the same frequency and behaved as receivers. The location of the moving beacons was calculated based on triangulating the distances from stationary beacons, and the current location was updated at a frequency of 16 Hz. With an accuracy of up to 2 cm and a bandwidth accommodating up to six mobile devices seamlessly, the beacon system implemented was used for false ceiling deterioration mapping and localization.

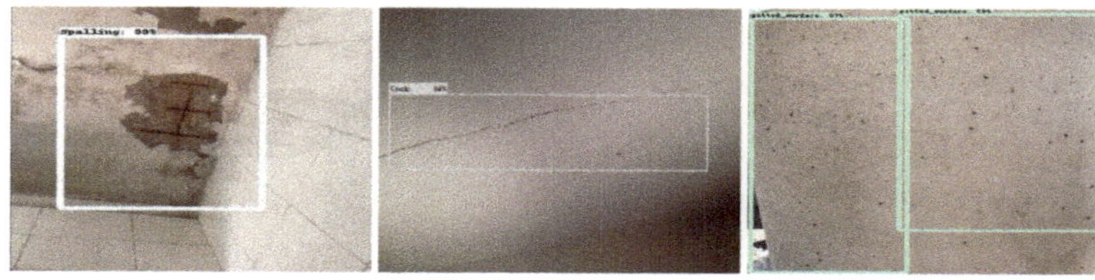

(**a**) Spalling (**b**) Crack (**c**) Pitted Surface

(**d**) Water Damage

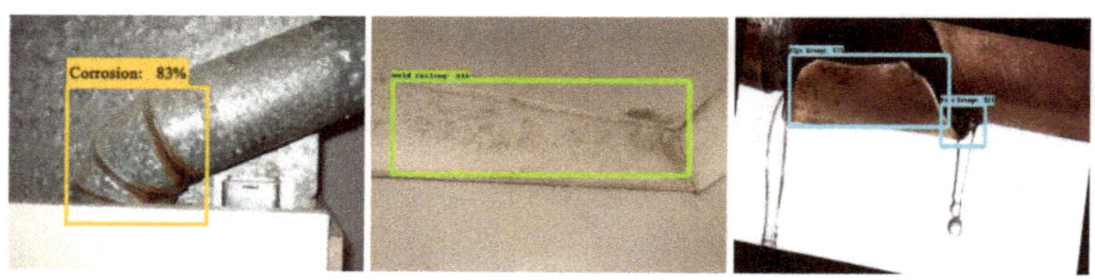

(**e**) Corrosion (**f**) Molding (**g**) Pipe Damage

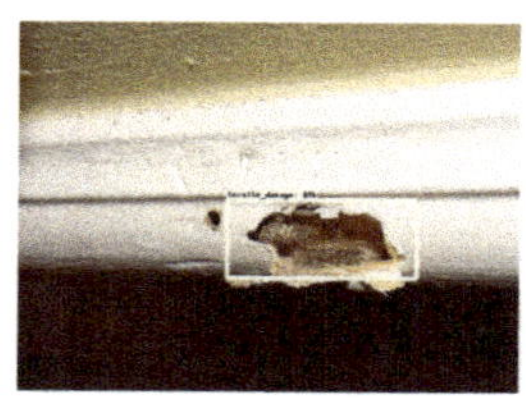

(**h**) Termite

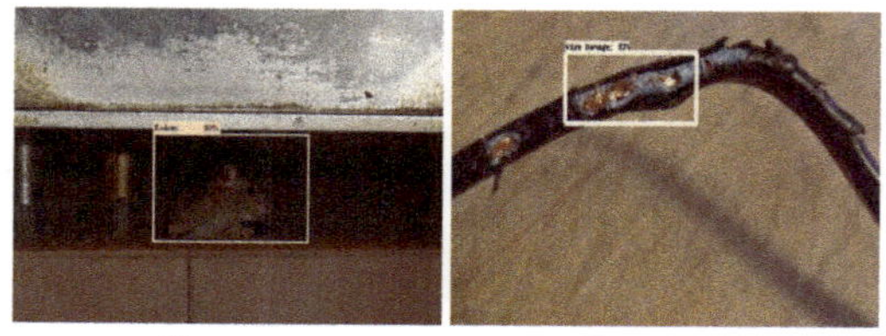

(**i**) Rodent (**j**) Fray Wire

Figure 7. Structural Defects (**a**–**d**), degradation in HVAC system (**e**–**g**), infestation (**h**,**i**), electrical damage (**j**) during Offline Testing.

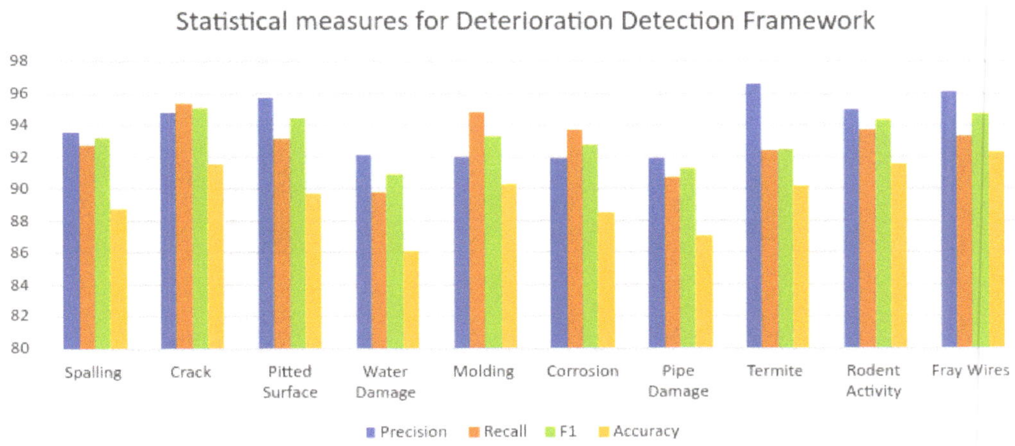

Figure 8. Graphical representation of the statistical measures of the proposed framework.

Figures 9 and 10 show the Falcon robot in the prototype of the false ceiling (Oceania Robotics test bed), while Figure 11 shows the robot in a real false ceiling environment (SUTD ROAR Laboratory). During the inspection, the robot was controlled by a mobile GUI interface, and the robot's position and the defect region were localized through UWB modules fixed in the false ceiling environment. The robot was paused at each stage for a few seconds to capture better quality images in these real-time field experiments.

(**a**) False Ceiling Testbed (**b**) Robot Falcon in Testbed

Figure 9. Falcon and the false ceiling testbed prototype.

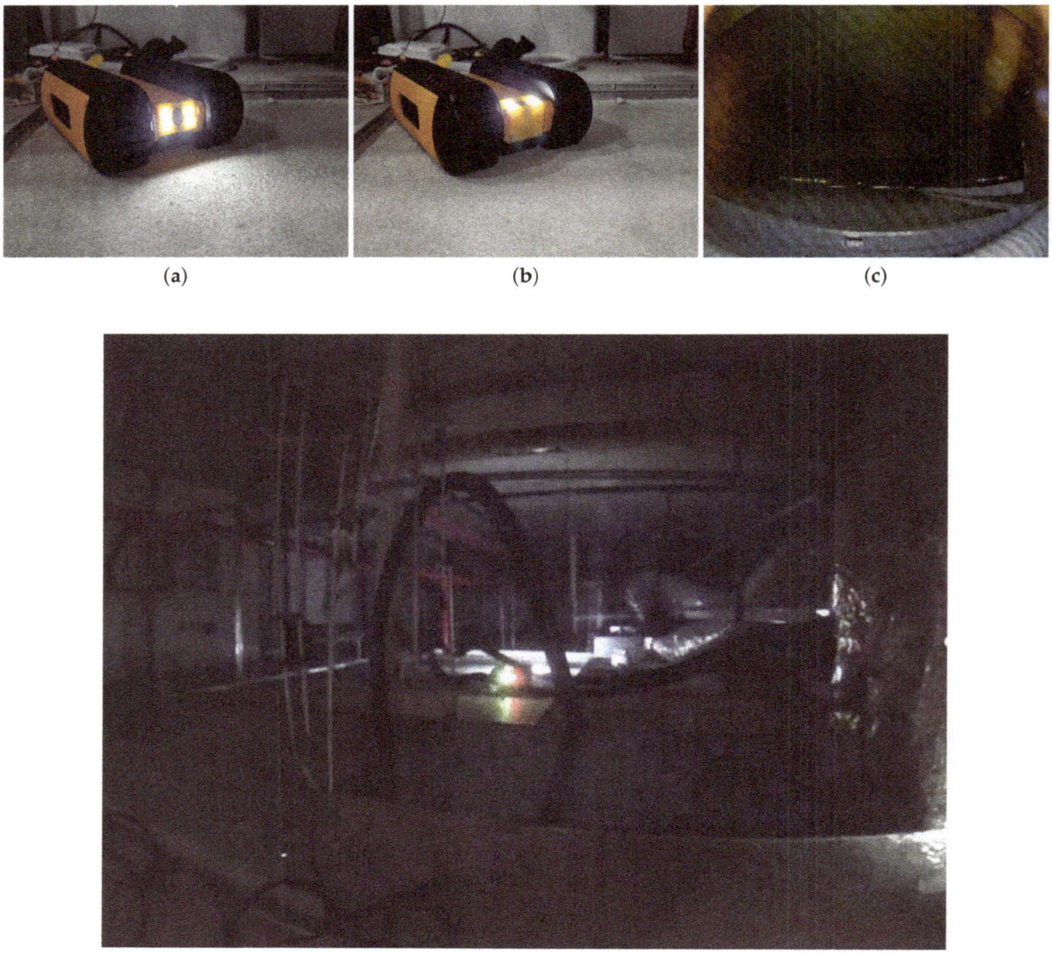

(a) (b) (c)

(d)

Figure 10. Falcon's performance on the false ceiling prototype at Oceanica Robotics. (**a**) Falcon in Prototype of False Ceiling (Camera at 0 degree); (**b**) Falcon in Prototype of False Ceiling (Camera at 90 degree); (**c**) Image collected by Falcon; (**d**) Falcon in Prototype of False Ceiling (Zoomed out).

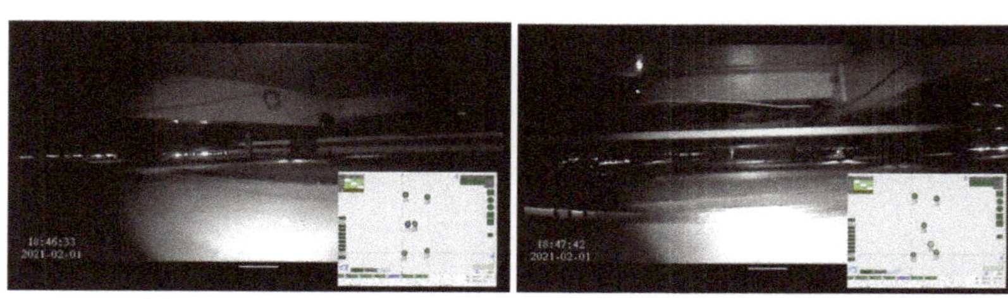

(**a**) Falcon at Location 1 (**b**) Falcon at Location 2

Figure 11. Falcon's performance on the false ceiling at the SUTD ROAR Laboratory.

The captured images were transferred to a high-powered GPU-enabled local server for false inspection tasks via WiFi communication. Figure 12 depicts the real-time filed trial deterioration detection results of the false ceiling testbed, and its localization results are shown in Figure 13. These deterioration-detected image frame locations were identified by fusing the beacon coordinates, wheel decoder data, and IMU sensor data on the Marvel Mind Dashboard tracking software. Figure 13 also shows the deterioration location mapping results for the real-time field trials, where the color codes indicate the class of deterioration.

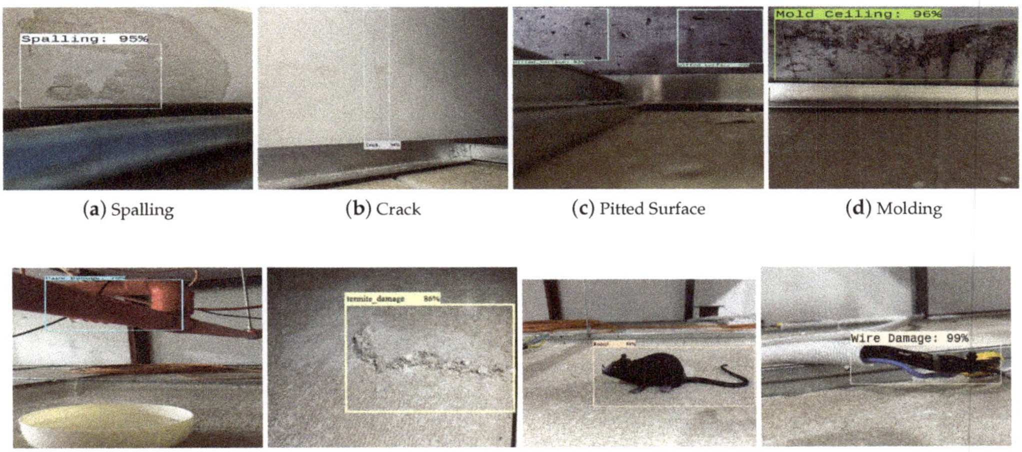

(**a**) Spalling (**b**) Crack (**c**) Pitted Surface (**d**) Molding

(**e**) Pipe Damage (**f**) Termite (**g**) Rodent (**h**) Fray Wire

Figure 12. Structural defects (**a**–**c**), degradation in the HVAC system (**d**,**e**), infestation (**f**,**g**), and electrical damage (**h**) during online testing.

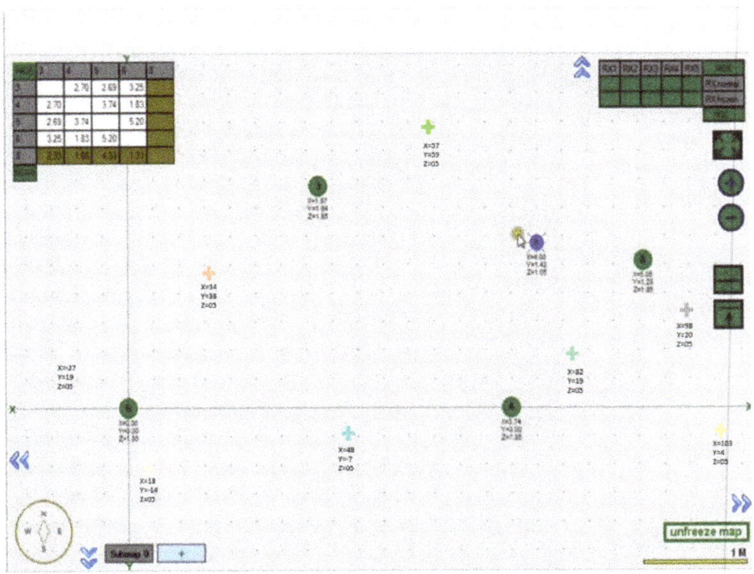

Figure 13. Beacon maps with static beacons and a mobile beacon.

The findings of the experiment reveal that the Falcon robot's maneuverability was stable. It could move around a complex false ceiling environment and accurately capture it for false ceiling deterioration identification. The detection algorithm detected most of the false ceiling deterioration in the real-time field trial with a good confidence level and scored an 88% mean detection accuracy. Furthermore, the Falcon robot's position on the false ceiling could be reliably tracked using the UWB localization results. This will further help inspection teams to identify defects and degradation efficiently.

5. Discussion

The proposed system's performance is discussed in this section by a comparison with two models (Faster Inception ResNet and Faster Resnet 152) and other existing studies. The comparison analysis findings are shown in Tables 6–8. The three detection frameworks were trained on the same image dataset and with the same number of epochs. Here, overall detection accuracies of 86.8% for the Faster Resnet 152 and 86.53% for the Faster Inception Resnet were observed. The detection accuracy of these two models was relatively low due to a high false-positive rate and misclassification issues due to similar deterioration factors and the impact of object illumination. These issues can be further resolved by retraining the algorithm with misclassified classes and applying nonlinear detrending techniques [32]. Further, Figure 14 shows a graphical representation of Table 8 for improved visualisation.

Table 6. Statistical measures for the deterioration detection framework (Faster Resnet 152).

Category	Class	Precision	Recall	F1	Accuracy
Structural Defect	Spalling	87.12	86.11	86.61	86.4
	Crack	90.87	90.47	90.67	90.2
	Pitted Surface	85.61	84.8	85.2	84.2
	Water Damage	86.79	87	86.89	84.6
Degradation in HVAC	Molding	90.69	91.09	90.89	89.8
	Corrosion	88.86	89.27	89.07	87.6
	Pipe Damage	85.51	84.89	85.2	83.4
Infestation	Termite	85.55	86.56	86.05	84.8
	Rodent Activity	90.38	88.99	89.68	90.8
Electrical Damage	Fray wires	90.54	88.86	89.7	86.2

Table 7. Statistical measures for the deterioration detection framework (Faster Inception Resnet).

Category	Class	Precision	Recall	F1	Accuracy
Structural Defect	Spalling	88.19	88.81	88.5	85.8
	Crack	92.51	92.92	92.72	90.4
	Pitted Surface	86.57	85.95	86.26	83.9
	Water Damage	87.8	86.97	87.38	84.3
Degradation in HVAC	Molding	92.05	92.46	92.27	90.2
	Corrosion	88.84	87.61	88.22	87.2
	Pipe Damage	86.76	87.17	86.97	84.1
Infestation	Termite	85.16	84.13	84.64	83.2
	Rodent Activity	93.16	92.54	92.85	91.1
Electrical Damage	Fray wires	87.21	88.03	87.62	85.1

Table 8. Comparison with other object detection frameworks.

Class	Precision	Recall	F1	Overall Accuracy
Faster Inception Resnet	94.92	93.3	94.11	86.5
Faster Resnet 152	92.9	93.21	93.05	86.8
Faster Inception (Proposed Model)	93.96	92.93	93.43	89.53

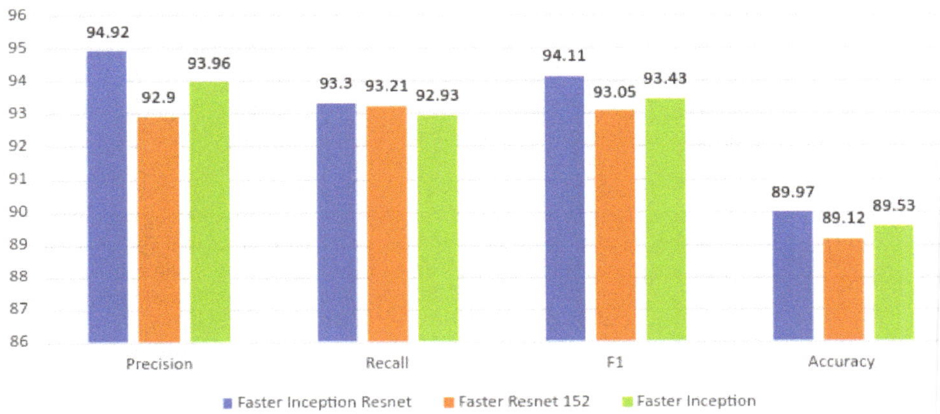

Figure 14. Graphical representation of the comparison with other detection frameworks.

The cost of training and testing is shown in Table 9. In that analysis, we found that the proposed model also had a lower execution time compared to the Faster Inception Resnet and Faster Resnet 152 models. Because of this, the framework that has been proposed is better suited for false ceiling deterioration detection tasks.

Table 9. Computational cost analysis.

Algorithm	Training Time (Hours:Minutes)	Speed (Milliseconds)
Faster Inception Resnet	23:20	647
Faster Resnet 152	19:48	135
Faster Inception (Proposed Model)	16:18	68

Table 10 shows the accuracy of various defect detection algorithms based on different classes. However, a fair comparison is lacking because their algorithm, datasets, and training parameters are not the same. Finally, the proposed method involves robotic inspection, which is another contribution with respect to the state of the art.

Table 10. Comparison of results with other methodologies in related work.

Reference	Algorithm	Accuracy
Gary et al. [5]	Mask-RCNN	80.00%
Hung La et al. [15]	CNN-based image stitching and 3D registration	93.10 %
Gui et al. [16]	CNN-based	70.00 %
Perez et al. [17]	VGG-16	87.50 %
Xing et al. [18]	CNN model (SCN) and optimised IoU	85.84%
Xian et al. [19]	Cascaded encoder (CASAE)	89.60%
Cheon et al. [20]	Single CNN	96.20 %
Proposed Method	Faster R-CNN Inception	89.53%

6. Conclusions

False ceiling defect detection and mapping were presented using our in-house-developed Falcon robot and the Faster Inception object detection algorithm. The efficiency of the proposed system was tested through a robot maneuverability test and showed defect detection accuracy in offline and real-time field trials. The robot's maneuverability was tested in two different false ceiling environments: the Oceania Robotics prototype false ceiling testbed and the SUTD ROAR laboratory real false ceiling. The experimental results proved that the Falcon robot's maneuverability was stable and that its defect mapping was accurate in a complex false ceiling environment. Further, the defect detection algorithm was tested on a test dataset, and real-time false ceiling images were collected by the Falcon robot. The experimental results show that Faster Inception has a good trade-off between detection accuracy and computation time, with a detection accuracy of 89.53% for detecting deterioration in real-time Falcon-collected false-ceiling-environment video streams, whereas the average detection accuracies of Faster Resnet 152 and Faster Inception Resnet were 86.8% and 86.53%, respectively. Further, Faster Inception required only 68 ms to process one image on the local server, which is lower compared with other algorithms, including Faster Inception Resnet and Faster Resnet 152. Further, the mapping results precisely indicated the location of deterioration on the false ceiling. Thus, it can be concluded that the suggested method is more suited for defect detection in false ceiling environments and can improve inspection services. In our future work, we plan to add more features to the false ceiling inspection framework, such as olfactory contamination detection.

Author Contributions: Conceptualization, B.R. and R.E.M.; methodology, A.S., R.E.M. and B.R.; software, A.S., L.M.J.M. and C.B.; validation, L.Y. and S.P.; formal analysis, A.S., S.P. and C.B.; investigation, R.E.M. and B.R.; resources, R.E.M.; data, A.S., L.Y., P.P. and L.M.J.M.; writing—original draft preparation, A.S., B.R., L.M.J.M., P.P., S.P. and C.B.; supervision, R.E.M.; project administration, R.E.M.; funding acquisition, R.E.M. All authors have read and agreed to the published version of the manuscript.

Funding: This research was supported by the National Robotics Programme under Robotics Enabling Capabilities and Technologies (Funding Agency Project No. 192 25 00051) and the National Robotics Programme under its Robot Domain (Funding Agency Project No. 192 22 00108) and was administered by the Agency for Science, Technology and Research.

Institutional Review Board Statement: Not applicable.

Informed Consent Statement: Not applicable.

Data Availability Statement: Not applicable.

Acknowledgments: The authors would like to thank the National Robotics Programme, the Agency for Science, Technology and Research, and SUTD for their support.

Conflicts of Interest: There are no conflicts of interest.

References

1. Edgemon, G.L.; Moss, D.; Worland, W. Condition Assessment of the Los Alamos National Laboratory Radioactive Liquid Waste Collection System. In *CORROSION 2005*; OnePetro: Los Alamos, NM, USA, 2005.
2. Henry, R.S.; Dizhur, D.; Elwood, K.J.; Hare, J.; Brunsdon, D. Damage to concrete buildings with precast floors during the 2016 Kaikoura earthquake. *Bull. N. Z. Soc. Earthq. Eng.* **2017**, *50*, 174–186. [CrossRef]
3. Jiang, Y.; Han, S.; Bai, Y. Building and Infrastructure Defect Detection and Visualization Using Drone and Deep Learning Technologies. *J. Perform. Constr. Facil.* **2021**, *35*, 04021092. [CrossRef]
4. Aliyari, M.; Droguett, E.L.; Ayele, Y.Z. UAV-Based Bridge Inspection via Transfer Learning. *Sustainability* **2021**, *13*, 11359. [CrossRef]
5. Atkinson, G.A.; Zhang, W.; Hansen, M.F.; Holloway, M.L.; Napier, A.A. Image segmentation of underfloor scenes using a mask regions convolutional neural network with two-stage transfer learning. *Autom. Constr.* **2020**, *113*, 103118. [CrossRef]
6. Ramalingam, B.; Tun, T.; Mohan, R.E.; Gómez, B.F.; Cheng, R.; Balakrishnan, S.; Mohan Rayaguru, M.; Hayat, A.A. AI Enabled IoRT Framework for Rodent Activity Monitoring in a False Ceiling Environment. *Sensors* **2021**, *21*, 5326. [CrossRef]
7. Protopapadakis, E.; Voulodimos, A.; Doulamis, A.; Doulamis, N.; Stathaki, T. Automatic crack detection for tunnel inspection using deep learning and heuristic image post-processing. *Appl. Intell.* **2019**, *49*, 2793–2806. [CrossRef]
8. Menendez, E.; Victores, J.G.; Montero, R.; Martínez, S.; Balaguer, C. Tunnel structural inspection and assessment using an autonomous robotic system. *Autom. Constr.* **2018**, *87*, 117–126. [CrossRef]
9. Palanisamy, P.; Mohan, R.E.; Semwal, A.; Jun Melivin, L.M.; Félix Gómez, B.; Balakrishnan, S.; Elangovan, K.; Ramalingam, B.; Terntzer, D.N. Drain Structural Defect Detection and Mapping Using AI-Enabled Reconfigurable Robot Raptor and IoRT Framework. *Sensors* **2021**, *21*, 7287. [CrossRef] [PubMed]
10. Melvin, L.M.J.; Mohan, R.E.; Semwal, A.; Palanisamy, P.; Elangovan, K.; Gómez, B.F.; Ramalingam, B.; Terntzer, D.N. Remote drain inspection framework using the convolutional neural network and re-configurable robot Raptor. *Sci. Rep.* **2021**, *11*, 22378. [CrossRef] [PubMed]
11. Kouzehgar, M.; Tamilselvam, Y.K.; Heredia, M.V.; Elara, M.R. Self-reconfigurable façade-cleaning robot equipped with deep-learning-based crack detection based on convolutional neural networks. *Autom. Constr.* **2019**, *108*, 102959. [CrossRef]
12. Pan, Z.; Yang, J.; Wang, X.; Wang, F.; Azim, I.; Wang, C. Image-based surface scratch detection on architectural glass panels using deep learning approach. *Constr. Build. Mater.* **2021**, *282*, 122717. [CrossRef]
13. Jamil, M.; Sharma, S.K.; Singh, R. Fault detection and classification in electrical power transmission system using artificial neural network. *SpringerPlus* **2015**, *4*, 334. [CrossRef] [PubMed]
14. Tun, T.T.; Elara, M.R.; Kalimuthu, M.; Vengadesh, A. Glass facade cleaning robot with passive suction cups and self-locking trapezoidal lead screw drive. *Autom. Constr.* **2018**, *96*, 180–188. [CrossRef]
15. La, H.M.; Dinh, T.H.; Pham, N.H.; Ha, Q.P.; Pham, A.Q. Automated robotic monitoring and inspection of steel structures and bridges. *Robotica* **2019**, *37*, 947–967. [CrossRef]
16. Gui, Z.; Li, H. Automated defect detection and visualization for the robotic airport runway inspection. *IEEE Access* **2020**, *8*, 76100–76107. [CrossRef]
17. Perez, H.; Tah, J.H.; Mosavi, A. Deep learning for detecting building defects using convolutional neural networks. *Sensors* **2019**, *19*, 3556. [CrossRef]
18. Xing, J.; Jia, M. A convolutional neural network-based method for workpiece surface defect detection. *Measurement* **2021**, *176*, 109185. [CrossRef]
19. Tao, X.; Zhang, D.; Ma, W.; Liu, X.; Xu, D. Automatic metallic surface defect detection and recognition with convolutional neural networks. *Appl. Sci.* **2018**, *8*, 1575. [CrossRef]
20. Cheon, S.; Lee, H.; Kim, C.O.; Lee, S.H. Convolutional neural network for wafer surface defect classification and the detection of unknown defect class. *IEEE Trans. Semicond. Manuf.* **2019**, *32*, 163–170. [CrossRef]
21. Ladig, R.; Shimonomura, K. High precision marker based localization and movement on the ceiling employing an aerial robot with top mounted omni wheel drive system. In Proceedings of the 2016 IEEE/RSJ International Conference on Intelligent Robots and Systems (IROS), Daejeon, Korea, 9–14 October 2016; pp. 3081–3086.
22. Zhang, Y.; Dodd, T.; Atallah, K.; Lyne, I. Design and optimization of magnetic wheel for wall and ceiling climbing robot. In Proceedings of the 2010 IEEE International Conference on Mechatronics and Automation, Xi'an, China, 4–7 August 2010; pp. 1393–1398.
23. Unver, O.; Sitti, M. A miniature ceiling walking robot with flat tacky elastomeric footpads. In Proceedings of the 2009 IEEE International Conference on Robotics and Automation, Kobe, Japan, 12–17 May 2009; pp. 2276–2281.
24. Hayat, A.A.; Ramanan, R.K.; Abdulkader, R.E.; Tun, T.T.; Ramalingam, B.; Elara, M.R. Robot with Reconfigurable Wheels for False-ceiling Inspection: Falcon. In Proceedings of the 5th IEEE/IFToMM International Conference on Reconfigurable Mechanisms and Robots (ReMAR), Toronto, ON, Canada, 12–14 August 2021; pp. 1–10.
25. Tan, N.; Hayat, A.A.; Elara, M.R.; Wood, K.L. A framework for taxonomy and evaluation of self-reconfigurable robotic systems. *IEEE Access* **2020**, *8*, 13969–13986. [CrossRef]
26. Manimuthu, M.; Hayat, A.A.; Elara, M.R.; Wood, K. Transformation design Principles as enablers for designing Reconfigurable Robots. In Proceedings of the International Design Engineering Technical Conferences and Computers and Information in Engineering Conference, St. Louis, MO, USA, 17–20 August 2021; pp. 1–12.

27. Caesar, H.; Uijlings, J.; Ferrari, V. Coco-stuff: Thing and stuff classes in context. In Proceedings of the IEEE Conference on Computer Vision and Pattern Recognition, Salt Lake City, UT, USA, 18–23 June 2018; pp. 1209–1218.
28. Ren, S.; He, K.; Girshick, R.; Sun, J. Faster r-cnn: Towards real-time object detection with region proposal networks. *arXiv* **2015**, arXiv:1506.01497.
29. Guan, S.; Lei, M.; Lu, H. A steel surface defect recognition algorithm based on improved deep learning network model using feature visualization and quality evaluation. *IEEE Access* **2020**, *8*, 49885–49895. [CrossRef]
30. Shi, Y.; Cui, L.; Qi, Z.; Meng, F.; Chen, Z. Automatic road crack detection using random structured forests. *IEEE Trans. Intell. Transp. Syst.* **2016**, *17*, 3434–3445. [CrossRef]
31. Cui, L.; Qi, Z.; Chen, Z.; Meng, F.; Shi, Y. Pavement Distress Detection Using Random Decision Forests. In *International Conference on Data Science*; Springer: Berlin/Heidelberg, Germany, 2015; pp. 95–102.
32. Civera, M.; Fragonara, L.Z.; Surace, C. Video processing techniques for the contactless investigation of large oscillations. In *Journal of Physics: Conference Series*; IOP Publishing: Bristol, UK, 2019; Volume 1249, p. 012004.

sensors

MDPI

Article

Smooth Complete Coverage Trajectory Planning Algorithm for a Nonholonomic Robot

Ana Šelek, Marija Seder *, Mišel Brezak and Ivan Petrović

Laboratory for Autonomous Systems and Mobile Robotics (LAMOR), Faculty of Electrical Engineering and Computing, University of Zagreb, 10000 Zagreb, Croatia
* Correspondence: marija.seder@fer.hr

Abstract: The complete coverage path planning is a process of finding a path which ensures that a mobile robot completely covers the entire environment while following the planned path. In this paper, we propose a complete coverage path planning algorithm that generates smooth complete coverage paths based on clothoids that allow a nonholonomic mobile robot to move in optimal time while following the path. This algorithm greatly reduces coverage time, the path length, and overlap area, and increases the coverage rate compared to the state-of-the-art complete coverage algorithms, which is verified by simulation. Furthermore, the proposed algorithm is suitable for real-time operation due to its computational simplicity and allows path replanning in case the robot encounters unknown obstacles. The efficiency of the proposed algorithm is validated by experimental results on the Pioneer 3DX mobile robot.

Keywords: mobile robot; complete coverage; path planning; path smoothing; velocity profile optimization

Citation: Šelek, A.; Seder, M.; Brezak, M.; Petrović, I. Smooth Complete Coverage Trajectory Planning Algorithm for a Nonholonomic Robot. *Sensors* **2022**, *22*, 9269. https://doi.org/10.3390/s22239269

Academic Editor: Andrey V. Savkin

Received: 19 October 2022
Accepted: 23 November 2022
Published: 28 November 2022

1. Introduction

The task of a Complete Coverage Path Planning (CCPP) algorithm is to generate such a path for a mobile robot that ensures that the robot completely covers the entire environment while following the planned path. There are many real-world applications that require a CCPP algorithm, such as floor cleaning [1–3], demining [4,5], automated harvesting [6], lawn mowing [7], autonomous underwater exploration [8,9], etc. To achieve efficient coverage in these applications, the planned path must satisfy some requirements, where the most important ones are maximizing the coverage rate, minimizing the complete coverage time, path length, overlap area, and energy consumption of the robot, and rapid path replanning if the environment changes during the execution of the coverage task. Many CCPP algorithms have been developed, but none of them meet all the above requirements, so there is still room for improvement.

In this paper, we propose a CCPP algorithm which generates smooth complete coverage paths that allow nonholonomic mobile robots to move in optimal time while following them (we called it SCCPP). Moreover, the proposed SCCPP algorithm is suitable for real-time operation due to its computational simplicity and allows path replanning in case the robot encounters unknown obstacles. The existing coverage algorithms in the literature are either non-smooth so they have increased coverage redundancy due to the non-ideal path following, or they have slow path planning and replanning. The SCCPP algorithm combines two of our previous works: the fast coverage planning algorithm [10] with the fast clothoid calculation [11]. The first algorithm is our replanning spanning tree coverage (RSTC) algorithm that generates a path in a low-resolution occupancy grid map to reduce the computational complexity and minimize the overlap rate. The path is the shortest possible coverage path in the corresponding graph, which contains sharp 90° turns. To avoid stopping and rotating the robot at turning points, we used clothoids to smooth the path generated by the RSTC algorithm. The main advantage of the clothoids is the linear change in their curvature, which allows the nonholonomic robot to move in a time-optimal manner while following such a smoothed path.

The SCCPP is the real-time traversable collision-free complete coverage path planning algorithm based on clothoids, which gives minimal path length, the coverage time, and overlap area and maximal coverage rate compared to the state-of-the-art coverage algorithms. Such a path is suitable and feasible for nonholonomic mobile robots.

The remaining of the paper is structured as follows. Section 2 presents the overview of the complete coverage path planning algorithms. The proposed smooth complete coverage path planning algorithm is described in Section 3. The simulation and experimental results are given in Sections 4 and 5, respectively, followed by the conclusion of the paper.

2. Related Work

One of the earliest works on complete coverage path planning is presented in [12], which defines requirements for covering a continuous two-dimensional (2D) environment. Typically, methods for CCPP tasks assume that the environment is known, including the obstacle configuration. A number of methods perform an exact cell decomposition of the environment, e.g., [13–15], where free space is decomposed into trapezoidal, rectangular, or triangular cells. The path that completely covers each cell (regarding their shape) can lead to increased coverage redundancy and is time-consuming when the environment changes. Simpler and faster are methods that perform an approximate cell decomposition where the environment is represented with a 2D occupancy grid map, i.e., equally distributed square grid of discrete cells [16–19].

The size of the square grid cells directly affects the replanning rate and the coverage rate. These two requirements are opposite, i.e., larger cell sizes allow real-time replanning due to lower computational complexity, while smaller ones ensure higher coverage rate at the cost of higher computational complexity. Fast replanning algorithms usually select the cell size equal to the footprint of the robot, while the cell size is much smaller in the algorithms that ensure a high coverage rate, usually from 2 to 10 cm for the cell side [20,21].

A graph can be constructed from the occupancy grid map, where the grid cells are the nodes and the connections between adjacent grid cells are the edges. An optimal CCPP would ensure that the robot completely covers the entire environment by visiting all nodes in the graph only once, but this is a NP -hard problem, known as the Traveling Salesman Problem (TSP) [22]. Therefore, CCPP algorithms use approximate or heuristic solutions (see Figure 1). The complete coverage path planning algorithm based on a numeric potential field approach called the path transform (PT) can be used for the TSP solution [16]. Ant colony optimization (ACO) [23,24] is based on artificial ants colony that are able to generate the shortest feasible path of the TSP successively. Another solution for TSP can be determined by using the Hopfield neural network (HNN) for an optimization that consists of a single layer containing one or more fully connected recurrent neurons [25–27]. The Spanning Tree Coverage (STC) algorithm [28] and the Replanning Spanning Tree Coverage (RSTC) algorithm [10] produce the optimal coverage path in linear time. The CCPP algorithms on a high-resolution occupancy grid map have increased coverage rate.

The Complete Coverage D* (CCD*) algorithm [2] has a high coverage rate, which leads to increased coverage redundancy when the environment changes. The hex-decomposition-based coverage planning (HDCP) algorithm [29] for unknown environments plans smooth paths for Dubins vehicles to follow at constant velocity in real-time. Adaptive coverage path planning [30] is aiming to achieve complete coverage with minimal path length and it is efficient in dynamic environments. Complete coverage path planning based on biologically inspired neural network [31] plans collision-free trajectory for real-time coverage task in dynamic environments. The energy constrained online coverage path planning [32] is based on contour following, which causes sharp turns in rectangular environments. The drawbacks of the above algorithms are that the coverage redundancy is increased when dynamic obstacles are present, and a trade-off exists between a higher coverage rate and higher coverage redundancy. Moreover, the energy consumption and the coverage time are high because most of these algorithms generate the path with straight lines that form sharp turns. When the nonholonomic robot follows such a path, it must stop and reorient itself so that the heading of the robot matches the path direction [33]. Authors in [34] do a complete 2D sweep coverage, but their approach does not produce smooth paths, is not computationally efficient, and does not consider dynamic obstacles but only static environments.

To provide optimal and feasible paths with curvature continuity that are easy to follow by nonholonomic mobile robots, path smoothing algorithms are used. A smoothing algorithm provides motion continuity and reduces the execution time of coverage tasks. The first research on finding the shortest curvature constrained smooth paths consisting of straight lines and arcs was done by Dubin in [35]. The presence of discontinuities at the intersection of the straight lines and circular arcs makes them infeasible for real applications. Backman et al. [36] presented a curvature continuous CCPP algorithm using the Dubin curve for agricultural vehicles. The generated path is easy to follow for real-world applications and the algorithm is efficient in coverage time but computationally complex. Yu et al. [37] proposed Dubin's vehicle-based coverage algorithm, which minimizes the non-working distance in CCPP, reduces the number of turns of the vehicle, maximizes the coverage area, but increases the overlap area. Jin and Tang [38] developed a coverage path planning algorithm which respects the kinematic constraints of the robot, but it is specialized for agricultural or similar purposes and is computationally complex. Lee et al. [39] presented a smooth CCPP algorithm based on Bézier curve, wall following, and high-resolution grid map representation. The algorithm maximizes coverage rate and minimizes energy consumption, but has increased coverage redundancy. All the above smoothing algorithms are not suitable for changing environments due to high computational complexity.

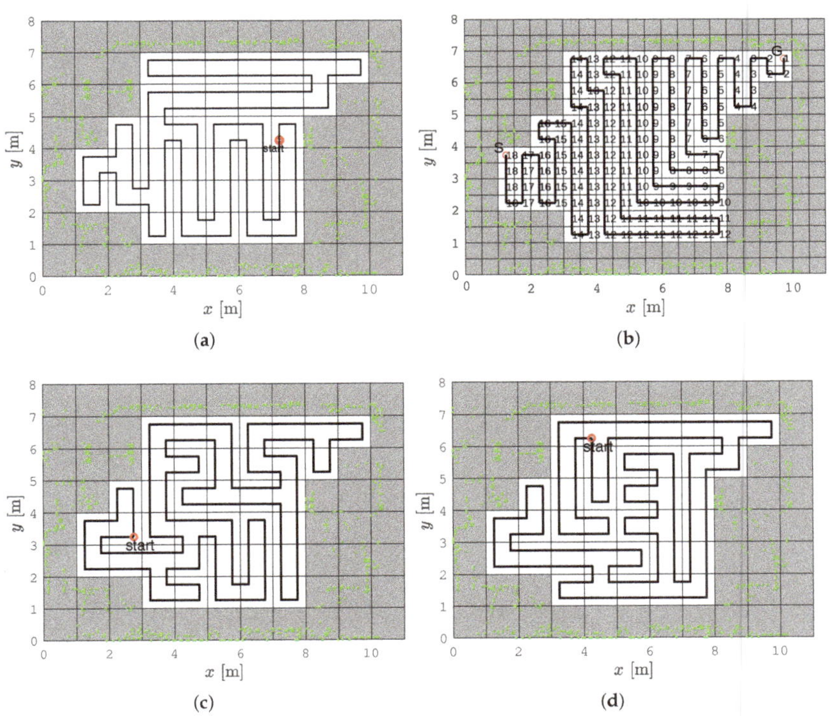

Figure 1. Solution examples of different methods for the TSP problem. (**a**) The RSTC algorithm, (**b**) The PT algorithm, (**c**) The ACO algorithm, (**d**) The HNN algorithm.

3. The Proposed Smooth Complete Coverage Path Planning Algorithm

The pipeline of the SCCPP algorithm, shown in Figure 2, consists of four sequentially computed modules. The first module is the replanning spanning tree coverage algorithm (RSTC), which generates a path composed of straight-line path segments perpendicular to each other (Section 3.1). The algorithm initially requires the map of the environment and the starting pose of the robot. The initial map can be a CAD map or a map created with a simultaneous localization and mapping algorithm (SLAM). The second module is the path smoothing algorithm, which smooths the path generated by the RSTC algorithm using clothoids (Section 3.2). The third module is the velocity profile optimization algorithm, which computes the highest admissible velocity profile along the smoothed path accounting for kinematic and dynamic constraints of the robot (Section 3.3). The fourth module is the trajectory tracking algorithm, which ensures that the robot tracks the optimal 5D trajectory generated by the third module (Section 3.3).

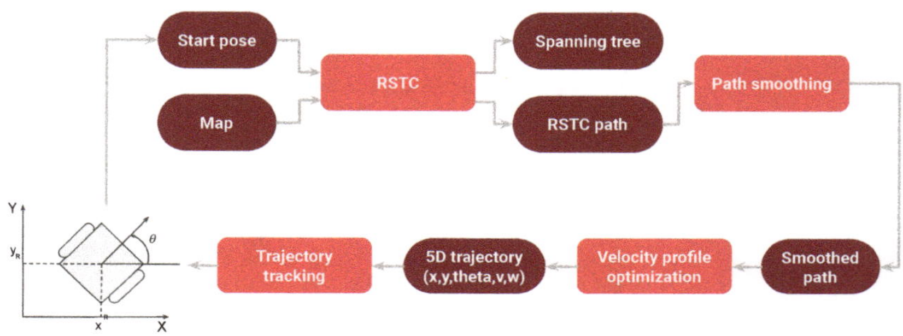

Figure 2. The pipeline of the smooth complete coverage path planning algorithm.

3.1. The Replanning Spanning Tree Coverage Algorithm

We use the replanning spanning tree coverage (RSTC) algorithm [10], which we briefly describe here. A map of the environment, given in the Portable Network Graphics (PNG) format, was edited to align the walls with the grid for better performance of the coverage algorithm and downsampled to a low-resolution occupancy grid with equally sized square cells (Figure 3a). Each cell contains occupancy information depending on the obstacle position. Only completely free cells were considered in the path planning and they are divided into four equal subcells of size D, where D is the diameter of the circumscribed circle around the robot's footprint (Figure 3b).

To create an optimal path, which visits each subcell exactly once, a spanning tree is constructed (Figure 3c). Free $2D$-size cells induce an undirected graph structure whose nodes are the centers of the cells, while the edges are the connecting lines between centers of the adjacent cells that share a common cell side. A spanning tree of an undirected graph is a subgraph that includes all nodes of the graph, with a minimum possible number of edges. A spanning tree is found in linear time by using the Breadth First Search (BFS) algorithm, meaning that the candidate neighbor nodes are kept in a queue. The pseudocode for spanning tree determination is given by Algorithm 1. The construction of the spanning tree starts from the node which contains the robot's position. All its neighbor nodes are inserted into the queue. The second node of the tree defines the first edge of the spanning tree, which has a direction aligned with the robot's orientation, if possible. The spanning tree continues to grow by removing from the queue and connecting to the last added node of the tree the node that is its neighbor. If this node has no neighbors in the queue, the node added one step earlier is compared to the queue, and so on. The tree is constructed when there are no nodes in the queue.

Algorithm 1 Pseudocode for the spanning tree

```
Input: occupancyGridMap, currentPose
Output: spanningTree
insert starting cell which contains the robot's position in the queue spanningTree
while all free cells in occupancyGridMap have been visited do
    determine all orthogonal and unvisited neighbors of the current cell moving counterclockwise and add them to the queue neighbor
    for all cells in queue neighbor do
        if distance of current cell in spanningTree and neighbor ≤ 1 then
            add current cell to the queue spanningTree
            set current cell in occupancyGridMap as visited
            go to the neighbor cell
        end if
    end for
    if distance of current cell in spanningTree and neighbor > 1 then
        for cells in queue spanningTree do
            if distance of current cell in spanningTree and neighbor ≤ 1 then
                go to the neighbor cell
            else
                move to the previous cell in queue spanningTree
            end if
        end for
    end if
end while
return the spanning tree
```

Once the spanning tree is created, the coverage path computation begins. The pseudocode for the path planning is given by Algorithm 2. The robot is assumed to follow the path around the spanning tree, always on the right side, until it completely covers all subcells. The coverage ends when the robot gets to the start subcell. The result is the complete coverage path, which consists of a series of connected lines (Figure 3d).

Algorithm 2 Pseudocode for the path planning

```
Input: spanningTree, currentPose
Output: RSTCpath
while it reaches the starting subcell again do
    calculate the direction of the spanning tree form current cell to the next first neighbor which is connected with the edge in the spannning tree
    mark current subcell as visited
    add subcell center coordinates in the queue RSTCpath
    go to neighbor cell
    if dynamic obstacle is detected then
        end the path planning algorithm
    end if
end while
return the RSTC path
```

Figure 1 shows a comparison of the RSTC algorithm with heuristic methods that solve TSP: the PT, ACO, and HNN algorithms. The length of the path for all methods is 76 m. The RSTC path has the smallest number of turning points (42 turning points compared to 48 for PT, 56 for ACO, and 50 for HNN). The higher number of turns results with longer coverage execution time. Therefore, our algorithm produces the coverage path more suitable for nonholonomic robots than other TSP solutions. The RSTC path has the lowest computational time (measured in Matlab/Python implementation) (0.32 s compared to 1.56 s for PT, 124 s for ACO, and 0.6 s for HNN).

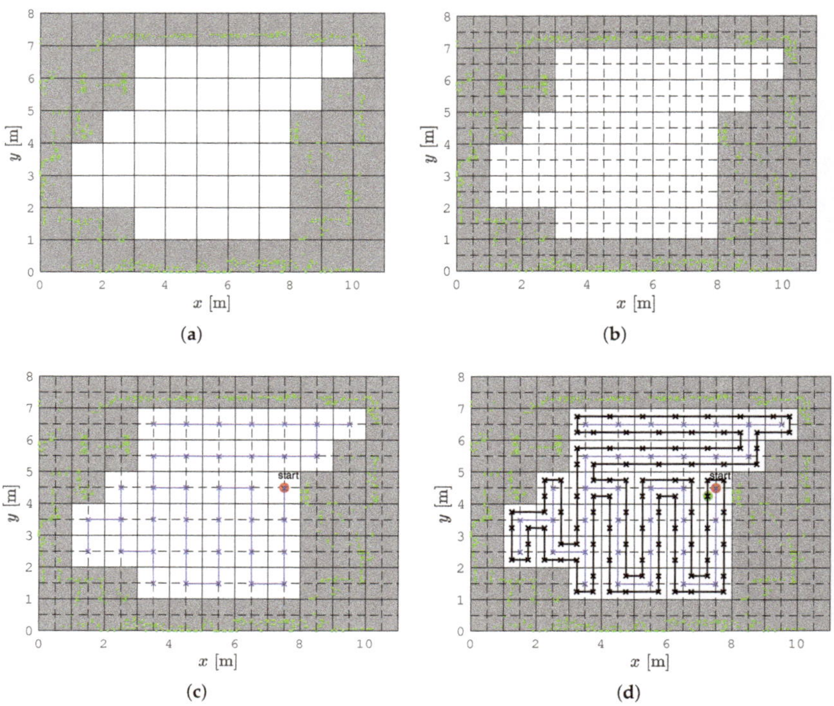

Figure 3. Example of the complete coverage path planning for the known environment step-by-step. The real obstacles collected by SLAM algorithm are presented with green dots. (**a**) Low resolution occupancy grid map, (**b**) Subdividing of grid cells, (**c**) The spanning tree (blue line), (**d**) Complete coverage path (black line).

While the robot follows the planned path and visits subcells one by one, it may encounter an unknown obstacle (Algorithm 3). In such a case, the new spanning tree is created for the rest of the unvisited grid cells and the path is recomputed. The robot continues to follow the new path from the right side of the spanning tree until it returns to the cell where the replanning started. It then continues to follow the previously planned path. If all cells are visited then the shortest path around the obstacle is determined and connected with the previously planned path. If the unknown obstacles free occupied cells, set these cells as free in the occupancy grid map. Add these cells to the previously determined spanning tree if the current cell is the diagonal neighbor of the occupied cells. The new path around this spanning tree is determined. The complete coverage algorithm ends when the robot returns to the start subcell of the initial path. Since only completely free cells are considered, the coverage path is optimal and minimizes the overlap area.

In summary, the main advantages of the RSTC algorithm are that it generates the shortest path, minimizes the overlap area, and allows real-time path replanning in changing environments. However, its drawbacks are sharp turns of the planned path where a robot has to stop and reorient itself to continue, which is inefficient regarding the task duration and energy consumption.

Algorithm 3 Pseudocode for the complete coverage path planning

```
Input: Map, currentPose
Output: RSTCpath
determine occupancy grid map based on png map image
determine spanning tree based on Algorithm 1
determine the RSTC path based on Algorithm 2
while it reaches the starting subcell again do
    if unknown obstacle is detected then
        occupy cells in which unknown obstacle is detected
        if exist the unvisited grid cells then
            determine the new spanning tree based on Algorithm 1 for the rest unvisited grid cells
            determine the new RSTC path based on Algorithm 2 for the rest unvisited grid cells
        else
            determine the RSTC path around the obstacle so that the minimum number of double-covered subcells is obtained and connect it with previously planned path
        end if
    end if
    if unknown obstacle cleared occupied cells then
        set these cells as free in the occupancy grid map
        if current cell is diagonal neighbor of the occupied cell then
            add these cells in the previously determined spanning tree
            determine the new RSTC path based on Algorithm 2
        end if
    end if
end while
return the RSTC path
```

3.2. Path Smoothing

The task of the path smoothing algorithm is to smooth the path generated by the RSTC algorithm at the sharp turns to allow continuous motion of the robot without stopping. The path smoothing algorithm [11] is used, which uses clothoids to smoothly connect adjacent lines of the path.

The coordinates of a general clothoid are:

$$x(s) = x_0 + \int_0^s \cos\left(\theta_0 + \kappa_0\xi + \frac{1}{2}c\xi^2\right)d\xi, \tag{1a}$$

$$y(s) = y_0 + \int_0^s \sin\left(\theta_0 + \kappa_0\xi + \frac{1}{2}c\xi^2\right)d\xi, \tag{1b}$$

where (x_0, y_0, θ_0) is the initial robot pose, the parameter κ_0 is the initial curvature at $s \geq 0$, which denotes the arc length, and the parameter c is the sharpness, which describes how much the curvature changes with traveled distance. Curvature κ can be expressed as a linear function of the arc length s:

$$\kappa(s) = \kappa_0 + cs. \tag{2}$$

The Equation (1) contain Fresnel integrals, which are transcendental functions that cannot be solved analytically, making them difficult to use in real-time applications. To enable the use of clothoids in real-time, various methods have been developed and we use the one that is particularly fast and thus very suitable for real-time applications [11]. It is based on a so-called basic clothoid with bounded approximation error [40] whose coordinates can be stored in a lookup table, and the coordinates of any other clothoid can be determined based on the points of the basic clothoid by rescaling, rotating, and translating.

An example of the smoothed sharp turn is shown in Figure 4, where the points S, T, and G denote the start, sharp turn, and the goal position, respectively. The smooth path is shown as a red curve consisting of four parts: a straight line $\overline{SC_1}$, the first clothoid from C_1 to C_2, the second clothoid from C_2 to C_3, and a straight line $\overline{C_3G}$. The two clothoids are symmetric–the second one is obtained by mirroring the first one at the bisector of the angle spanned by the lines $\overline{ST}$ and $\overline{TG}$.

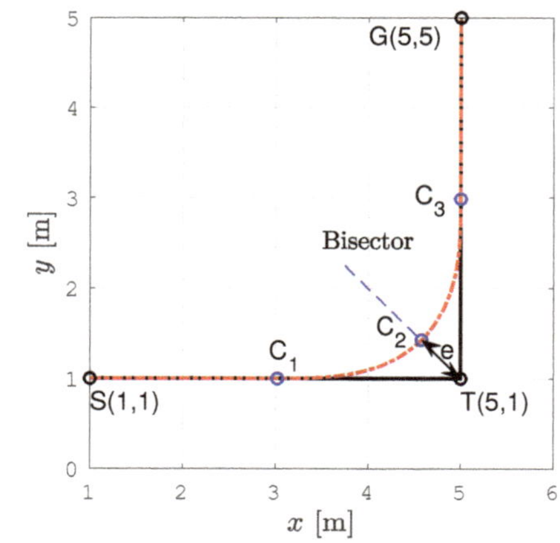

Figure 4. Illustration of the sharp turn smoothing by two clothoids.

The parameter e is the deviation from the planned path. If it is high, the path curvature is low and the robot can drive at a higher velocity. Since for our problem, each sharp turn is at a 90° angle, the same deviation parameter e, and consequently the same pair of symmetric clothoids can be applied for all of them, which additionally simplifies the smoothing.

We determined the maximum deviation e_{max} to ensure that the robot's footprint does not collide with the obstacle cell at the point of maximal deviation, i.e.:

$$e_{\text{max}} \leq \frac{D\sqrt{2} - D}{2}, \tag{3}$$

where $D\sqrt{2}$ is the subcell diagonal, and D is the diameter of the circumscribed circle around the robot's footprint. In our setup it is $D = 0.5$ m, so we choose $e_{\text{max}} = 0.1$ m. From e_{max} it is possible to determine the maximum curvature in point C_2 (see [11]), so we obtain $\kappa_{\text{max}} = 6.4\ \text{m}^{-1}$.

From the radial acceleration, which is $a_{\text{radial}} = v \times \omega = v^2\kappa$, where v and ω are the linear and angular velocity of the robot, respectively, we can get the velocity limit at the curvature extrema as:

$$v_{\text{min}} = \sqrt{\frac{a_{\text{radial}_{\text{max}}}}{\kappa_{\text{max}}}}. \tag{4}$$

Since $a_{\text{radial}_{\text{max}}}$ is characteristic of the mobile robot, and in our setup it is $a_{\text{radial}_{\text{max}}} = 0.1\ \text{m}/\text{s}^2$, this gives the velocity limit $v_{\text{min}} = 0.125\ \text{m}/\text{s}$, which is used later in the velocity profile optimization module.

Figure 5 shows a part of the original and smoothed path with two positions of the robot. It can be noticed that the robot's circular footprint when moving along the smoothed path is always within the free subcells of size D. It does not collide with the obstacle cell even at points where a deviation of the smoothed path from the original path is the largest (noted with e).

As described in Section 3.1, the RSTC algorithm recalculates the path for the unvisited grid subcells when the robot encounters an unknown obstacle in the neighboring cells. Due to the computational simplicity of the proposed path smoothing algorithm, the smoothing of the replanned path can be conducted online so that the robot does not need to slow down or stop.

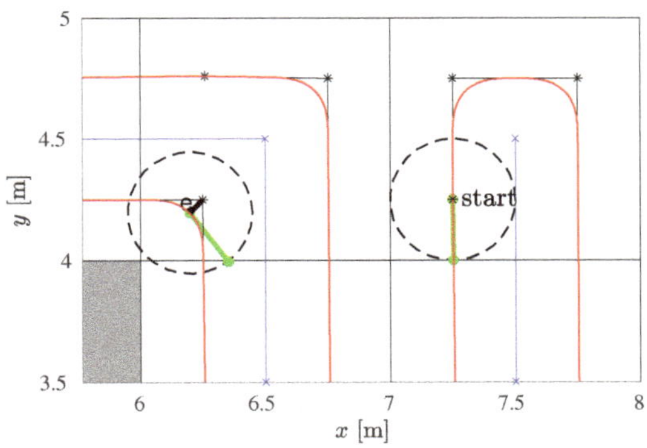

Figure 5. The enlarged part of the coverage path with two positions of the robot: (1) the robot's starting position and (2) the robot's position at the obstacle corner. The obstacle cell is shown in gray, the robot's positions in dashed circles, the robot's orientation in green, the spanning tree in blue, the RSTC path in black, and the smoothed path in red.

3.3. Velocity Profile Optimization and Trajectory Tracking

Although the path smoothing algorithm takes into account the kinematic and dynamic constraints of the robot, it only outputs x and y coordinates and curvature of the smooth complete coverage path. To ensure time-optimal robot motion along the smooth path, the velocity profile with maximum allowable velocities should be calculated. The turning points of the smoothed path (e.g., as point C_2 in Figure 4) are the local extrema points of the smooth path curvature, i.e., at these points, the turning radius reaches a local minimum, and consequently, the velocity is locally lowest. The maximum allowable velocity of the robot at a turning point is determined by the radial acceleration limit a_{radial}. The robot must decelerate and accelerate before and after the turning point as much as maximally tangentially allowed by the acceleration limit. The robot should always remain within its acceleration limits. The overall acceleration is calculated as:

$$a = \sqrt{a_{\text{radial}}^2 + a_{\text{trans}}^2} \tag{5}$$

where $a_{\text{radial}} = v^2\kappa$ is the radial acceleration of the robot and $a_{\text{trans}} = \frac{dv}{dt}$ is the translation acceleration.

We used the velocity profile optimization algorithm described in detail in [41]. It outputs the optimal linear and angular velocities of the robot based on the points of the smooth complete coverage path and the driving limitations. Linear and angular velocities, together with the robot position and orientation, result in the 5D trajectory $(x, y, \theta, v, \omega)^T$. This trajectory is the input to the trajectory tracking algorithm based on the Kanayama controller [42]. This controller ensures that the robot tracks the optimal 5D trajectory by dumping perturbations of the robot pose provided by the localization module and velocity measurements from the wheel encoders.

The kinematic model of the differential drive mobile robot can be represented as follows:

$$\begin{bmatrix} \dot{x}(t) \\ \dot{y}(t) \\ \dot{\theta}(t) \end{bmatrix} = \begin{bmatrix} cos\theta(t) & 0 \\ sin\theta(t) & 0 \\ 0 & 1 \end{bmatrix} \begin{bmatrix} v(t) \\ \omega(t) \end{bmatrix}, \tag{6}$$

where t is time, $(x(t), y(t))$ is position, $\theta(t)$ is orientation, $v(t)$ is linear velocity, and $\omega(t)$ is the angular velocity of the mobile robot. For the system (6), the nonholonomic constraint is:

$$-\dot{x}(t) \cdot sin\theta(t) + \dot{y}(t) \cdot cos\theta(t) = 0, \tag{7}$$

which constrains the drive wheels to roll and prevents slippage. A path is feasible for nonholonomic mobile robot if it can follow the path with the reference velocity commands.

The main task of the tracking control for mobile robots is to find appropriate velocities $v(t)$ and $\omega(t)$ to achieve the control objective. Inputs for the Kanayama controller are the reference configuration $\mathbf{p_r} = (x_r, y_r, \theta_r)^T$, which is taken from the 5D trajectory at the current time instant, and the measured current configuration $\mathbf{p_c} = (x_c, y_c, \theta_c)^T$. The Kanayama controller outputs are the corrected reference linear and angular velocities. The first step of the control law is to compute the error configuration $\mathbf{p_e}$ as the difference between $\mathbf{p_r}$ and $\mathbf{p_c}$, which must converge to zero. For the kinematic model given by (6) and (7), the expression for the error configuration is:

$$\mathbf{p_e} = \begin{bmatrix} x_e \\ y_e \\ \theta_e \end{bmatrix} = \begin{bmatrix} cos\theta_c & sin\theta_c & 0 \\ -sin\theta_c & cos\theta_c & 0 \\ 0 & 0 & 1 \end{bmatrix} (\mathbf{p_r} - \mathbf{p_c}). \tag{8}$$

Using the error configuration $\mathbf{p_e}$, the reference linear and angular velocities v_r and ω_r, the corrected linear and angular velocities v and ω can be calculated as follows:

$$\begin{bmatrix} v \\ w \end{bmatrix} = \begin{bmatrix} v_r cos\theta_e + K_x x_e \\ \omega_r + v_r(K_y y_e + K_\theta sin\theta_e) \end{bmatrix}, \tag{9}$$

where K_x, K_y and K_θ are the positive constant parameters of the controller.

Figure 6 shows the smoothed path and the tracked trajectory for the example shown in Figure 4. There is no significant deviation of the tracked trajectory from the smoothed path and the mobile robot moves smoothly from one segment of the path to another.

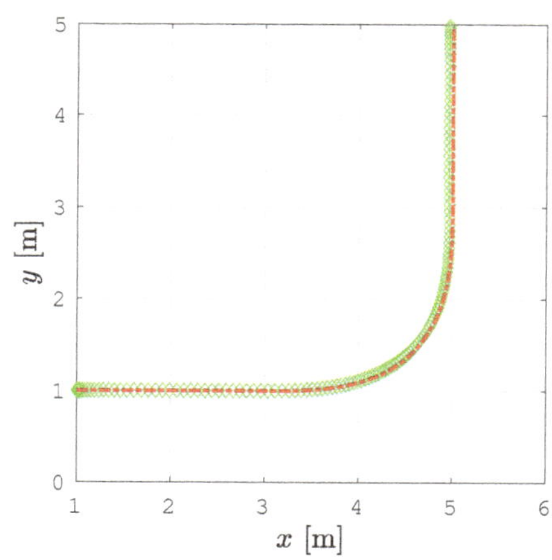

Figure 6. The smoothed path from Figure 4 (red line) and the tracked trajectory (green diamonds).

Figure 7 shows linear and angular velocities for the example shown in Figure 4 calculated by the velocity profile optimization procedure (blue line) and the real linear and angular velocity profile, while the robot tracked the trajectory (magenta diamonds). The corrected reference velocities calculated by Kanayama's algorithm are shown with black diamonds. The robot in the simulation reached exactly the corrected reference velocity with a delay of one control time step.

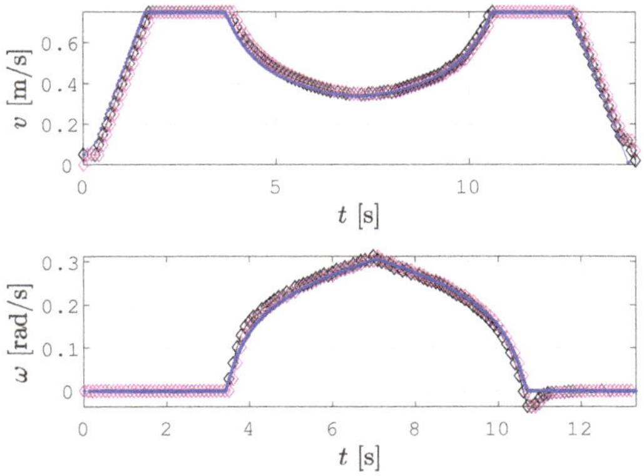

Figure 7. The calculated linear and angular velocities (blue line) and real linear and angular velocities (magenta diamonds) at which the robot tracked the trajectory.

4. Simulation Results

We tested the proposed motion planning approach in three scenarios: the *Lab scenario* (Section 4.1), *Aula scenario* (Section 4.2), and *Gallery scenario* (Section 4.3). For each scenario, the proposed SCCPP algorithm is compared with the original RSTC algorithm without smoothing (called the CCPP algorithm in the following), and the results are discussed in Section 4.4. We also compared SCCPP with CCD* and HDCP algorithms for the *Gallery scenario*. CCD* was selected for comparison because it is a graph search method based on D* search of the high resolution grid map of the environment. HDCP was selected for comparison because it plans smooth paths based on Dubins curve, which is followed at constant velocity in real-time.

We used a receding horizon control (RHC) algorithm developed within our research group [43] for the path following of the the CCPP without smoothing. This algorithm can adapt to dynamic changes in the environment. If the selected subgoal is too close to the detected obstacle, a new subgoal is chosen in the local vicinity from the critical one. With this procedure, the robot follows the planned path with a minimal drift.

The simulations were performed on an Asus ROG Strix SCAR II (Intel i7-8750H, 16 GB DDR4 RAM). The algorithms were implemented and tested in the Robot Operating System (ROS). The stage simulator was used for the simulations.

To create the environment map, for each test scenario the environment was explored through sensors on the Pioneer 3DX robot and data was collected by the *slam_gmapping* ROS package. Then, the map was created and post-processed to align walls to the grid for better performance of the coverage algorithm. The post-processed map is resampled to an occupancy grid map with low resolution. The size of the cells is 1×1 m and each cell is divided into four 0.5×0.5 m sub-cells, which are approximately of the same size as the robot's footprint. The *amcl* ROS package was used for the robot localization in created maps.

The *Lab scenario* tests how the algorithm works in a small room with static and dynamic obstacles. The *Aula scenario* tests how the algorithm works in a large hallways with a large number of static obstacles. The third scenario is the *Gallery scenario*, which tests the algorithm performances in narrow spaces. To validate the performances of the SCCPP algorithm and to compare it with other algorithms, the path length and execution time were measured, and the coverage rate and coverage redundancy were calculated.

The coverage rate is calculated as:

$$\text{Coverage rate} = \mathrm{A_p}/(\mathrm{A_t} - \mathrm{A_o}) \cdot 100\%, \tag{10}$$

where A_t is the total area of the environment, A_o is the area of obstacles, and A_p is the area covered by the driven trajectory of the robot. The coverage redundancy is calculated as follows:

$$\text{Coverage redundancy} = A_r / A_p \cdot 100\%, \tag{11}$$

where A_r is the overlap area, calculated as the sum of all subcells visited more than once.

In all figures below that show the paths or trajectories, static obstacles are shown with green dots, and if they partially occupy a cell, the entire cell is shown as occupied (gray cells), the spanning tree is shown as a blue line, the RSTC path as a black line, the smoothed RSTC path as a red line, and the tracked trajectory as a green line.

4.1. The Lab Scenario

The coverage starts at cell (7, 4) which is the starting cell for the spanning tree construction and the path circumnavigates around the constructed spanning tree (see Figure 8a). The robot follows the path from the right side of the spanning tree and covers each subcell exactly once. The coverage is complete when the robot returns to the start subcell.

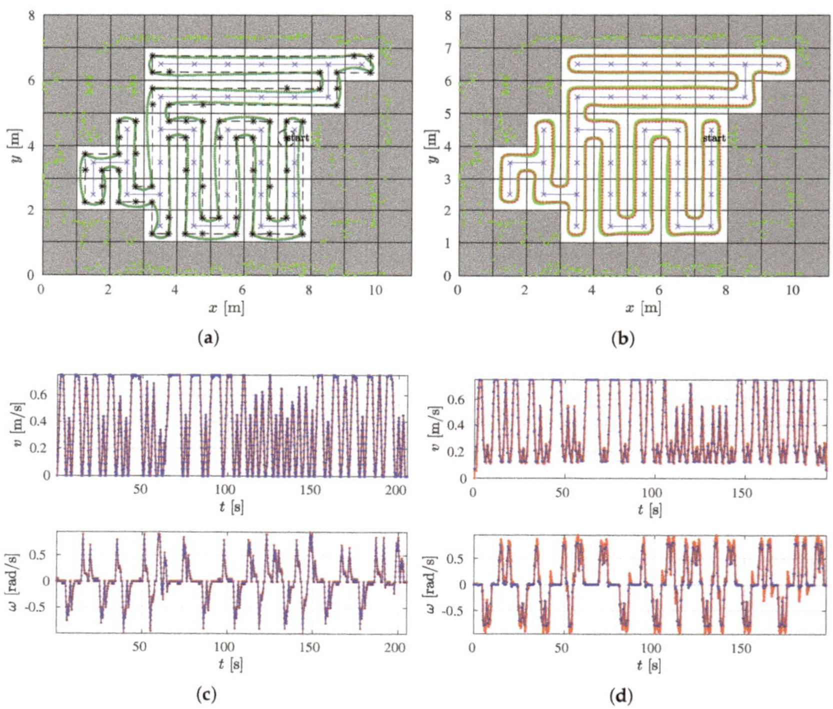

Figure 8. The comparison of the CCPP and SCCPP algorithms in the *Lab scenario* with the presented RSTC spanning tree (blue line), the RSTC path (black line), the smoothed path by the SCCPP algorithm (red line), the executed path by the CCPP and SCCPP algorithm (green line), the reference velocity profile (blue line), and the actual velocity profile (red line). (**a**) The CCPP path in the *Lab scenario*. (**b**) The SCCPP path in the *Lab scenario*. (**c**) The CCPP velocity profile in the *Lab scenario*. (**d**) The SCCPP velocity profile in the *Lab scenario*.

The path is created with straight lines that form sharp turns. In each turn for the CCPP variant, the robot must stop and reorient itself so that the robot heading is equal to the path direction. For this reason, the linear velocity is zero and the angular velocity is close to the maximal value. On straight sections of the path, the linear velocity is maximal and the angular velocity is zero. Figure 8c shows the calculated linear and angular velocities (blue line) and the real linear and angular velocities of the robot executed by the motors (red line).

To improve the coverage and reduce the execution time, the smoothed variant–the SCCPP algorithm is used. The output of this algorithm is the smoothed path that circumnavigates around the constructed spanning tree (see Figure 8b). In this way, sharp turns are now smoothed and the robot can travel without stopping.

The execution of the SCCPP algorithm can be examined from the linear and angular velocities shown in Figure 8d. The blue lines are the linear and angular velocities calculated by the optimization process, and the red lines show how the linear and angular velocities changed as the robot tracked the trajectory. It can be observed that the linear velocity is not zero while the robot has to overcome the turn as opposed to zero velocities in Figure 8c, where the smoothing algorithm is not used.

The replanning SCCPP algorithm is executed in a dynamic environment. Dynamic changes can be detected by the robot in neighboring cells of the current cell where the robot is currently located; see Figure 9a for an example where the obstacle is detected in cell (4, 4). All unvisited cells are considered as nodes for the new spanning tree calculation and the visited subcells are not part of the new path coverage, see Figure 9b. After the new spanning tree is determined for the rest of the unvisited grid cells, the new path is determined. The path always follows the spanning tree counterclockwise. When the robot returns to the cell where the replanning started, it continues to follow the path from the right side of the spanning tree that was already formed during the previous path planning. Therefore, this part of the spanning tree is not shown in Figure 9b from cell (4, 4) to starting cell (7, 5). The robot traverses these visited grid cells by visiting only subcells that are not yet visited. The complete coverage algorithm ends when the robot gets to the starting subcell. The red dots in Figure 9 are the position of the robot where the replanning algorithm is executed. In this example, the replanning process is fast enough that the robot does not need to stop for the recalculation of the smoothed path. The recalculation of the smoothed RSTC path takes less than 1 ms on average.

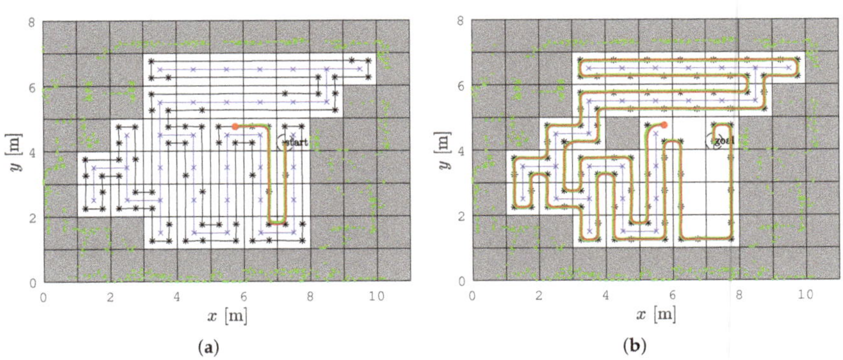

Figure 9. The RSTC algorithm in the *Lab scenario* with the presented RSTC spanning tree (blue line), the RSTC path (black line), the smoothed path by the SCCPP algorithm (red line), and the executed path by the smoothed RSTC algorithm (green line). Red points are the robots' position where the replanning algorithm is executed. (**a**) The smoothed RSTC path in the *Lab scenario* before replanning. (**b**) The smoothed RSTC path in the *Lab scenario* after replanning.

4.2. The Aula Scenario

Similar behavior of the CCPP and SCCPP algorithms can be observed in the example with few hallways and more static obstacles; see Figure 10. Static obstacles are shown with green dots and if they partially or fully occupy a cell, the cell is presented as occupied (gray cells). Some static obstacles are detected by the laser outside the building boundaries because there are windows in parts of the hallway and these grid cells are represented as occupied. The direction of the first segment of the complete coverage path is −90° due to the initial orientation of the robot. The coverage starts at cell (6, 32), which is the starting cell for the spanning tree construction and the path circumnavigates around the constructed spanning tree, but this is not represented in Figure 10. Due to the transparency of the figure we presented, only a smoothed path (red line) and driven trajectory (green line). The smoothed path does not contain sharp turns and the robot can drive the trajectory without stopping and reorienting itself to align heading with the path direction. This improves the coverage rate and reduces the execution time of the coverage task. The SCCPP variant outperforms the CCPP algorithm here as well.

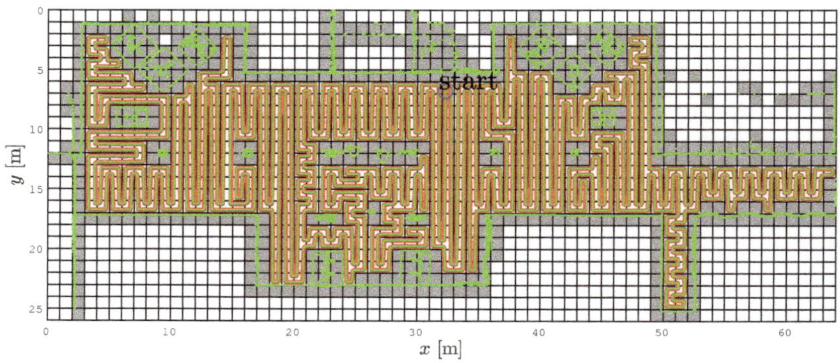

Figure 10. The SCCPP algorithm in the *Aula scenario*, which presents the smoothed path with red line and the executed path with green line.

4.3. The Gallery Scenario

This scenario has the most turns due to the narrow dimensions of the environment. The starting cell for the spanning tree construction and the RSTC path is (11, 3), see Figure 11. Again, the SCCPP algorithm variant is faster than the CCPP variant.

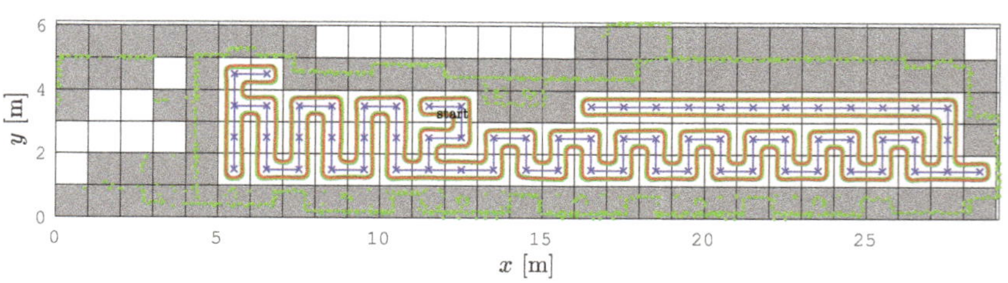

Figure 11. The SCCPP algorithm in the *Gallery scenario*: the RSTC spanning tree (blue line), the smoothed path (red line), and the executed path (green line).

The SCCPP algorithm is compared to the CCD* and HDCP algorithms in this scenario. The complete coverage path of the CCD* algorihm is shown in Figure 12 with the black line, and visited cells are marked by different colors according to the number of visits, so some parts of the map are visited nine times.

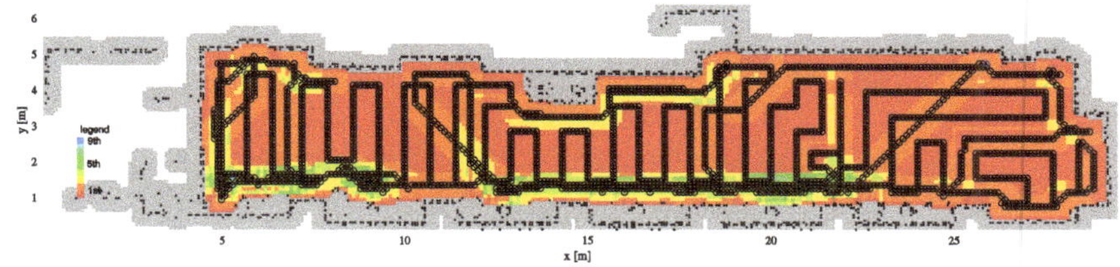

Figure 12. The complete coverage of the CCD* algorithm with the path and noted redundant numbers of cell visits (from 1 to 9).

The complete coverage path of the HDCP algorithm is shown in Figure 13 with the green line. Partially occupied cells are shown in gray hexagons and only completely free hexagonal cells are used for the coverage task. The path begins in the cell (11, 3) and ends in the cell (6, 2). Some cells are visited more than once due to the narrow environment and low resolution occupancy grid map.

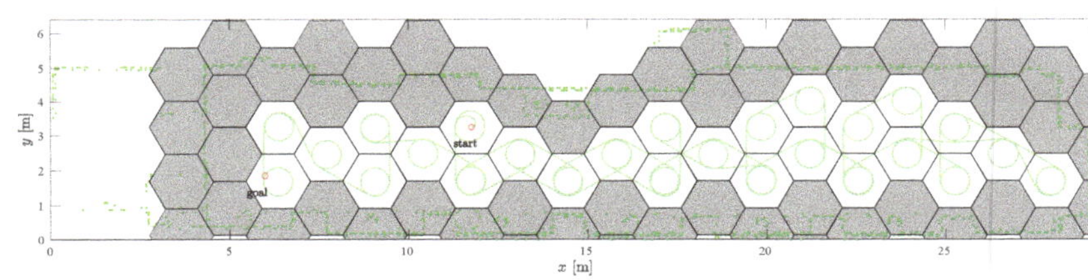

Figure 13. The complete coverage of the HDCP algorithm with hexagonal cell grid.

4.4. Discussion

The results of the CCPP and SCCPP comparison in all three scenarios are given in Table 1. When the smoothing algorithm is used, the length of the tracked trajectory is shorter and the execution time is reduced compared to the results obtained without smoothing. In the *Gallery scenario*, the RSTC algorithm provides a path with many sharp turns and because of the discontinuity during the motion, the execution time for the complete coverage is about 10% longer for CCPP than for SCCPP. It can be noted that the environment configuration and the position of the obstacles can affect the execution time and the total length of the tracked trajectory. In addition, the CCPP algorithm has higher localization errors at points where the robot rotates in place. These cause higher uncertainty in the robot's pose and the deviation from the tracked trajectory is larger than in the SCCPP examples. This also leads to an increase in the coverage redundancy.

Table 1. Table of comparison of CCPP and SCCPP algorithms.

	Lab CCPP	Lab SCCPP	Gallery CCPP	Gallery SCCPP	Aula CCPP	Aula SCCPP
Coverage path length	77.22 m	**72.46 m**	141.04 m	**129.44 m**	1255.91 m	**1191.22 m**
Coverage time	217.30 s	**193.26 s**	453.99 s	**409.58 s**	3479.51 s	**2834.23 s**
Coverage rate	71.48%	**74.42%**	66.37%	**72.11%**	79.03%	**79.84%**
Coverage redundancy	35.53%	**7.89%**	48.19%	**5.79%**	39.23%	**3.37%**
Nodes number	61	61	97	97	744	744
Path calculation time	**0.5 ms**	0.9 ms	**0.6 ms**	1.2 ms	**5 ms**	10 ms

From these three scenarios, it can be observed that the SCCPP algorithm has, on average, a 6.5% shorter tracked trajectory, an 8.8% reduction in coverage execution time, a 4.5% better coverage rate, and an 82.34% lower coverage redundancy than the original CCPP algorithm. The better coverage rate of the SCCPP compared to the CCPP can be explained by higher path following accuracy, while accuracy of the CCPP is worsened by sharp turns. To provide the scalability of the proposed algorithm, each scenario has a different size of the environment, which leads to different number of nodes (cells) in a spanning tree construction. Because the number of nodes is proportional to the computation time, *Lab scenario* has the smallest and *Aula scenario* has the biggest elapsed time required to calculate the complete coverage path. For each scenario, SCCPP has a larger time than CCPP for the time required to calculate the smooth path. The time increases linearly with the number of nodes, while the number of nodes increases quadratically with the map dimensions.

The SCCPP algorithm is compared to the CCD* and HDCP algorithms, and the results are shown in Table 2. Compared to CCD* and HDCP, SCCPP has the shortest coverage path length (42% shorter than CCD* and 21% shorter than HDCP), the shortest coverage time (47% shorter than CCD* and 53% shorter than HDCP), and the smallest coverage redundancy (82% smaller than CCD* and 40% smaller than HDCP). However, SCCPP has worse coverage rate compared to CCD* (27% lower), and better coverage rate compared to HDCP (15% higher). The reason of the lower coverage rate of both the SCCPP and HDCP algorithm is the used low resolution occupancy grid map. However, the high resolution occupancy grid map results with higher coverage redundancy in CCD*.

Table 2. Comparisons of the SCCPP, CCD* and HDCP algorithms for the *Gallery scenario*.

	SCCPP	CCD*	HDCP
Coverage path length	**129.44 m**	222.79 m	163.48 m
Coverage time	**409.58 s**	768.22 s	879.65 s
Coverage rate	72.11%	**98.85%**	57.38%
Coverage redundancy	**5.79%**	87.48%	46.15%

The coverage rate for the SCCPP algorithm can be increased if a wall following method is used, but this also increases the redundancy. The wall following algorithm used after SCCPP is presented in Figure 14 with driven trajectory (red line) and robot's footprint at every point on the trajectory (degradation from black to gray). The coverage rate is increased to 97.64% with the use of the wall following, and the coverage redundancy is increased to 24.64%.

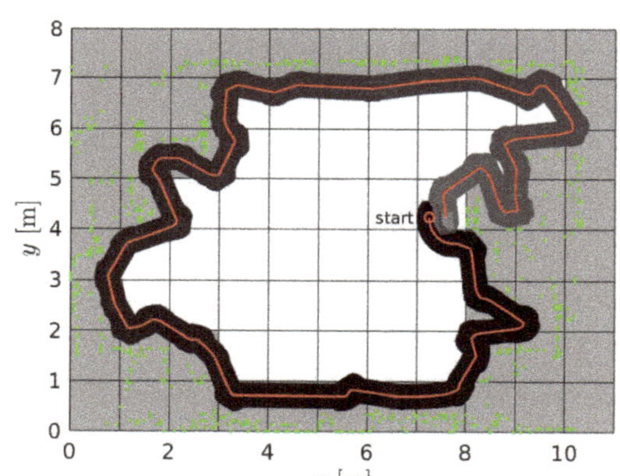

Figure 14. The wall following algorithm after SCCPP: driven trajectory (red line), and robot's footprint in every point on the trajectory is presented degraded from black to gray.

The advantages of our SCCPP algorithm are completeness of coverage, robustness to environmental shape and initial robot pose, optimal path that visits all subcells exactly once, time efficiency, low coverage redundancy, and fast replanning. The coverage rate can be significantly increased if the wall following method is used. The limitation is that the algorithm requires a priori knowledge about the workspace. The execution of the proposed smooth complete coverage path planning algorithm is around 10 ms and it is suitable for real-time operation due to its computational simplicity.

5. Experiments on a Real Robot

The experiment (the experiments are demonstrated in the accompanying video available here: https://youtu.be/VEAGHIAIpRA accessed on 18 October 2022) was performed on the Pioneer 3DX robot with a SICK laser sensor LMS200. To have a similar scenario as in the simulation, the experiment was performed in the *Lab scenario*, where a new map had to be created due to some minor changes in the placement of the furniture. First, the *Lab* was explored using the laser sensor on the robot, and the data of the environment was collected using the *slam_gmapping* ROS package. After the environment was explored, the map was created and the package *amcl* ROS was used for robot localization in the created map.

The SCCPP algorithm was executed and the results are shown in Figure 15. The static obstacle configuration is represented as green dots, which are shown as occupied cells (gray cells in Figure 15) in the occupancy grid map. The initial position of the robot ($x = 5.25$ [m], $y = 2.75$ [m]) is marked with a blue star and this is the starting position for the coverage of the environment. The blue line is the spanning tree determined by RSTC algorithm, the red line is the smoothed RSTC path, and the green line is the driven trajectory by the Pioneer 3DX robot.

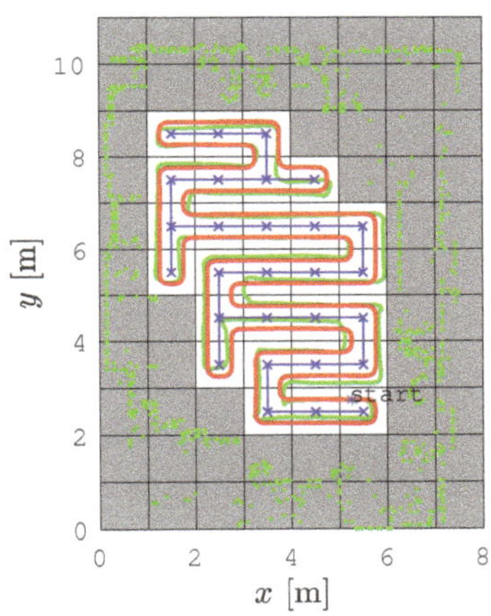

Figure 15. The calculated SCCPP path (red line) and trajectory tracked by the Pioneer 3DX robot (green line).

There are some deviations between the smoothed path determined by the SCCPP algorithm (red line) and the real trajectory tracked by the robot (green line), especially when dealing with parts of the path that are curved. This problem is due to the inaccurate and noisy localization of the robot. Another problem is the hardware setup. We used serial communication between the robot and the laptop with ROS, which caused a delay of three cycles in sending the calculated velocities to the robot. In Figure 16, the calculated linear and angular reference velocity profile are shown with blue lines and the actual linear and angular velocity profile of the robot during trajectory tracking is shown with magenta lines. For the experiment with the real robot, we used 0.5 m/s and 0.75 rad/s as the maximal linear and angular velocities, respectively. The delay of three cycles caused the corrected reference velocity by the Kanayama tracking algorithm to deviate from the optimized velocity profile of the smoothed path. This is shown in Figure 17, where the first 1.5 s of the linear velocity profile from Figure 16 are shown magnified.

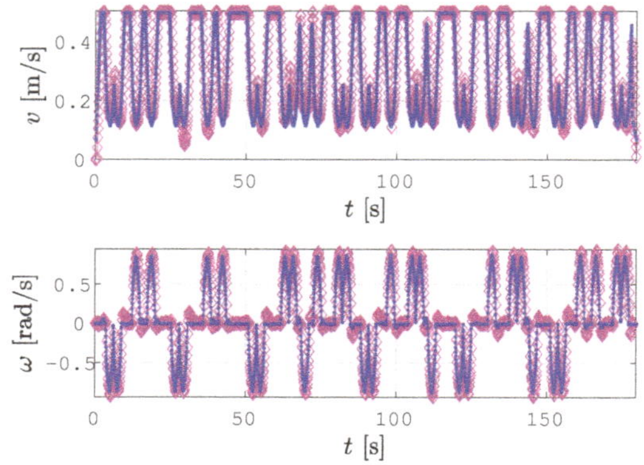

Figure 16. Calculated linear and angular velocities (blue line) and linear and angular velocities, while the Pioneer 3DX robot is tracking the trajectory (magenta line).

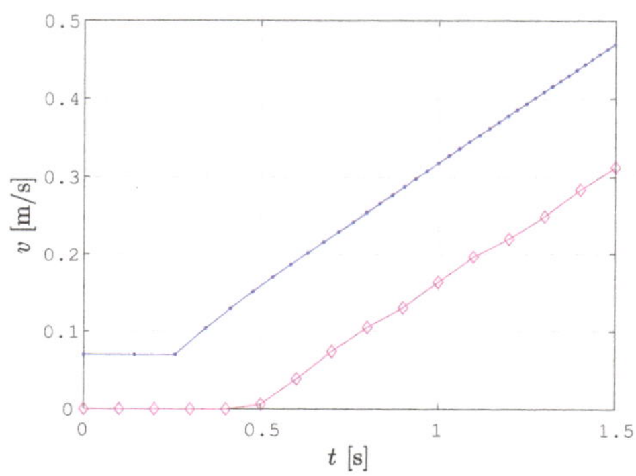

Figure 17. Real linear velocity, while the robot Pioneer 3DX is following the trajectory (magenta line), deviates from the applied linear velocity (blue line).

Table 3 presents the experimental evaluation of the SCCPP algorithm efficiency on the Pioneer 3DX robot for the *Lab scenario*. The coverage path length is 53.85 m, the coverage execution time is 179.5 s, the coverage rate is 77.7%, and the coverage redundancy is 12.96%. The trajectory tracking error is determined as the difference of the calculated trajectory (Figure 15 red dotted line) and tracked trajectory (Figure 15 green line) and it is 6.43 m^2. When compared with simulation results in Table 1 for the *Lab scenario*, although the maps are slightly different, it can be observed that the coverage rate in both cases is approximately equal, which confirms that our algorithm is equally efficient in the real robot setup. The coverage redundancy is slightly worse on a real robot, which is mainly caused by noisy localization and delay in the control loop.

Table 3. SCCPP algorithm on the Pioneer 3DX robot for the *Lab scenario*.

	SCCPP
Coverage path length	53.85 m
Coverage time	179.5 s
Coverage rate	77.7%
Coverage redundancy	12.96%
Tracking error	6.43 m^2

6. Conclusions

The proposed SCCPP algorithm is the online algorithm that generates a traversable collision-free trajectory based on clothoids with low computational cost. Such a path is suitable and feasible for nonholonomic mobile robots since it does not contain sharp turns. By using a smoothing technique on the proposed coverage path, the coverage efficiency can be significantly improved in terms of the time required and energy consumption during the coverage tasks and has very low overlap redundancy. The complexity of the environment affects the coverage efficiency, and the experimental results evaluated the efficiency of the CCPP algorithms on maps with different complexity levels. By using the SCCPP algorithm, the trajectory followed by the robot can be executed faster and with higher accuracy than without the smoothing algorithm. The SCCPP algorithm produces the shortest coverage path, takes the shortest time for coverage execution, and has the smallest coverage redundancy compared to the CCD* and HDCP algorithms. The SCCPP algorithm takes advantage of the large size of the grid cells to ensure real-time operation and minimization of overlapping areas, but at the cost of a lower coverage rate due to the uncovered areas around obstacles and walls. However, the coverage rate could be easily increased by simply combining our SCCPP algorithm with a wall following algorithm.

As future work, more experiments are planned for other robot designs such as omnidirectional mobile robots and Ackermann steering vehicles. Furthermore, we consider the extension of this work to multiple robots in the form of a decentralized solution for the coordinated multi-robot complete coverage task. This will decrease the total task time significantly due to the division of workload overall robots, while decentralization will prevent a single point of failure.

Author Contributions: Conceptualisation, A.Š., M.S., M.B. and I.P.; methodology, A.Š., M.S. and M.B.; software, A.Š. and M.B.; validation, A.Š; formal analysis, A.Š.; investigation, A.Š.; resources, A.Š. and M.B.; data curation, A.Š.; writing—original draft preparation, A.Š., M.S., M.B. and I.P.; writing–review and editing, A.Š., M.S. and M.B.; visualisation, A.Š.; supervision, M.S., M.B. and I.P.; project administration, M.S. and I.P.; funding acquisition, I.P. All authors have read and agreed to the published version of the manuscript.

Funding: This work has been supported by the European Regional Development Fund under the grant KK.01.2.1.01.0138–Development of a multi-functional anti-terrorism system.

Institutional Review Board Statement: Not applicable.

Informed Consent Statement: Not applicable.

Data Availability Statement: Not applicable.

Conflicts of Interest: The authors declare no conflict of interest.

References

1. Gao, X.S.; Xu, D.G.; Wang, Y. Omni-directional mobile robot for floor cleaning. *Chin. J. Mech. Eng.* **2008**, *44*, 228–233. [CrossRef]
2. Dakulović, M.; Horvatić, S.; Petrović, I. Complete Coverage D* Algorithm for Path Planning of a Floor-Cleaning Mobile Robot. In Proceedings of the Preprints of the 18th IFAC World Congress, Milano, Italy, 28 August–2 September 2011; pp. 5950–5955. [CrossRef]
3. Yang, S.X.; Luo, C. A neural network approach to complete coverage path planning. *IEEE Trans. Syst. Man Cybern. Part B (Cybern.)* **2004**, *34*, 718–724. [CrossRef] [PubMed]
4. Acar, E.; Choset, H.; Zhang, Y.; Schervish, M. Path Planning for Robotic Demining: Robust Sensor-Based Coverage of Unstructured Environments and Probabilistic Methods. *Int. J. Robot. Res.* **2003**, *22*, 441–466. [CrossRef]
5. Dakulović, M.; Petrović, I. Complete coverage path planning of mobile robots for humanitarian demining. *Ind. Robot. Int. J.* **2012**, *39*, 484–493. [CrossRef]
6. Ollis, M.; Stentz, A. First results in vision-based crop line tracking. In Proceedings of the IEEE International Conference on Robotics and Automation, Minneapolis, MN, USA, 22–28 April 1996; pp. 951–956. [CrossRef]
7. Weiss-Cohen, M.; Sirotin, I.; Rave, E. Lawn Mowing System for Known Areas. In Proceedings of the 2008 International Conference on Computational Intelligence for Modelling Control & Automation, Vienna, Austria, 10–12 December 2008; pp. 539–544. [CrossRef]

8. Kapoutsis, A.C.; Chatzichristofis, S.A.; Doitsidis, L.; de Sousa, J.B.; Kosmatopoulos, E.B. Autonomous navigation of teams of Unmanned Aerial or Underwater Vehicles for exploration of unknown static dynamic environments. In Proceedings of the 21st Mediterranean Conference on Control and Automation, Platanias, Greece, 25–28 June 2013; pp. 1181–1188. [CrossRef]
9. Kapoutsis, A.; Chatzichristofis, S.; Doitsidis, L.; Sousa, J.; Pinto, J.; Braga, J.; Kosmatopoulos, E. Real-time adaptive multi-robot exploration with application to underwater map construction. *Auton. Robot.* **2015**, *40*, 987–1015. [CrossRef]
10. Šelek, A.; Seder, M.; Petrović, I. Mobile robot navigation for complete coverage of an environment. *IFAC-PapersOnLine* **2018**, *51*, 512–517. [CrossRef]
11. Brezak, M.; Petrović, I. Path smoothing using clothoids for differential drive mobile robots. *IFAC Proc. Vol.* **2011**, *44*, 1133–1138. [CrossRef]
12. Cao, Z.L.; Huang, Y.; Hall, E.L. Region filling operations with random obstacle avoidance for mobile robotics. *J. Robot. Syst.* **1988**, *5*, 87–102. [CrossRef]
13. Galceran, E.; Carreras, M. A Survey on Coverage Path Planning for Robotics. *Robot. Auton. Syst.* **2013**, *61*, 1258–1276. [CrossRef]
14. An, V.; Qu, Z.; Roberts, R. A Rainbow Coverage Path Planning for a Patrolling Mobile Robot With Circular Sensing Range. *IEEE Trans. Syst. Man Cybern. Syst.* **2017**, *48*, 1238–1254. [CrossRef]
15. An, V.; Qu, Z.; Crosby, F.; Roberts, R.; An, V. A Triangulation-Based Coverage Path Planning. *IEEE Trans. Syst. Man Cybern. Syst.* **2020**, *50*, 2157–2169. [CrossRef]
16. Zelinsky, A.; Jarvis, R.; Byrne, J.C.; Yuta, S. Planning Paths of Complete Coverage of an Unstructured Environment by a Mobile Robot. In Proceedings of the International Conference on Advanced Robotics, Tokyo, Japan, 26–30 July 1993; pp. 533–538.
17. Gabriely, Y.; Rimon, E. Competitive online coverage of grid environments by a mobile robot. *Comput. Geom.* **2003**, *24*, 197–224. [CrossRef]
18. Dakulović, M.; Čikeš, M.; Petrović, I. Efficient Interpolated Path Planning of Mobile Robots based on Occupancy Grid Maps. *IFAC Proc. Vol.* **2012**, *45*, 349–354. [CrossRef]
19. Klančar, G.; Seder, M.; Blažič, S.; Škrjanc, I.; Petrović, I. Drivable Path Planning Using Hybrid Search Algorithm Based on E* and Bernstein-Bézier Motion Primitives. *IEEE Trans. Syst. Man Cybern. Syst.* **2019**, 51, 4868–4882. [CrossRef]
20. Le, A.V.; Prabakaran, V.; Sivanantham, V.; Mohan, R.E. Modified A-Star Algorithm for Efficient Coverage Path Planning in Tetris Inspired Self-Reconfigurable Robot with Integrated Laser Sensor. *Sensors* **2018**, *18*, 2585. [CrossRef] [PubMed]
21. Lui, Y.T.; Sun, R.Z.; Zhang, T.Y.; Zhang, X.N.; Li, L.; Shi, G.Q. Warehouse-Oriented Optimal Path Planning for Autonomous Mobile Fire-Fighting Robots. *Secur. Commun. Netw.* **2020**, *2020*, 6371814. [CrossRef]
22. LaValle, S.M. *Planning Algorithms*; Cambridge University Press: Cambridge, MA, USA, 2006; p. 1007.
23. Shrivastava, K.; Kumar, S. The Effectiveness of Parameter Tuning on Ant Colony Optimization for Solving the Travelling Salesman Problem. In Proceedings of the 8th International Conference on Communication Systems and Network Technologies, Bhopal, India, 24–26 November 2018. [CrossRef]
24. Shweta, K.; Singh, A. An Effect and Analysis of Parameter on Ant Colony Optimization for Solving Travelling Salesman Problem. *Int. J. Comput. Sci. Mob. Comput.* **2013**, *2*, 222–229.
25. Hopfield, J.; Tank, D. Neural Computation of Decisions in Optimization Problems. *Biol. Cybern.* **1985**, *52*, 141–152. [CrossRef]
26. García, L.; Talaván, P.; Yáñez, J. Improving the Hopfield model performance when applied to the traveling salesman problem. *Soft Comput.* **2017**, *21*, 3891–3905. [CrossRef]
27. Shi, Y.; Zhang, Y. The neural network methods for solving Traveling Salesman Problem. *Procedia Comput. Sci.* **2022**, *199*, 681–686. [CrossRef]
28. Gabriely, Y.; Rimon, E. Spiral-STC: An On-Line Coverage Algorithm of Grid Environments by a Mobile Robot. In Proceedings of the IEEE International Conference on Robotics and Automation, ICRA'02, Washington, DC, USA, 11–15 May 2002; Volume 1, pp. 954–960. [CrossRef]
29. Kan, X.; Teng, H.; Karydis, K. Online Exploration and Coverage Planning in Unknown Obstacle-Cluttered Environments. *IEEE Robot. Autom. Lett.* **2020**, *5*, 5969–5976. [CrossRef]
30. Hassan, M.; Liu, D. PPCPP: A Predator–Prey-Based Approach to Adaptive Coverage Path Planning. *IEEE Trans. Robot.* **2020**, *36*, 284–301. [CrossRef]
31. Yang, C.; Tang, Y.; Zhou, L.; Ma, X. Complete Coverage Path Planning Based on Bioinspired Neural Network and Pedestrian Location Prediction. In Proceedings of the 2018 IEEE 8th Annual International Conference on CYBER Technology in Automation, Control, and Intelligent Systems (CYBER), Tianjin, China, 19–23 July 2018; pp. 528–533. [CrossRef]
32. Dogru, S.; Marques, L. ECO-CPP: Energy constrained online coverage path planning. *Robot. Auton. Syst.* **2022**, *157*, 104242. [CrossRef]
33. Di Franco, C.; Buttazzo, G. Energy-Aware Coverage Path Planning of UAVs. In Proceedings of the 2015 IEEE International Conference on Autonomous Robot Systems and Competitions, Vila Real, Portugal, 8–10 April 2015; pp. 111–117. [CrossRef]
34. Savkin, A.V.; Huang, H. Asymptotically Optimal Path Planning for Ground Surveillance by a Team of UAVs. *IEEE Syst. J.* **2022**, *16*, 3446–3449. [CrossRef]
35. Dubin, L.E. On curves of minimal length with constraint on average curvature, and with prescribed initial and terminal positions and tangents. *Am. J. Math.* **1957**, *79*, 497–516. [CrossRef]
36. Backman, J.; Piirainen, P.; Oksanen, T. Smooth turning path generation for agricultural vehicles in headlands. *Biosyst. Eng.* **2015**, *139*, 76–86. [CrossRef]

37. Yu, X.; Roppel, T.A.; Hung, J.Y. An Optimization Approach for Planning Robotic Field Coverage. In Proceedings of the 41st Annual Conference of the IEEE Inductrial Electronics Society, Yokohama, Japan, 9–12 November 2015. [CrossRef]
38. Jin, J.; Tang, L. Optimal Coverage Path Planning for Arable Farming on 2D Surfaces. *Trans. ASABE* **2010**, *53*, 283–295. [CrossRef]
39. Lee, T.K.; Baek, S.; Choi, Y.H.; Oh, S.Y. Smooth coverage path planning and control of mobile robots based on high-resolution grid map representation. *Robot. Auton. Syst.* **2011**, *59*, 801–812. [CrossRef]
40. Brezak, M.; Petrović, I. Real-time Approximation of Clothoids With Bounded Error for Path Planning Applications. *IEEE Trans. Robot.* **2014**, *30*, 507–515. [CrossRef]
41. Lepetič, M.; Klančar, G.; Škrjanc, I.; Matko, D.; Potočnik, B. Time optimal path planning considering acceleration limits. *Robot. Auton. Syst.* **2003**, *45*, 199–210. [CrossRef]
42. Kanayama, Y.; Kimura, Y.; Miyazaki, F.; Noguchi, T. A stable tracking control method for an autonomous mobile robot. In Proceedings of the IEEE International Conference on Robotics and Automation, Cincinnati, OH, USA, 13–18 May 1990; Volume 1, pp. 384–389. [CrossRef]
43. Seder, M.; Baotić, M.; Petrović, I. Receding Horizon Control for Convergent Navigation of a Differential Drive Mobile Robot. *IEEE Trans. Control. Syst. Technol.* **2017**, *25*, 653–660. [CrossRef]

sensors

MDPI

Article

Minimum-Time Trajectory Generation for Wheeled Mobile Systems Using Bézier Curves with Constraints on Velocity, Acceleration and Jerk

Martina Benko Loknar, Gregor Klančar and Sašo Blažič *

Faculty of Electrical Engineering, University of Ljubljana, Tržaška cesta 25, SI-1000 Ljubljana, Slovenia
* Correspondence: saso.blazic@fe.uni-lj.si

Abstract: This paper considers the problem of minimum-time smooth trajectory planning for wheeled mobile robots. The smooth path is defined by several Bézier curves and the calculated velocity profiles on individual segments are minimum-time with continuous velocity and acceleration in the joints. We describe a novel solution for the construction of a 5th order Bézier curve that enables a simple and intuitive parameterization. The proposed trajectory optimization considers environment space constraints and constraints on the velocity, acceleration, and jerk. The operation of the trajectory planning algorithm has been demonstrated in two simulations: on a racetrack and in a warehouse environment. Therefore, we have shown that the proposed path construction and trajectory generation algorithm can be applied to a constrained environment and can also be used in real-world driving scenarios.

Keywords: wheeled mobile robots; trajectory generation; velocity profile; trajectory optimization; Bézier curves

Citation: Benko Loknar, M.; Klančar, G.; Blažič, S. Minimum-Time Trajectory Generation for Wheeled Mobile Systems Using Bézier Curves with Constraints on Velocity, Acceleration and Jerk. *Sensors* **2023**, 23, 1982. https://doi.org/10.3390/s23041982

Academic Editor: Cosimo Distante

Received: 18 January 2023
Revised: 3 February 2023
Accepted: 7 February 2023
Published: 10 February 2023

1. Introduction

Path planning and trajectory planning are fundamental topics in autonomous mobile robotics and even more broadly in the context of automation [1]. Path planning algorithms generate a path through predefined points with the main goal of finding a continuous path that minimizes the distance between the starting point and an end point without colliding with obstacles [2,3]. While path planning is a geometric problem, trajectory planning additionally parameterizes the resulting path by time. Consequently, defining time moments along a path affects the kinematic and dynamic properties of the motion of a mobile system. Forces and torques depend on acceleration along a trajectory, while vibrations of its mechanical structure are mainly determined by values of jerk, the time derivative of acceleration [4].

The aim of our work was to solve the problem of minimum-time trajectory generation for wheeled mobile systems with constraints on velocity, acceleration, and jerk in a limited planar space without obstacles. The idea we propose is to apply an optimization method to determine the construction parameters of a Bézier curve primitive such that the resulting travel time on a complete smooth path is minimal. The algorithm we use to compute the minimum-time velocity profile is presented in Ref. [5]. It computes the velocity profile on a predefined path under specified constraints on velocity, radial and tangential acceleration, and radial and tangential jerk.

This paper is organized as follows. Section 2 gives a general overview of the related work. Section 3 introduces the research problem and our main objectives, and Section 4 briefly lists the main contributions. The novel methodology of constructing and parameterizing the fifth-order Bézier curves that make up the resulting geometric path is detailed in Section 5.1. In Section 6, we present two applications of our proposed trajectory generation algorithm, namely the computation of the minimum-time trajectory of a wheeled mobile system on a racetrack (Section 6.1) and in a warehouse (Section 6.2). Our conclusions are drawn in the last section.

2. Related Work

The problem of minimum-time trajectory planning remains relevant due to the growing demands for optimal operation of mobile systems, robots, and automated machines. Trajectory planning or, more generally, the planning of the motion of mobile systems can be divided into two parts: velocity profile optimization and path search [6].

The problem of velocity profile optimization is to determine the time-optimal speed law that satisfies certain kinematic or dynamic constraints, and was considered in Refs. [7,8]. The authors in Ref. [9] have provided a comprehensive review of the consideration of jerk in science and engineering, where it is used as a design factor to ensure ride comfort (e.g., amusement park rides, watercraft, elevators, and autonomous buses), and also reference jerk-related ISO standards. As a result, jerk has established its relevance in numerous scientific and engineering applications. Much of the research is dedicated to limiting or minimizing jerk to reduce vibration, decrease positional errors, or improve the overall performance of machine tools [10], robotic manipulators [11–15], and autonomous mobile robots [5,16–19].

Numerous path planning strategies have been designed and implemented in the literature [20–22]. To meet the kinematic limits of the vehicle and successfully transport a hazardous, fragile, or valuable load, the resulting path must be smooth [23]; it must be feasible at high speeds while being harmless to the mechanical system by avoiding vibration and excessive acceleration of the actuators. Often, path planning techniques must also comply with geometric constraints [3,24]. A significant part of path planning methods is the choice of geometric curves, which can be polynomials [25], Bézier curves [6,26–28], cubic splines [29], B-splines [30], Dubins curves, clothoids [31], hypocycloids [32], and others, as presented in Ref. [23].

In this work, we have utilized Bézier curves due to their favorable properties, including low computational cost and flexibility. The authors in Refs. [33–36] also used quintic Bézier curves and various optimization approaches in an attempt to improve the efficiency and accuracy of path planning for autonomous vehicles. In Ref. [33], the author described the cubic and quintic (trigonometric) Bézier curves using a few shape parameters, which makes the method flexible for use in cluttered environments. However, the author only evaluated and compared the values of velocity, radial acceleration, longitudinal and radial jerks on given unit speed curves. In Ref. [34], the authors proposed a real-time motion planning approach for automated driving in urban environments. Similar to our case, they used a decoupled method by separating path and speed planning. While their trajectory generation approach is suitable for environments with obstacles, the generated velocity profiles do not include jerk constraints. In Ref. [35], the presented method combines jump point search with Bézier curves. However, their approach only ensures $\mathcal{C}^2$ continuity and considers velocity and acceleration constraints. In Ref. [36], the authors proposed an optimization approach for path planning for driverless vehicles in parallel parking using a radial basis function neural network. The authors optimized performance to ensure curve continuity, safety, and compliance with curvature constraints, but did not address the problem of velocity planning or compliance with other dynamic constraints.

Mobile robots are finding broader application and have become an integral part of a variety of environments: in manufacturing, medicine, and many other robotics-based services, including automated warehouses [37–40]. In work environments where simple and labor-intensive tasks of workers are replaced by mobile robots, labor efficiency, scalability, adaptability, and warehouse visibility increase, and turnaround time decreases.

3. Problem Formulation

Let the motion of a mobile system along a three times continuously differentiable plane curve $\mathcal{C}$ be described as a function of time $t \in \left[0, t_f\right]$ by the position vector $\mathbf{r}(t)$

measured from a given fixed origin. The velocity vector $\mathbf{v}(t)$, the acceleration vector $\mathbf{a}(t)$, and the jerk vector $\mathbf{j}(t)$ in the tangential-normal form are:

$$\mathbf{v}(t) = v(t) \cdot \hat{\mathbf{T}}, \tag{1a}$$

$$\mathbf{a}(t) = a_{\mathrm{T}}(t) \cdot \hat{\mathbf{T}} + a_{\mathrm{R}}(t) \cdot \hat{\mathbf{N}} = \dot{v} \cdot \hat{\mathbf{T}} + v^2\kappa \cdot \hat{\mathbf{N}}, \tag{1b}$$

$$\mathbf{j}(t) = j_{\mathrm{T}}(t) \cdot \hat{\mathbf{T}} + j_{\mathrm{R}}(t) \cdot \hat{\mathbf{N}} = \left(\ddot{v} - v^3\kappa^2\right) \cdot \hat{\mathbf{T}} + \frac{1}{v}\left(\frac{\mathrm{d}}{\mathrm{d}t}(v^3\kappa)\right) \cdot \hat{\mathbf{N}}, \tag{1c}$$

where $\hat{\mathbf{T}}$ and $\hat{\mathbf{N}}$ are the unit tangential and the unit normal vector, respectively, and κ is the curvature of the path at time t. In Equation (1a), v is called speed, and the tangential and normal components of the acceleration (Equation (1b)) are called acceleration along the path and centripetal acceleration (also called radial acceleration), respectively. The expression (1c) is obtained by differentiating Equation (1b) and applying the Frenet–Serret formulas for movement in two-dimensional Euclidean space $\mathbb{R}^2$ [41].

The maximum allowable value of velocity v_{MAX} is determined by the capabilities of the robot actuators and also the environmental conditions (e.g., surface type). Driving in the reverse direction is not permitted. The maximum values of radial acceleration $a_{\mathrm{R_{MAX}}}$ and tangential acceleration $a_{\mathrm{T_{MAX}}}$ can be set based on the dynamic constraints of a mobile robot (e.g., maximum centripetal force and rolling resistance in a turn) [6]. Similarly, we imposed third-order constraints $j_{\mathrm{R_{MAX}}}$ and $j_{\mathrm{T_{MAX}}}$, whose values can be derived from ride comfort criteria [9]. Although we could limit the acceleration and jerk components separately (and treat them individually), we additionally restricted their values to the range within an ellipse (similar to researchers in [5,6,35,42]:

$$0 \leq |v(t)| \leq v_{\mathrm{MAX}}, \tag{2a}$$

$$\frac{a_{\mathrm{R}}^2(t)}{a_{\mathrm{R_{MAX}}}^2} + \frac{a_{\mathrm{T}}^2(t)}{a_{\mathrm{T_{MAX}}}^2} \leq 1, \tag{2b}$$

$$\frac{j_{\mathrm{R}}^2(t)}{j_{\mathrm{R_{MAX}}}^2} + \frac{j_{\mathrm{T}}^2(t)}{j_{\mathrm{T_{MAX}}}^2} \leq 1. \tag{2c}$$

Treating the tangential and radial components of acceleration (Equation (2b)) and jerk (Equation (2c)) together is more rigorous than limiting the individual components. It also provides greater ride comfort by constraining the overall norms. The goal of this research was to develop a trajectory planning method for a mobile system operating in a constrained, obstacle-free, planar environment while subject to kinematic constraints. Although motion planning algorithms have been the subject of extensive research, dealing with third-order constraints still proves challenging.

4. Contributions

The main contributions of this paper can be summarized as follows:

- We describe an innovative construction method for 5th order Bézier curves. The proposed parameterization is simple and intuitive, yet effective for generating smooth paths consisting of multiple splines (Section 5);
- The above smooth path generation basis is coupled with an algorithm that computes a minimum-time velocity profile with velocity, acceleration, and jerk constraints on a predefined path (see Ref. [5]). Together they form a powerful trajectory generation algorithm (Section 6). The resulting trajectories thus provide continuous velocity and acceleration profiles;
- To prove the applicability of our approach to trajectory optimization, we performed simulation experiments on a racetrack and in a warehouse environment (Sections 6.1 and 6.2).

In the warehouse simulation, we identified and analyzed realistic situations with different dynamic constraints to investigate and propose the most appropriate driving scenarios.

5. Curve Primitives

A Bernstein–Bézier curve (or Bézier curve) is defined by a set of control points $P_0, P_1, \dots P_b, b \in \mathbb{N}$:

$$\boldsymbol{r}_b(\lambda) = \sum_{i=0}^{b} \boldsymbol{p}_i B_{i,b}(\lambda), \tag{3}$$

where λ is a normalized time variable ($0 \leq \lambda \leq 1$) and $\boldsymbol{p}_i$ denotes the position vector of a control point P_i. The polynomials $B_{i,b}(\lambda)$:

$$B_{i,b}(\lambda) = \binom{b}{i} \lambda^i (1-\lambda)^{b-i} = \frac{b!}{i!(b-i)!} \lambda^i (1-\lambda)^{b-i}, \tag{4}$$

are known as Bernstein basis polynomials of degree b. Bézier curves can be defined for N-dimensional space, $N \in \mathbb{N}$. In planar space, the curve $\boldsymbol{r}_b(\lambda)$ and the vectors $\boldsymbol{p}_i$ are two-element vectors: $\boldsymbol{r}_b(\lambda) = [X(\lambda), Y(\lambda)]^T$ and $\boldsymbol{p}_i = [x_i, y_i]^T$. These curves have several useful properties for path planning. The first and last points of the Bézier curves introduced in Equation (3) are their endpoints:

$$\boldsymbol{r}_b(0) = \boldsymbol{p}_0 \quad \text{and} \quad \boldsymbol{r}_b(1) = \boldsymbol{p}_b. \tag{5}$$

The N-dimensional, b-th order Bézier curve also lies within the convex hull defined by its control points. Furthermore, the beginning and the end of the curve are tangent to the first and the last section of the convex polygon, respectively (Figure 1).

$$\left.\frac{\mathrm{d}\boldsymbol{r}_b}{\mathrm{d}\lambda}\right|_{\lambda=0} = b(\boldsymbol{p}_1 - \boldsymbol{p}_0), \tag{6}$$

$$\left.\frac{\mathrm{d}\boldsymbol{r}_b}{\mathrm{d}\lambda}\right|_{\lambda=1} = b(\boldsymbol{p}_b - \boldsymbol{p}_{b-1}). \tag{7}$$

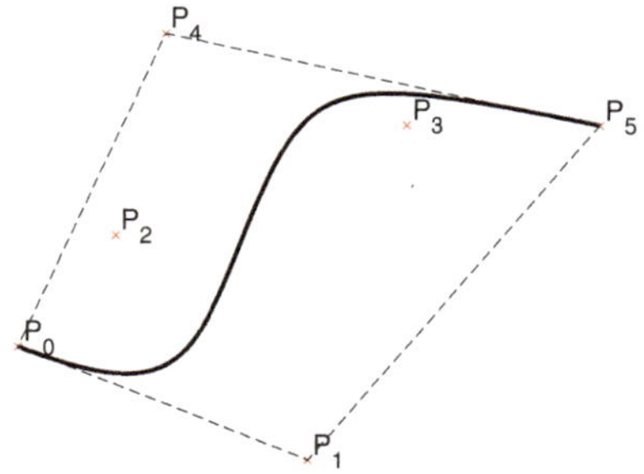

Figure 1. Fifth order Bernstein–Bézier curve within its convex hull (dashed lines). The curve is tangent to the sides of the convex hull, line segments $\overline{P_0P_1}$ and $\overline{P_4P_5}$.

Other properties of Bernstein polynomials (derivatives, calculation of definite integrals, the de Casteljau algorithm, degree elevation, etc.) fall outside the scope of this article; more details on this topic can be found in Ref. [28].

Bézier curves constructed by a large number of control points are computationally intensive. For this reason, in path planning, it is desirable to construct a smooth path by connecting low-degree Bézier curves [6]. The authors in Ref. [43] proposed a new parameterization of motion primitives based on Bézier curves for path planning applications of wheeled mobile robots. However, the method was presented for third-order polynomials and the algorithm does not guarantee the existence of the curve for all possible parameterizations. We used fifth-order Bézier curves because this is the degree of Bézier curves that always satisfies the curvature continuity requirement (C^2) in the joints. The 5th order Bézier curve $\boldsymbol{r}_5(\lambda)$ is defined by six control points P_i:

$$\boldsymbol{r}_5(\lambda) = (1-\lambda)^5\boldsymbol{p}_0 + 5\lambda(1-\lambda)^4\boldsymbol{p}_1 + 10\lambda^2(1-\lambda)^3\boldsymbol{p}_2 + 10\lambda^3(1-\lambda)^2\boldsymbol{p}_3 + 5\lambda^4(1-\lambda)\boldsymbol{p}_4 + \lambda^5\boldsymbol{p}_5. \tag{8}$$

5.1. Construction of 5th Order Bézier Curves

It is very important to choose the appropriate construction parameters that would efficiently define the Bézier curves and facilitate the search for the minimum-time trajectory.

With the above notation, let us mark the distances between consecutive control points $d(P_i, P_{i+1})$ as d_{i+1} and the angles between (P_i, P_{i+1}) and the positive direction of the x-axis as φ_{i+1} (Figure 2), $i = \{0, 1 \ldots, 4\}$. For the coordinates of two consecutive control points, it follows that:

$$x_{i+1} - x_i = d_{i+1}\cos\varphi_{i+1}, \tag{9a}$$

$$y_{i+1} - y_i = d_{i+1}\sin\varphi_{i+1}. \tag{9b}$$

We evaluate (9a) and (9b) for $i \in \{0, 1\}$ using the sum and difference formulas for sine and cosine. This gives the following expression for the value of the curvature in P_0:

$$\lim_{\lambda\to 0}\kappa(\lambda) = \kappa_0 = \frac{4}{5}\frac{d_2}{d_1^2}\sin(\varphi_2 - \varphi_1). \tag{10}$$

The derivative of curvature κ_0 in P_0 with respect to λ is:

$$\lim_{\lambda\to 0}\frac{\mathrm{d}\kappa}{\mathrm{d}\lambda} = \frac{12}{5}\frac{1}{d_1^2}d_3\sin(\varphi_3 - \varphi_1) + \kappa_0\left(-12\frac{d_2}{d_1}\cos(\varphi_2 - \varphi_1) + 6\right). \tag{11}$$

We choose the parameters of the curve so that the second term in Equation (11) becomes 0. This happens when:

$$\frac{d_2}{d_1} = \frac{1}{2\cos(\varphi_2 - \varphi_1)}. \tag{12}$$

The curvature κ_0 from Equation (10) and its derivative κ_0' in P_0 from Equation (11) then become:

$$\kappa_0 = \frac{4}{10}\frac{\tan(\varphi_2 - \varphi_1)}{d_1}, \tag{13a}$$

$$\kappa_0' = \frac{12}{5}\frac{d_3\sin(\varphi_3 - \varphi_1)}{d_1^2}. \tag{13b}$$

The purpose of introducing notations for d_i and φ_i and deriving expressions for κ_0 and κ_0' is to make the process of path construction as efficient and intuitive as possible. This also takes into account that the path ultimately consists of several Bézier curves. Thus, the parameters needed to generate a 5th order Bézier curve are $P_0, P_5, \varphi_1, \varphi_5, \kappa_0, \kappa_5, d_1, d_5$, and κ_0'. However, how would one set the value of κ_0'? It could be simply set to zero, but perhaps it is also useful to examine Equation (1c) and choose such a value for κ'_0 that the value of the radial component (in P_0) of the jerk vector is zero.

A 5th order Bézier curve is therefore constructed in the following steps (Figure 2):

1. Outline the first control point and mark it as P_0. In the direction of φ_1, measure out the distance d_1 and mark the second point as P_1.
2. In the direction φ_1, measure out the distance $d_2^{||}$ (from Equation (12)):

$$d_2^{||} = d_2 \cos(\varphi_2 - \varphi_1) = \tfrac{1}{2} d_1. \tag{14}$$

3. Measure in the perpendicular direction the distance $d_2^{\perp}$ (from Equation (10)):

$$d_2^{\perp} = \tfrac{5}{4} d_1^2 \kappa_0. \tag{15}$$

   and mark the third point as P_2.
4. All points away from P_2 for $d_3^{\perp}$ (Equation (11)) in the same direction (perpendicular to the line segment $\overline{P_0P_1}$) lie on the red dashed line.

$$d_3^{\perp} = \tfrac{5}{12} d_1^2 \kappa'_0. \tag{16}$$

5. Mark the last point as P_5. Measure out the distance d_5 in the opposite direction from φ_5 and mark the fifth point as P_4.
6. All points away from P_4 for $d_4^{\perp}$ (Equations (9a), (9b) and (10) for $i = 4$) in the same direction (perpendicular to the line segment $\overline{P_4P_5}$) lie on the green dashed line:

$$d_4^{\perp} = \tfrac{5}{4} d_5^2 \kappa_5. \tag{17}$$

7. The fourth control point P_3 lies on the intersection of the red and green dashed lines. The Bézier curve is now completely defined.

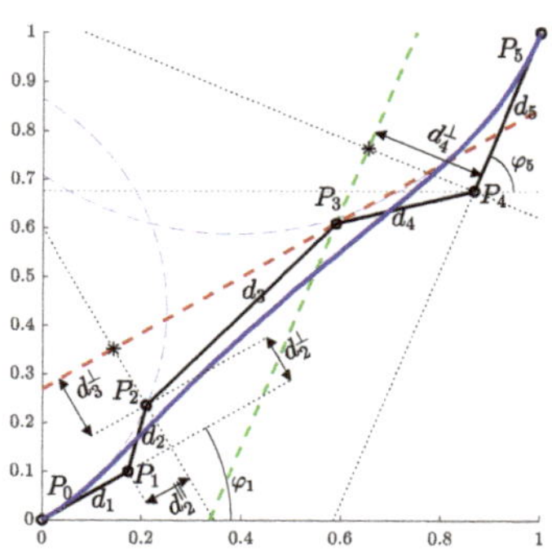

Figure 2. The proposed construction of a Bézier curve that enables efficient parameterization.

6. Generation of Minimum-Time Trajectories

We have shown the use of the proposed trajectory planning algorithm in two environments. On a racetrack, the focus was on demonstrating path construction and ensuring that it is within the corridor boundaries. In a warehouse, we demonstrated the benefits of our proposed methods in a real-world application. All simulations were performed using the Matlab programming environment on a computer with an Intel(R) Core(TM) i7-8700 CPU 3.2 GHz processor with 16 GB RAM memory.

A minimum-time trajectory is computed by applying an algorithm that generates a minimum-time velocity profile (proposed in Ref. [5]) to Bézier curve splines. The algorithm, which considers velocity, acceleration, and jerk constraints along a given path, consists of two steps. In the first step of the algorithm, the velocity and acceleration constraints are considered. In the second step, the algorithm modifies the original velocity profile to include the jerk constraints, with the process varying depending on the type of violation (single-point or interval jerk violations). The simulation methodology for computing the minimum-time velocity profile, described in detail in Ref. [5], can essentially be described as solving the presented ordinary differential equation with a given initial value. In our own implementation, numerical methods (Euler's method and trapezoidal integration) were used to calculate the required values in the discrete time samples.

We then used a nonlinear gradient optimization method, an optimization routine built into Matlab, to change the construction parameters of the Bézier curves. Using this method, we found a solution where the travel time reached a minimum. Since the simulated environments were static and free of obstacles, we divided the environments into several individual sections. This was done primarily to reduce the number of optimization parameters and consequently speed up the minimization process.

6.1. Racetrack Environment

The model of a racetrack that we used in our simulations is shown in Figure 3. It is defined by the centerline. The left and right edges of the racetrack are at a distance $w/2$ from the centerline and represent the corridor boundaries. The shape of the racetrack can in general be arbitrary complex and is therefore divided into segments. This is done by analyzing the curvature of the centerline. Points where the curvature reaches local extrema are denoted by the sequence C_i, $i \in \{1, \ldots, N_{\text{parts}} + 1\}$ where N_{parts} is the number of segments. Then perpendicular lines to the centerline are drawn in C_i. These lines represent the edges of individual segments. The first and last control points of the Bézier curves lie somewhere on the segment edges. For simplicity, we represent the positions of P_0^i and P_5^i, $i \in \{1, \ldots, N_{\text{parts}}\}$, by the parameter $p \in [-1, 1]$. The sign of p indicates whether the control point lies somewhere between the centerline and the right (+) or left (−) edge of the racetrack.

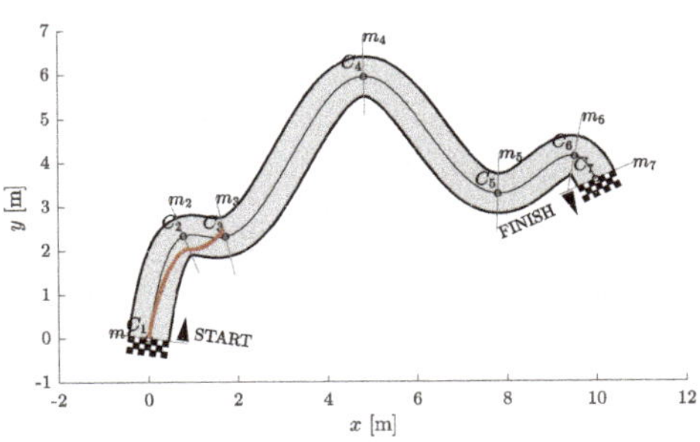

Figure 3. The racetrack model we used for the simulations. Shown are points C_i on the centerline where the curvature is locally highest, and lines m_i where P_1^i and P_5^i lie.

Thus, each curve in a segment is completely defined by the positions, angles, and curvatures in the first and last control points (p_0, φ_1, κ_0, p_5, φ_5, κ_5), d_1, and d_5. Note that the angles are measured from a tangent to the centerline in C_i (Figure 4). To find a reasonable starting point for the optimization, we devised a simple heuristics. In each segment, a line ℓ is drawn from P_0^i through the outermost edge of the inner side of the corridor at the end of the $(i+1)$th segment. The intersection of lines ℓ and m_i is the initial estimate for the position $p_{5,\text{init}}^i$ of the last control point $P_5^{\prime i}$. If the intersection point is outside the corridor (as in Figure 5), $p_{5,\text{init}}^i$ is set to its edge. The initial estimate for the angle $\varphi_{5,\text{init}}^i$ is the angle between the line ℓ and the line perpendicular to m_{i+1}, while $\kappa_{5,\text{init}}^i$ is the curvature of the centerline in C_{i+1}. The values of d_1^i and d_5^i were set to the value of a certain fraction of the distance between P_0^i and $P_5'^i$. The heuristic procedure described is shown in Figure 5. To simplify the notation, we will from now on omit the superscript i.

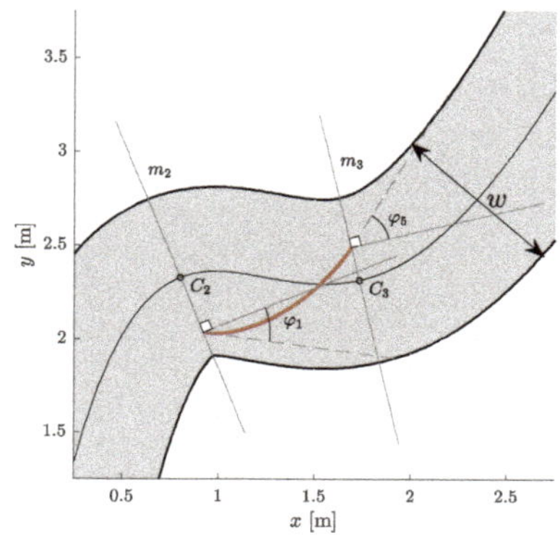

Figure 4. A segment with a Bézier curve. φ_1 and φ_5 are measured from the line perpendicular to m_i.

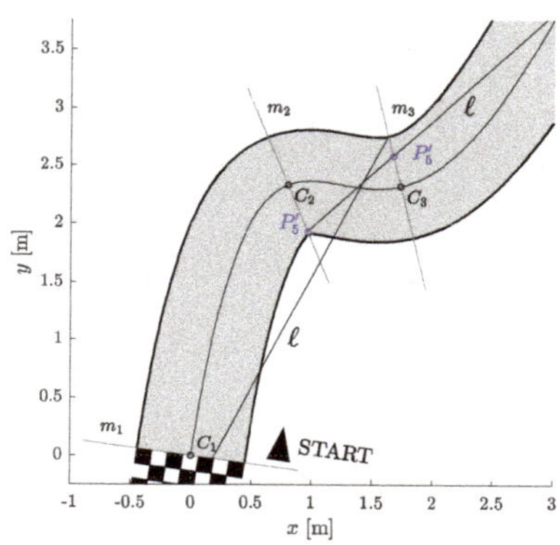

Figure 5. An example of heuristic determination of initial guesses for construction of Bézier curve(s).

We devised a series of separate simulation experiments to demonstrate the operation and efficiency of the proposed Bézier curve construction and trajectory planning algorithm.

Let N_{free} denote the number of optimization parameters on each corridor segment. It is expected that the higher the number of (free) curve construction parameters N_{free}, the more versatile a curve is. Thus a better solution can be provided. However, in this way the optimization problem becomes more computationally expensive and the solution more difficult to obtain due to the complex form of the objective function.

Another problem is the gradual generation of the final trajectory. A Bézier curve can be constructed for each segment separately and have the criterion assigned to it. Alternatively, multiple Bézier curves spanning several segments can be constructed together, and the objective is the travel time along all of them. Then only the solution on the first segment is kept and this procedure is repeated in a receding horizon manner. Let us denote by N_{seg} the number of segments that are treated simultaneously. If only a one-segment optimization is performed ($N_{\text{seg}} = 1$), the solution does not take into account the corridor shape in the following segment, e.g., when a sharp turn follows. On the other extreme, the complete curve (on all corridor segments) can be generated in each run of the optimization, but since the dimension of the optimization problem is the product of the construction parameters, namely $N_{\text{free}} \times N_{\text{seg}}$, it is reasonable not to exaggerate the two values and to find a sensible compromise.

Simulations were performed for $N_{\text{free}} \in \{2, 5\}$ and $N_{\text{seg}} \in \{1, 2, 3\}$. Therefore, in one group of experiments, the two optimization parameters are d_1 and d_5, while the other three necessary parameters for curve construction (p_5, φ_5, and κ_5) are given by the heuristics discussed above and shown in Figure 5. In the other set of simulations, there are five optimization parameters, namely d_1, d_5, p_5, φ_5, and κ_5. Certainly, both simulation scenarios also include cases with different N_{seg}. The data obtained from the simulation experiments are compiled in Table 1, where t_i represents the travel time at the end of a given corridor segment and $\sum_{i=1}^{6} t_i$ is the total travel time on a resulting path.

Table 1. Resulting travel times on segments and total travel times. Simulations were performed for different numbers of optimization parameters $N_{\text{free}} \in \{2, 5\}$, and different numbers of segments $N_{\text{seg}} \in \{1, 2, 3\}$, whose geometry was taken into account when calculating the solution for the current segment.

N_{seg}	N_{free}	t_1 [s]	t_2 [s]	t_3 [s]	t_4 [s]	t_5 [s]	$\sum_{i=1}^{6} t_i$ [s]
1	2	2.91	4.20	7.18	9.73	11.24	12.11
	5	2.34	2.87	5.44	7.89	9.42	10.20
2	2	2.87	4.15	7.14	9.69	11.19	12.03
	5	2.34	2.88	5.54	7.90	9.27	9.66
3	2	2.89	4.16	7.13	9.68	11.22	12.07
	5	2.34	2.90	5.52	7.68	8.87	9.17

Figures 6–11 show the resulting paths in the racetrack and the corresponding velocity profiles. We imposed the following constraints: $v_{\text{MAX}} = 1.75$ m/s (represented by the dashed horizontal lines), $a_{\text{R}_{\text{MAX}}} = 1.6$ m/s^2, $a_{\text{T}_{\text{MAX}}} = 0.8$ m/s^2, $j_{\text{R}_{\text{MAX}}} = 16$ m/s^3, $j_{\text{T}_{\text{MAX}}} = 12$ m/s^3.

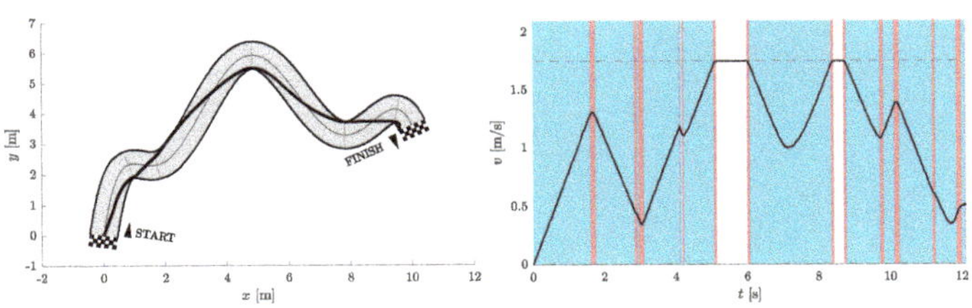

Figure 6. Resulting path as optimization result for $N_{\text{free}} = 2$ and $N_{\text{seg}} = 1$ (**left**) with the corresponding velocity profile (**right**). Blue vertical bands indicate intervals where the acceleration reaches its maximum allowable values. Similarly, red vertical bands indicate intervals where the jerk reaches its maximum allowable value.

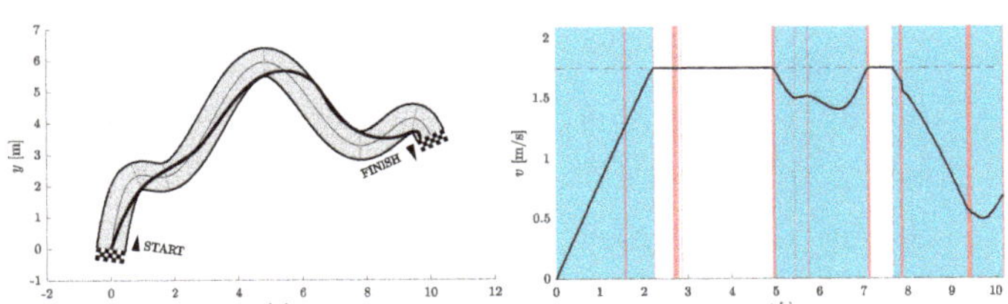

Figure 7. Resulting path as optimization result for $N_{\text{free}} = 5$ and $N_{\text{seg}} = 1$ (**left**) with the corresponding velocity profile (**right**). Blue vertical bands indicate intervals where the acceleration reaches its maximum allowable values. Similarly, red vertical bands indicate intervals where the jerk reaches its maximum allowable value.

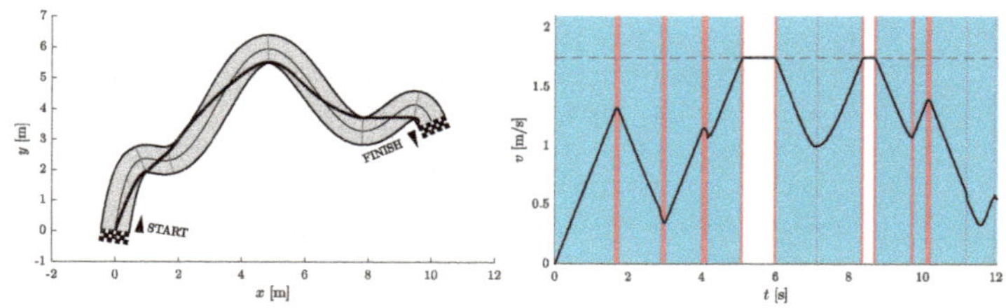

Figure 8. Resulting path as optimization result for $N_{\text{free}} = 2$ and $N_{\text{seg}} = 2$ (**left**) with the corresponding velocity profile (**right**). Blue vertical bands indicate intervals where the acceleration reaches its maximum allowable values. Similarly, red vertical bands indicate intervals where the jerk reaches its maximum allowable value.

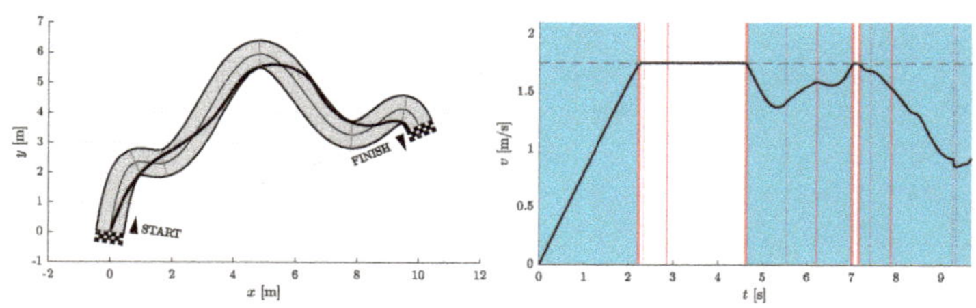

Figure 9. Resulting path as optimization result for $N_{\text{free}} = 5$ and $N_{\text{seg}} = 2$ (**left**) with corresponding velocity profile (**right**). Blue vertical bands indicate intervals where the acceleration reaches its maximum allowable values. Similarly, red vertical bands indicate intervals where the jerk reaches its maximum allowable value.

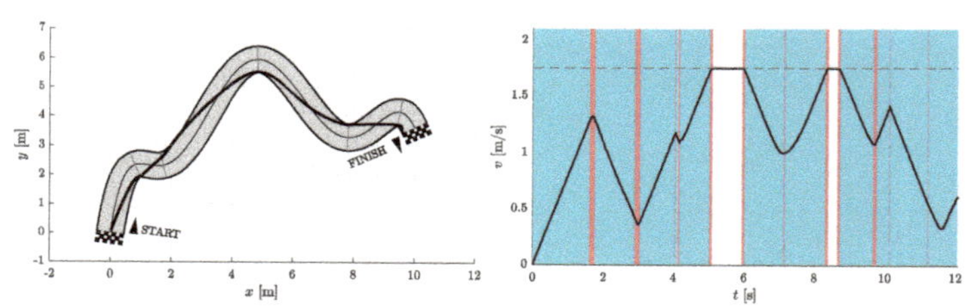

Figure 10. Resulting path as optimization result for $N_{\text{free}} = 2$ and $N_{\text{seg}} = 3$ (**left**) with the corresponding velocity profile (**right**). Blue vertical bands indicate intervals where the acceleration reaches its maximum allowable values. Similarly, red vertical bands indicate intervals where the jerk reaches its maximum allowable value.

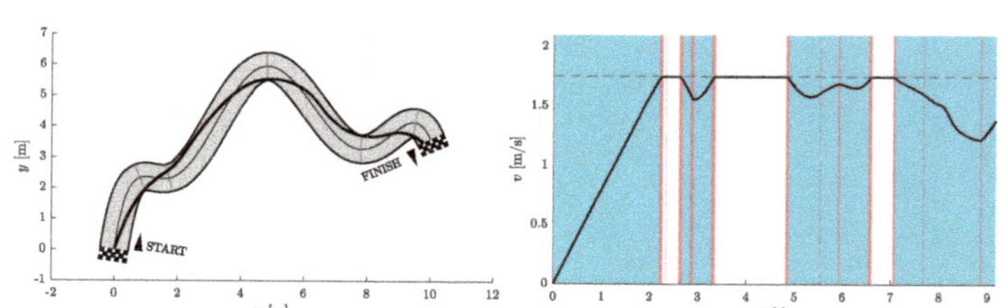

Figure 11. Resulting path as optimization result for $N_{\text{free}} = 5$ and $N_{\text{seg}} = 3$ (**left**) with the corresponding velocity profile (**right**). Blue vertical bands indicate intervals where the acceleration reaches its maximum allowable values. Similarly, red vertical bands indicate intervals where the jerk reaches its maximum allowable value.

As expected, the results in Table 1 show that the travel times decrease when either N_{seg} or N_{free} increases. The shortest travel time is calculated for the case where $N_{\text{seg}} = 3$ and $N_{\text{free}} = 5$.

6.2. Warehouse Environment

The enormous technological capabilities of automated guided vehicles (AGVs) and other autonomous mobile robots (AMRs) are facilitating the launch of fully automated warehouses. Common warehouse tasks performed by mobile robots include loading and unloading goods, stacking and retrieving items, picking and sorting orders, inventory tracking, and warehouse maintenance.

We tested the proposed trajectory planning algorithm in a simple warehouse environment by simulating the task of moving between three rows of storage racks, picking up and dropping off loads from specific locations (Figure 12). Warehouses are usually very confined environments, so we assumed that movement in the two aisles is restricted to a straight line. To avoid collisions of AGVs with storage racks, the straight segments on both sides protrude slightly beyond the edges (black solid dots in Figure 13). The optimization problem is to find the most suitable path shape between the aisles.

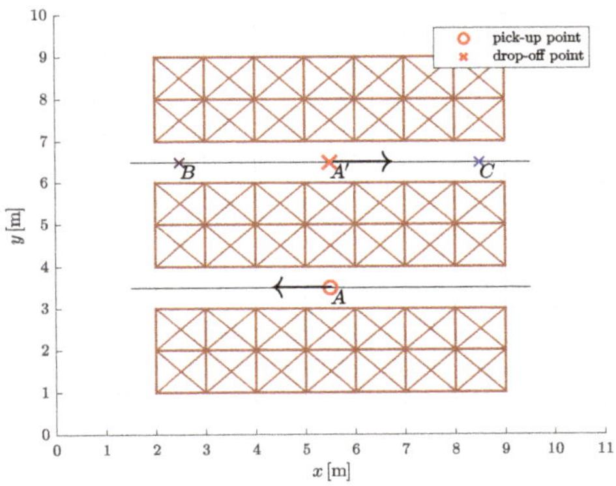

Figure 12. The warehouse floor plan with three pairs of pick-up and drop-off points: A and A', A and B, A and C.

The simulation experiment was designed as follows. An AGV travels clockwise along a circular route from the **p**ick-**u**p **p**oint (PUP) to the **d**rop-**o**ff **p**oint (DOP) and back to the starting point. At the pick-up and drop-off points, the speed is set to zero. As the load is delicate, the dynamic constraints on the mobile system are more severe in the first part of the path, as shown in Table 2.

Table 2. Constraints on velocity, acceleration, and jerk for a fully loaded (✓) and an unloaded (×) mobile system.

Load	v_{MAX} [m/s]	$a_{R_{\mathrm{MAX}}}$ [m/s^2]	$a_{T_{\mathrm{MAX}}}$ [m/s^2]	$j_{R_{\mathrm{MAX}}}$ [m/s^3]	$j_{T_{\mathrm{MAX}}}$ [m/s^3]
✓	1.0	2.0	1.0	4.0	4.0
×	2.25	4.0	2.0	16	16

We proposed that the continuous curvature path between a pick-up point and a drop-off point (and vice versa) consists of two straight lines and two 5th order Bézier curves. The coordinates of the control points were determined by an optimization process that minimizes travel time. Since the velocity is set to zero at the symmetrically placed drop-off point A', the optimizations can be performed only on one half of the circular route (thus on four segments instead of eight). The free optimization parameters for the construction of the Bézier curves were d_1 and d_5 (Section 5.1 for a full explanation) and the x coordinate of the joint between them, which is P_5 of the first Bézier curve and P_0 of the second Bézier curve.

First, we computed the minimum-time trajectories for symmetrically placed pair of pick-up and drop-off point (A and A'). Let us denote the path representing the full-load optimization solution for the symmetrically placed pair A and A' by $\mathcal{F}$. Similarly, let $\mathcal{N}$ be the path representing the no-load optimization solution for the symmetrically placed pair A and A' (Figure 13).

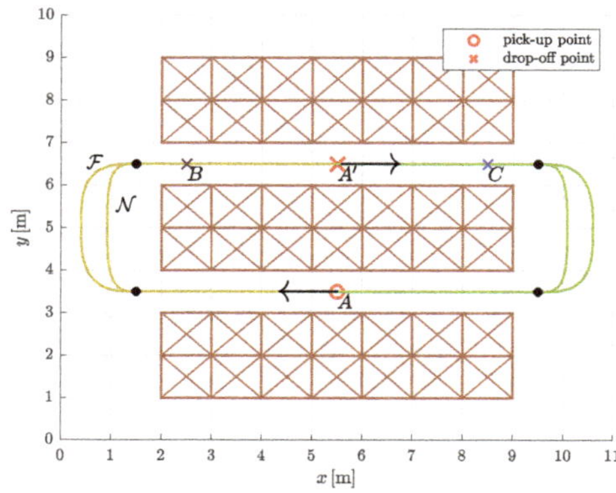

Figure 13. The drawn paths $\mathcal{F}$ and $\mathcal{N}$ are the result of an optimization that minimizes travel time. The filled dots mark the points where the straight segments meet the Bézier curves.

We then conducted a comparative travel time analysis. The main question was whether travel times differ in cases where a fully loaded/unloaded AGV travels along a path that is not optimized for a load of the same type. Normally, AGVs in warehouses travel along predefined trajectories. So with the simulation experiment, we wanted to test whether it is possible to reduce travel time if each curve segment is optimized for the actual load being carried. We also included examples with different ratios of travel times (or path lengths) of fully loaded or unloaded mobile systems by adding drop points B and C. Essentially, we calculated travel times for the three pick-up and drop-off pairs where the AGV was fully loaded on the first part (PUP$\rightarrow$DOP) and unloaded on the second part (DOP$\rightarrow$PUP) of the circular path, but traveling on either $\mathcal{F}$ or $\mathcal{N}$. The travel times are given in Table 3, where the subscripts indicate the load type of the AGV. By μ, we denote the relative increase (in percent) in travel time in a given route case scenario compared to the travel time when the AGV travels on a path optimized for a load of the same type (the last three rows in Table 3).

The results in Table 3 show that the travel time is indeed the shortest when the mobile system travels along the route optimized for the actual load (PUP$\rightarrow$DOP: $\mathcal{F}_{\mathrm{F}}$, and DOP$\rightarrow$PUP: $\mathcal{N}_{\mathrm{N}}$). Moreover, it can be seen that when the default path is $\mathcal{F}$ (rows 4–6 in Table 3), the corresponding travel times are always shorter than in the case when the default path is $\mathcal{N}$ (rows 1–3 in Table 3). However, generally, the travel times are not radically different and this observation is not entirely unexpected. We could achieve more obvious travel time differences if we increased the ratio of fully loaded to unloaded constraint values (see Table 2) or chose a more complex arrangement of pick-up and drop-off points spanning multiple rows of storage racks. Nevertheless, the selected values for velocity, acceleration, and jerk constraints (and the ratio between the two load types) should reflect reality. Additionally, the examples presented can be viewed as the smallest units for which this analysis can be performed. These subtle differences in travel times (approximately 1% reduction) imply significant absolute time differences when the presented trajectories are combined into larger trajectories. Or if one considers that a warehouse robot would traverse the same trajectories over and over again during its entire operation.

Table 3. Travel times of the AGV on a circular route for different placements of pick-up (PUP) and drop-off points (DOP). For the symmetrically placed pair $A - A'$, $\mathcal{F}$ and $\mathcal{N}$ denote the paths representing the optimization solutions full-load and no-load, respectively. Similarly, the indices F and N denote the type of load on the AGV. We write μ for the increase in travel time according to the last three rows, expressed as a percentage.

Circular Route Case		Pick Up Point	Drop off Point	Travel Time	μ
PUP → DOP	**DOP → PUP**	**(PUP)**	**(DOP)**	**[s]**	**[%]**
		A	A'	20.74	1.49
$\mathcal{N}_{\text{F}}$	$\mathcal{N}_{\text{N}}$	A	B	19.08	1.62
		A	C	22.40	1.38
		A	A'	20.65	1.04
$\mathcal{F}_{\text{F}}$	$\mathcal{F}_{\text{N}}$	A	B	18.99	1.13
		A	C	22.25	0.66
		A	A'	20.44	0
$\mathcal{F}_{\text{F}}$	$\mathcal{N}_{\text{N}}$	A	B	18.78	0
		A	C	22.10	0

Figure 14 shows the velocity profiles for all three pairs of pick-up and drop-off points for the case where $\mathcal{F}_{\text{F}}$ is on the first part and $\mathcal{N}_{\text{N}}$ is on the second part of the circular route.

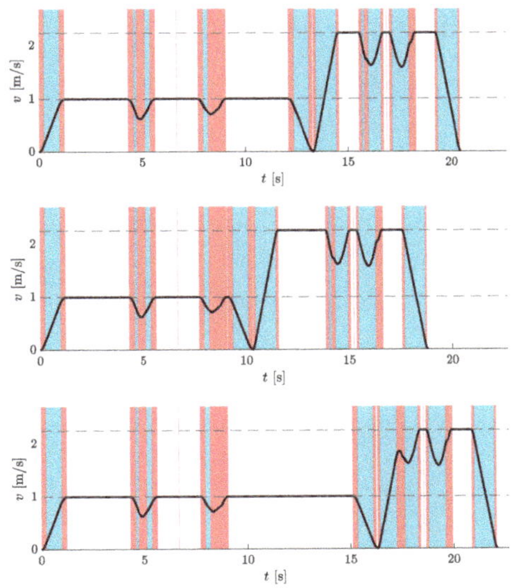

Figure 14. Velocity profiles of AGV traveling on a circular route for different placements of pick-up and drop-off points: A and A' (**top**), A and B (**middle**), A and C (**bottom**). The graphs show the results for PUP→DOP: $\mathcal{F}_{\text{F}}$, and DOP→PUP: $\mathcal{N}_{\text{N}}$. Dashed horizontal lines represent the velocity limit values. Blue vertical bands indicate intervals where the acceleration reaches its maximum allowable values. Similarly, red vertical bands indicate intervals when the jerk reaches its maximum allowable values.

7. Conclusions

In this paper, we present a new minimum-time optimization-based approach for planning the trajectory of a mobile robot in a planar constrained environment. We assumed that a mobile system has constraints on velocity, acceleration, and jerk. The resulting smooth path consists of 5th order Bézier curves, for whose construction we propose a new method that allows efficient parameterization.

We analyzed the results of the proposed approach for generating trajectories in a simulated automated warehouse. Different sets of dynamic constraints led to different solutions for trajectories. We have shown that it is possible to achieve noticeable improvements in travel times by choosing the appropriate trajectories. The approach is applicable for trajectory and velocity planning of a single wheeled robot, but could be extended for the use of multiple robots to take into account evasive maneuvers or cooperation on a given task. It can also be used by various other mobile systems moving in a plane (e.g., track robots, robotic manipulators), especially non-holonomic systems.

Our findings may be especially useful and have great potential for determining minimum-time trajectories in automated warehouses, where the dynamic constraints imposed on autonomous mobile robots may depend on the type of load the mobile system is transporting. Our approach could also be applied to other planar environments with similar requirements.

The values of the constraints in the warehouse environment were conservatively estimated to ensure the vertical stability of a mobile system. However, the stability of the system (mobile robot with load) itself was not the subject of our research. Future studies should aim to describe the characteristics of the load in more detail, as this could impose additional or more demanding constraints on a mobile system. For a specific mobile system with known load characteristics (mass, mass distribution, dimensions, contact area conditions) it would be possible to calculate the tipping angle and consequently determine the allowable accelerations. The use of higher order Bézier curves or other curve primitives would also be of particular interest. More broadly, research is needed to apply the proposed trajectory planning approach to environments with static or dynamic obstacles to demonstrate the proposed idea using global or local path planning methods.

Author Contributions: Conceptualization and methodology, M.B.L., G.K. and S.B.; software, M.B.L.; writing—original draft preparation, M.B.L.; writing—review and editing, M.B.L., G.K. and S.B. All authors have read and agreed to the published version of the manuscript.

Funding: This work was funded by the Slovenian Research Agency under Grants P2-0219 and L2-3168.

Institutional Review Board Statement: Not applicable.

Informed Consent Statement: Not applicable.

Data Availability Statement: Not applicable.

Conflicts of Interest: The authors declare no conflict of interest.

References

1. Gasparetto, A.; Boscariol, P.; Lanzutti, A.; Vidoni, R., Path Planning and Trajectory Planning Algorithms: A General Overview. In *Motion and Operation Planning of Robotic Systems: Background and Practical Approaches*; Carbone, G., Gomez-Bravo, F., Eds.; Springer International Publishing: Cham, Switzerland, 2015; pp. 3–27. [CrossRef]
2. Tharwat, A.; Elhoseny, M.; Hassanien, A.E.; Gabel, T.; Kumar, A. Intelligent Bézier curve-based path planning model using Chaotic Particle Swarm Optimization algorithm. *Clust. Comput.* **2019**, *22*, 4745–4766. [CrossRef]
3. Zhang, H.; Yang, S. Smooth path and velocity planning under 3D path constraints for car-like vehicles. *Robot. Auton. Syst.* **2018**, *107*, 87–99. [CrossRef]
4. Tsirlin, M. Jerk by axes in motion along a space curve. *J. Theor. Appl. Mech.* **2017**, *55*, 1437–1441. [CrossRef]
5. Benko Loknar, M.; Blažič, S.; Klančar, G. Minimum-time velocity profile planning for planar motion considering velocity, acceleration and jerk constraints. *Int. J. Control.* **2023**, *96*, 251–265. [CrossRef]
6. Zdešar, A.; Škrjanc, I. Optimum Velocity Profile of Multiple Bernstein-Bézier Curves Subject to Constraints for Mobile Robots. *ACM Trans. Intell. Syst. Technol.* **2018**, *9*, 1–23. [CrossRef]

7. Consolini, L.; Locatelli, M.; Minari, A.; Nagy, A.; Vajk, I. Optimal Time-Complexity Speed Planning for Robot Manipulators. *IEEE Trans. Robot.* **2019**, *35*, 790–797. [CrossRef]
8. Lima, P.F.; Trincavelli, M.; Martensson, J.; Wahlberg, B. Clothoid-Based Speed Profiler and Control for Autonomous Driving. In Proceedings of the 2015 IEEE 18th International Conference on Intelligent Transportation Systems, Gran Canaria, Spain, 15–18 September 2015; pp. 2194–2199. [CrossRef]
9. Hayati, H.; Eager, D.; Pendrill, A.M.; Alberg, H. Jerk within the Context of Science and Engineering—A Systematic Review. *Vibration* **2020**, *3*, 371–409. [CrossRef]
10. Sato, R.; Shirase, K. Analytical time constant design for jerk-limited acceleration profiles to minimize residual vibration after positioning operation in NC machine tools. *Precis. Eng.* **2021**, *71*, 47–56. [CrossRef]
11. Ma, J.w.; Gao, S.; Yan, H.t.; Lv, Q.; Hu, G.q. A new approach to time-optimal trajectory planning with torque and jerk limits for robot. *Robot. Auton. Syst.* **2021**, *140*, 103744. [CrossRef]
12. Rout, A.; BBVL, D.; Biswal, B.B.; Mahanta, G.B. Optimal trajectory planning of industrial robot for improving positional accuracy. *Ind. Robot. Int. J. Robot. Res. Appl.* **2021**, *48*, 71–83. [CrossRef]
13. Palleschi, A.; Garabini, M.; Caporale, D.; Pallottino, L. Time-optimal path tracking for jerk controlled robots. *IEEE Robot. Autom. Lett.* **2019**, *4*, 3932–3939. [CrossRef]
14. Lin, J.; Somani, N.; Hu, B.; Rickert, M.; Knoll, A. An Efficient and Time-Optimal Trajectory Generation Approach for Waypoints under Kinematic Constraints and Error Bounds. In Proceedings of the IEEE International Conference on Intelligent Robots and Systems, Madrid, Spain, 1–5 October 2018; pp. 5869–5876. [CrossRef]
15. Pham, H.; Pham, Q.C. On the structure of the time-optimal path parameterization problem with third-order constraints. In Proceedings of the IEEE International Conference on Robotics and Automation, Singapore, 29 May–3 June 2017; pp. 679–686. [CrossRef]
16. Consolini, L.; Locatelli, M.; Minari, A. A Sequential Algorithm for Jerk Limited Speed Planning. *IEEE Trans. Autom. Sci. Eng.* **2021**, *19*, 3192–3209. [CrossRef]
17. Raineri, M.; Bianco, C.G.L. Jerk limited planner for real-time applications requiring variable velocity bounds. In Proceedings of the IEEE International Conference on Automation Science and Engineering, Vancouver, BC, Canada, 22–26 August 2019; pp. 1611–1617. [CrossRef]
18. Lima, P.F.; Nilsson, M.; Trincavelli, M.; Martensson, J.; Wahlberg, B. Spatial Model Predictive Control for Smooth and Accurate Steering of an Autonomous Truck. *IEEE Trans. Intell. Veh.* **2017**, *2*, 238–250. [CrossRef]
19. Lu, S.; Zhao, J.; Jiang, L.; Liu, H. Solving the Time-Jerk Optimal Trajectory Planning Problem of a Robot Using Augmented Lagrange Constrained Particle Swarm Optimization. *Math. Probl. Eng.* **2017**, *2017*, 1–10. [CrossRef]
20. Zhang, H.Y.; Lin, W.M.; Chen, A.X. Path planning for the mobile robot: A review. *Symmetry* **2018**, *10*, 450. [CrossRef]
21. Patle, B.K.; Babu L, G.; Pandey, A.; Parhi, D.R.; Jagadeesh, A. A review: On path planning strategies for navigation of mobile robot. *Def. Technol.* **2019**, *15*, 582–606. [CrossRef]
22. Abdallaoui, S.; Aglzim, E.H.; Chaibet, A.; Kribèche, A. Thorough Review Analysis of Safe Control of Autonomous Vehicles: Path Planning and Navigation Techniques. *Energies* **2022**, *15*, 1358. [CrossRef]
23. Ravankar, A.; Ravankar, A.; Kobayashi, Y.; Hoshino, Y.; Peng, C.C. Path Smoothing Techniques in Robot Navigation: State-of-the-Art, Current and Future Challenges. *Sensors* **2018**, *18*, 3170. [CrossRef]
24. Kim, Y.; Kim, B.K. Time-Optimal Trajectory Planning Based on Dynamics for Differential-Wheeled Mobile Robots with a Geometric Corridor. *IEEE Trans. Ind. Electron.* **2017**, *64*, 5502–5512. [CrossRef]
25. Ghazaei, M.; Robertsson, A.; Johansson, R. Online Minimum-Jerk Trajectory Generation. *Proc. IMA Conf. Math. Robot.* **2015**.
26. Renny Simba, K.; Uchiyama, N.; Sano, S. Real-time smooth trajectory generation for nonholonomic mobile robots using Bézier curves. *Robot. Comput. Integr. Manuf.* **2016**, *41*, 31–42. [CrossRef]
27. Song, B.; Wang, Z.; Zou, L. An improved PSO algorithm for smooth path planning of mobile robots using continuous high-degree Bezier curve. *Appl. Soft Comput.* **2021**, *100*, 106960. [CrossRef]
28. Kielas-Jensen, C.; Cichella, V.; Berry, T.; Kaminer, I.; Walton, C.; Pascoal, A. Bernstein Polynomial-Based Method for Solving Optimal Trajectory Generation Problems. *Sensors* **2022**, *22*, 1869. [CrossRef] [PubMed]
29. Li, W.; Tan, M.; Wang, L.; Wang, Q. A cubic spline method combing improved particle swarm optimization for robot path planning in dynamic uncertain environment. *Int. J. Adv. Robot. Syst.* **2020**, *17*, 172988141989166. [CrossRef]
30. Eshtehardian, S.A.; Khodaygan, S. A continuous RRT*-based path planning method for non-holonomic mobile robots using B-spline curves. *J. Ambient. Intell. Humaniz. Comput.* **2022**. [CrossRef]
31. Kang, C.M.; Lee, S.H.; Chung, C.C. On-Road Path Generation and Control for Waypoints Tracking. *IEEE Intell. Transp. Syst. Mag.* **2017**, *9*, 36–45. [CrossRef]
32. Ravankar, A.; Ravankar, A.A.; Kobayashi, Y.; Emaru, T. SHP: Smooth Hypocycloidal Paths with Collision-Free and Decoupled Multi-Robot Path Planning. *Int. J. Adv. Robot. Syst.* **2016**, *13*, 133. [CrossRef]
33. Bulut, V. Path planning for autonomous ground vehicles based on quintic trigonometric Bézier curve: Path planning based on quintic trigonometric Bézier curve. *J. Braz. Soc. Mech. Sci. Eng.* **2021**, *43*, 1–14. [CrossRef]
34. Artunedo, A.; Villagra, J.; Godoy, J. Real-Time Motion Planning Approach for Automated Driving in Urban Environments. *IEEE Access* **2019**, *7*, 180039–180053. [CrossRef]

35. Zhang, B.; Zhu, D. A new method on motion planning for mobile robots using jump point search and Bezier curves. *Int. J. Adv. Robot. Syst.* **2021**, *18*, 1–11. [CrossRef]
36. Yu, L.; Wang, X.; Hou, Z.; Du, Z.; Zeng, Y.; Mu, Z. Path planning optimization for driverless vehicle in parallel parking integrating radial basis function neural network. *Appl. Sci.* **2021**, *11*, 8178. [CrossRef]
37. Kumar, N.V.; Kumar, C.S. Development of collision free path planning algorithm for warehouse mobile robot. *Procedia Comput. Sci.* **2018**, *133*, 456–463. [CrossRef]
38. Frego, M.; Bevilacqua, P.; Divan, S.; Zenatti, F.; Palopoli, L.; Biral, F.; Fontanelli, D. Minimum Time—Minimum Jerk Optimal Traffic Management for AGVs. *IEEE Robot. Autom. Lett.* **2020**, *5*, 5307–5314. [CrossRef]
39. Kim, C.; Suh, J.; Han, J.H. Development of a hybrid path planning algorithm and a bio-inspired control for an omni-wheel mobile robot. *Sensors* **2020**, *20*, 4258. [CrossRef]
40. Lee, H.; Hong, J.; Jeong, J. MARL-Based Dual Reward Model on Segmented Actions for Multiple Mobile Robots in Automated Warehouse Environment. *Appl. Sci.* **2022**, *12*, 4703. [CrossRef]
41. Schot, S.H. Jerk: The time rate of change of acceleration. *Am. J. Phys.* **1978**, *46*, 1090–1094. [CrossRef]
42. Lepetič, M.; Klančar, G.; Škrjanc, I.; Matko, D.; Potočnik, B. Time optimal path planning considering acceleration limits. *Robot. Auton. Syst.* **2003**, *45*, 199–210. [CrossRef]
43. Blažič, S.; Klančar, G. Effective Parametrization of Low Order Bézier Motion Primitives for Continuous-Curvature Path-Planning Applications. *Electronics* **2022**, *11*, 1709. [CrossRef]

sensors

MDPI

Article

Robot Navigation Based on Potential Field and Gradient Obtained by Bilinear Interpolation and a Grid-Based Search

Gregor Klančar [1] and Andrej Zdešar [1] and Mohan Krishnan [2,*]

[1] Faculty of Electrical Engineering, University of Ljubljana, Tržaška 25, 1000 Ljubljana, Slovenia; gregor.klancar@fe.uni-lj.si (G.K.); andrej.zdesar@fe.uni-lj.si (A.Z.)
[2] Electrical & Computer Engineering and Computer Science Department, University of Detroit Mercy, Detroit, MI 48208, USA
* Correspondence: mohank@udmercy.edu

Abstract: The original concept of the artificial potential field in robot path planning has spawned a variety of extensions to address its main weakness, namely the formation of local minima in which the robot may be trapped. In this paper, a smooth navigation function combining the Dijkstra-based discrete static potential field evaluation with bilinear interpolation is proposed. The necessary modifications of the bilinear interpolation method are developed to make it applicable to the path-planning application. The effect is that the strategy makes it possible to solve the problem of the local minima, to generate smooth paths with moderate computational complexity, and at the same time, to largely preserve the product of the computationally intensive static plan. To cope with detected changes in the environment, a simple planning strategy is applied, bypassing the static plan with the solution of the A* algorithm to cope with dynamic discoveries. Results from several test environments are presented to illustrate the advantages of the developed navigation model.

Keywords: robot navigation; path planning; potential field; bilinear interpolation; dynamic local re-planning

Citation: Klančar, G.; Zdešar, A.; Krishnan, M. Robot Navigation Based on Potential Field and Gradient Obtained by Bilinear Interpolation and a Grid-Based Search. *Sensors* **2022**, *22*, 3295. https://doi.org/10.3390/s22093295

Academic Editor: Andrey V. Savkin

Received: 4 April 2022
Accepted: 23 April 2022
Published: 25 April 2022

1. Introduction

The main goal of a navigation function is to create feasible, safe paths that avoid obstacles and allow a robot to move from its start configuration to its goal configuration [1]. Online robot navigation and path planning consists of two complementary aspects. In the global path-planning phase, the task is to find an optimal path to the intended goal, starting from the robot's starting position and using all the previous information about the environment. This plan takes into account the need to avoid obstacles but only those that are assumed to be present before the robot starts to move towards the goal. This is coupled with the local obstacle avoidance phase in which the robot avoids new obstacles detected by its sensors while navigating the planned path. The former can be thought of as proactive, while the latter is reactive. With this understanding comes the acceptance that the former is usually more optimal than the latter in some sense. The biggest challenge in real-world applications is the ability to handle unanticipated changes in both structured and unstructured environments. Because the discovery of new obstacles is an evolutionary process, it cannot be assumed that the overall path that is eventually completed will be as optimal as a path that was planned with the knowledge of all the aspects of an unchanged environment at the beginning. However, this is not a fair comparison because new map situations are usually discovered after the robot has begun to execute the original plan. Moreover, due to these new discoveries, the robot may find itself in situations where it seems to be trapped if it continues to follow the global plan while only imposing a requirement to avoid collisions with the new obstacles. The goal is then to develop strategies to overcome the new obstacles in an effective and situationally appropriate way while the robot continues to head towards its original goal.

It is important to understand the trade-off involved in the above situation, which can be thought of as a see-and-react strategy when dealing with new discoveries. In contrast to this approach, there is the alternative of re-implementing the global path-planning strategy at the point where new environmental discoveries are made. However, implementing a full-featured online global path-planning strategy is usually not feasible due to computational costs. Speed is essential because potential delays in reacting could affect the ability to deal with new discoveries safely and efficiently unless one is willing to slow down or stop until a new plan is available. This is particularly true when considering that changes in the environment relative to prior assumptions are quite likely in actual navigation applications. It is to address this need that several incremental methods have been developed [2–6], which reduce computational and storage costs by reusing existing planning information.

An extensive survey of path planning algorithms has been carried out in [7]. Algorithms are divided into categories and sub-categories within them, based on the modality of their development. This work falls in the sub-category of graph search , which supports a variety of path-planning approaches, but here specifically, a graph search based on the use of a uniform grid. It combines this with the adoption of a variant of the original artificial potential fields (APF) method [8]. According to [7], the APF is categorised as reactive manoeuvring. Thus, essentially, a uniform grid-based graph search is combined with a reactive manoeuvring technique to carry out global and local path planning.

The environment can be represented as a graph using cell decomposition or roadmaps. Examples of the latter approach are the Voronoi graph [9], which can produce optimum clearance from the obstacles, and the tangent graph [10], which contains the optimal solution and requires less memory than the visibility graph, its superset. In [11], a tangent graph is constructed for obstacles described with analytic curves in which a finite search algorithm can be used to find the optimal path. The optimal path found in the tangent graph may not be smooth. The authors in [11] combine the tangent graph with online reactive navigation to generate smooth paths for the unicycle drive. Although the algorithm always finds a path, the path may not be optimal. The computational complexity of roadmap-based path-planning algorithms depends on the number of obstacles and the complexity of the obstacle shapes (i.e., the number of the primitives describing all obstacles). The problem of reducing the computational complexity of constructing a tangent graph was addressed in [12]. Among the cell decomposition approaches, grid-based tessellation is the most common and is also used in this work. In grid-based approaches, the complexity mainly depends on the number of cells—a smaller cell size leads to a higher path resolution. The cell size must be small enough to describe the environment with sufficient detail, but must not be too small, as this significantly increases the computational complexity of the search algorithms. The proposed approach introduces an interpolation that can produce smooth paths even at a coarse map resolution.

The earliest graph-theoretic path-planning algorithm is arguably the one developed by Dijkstra [13], which inspired many subsequent variations. It has two aspects associated with its basic construction that could make it less suitable for use in some real-world robot navigation problems. Firstly, it finds the optimal paths between a source node and all destination nodes (or equivalently, between multiple source nodes and a single destination node). In many applications involving only a single robot and a single destination, it increases the computational burden in doing much more than is needed. Secondly, it does not accommodate new discoveries made as the robot's sensor horizon advances on the way to the goal. Yet, in other applications involving several missions originating at different locations, with the need to converge to a single destination such as in warehouses [14,15], or games [16,17], Dijkstra's algorithm matches the need. The computational burden associated with it then becomes justifiable, especially if it is required to be exercised occasionally and the results reused with different starting locations. The focus then shifts to the second issue mentioned above, as to how to make the Dijkstra paths viable even when dealing with environmental changes.

The A* algorithm [18] and its derivatives, such as D*Lite [4], are computationally efficient algorithms compared to Dijkstra in the specific task that they address of finding a path between a single source and goal nodes. This reduced task allows an informed search strategy in the form of a heuristic to be deployed, which leads to the computational savings. The one-source–one-goal paradigm can be identified with a single robot trying to get to a single destination. Both A* and D*Lite accommodate new obstacle discoveries, but the latter is an incremental algorithm which makes it suitable when there are continuing map changes while navigating to the same goal.

Some of the prior efforts of other researchers that use methods related to our path planning and navigation strategy are now discussed. The concept of artificial or virtual potential functions (APF) was first proposed in [8]. They are called artificial because these are not actual electric potentials but are only conceptualised as such. In the original formulation, as explained in Choset [19], the two attractive and repulsive potential functions were algebraically summed to obtain the overall potential function. In practice, the ad hoc parametric choices of the model could set up local minima at which the net force on the robot is zero, resulting in the robot being trapped on its way to the goal.

Let us now turn our attention from the core APF concept to how it was actually realised in prior robot navigation research. Ratering and Gini [20] proposed a hybrid potential field consisting of the combination of a global potential field calculated with a variant of Dijkstra's algorithm and a local potential field synthesised with the help of sonar measurements. Wang et al. [21] also constructed the global and local planners separately. Distance transformation, another variant of Dijkstra's algorithm, is used for the global planner, while an APF-based method is used for the local planner. The overall navigation strategy is characterised by a mediation between the strict need to achieve the subgoals of the global plan and the freedom of the APF-based local planner, so that local minima are avoided. Azmi and Ito [22] propose a technique to handle the local minima problem, in this case, a repetitive oscillatory excursion between two local minima. A map transformation operation was proposed that resulted in the stalemate being resolved through the rotation of the environment space. Lazarowska [23] devised the planning of trajectories for autonomous ships navigating amongst both static and dynamic obstacles. The static APF model accounted for the compliance of special maritime rules that prescribed deliberate actions to avoid collisions between ships. Similarly, Klančar and Seder [15] combined the static APF with local reactive model–predictive planning to avoid collisions among multiple robotic vehicles in warehouse navigation. Amiryan and Jamzad [24] used the APF to complement a pre-determined path generated by a sampling-based path planner such as Rapidly-exploring Random Tree (RRT) [6] to avoid local minima problems. In [25], a hybrid planning method is proposed that combines a particle swarm optimisation algorithm with the APF for static obstacles and the potential field prediction for dynamic obstacles. Several solutions have been proposed to overcome problems with local minima, such as representing concave obstacles by convex representations [26], adding virtual obstacles to move away from local minima [27] or by small perturbations of the APF based on the input-to-state stability property [28]. Alternatively, a robot navigation function can be determined using deep neural networks with reinforced learning as in [29], ant colony optimisation [30], simulated annealing, particle swarm optimisation [25], genetic algorithms and fuzzy logic [31] or the like.

The sampling of the APF-based literature discussed above indicates that the APF concept continues to play a role in robot navigation. Specific use cases range from the original concept of an integrated formulation premised on attractive/repulsive forces to separate formulations addressing the static and dynamic planning phases, to dealing with dynamic moving obstacles and other variations.

The main contributions of this paper are the following.

- A new navigation function is proposed that generates smooth and collision-proof paths by using the bilinear interpolation (BiLI) method to obtain an artificial potential field gradient-descent navigation function from a discrete cost-to-goal (CtG) map obtained by an optimal discrete grid-based search method. The approach is computationally efficient as it relies on a coarse discrete graph search that can be precomputed for static environments and known goals and can be easily reused for multiple missions from different parts of the environment that need to navigate to a common goal. The bilinear interpolation method implements a continuous potential field and driving direction from a discrete grid-based search.

- Although BiLI is commonly used in computer vision applications, its use for robot planning requires some enhancements, such as handling occupied cells whose values are not defined, interpolating at the environmental boundaries and ensuring continuous gradient descent, which are the main novelties of this work. The resulting path is collision-proof, continuous and close to the optimum, even at the course grid resolution used. It also avoids the problems with local minima that are common in general APF-based methods.
- BiLI can be applied in dynamic environments where incremental graph search methods such as D* or similar [2,4,5,32] can be used to efficiently account for the changes in the environment. In this work, a simple strategy is proposed using Dijkstra for the global CtG map planning and A* for dealing with locally sensed environment changes. When different types of map changes are detected that affect the CtG values, the proposed model uses the A* algorithm to find an emergency bypass path to areas of the environment where the old CtG values are still valid. The bypass path is followed by the determination of a final gradient-descent path segment to the overall goal. Then, a situational decision is made whether to take the additional path segments at the point of the map change or to keep the original path. Using A* to determine the diversion path is relatively fast compared to regenerating the CtG values with Dijkstra, as usually only a few cells need to be examined.

The remaining parts of the article are organised as follows. The first part of the path-planning model developed in this effort is explained in Sections 2 and 3, with the help of a test environment scenario. The reactive strategy to negotiate a blockage of the global path is formulated in Section 4 with the same environment. Section 5 contains a description of additional test scenarios formulated to evaluate this model and the attendant results. Finally, Section 6 contains a brief summarising discussion of the work and suggests further steps, while Section 7 draws some broad conclusions.

2. The Environment and APF Generation

Following a description of the environment, the formulation of the APF values for each cell of the grid-based discretisation is discussed in this section.

2.1. The Environment

A sample environment is shown in Figure 1 which represents the static map in one experiment. Appropriate modifications to it will be subsequently made that reflect the dynamic discovery of new obstacles. Moreover, other obstacle configurations will be created later that represent additional challenges addressed in this work.

The environment consists of a walled-off 10 by 10 m field of play. It can be scaled up to any size, as desired. The entire area is divided into 400 cells (20 by 20), with each cell being 0.5 m square. The environment features four prominent symmetrically positioned "obstacle islands" that, given the starting and goal locations (also shown), block line of sight to the goal for significant portions of a possible path.

The broad characteristics of the environment are:

- Even when an obstacle only overlaps part of a cell, the entire cell is considered occupied. Thus, the map of the environment is in the form of a binary occupancy grid, which also factors in obstacle inflation.
- Dynamic changes in the environment over the initial static knowledge are assumed to be small and localised.

- They can be in the form of additions or subtractions. That is, cells that were unoccupied might be occupied as well as the opposite.

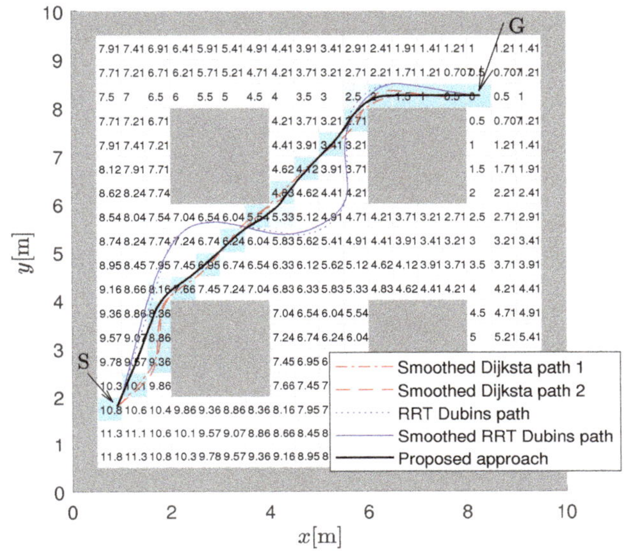

Figure 1. Static map of environment with start (S) and goal (G) locations, discrete CtG values assigned to the free cells (white cells, gray cells belong to obstacles) and some different smooth paths that connect the start and the goal locations.

2.2. Formulation of APF

A variant of the original APF (artificial or virtual potential functions) concept [8] is used to fulfill the global path-planning function. As pointed out earlier, the original APF was conceived as a function that is continuous with respect to space, that addressed both goal seeking and obstacle avoidance in an integrated manner. It followed from attractive and repulsive forces between artificial electrical charges. However, that approach is known to potentially spawn local minima trapping the robot.

As opposed to the classical approach, the floor space is tessellated into a suitable x–y grid to accommodate the use of a standardised discrete occupancy map to represent obstacles. The Dijkstra algorithm is used to generate the cost-to-goal (CtG) value for all cells in the environment, which constitutes the global cost map. In doing this, diagonal cell-to-cell transitions are given appropriate differential weights relative to horizontal and vertical transitions. The CtG values are shown overlaid in Figure 1 within each cell for the assumed environment. For example, the CtG is zero for the goal cell and 10.8 m for the starting cell.

Within the grid map, the Dijkstra algorithm can be used to obtain the shortest path. In Figure 1, the cells that lead from the start to the goal cell are shaded with a light blue colour. A smooth path can then be obtained if a spline (using, e.g., Bézier curves or clothoid curves) is fitted over the centres of the cells that comprise the path. In Figure 1, the Automated Driving Toolbox in Matlab is used to smooth the paths using cubic splines as shown in two examples (Smoothed Dijkstra path 1 and Smoothed Dijkstra path 2). This approach does not ensure that the smooth path does not go too close or even over the obstacles unless collision checking is also made during spline fitting. In Figure 1 is also an example of the path obtained by the approach proposed in this paper. This is a smooth path that is obtained based on the interpolated gradient of the APF that takes obstacles into account implicitly. Figure 1 also presents a smooth path that is obtained with Rapidly-exploring Random Tree (RRT*) [33]. In this case, RRT* uses Dubins' curves [1] to obtain the path from the start to the goal, and the obtained path is further smoothed by fitting a cubic Bézier spline. The shape of the curve depends on many constraints (e.g., path curvature, segment length, safety distance, vehicle constraints, etc.) that can be given to the algorithm to optimise; therefore, paths with different shapes and smoothness can be obtained. Moreover, the RRT approach is stochastic; therefore, a completely different solution can be obtained in every run of the algorithm even if the input conditions do not vary. Some smoothed paths can be very oscillatory or make large turns around the obstacles, or path smoothing can also produce infeasible paths that collide with obstacles. The proposed approach in this paper is deterministic and produces smooth and near optimal paths that ensure a minimum safe distance from the obstacles, and it is also computationally efficient because it produces satisfactory results even when the cell size is relatively large.

The CtG numbers serve as the classical APF values that are the basis of the global planning within the environment, after some refinement discussed shortly. Just as in the classic APF method, a gradient descent determines the direction of motion. If the static map does not change, a gradient-descent approach based on the CtG values can be used to move the robot all the way to the goal. However, details need to be addressed, such as how the gradients are calculated and smoothed for a function that is discrete over the floor space, as well as how dynamic discoveries in the environment are handled, etc.

3. Interpolation and Smoothing of Potentials and Gradients for Path Planning

The discretisation of the floor—a decision made to contain the computational cost as well as to simplify the consideration of obstacle occupancy—correspondingly creates a discrete CtG surface. This then restricts the resolution of the gradient determination which affects the smooth navigation. To address this, a refinement is introduced through the use of a well-known technique of image resampling known as bilinear interpolation (BiLI) [34,35], which operates on pixels that are like the cells in our environment, as discussed below.

3.1. Bilinear Interpolation

The conceptual basis of BiLI is explained using Figure 2 [35]. BiLI uses a 2 by 2 cell window to interpolate CtG values within the centred unit square region within this window (dashed square in Figure 2), thus creating new data points in an educated manner. It does so by using linear functions to perform the interpolation in what is essentially a planar extension of 1D linear interpolation. The spline-based function representing the interpolating surface is associated with 4 parameters whose values need to be estimated. This is achieved using the CtG cell values at the corners of the unit square.

According to location of a point $[x,y]^T$ in a cell $\mathbf{M}$, a 2 by 2 region of cells around it is chosen for the interpolation, whose centres are connected by a dashed square in Figure 2. Normalised coordinates are found by centres of these cells as:

$$x_n = \frac{x-x_0}{d_c} \quad , \quad y_n = \frac{y-y_0}{d_c} \quad , \tag{1}$$

where $[x_0, y_0]$ is the origin of the normalised coordinates defined by the lower left corner of the dashed square and d_c is the cell size. The interpolated and discrete CtG value (potential in the sequel) in normalised coordinates are expressed as $P_n(x_n, y_n) = P(x,y)$ and $U_n(x_n, y_n) = U(x,y)$, respectively. The potential for the four adjacent cell centres (corners of dashed square in Figure 2) are

$$p_{cr} = U_n(x_n, y_n)\Big|_{x_n=c,\ y_n=r} , \tag{2}$$

where $c, r \in \{0,1\}$ and $U_n(x_n, y_n) = U(x,y)$.

The interpolated potential $U_n(x_n, y_n)$ at any given normalised position $[(x_n, y_n)]^T$ inside the quadrant of cell $\mathbf{M}$ delineated by the unit square is defined as [35]:

$$P_n(x_n, y_n) = w_{00}p_{00} + w_{01}p_{01} + w_{10}p_{10} + w_{11}p_{11}, \tag{3}$$

where the BiLI weights are given by: $w_{00} = (1 - x_n)(1 - y_n)$, $w_{01} = (1 - x_n)y_n$, $w_{10} = x_n(1 - y_n)$ and $w_{11} = x_n y_n$.

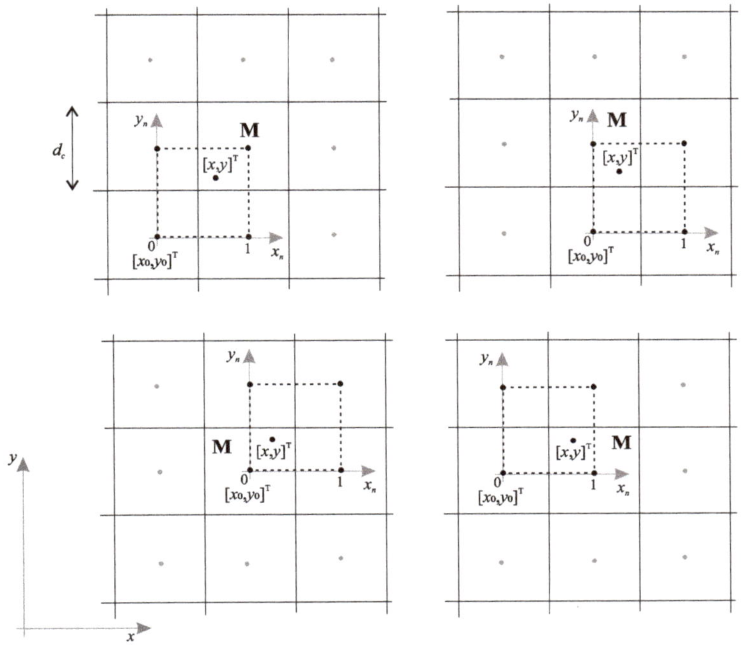

Figure 2. Basis of bilinear interpolation. Interpolated potential at a given point $[x, y]^T$ is defined by the discrete potential at centres (black dots) of four cells connected by the dashed square. Gray dots denote centres of cells.

By following the negative gradient of interpolated potential $P(x, y) = P_n(x_n, y_n)$, the safe path from everywhere in the environment towards the goal location (with potential 0) can be obtained. The negative gradient of $P(x, y)$ in $[x, y]^T$ can be obtained as :

$$\begin{aligned} \mathbf{g}(x,y) &= -\nabla P(x,y) = -\left[\frac{\partial P(x,y)}{\partial x}, \frac{\partial P(x,y)}{\partial y}\right]^T - \frac{1}{d_c}\left[\frac{\partial P_n(x_n,y_n)}{\partial x_n}, \frac{\partial P_n(x_n,y_n)}{\partial y_n}\right]^T \\ &= -\frac{1}{d_c}\begin{bmatrix} p_{11}y_n - p_{01}y_n + p_{00}(y_n - 1) - p_{10}(y_n - 1) \\ p_{11}x_n - p_{10}x_n + p_{00}(x_n - 1) - p_{01}(x_n - 1) \end{bmatrix}. \end{aligned} \tag{4}$$

3.2. Adjustments of Bilinear Interpolation for Path Planning

Before applying the interpolation of Equation (3), a check needs to be performed if any of the three neighbour cells of the cell **M** (see Figure 2) involved in the interpolation are occupied. Note that cell **M** is never occupied as we are interpolating potential at a point $[x, y]^T$ inside it. For occupied cells, the potential is typically infinite or undefined, as motion over the obstacles towards the goal should not be possible/permitted. The potential for occupied cell $U(x_m, y_m)$ (with centre at x_m, y_m) is reconstructed from the eight-cell neighbourhood by finding the unoccupied cell with the largest potential, as:

$$\begin{aligned} \{c, r\} &= \underset{c,r}{\operatorname{argmax}}\{U(x_m + d_c c, y_m + d_c c) \neq \infty\} \\ U(x_m, y_m) &= U(x_m + d_c c, y_m + d_c r) + d_c\sqrt{c^2 + r^2} \end{aligned}, \tag{5}$$

where $c, r \in \{-1, 0, 1\}$.

Additionally, a check needs to be performed if any of the four cells needed for the interpolation (also interpolation cells, see Equation (3) and Figure 2) is outside the environment. A simple solution to this could be that the grid cell area is always at least one cell larger than the area we are interpolating. More general solution applies calculation of the potential and gradient for a nearby location $[x_t, y_t]^T$, where all four cells used in the interpolation are inside the environment. For a position $[x, y]^T$, where one or more interpolation cells are outside the environment, the nearby location is determined by translating the position from the border for $\frac{d_c}{2}$ in x and/or y direction towards the inside of the environment as follows:

$$x_t = \begin{cases} x & ; \ x_{min} \le x \le x_{max} \\ x_{min} + \frac{d_c}{2} & ; \ x < x_{min} \\ x_{max} - \frac{d_c}{2} & ; \ x > x_{max} \end{cases} \qquad y_t = \begin{cases} y & ; \ y_{min} \le y \le y_{max} \\ y_{min} + \frac{d_c}{2} & ; \ y < y_{min} \\ y_{max} - \frac{d_c}{2} & ; \ y > y_{max} \end{cases}, \tag{6}$$

where environment borders are defined by $x_{min}, x_{max}, y_{min}, y_{max}$. For translated nearby location, interpolated potential is computed from (3) noted as $P(x_t, y_t)$ and the gradient from (4) noted as $\mathbf{g}_t(x_t, y_t)$. Finally, the appropriate potential for each interpolating cell outside the environment (noted as P^*) are reconstructed using Lie derivative (also direction derivative):

$$P^* = P(x_t, y_t) + \mathbf{g}_t(x_t, y_t) \times [x - x_t, y - y_t]^T. \tag{7}$$

Figure 3 (top-left image) shows the resulting potential (CtG values) surface obtained through application of the BiLI technique, corresponding to the environment of Figure 1. Notice how the CtG surface slopes continuously downward from the start point to the goal point, with the four "islands" represented by infinite potentials. This follows from the fact that the CtG values will monotonically decrease from any cell in the environment towards the goal. Moreover, a few sample paths obtained by following the negative gradient (computed from Equation (4)) from different locations towards the goal are shown in the top-right image with blue line. Notice discontinuous change of the negative gradient direction near occupied cells, which can also be observed by checking gradients near obstacle (the cells near the central obstacle in the bottom-left image from Figure 3 with enlarged cell at obstacle corner). To improve this and obtain smoother paths (as illustrated in the top-right figure by purple line), we additionally propose interpolation of the negative gradients in Section 3.3.

3.3. Interpolation of Gradients

The resulting potential in Figure 3 is continuous, but its negative gradient may be discontinuous especially in the vicinity of obstacles, as seen from the bottom-left image in Figure 3 where, at the boundaries between the quadrants of a cell, the gradient shows discontinuities. The negative gradient indicates the required driving direction to reach the goal in optimal manner. Every discontinuity of the gradient is problematic, as a wheeled robot would need to stop and rotate on the spot to reliably follow the course of the negative gradient towards the goal. To improve this, we propose interpolation of the negative gradient, similar to what was performed for the potential field.

For chosen location $[x, y]^T$ in cell $\mathbf{M}$, we estimated the interpolated potential $P(x, y)$ and its negative gradient $\mathbf{g}$ using Equations (3) and (4). Interpolated potential is obtained from the four interpolation cells centre potentials as indicated by Figure 2. In the following, we will use the same interpolation principle to also obtain the interpolated negative gradient $\mathbf{h}(x, y)$ with continuous course as shown in the right images of Figure 3. The gradient for the centres of the four interpolating cells can be estimated from (4), which for each cell considers only the left neighbour ($x_n = 0$) to obtain x element of the gradient $\mathbf{g}$ or only the upper neighbour ($y_n = 0$) to obtain y element of $\mathbf{g}$. Therefore, we estimate the cell centre gradient considering the smallest discrete potential (valid for the cell centre) of both neighbour cells in x or y axis. Denote the interpolating cell centre gradients (similarly as their potential in (2)) by $\mathbf{h}_{cr} = [hx_{cr}, hy_{cr}]^T$ where $c, r \in \{0,1\}$. For a cell with centre coordinates $x_m = x_0 + d_c c$, $y_m = y_0 + d_c r$ the cell gradient reads

$$\mathbf{h}_{cr} = -\begin{cases} \frac{1}{d_c}\begin{bmatrix} e(U(x_m + d_c e, y_m) - p_{cr}) \\ f(U(x_m, y_m + d_c e) - p_{cr}) \end{bmatrix} & ; \quad U(x_m, y_m) < \infty, \\ \frac{1}{d_c}\begin{bmatrix} S_x(U(x_m + d_c S_x, y_m) - p^*_{cr}) \\ S_y(U(x_m, y_m + d_c S_y) - p^*_{cr}) \end{bmatrix} & ; \quad U(x_m, y_m) = \infty, \end{cases} \tag{8}$$

where

$$\begin{aligned} e &= \underset{s}{\operatorname{argmin}}\{U(x_m + d_c s, y_m)\} \ ; s \in \{-1, 0, 1\} \\ f &= \underset{s}{\operatorname{argmin}}\{U(x_m, y_m + d_c s)\} \ ; s \in \{-1, 0, 1\} \end{aligned} \tag{9}$$

and $S_x = \operatorname{sgn}(x - x_m)$, $S_y = \operatorname{sgn}(y - y_m)$ where sgn(.) denotes the sign function. The first part of (8) relates to the case where the cell is free and the second for the case when cell is occupied. If the cell is occupied ($U(x_m, y_m) = \infty$), then its potential is updated (noted as p^*_{cr} in (8)) from already known potential $P(x, y)$ and the gradient $\mathbf{g}(x, y)$ in current position $[x, y]^T$ as follows:

$$p^*_{cr} = P(x, y) + \mathbf{g}(x, y) \times [x_m - x, y_m - y]^T.$$

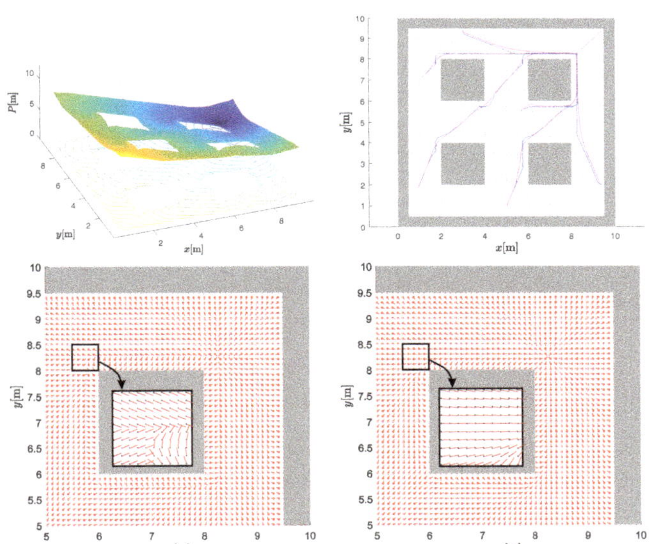

Figure 3. Interpolated potential surface $P_n(x_n, y_n)$ (CtG values, darker colours denote lower CtG values) obtained through bilinear interpolation with contours of equal potential, corresponding to the discrete CtG values of environment of Figure 1 (**top-left**). Obtained paths following the negative gradient (blue line) and interpolated negative gradient (purple line) are shown (**top-right**). Part of the environment near the goal with negative gradients (red lines going from black dots outwards) computed from (4) (**bottom-left**) and interpolated negative gradients (**bottom-right**).

Before computing the cell gradients in (8) and (9), a check needs to be made if any of the neighbour cells is outside the environment. If this is the case, the potential of this cell is reconstructed similarly as in (7) considering the known interpolated potential $P(x, y)$ and gradient $\mathbf{g}(x, y)$ for the point $[x, y]^T$.

From estimated cell gradients $\mathbf{h}_{00}$, $\mathbf{h}_{01}$, $\mathbf{h}_{10}$ and $\mathbf{h}_{11}$ (Equation (8)), the final interpolated gradient in current position is obtained by:

$$\mathbf{h}(x, y) = w_{00}\mathbf{h}_{00} + w_{01}\mathbf{h}_{01} + w_{10}\mathbf{h}_{10} + w_{11}\mathbf{h}_{11}, \tag{10}$$

where the same weights w_{00}, w_{01}, w_{10}, w_{11} as in (3) are used. The comparison of the gradient field **g** and the improved interpolated gradient **h** is shown in Figure 3 in the lower graphs. The obtained planned paths by following the interpolated gradient of the potential field result in collision safe and smooth paths as shown in the top-right graph of Figure 3.

Bilinear interpolation can therefore be elegantly used to obtain continuous potential field as well as appropriate desired driving directions (the interpolated negative gradients **h**) based on a discrete grid-based cost map (discrete CtG potential field). The obtained paths in path planning are collision safe, without local minima and with continuous course of driving direction, which is important for robotic vehicles. Note that Bicubic interpolation [15,36] could also produce smoother interpolations, but it requires 4 by 4 neighbourhood, which is problematic in the vicinity of obstacles or in narrow corridors, as the occupied cells have infinite CtG value. Occupied cells require special treatment before they are used in the interpolation. This becomes even more challenging for bigger neighbourhoods (e.g., 4 by 4 as opposed to 2 by 2). Therefore, for the path planning, we propose the use of bilinear interpolation with appropriate preprocessing of occupied cells and with additional gradient interpolation to obtain smooth paths. The path is smooth, even as it is intuitive and optimal. As mentioned earlier, no local minima will exist in the CtG contour, unlike in the classical APF formulation, because of the inherent manner of its construction. However, if the static map is augmented by new obstacles put in play after its creation, this could change. This eventuality is dealt with in the next section.

Let us analyse the computational complexity of the proposed smooth path planning. The path is obtained with a gradient-descent method. The number of steps that are required to reach the goal is dependent on the size of the sampling step. The sampling step can be lower bounded to prevent oversampling and upper bounded to obtain desired smoothness of the path. Consider that a particular path passes over M cells and that the sampling step is selected in a way that on average (or at maximum) L iterations of the gradient descent are made in each of the cells. This means that computational complexity of gradient descent is $\mathcal{O}(LM)$, and it is therefore dependent on the sampling step and path length. In each step of the gradient-descent calculation, an interpolated value of the gradient is obtained from the four nearest surrounding cells (see Equation (10)) around the current path point. The gradient in each of these four cells is calculated from the APF of the four neighbouring cells (Equation (8)). Therefore, to compute the interpolated gradient for a given cell, a neighbourhood of twelve cells is needed in total. Hence, the gradient does not need to be determined for every cell, but only for the cells that are along the path. We assume that the gradient calculations can be cached; therefore, the computational time and space complexity of obtaining the gradients in the cells around the path are $\mathcal{O}(M)$. Note that cell gradient calculations could also be made in parallel if calculation speed is crucial. We calculate the values of the APF for the entire map with N cells using Dijkstra algorithm, which in case of a grid map has a computational time complexity of $\mathcal{O}(N\log(N))$ and final space complexity of $\mathcal{O}(N)$. The values of the potential field could also be determined only for the cells in the vicinity of the path that are sufficient for gradient calculation. This is a viable option when we would like to quickly determine only a single path, especially if the map is very large. In this case, the Dijkstra algorithm can be stopped once the goal is reached (in this step, it is also beneficial to use A*) and then the Dijkstra algorithm is resumed only if the value of the potential for an unknown cell needs to be known and only until the final value of the cell potential is known. However, in our case, we calculate the APF for the entire map, as this enables fast recalculation of various paths that lead to a single goal (or begin at a common start), which is beneficial for the cases when part of the map changes, as presented in the next section.

4. Discovery of New Obstacles and Reactive Avoidance Manoeuvre

A change in the environment is shown in Figure 4 with the addition of an L-shaped obstacle (in Figure 4 (top-right)) cluster roughly halfway through the initially planned path in Figure 4 (top-left). The detection of these additional obstacles is assumed to take place when the robot's sensor horizon includes the region, through an iterative comparison between what it sees with its sensor and what it expects to see from the initial static map. The range of the sensor used will have some effect on when any reactive manoeuvre is initiated, but this detail is not critically relevant to our model development.

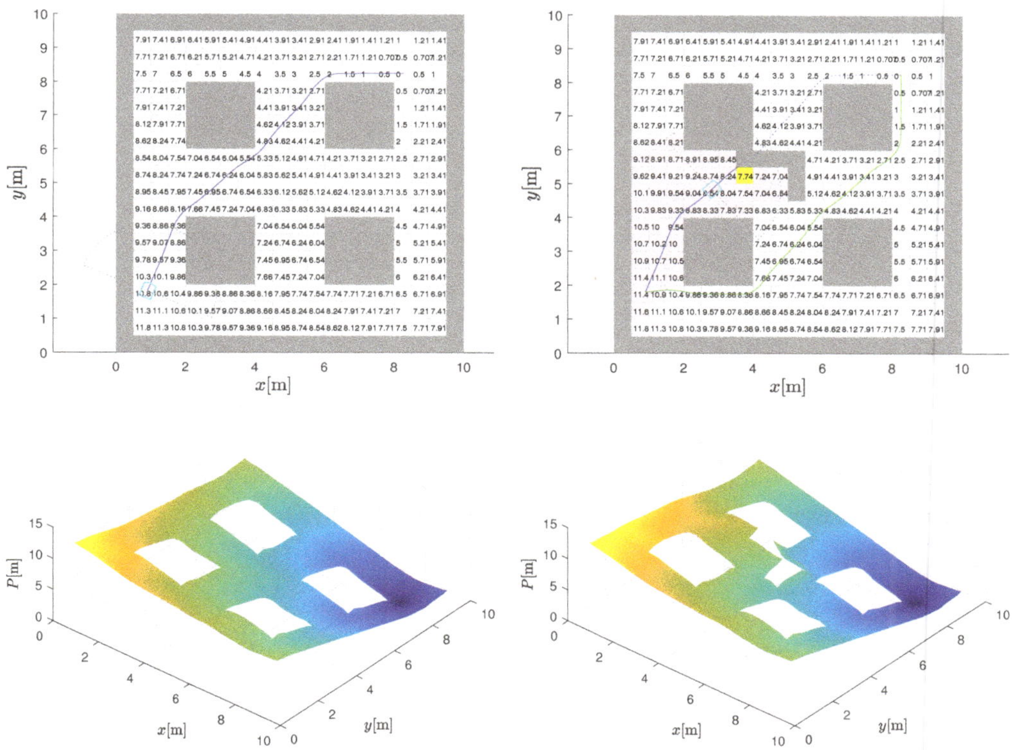

Figure 4. Augmented environment with new discoveries. Static map of environment with planned path in blue line towards goal (**top-left**). Change of environment with new L-shaped obstacle where robot (shown by the sky blue line and its sensor range by the gray dots) is blocked at stuck cell (highlighted by yellow) if continuing based on static CtG map. Replanned CtG values for new environment with highlighted light purple cells where CtG values have changed and new gradient-descent path in green (**top-right**). Interpolated potential (CtG, darker color denote lower values) surface before (**bottom-left**) and after the environment change (**bottom-right**).

It should be noted that if the new obstacle does not impede motion, then nothing needs to change. However, following the initially planned global path here will stop the robot at the "stuck cell" (see the global path in Figure 4 (top-left) in conjunction with the "stuck cell" shown in yellow in Figure 4 (top-right)), which is effectively a local minimum forced by the new obstacle. Essentially, the robot ends up in a trap in pursuing gradient descent based on the initial plan. In fact, the discovery—here, an addition to the static obstacle map—can be taken as an indication that the CtG values in the vicinity are no longer reliable.

The results of repeating the Dijkstra algorithm to refresh the CtG values of the entire environment with the new obstacles included are also shown in Figure 4 (top-right). The cells with modified CtG values are highlighted (in comparison to plot (top-left)), which clearly reveals that the new obstacles only influence the CtG values (increases them) of a small subset of the cells that are upstream of the new obstacles. The downstream CtG values are unaltered, as would be expected. The new interpolated CtG surface is shown in Figure 4 (bottom-right) and is in accordance with the updated map.

The new gradient-descent path from the original starting point is also shown in Figure 4 (top-right) with the green line, which confirms that if we had been aware of these new obstacles at the very beginning, the global path planning would have accounted for it and the CtG numbers would have been monotonic again. This recalculation could have also been performed from the trapped position of the robot to the goal. However, this is the calculation that is to be avoided because of the associated computational burden. The new path is just presented here to make a point.

Thus, the challenge is to come up with a reactive strategy that enables the robot to get around the obstruction through developing a bypass path that involves minimal computation effort. This path should lead the robot to an area where the old CtG values can be used again to continue travel. The flowchart shown in Figure 5 is a broad representation of the core method adopted. There are minor case-based variations stemming from the type of map change encountered, which are not included in this basic flowchart for simplicity. The explanation that follows is with reference to this flowchart as well as the two in Figures 6 and 7 that follow, dealing with lower-level steps in the algorithm.

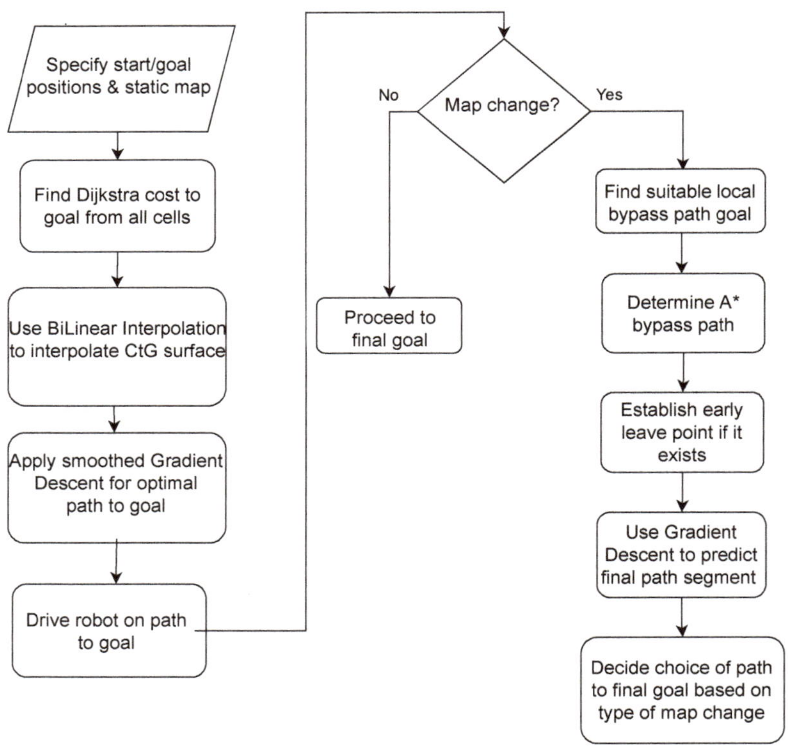

Figure 5. Core flowchart of algorithm (some variations based on case). Leave point determination applies only when current path is blocked.

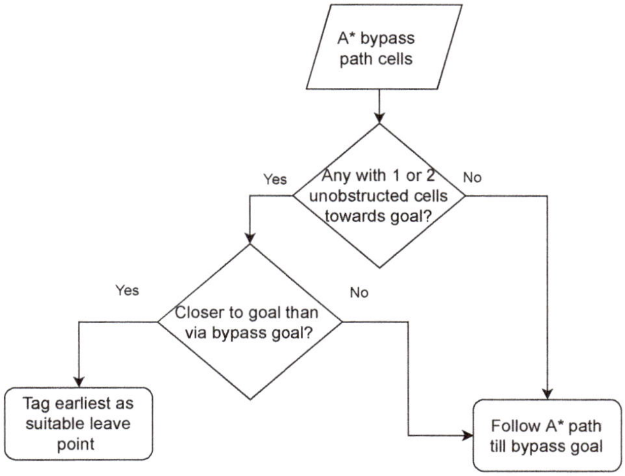

Figure 6. Finding suitable leave point. Leave point is established if cell on A* path has immediate unobstructed cells in original goal direction and is closer to it than path via bypass goal.

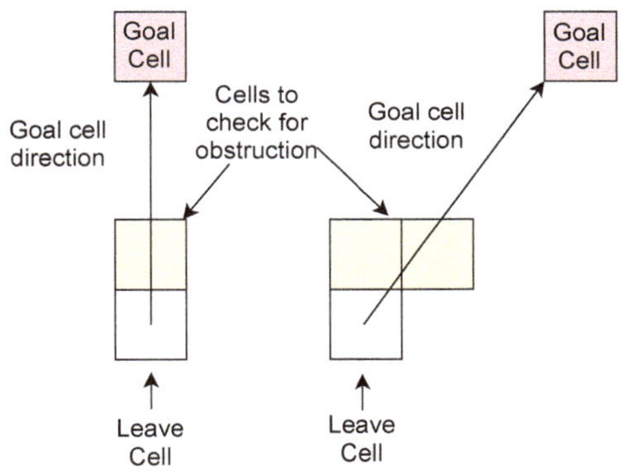

Figure 7. Checking blockage condition for leave cell candidates. Based on alignment of potential leave cell with original goal cell, occupancy status of either one or two cells is checked to determine whether obstruction is present.

A predefined, 5 by 5 search window of cells (see the yellow dashed square in Figure 8 (top-right)), which can be thought of as a "fishing net", is centred at the stuck robot cell (the yellow cell in Figure 8 (top-right)). The cells within the window are examined to find the lowest CtG value that is smaller than the CtG value at the stuck cell, as per the original static map, to use as a temporary intermediate goal (the green cell in Figure 8 (top-right) to help negotiate the blockage caused by the newly discovered obstacles with a sensor (e.g., LiDAR). An inherent assumption is that the size of the search window is sufficient to uncover a cell with a CtG value that is on the other side of the blockage, and hence unaffected by it. This is facilitated by centring the net at the stuck robot cell right next to the new obstacles blocking the path, even though the blockage may have been detected even further away by the robot's LiDAR. If this step does not uncover a satisfactory temporary local goal, larger nets are cast iteratively until a suitable cell is found. Moreover, it might be possible to get more creative with the footprint adopted for this search window, which is beyond the scope of this work.

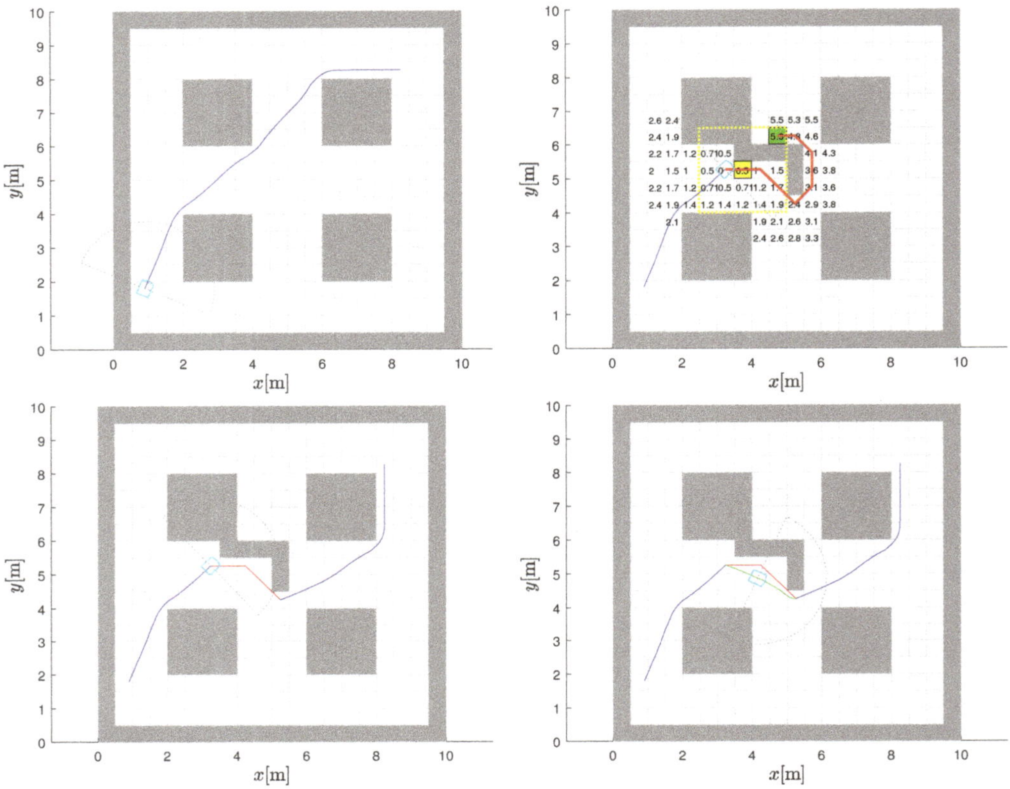

Figure 8. Strategy to negotiate dynamic obstacles. Original planned path based on the static environment map shown is shown with blue line (**top-left**). New obstacle blocking path initiates A* bypass (red line) calculation from stuck cell (highlighted in yellow) to the temporary intermediate goal cell (highlighted in green) (**top-right**). The leave point and final overall path is shown with A* bypass (**bottom-left**) and in by smoothed bypass (green line) using BiLI interpolation (**bottom-right**).

With the stuck cell and the temporary goal identified, the A* algorithm is then run to find a diversionary path from the robot's stuck location to the temporary local goal identified, as described above. This step should not result in an "unreachable goal" being returned, as the algorithm is executed with the known map at the time. The resulting path obtained is also shown in Figure 8 (bottom-left). It should be noted that using D*Lite instead of A* will not result in any computational savings, because the algorithm needs to be run just once, in which case there is no advantage of one over the other. Whether it is necessary to follow the A* bypass path all the way to completion would depend on the situation. The reason is that if in negotiating the obstacle via the A* path the robot finds itself in a cell that is closer to the final goal, completing the A* path and then proceeding to it is not as efficient compared to treating the cell as a leave point. This additional condition built into the algorithm is laid out in the flowchart extension presented in Figure 6.

A good leave cell must also satisfy another condition. It should also be one for which there is at least one unobstructed cell within its 8-cell neighbourhood in the direction of the goal. This is illustrated in Figure 7 for the two situations that represent all the possibilities that can occur. Essentially there are two cases, because of the quantisation imparted by the discretised cell structure. Either one or two cells need to be checked, depending on the relative positioning of the goal cell and the candidate leave cell. That is, are they vertically or horizontally aligned or at a different angle, in which case the line joining them will be straddled by two cells. In the latter case, only one needs to be free to meet the leave condition. If none of the leave cells pass the dual tests, the bypass path is followed all the way to the temporary bypass goal.

When a leave cell is established, it is taken to mean that the local obstruction has been satisfactorily bypassed and the old CtG values are valid again. The algorithm reverts to the gradient descent using the old CtG values. It should be noted that it is possible to determine a suitable leave point and the final gradient-descent segment even before the robot moves from the stuck cell. This enables the path to be evaluated before travel. The leave point and the final overall path of the robot between the original starting point and goal location are also shown in Figure 8 with A* bypass (bottom-left) and by the smoothed bypass path using BiLI interpolation (bottom-right).

The use of the CtG values and gradient descent as an overarching method to drive to the goal and the separate handling of unexpected obstructions through a bypass path ensures that, unlike in the classical formulation of the APF, local minima cannot be formed at the global path planning with static map phase. That is, the two-step approach results in local minima being caused only by newly discovered obstacles and transfers the burden of resolving them to the local path-planning phase. This is a key element of the path-planning strategy used here.

The planning model discussed here was evaluated in additional experiments. The environments used and the attendant results obtained are discussed in the next section.

5. Additional Experiments and Results

5.1. Obstacle Missing from Static Map within Sensor View

A new environment is shown in Figure 9 (top-left) with the start (lower left) and goal (upper right) locations, as well as the path resulting from the global path-planning phase. As the robot travels towards the goal and its sensor horizon advances, it discovers a change in the static map on which the planning was premised. A cell in the "wall" that was thought to be blocked is found to be clear from the robot's sensor view. Here, the change is a subtraction as compared to the earlier example. The cleared cell as well as the cell where the discovery was made are shown in Figure 9 (bottom-left).

This can be used to trigger an opportunistic strategy—when a blockage is removed, there is a possibility that a shorter path to the goal exists and needs to be investigated. The A* algorithm is invoked to determine a path to the region of the cell, which is now open, so that the old CtG surface can be used again. This is accomplished by setting the target cell for the A* bypass path to an unblocked cell beyond the cleared cell in the direction of the goal. The potential A* bypass path is shown in Figure 9 (bottom-left) in red and the corresponding smoothed version by the green line in Figure 9 (bottom-right). The remaining path from that cell to the original goal cell is again obtained through using the CtG values and gradient descent and is shown in Figure 9 (bottom-right) in which the overall path from the original starting position to the goal is also evident.

If there are multiple occupied cells that are now clear, all of them need to be assessed in the same manner as discussed above. Before taking a bypass path, a check needs to be conducted to see whether the modified path (made up of an A* segment followed by a gradient-descent segment) is shorter than the remaining current path to the goal. Only if the newer path is better should the robot use it to get to the goal instead of the original path. In the environment considered, this happens to be the case.

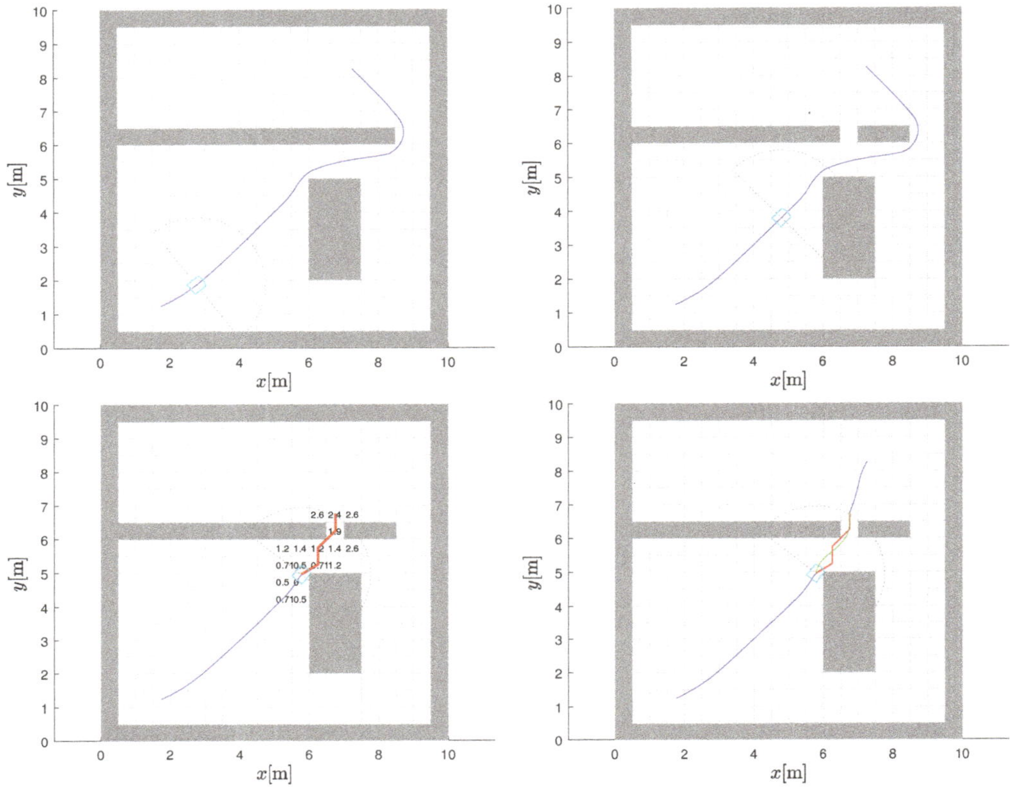

Figure 9. Path planning with obstacle clearance within sensor view. Initial path planned (blue line) based on knowledge of global static map is followed (**top-left** and **top-right**). Robot detects change in static map using sensor view (cleared cell) and computes A* bypass path (red line) to an unblocked cell beyond cleared cell (**bottom-left**). The smoothed bypass path (green line) connects to the gradient-descent path (blue line above the obstacle) based on CtG (**bottom-right**).

5.2. Obstacle Missing from Static Map Outside of Sensor View

It is also possible to conceptualise a situation where the robot comes to know about the removal of an obstacle from the static map in a region of the environment outside the sensor range of the robot while it is in motion on the current path to the goal. For example, this could happen when another robot passing by that region notices and relays the change(s) to a central station and/or all agents. While this could be a corrected error in the map creation, it could also be the result of a temporary obstacle (for example, a fallen tree) being cleared.

This situation can be handled in the same way as the previous one. A potential A* bypass path can be estimated from the robot's current position to a target cell just beyond the cell whose occupancy status has changed. This serves as a bridge to an area where the old CtG values are still valid. As before, the next segment of the path is established using the gradient descent from the temporary bypass goal on to the original destination. Then, based on whether this new alternate route is shorter than the remaining part of the current route, the robot can decide whether to switch to the alternate path or continue on the current one.

An example of this case is shown in the environment in Figure 10. The various parts of the figure are in line with Figure 9 and can be understood with the help of the caption.

In the environment of Figure 10, the information on the map change helps chart a shorter route to the goal.

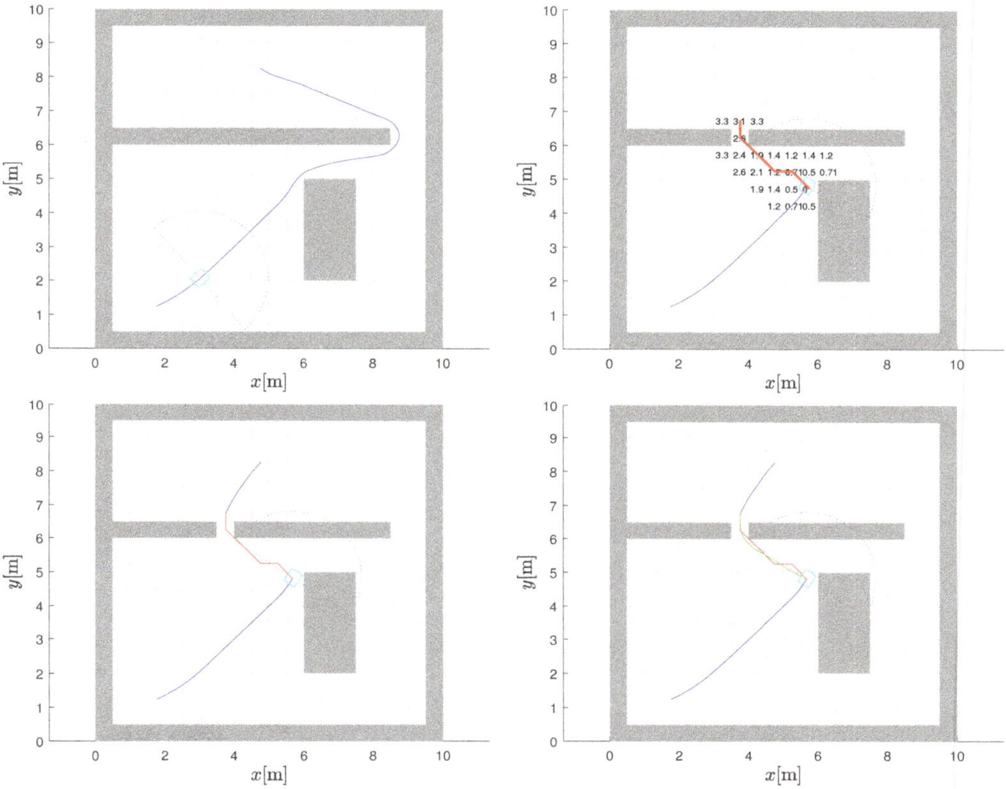

Figure 10. Path planning with obstacle cleared outside the robot sensor view. Initial path (blue line) planned based on knowledge of global static map is followed (**top-left**). Robot is informed of cleared obstacle cells (outside robot's sensor view) and A* bypass path (red line) to an unblocked cell beyond cleared cell is computed (**top-right**). The composite alternative path made up of final gradient-descent path (blue line above the obstacle) using the original CtG (**bottom-left**) and smoothed bypass version (green line) (**bottom-right**) is shorter than the one in the (**top-left**) figure and robot switches to it.

5.3. U-Shaped Trap

The last environment considered is one which incorporates the classic U-shaped concave trap. The static map is initially empty and the planned path between the start position and the goal is shown in Figure 11 (top-left), before the blockage is discovered, and is as expected. Even when the lower part of the U-shaped particle is discovered (Figure 11 (top-right)), the robot continues to proceed on the initial straight path recommended by global path planning. It does this until it encounters the core structure of the trap and the attendant local minima created (Figure 11 (bottom-left)). The ability to resolve an unexpected concave obstacle configuration on the planned path is a good test of the ability of an algorithm, because those that are purely combinatorial will fail this test.

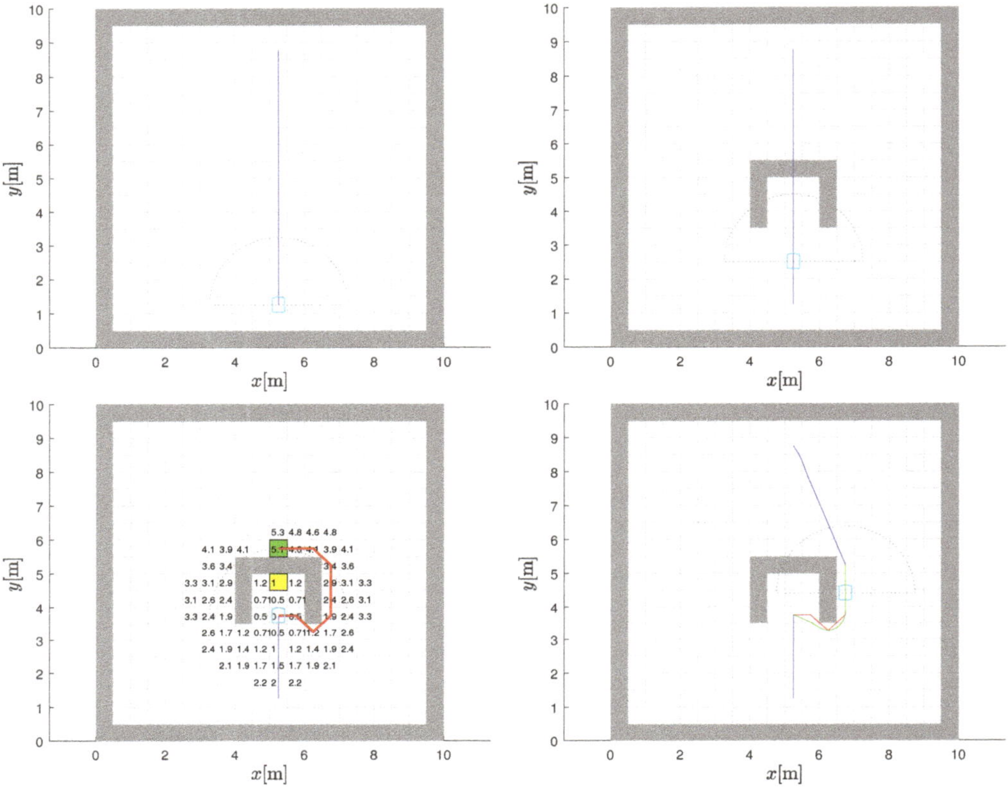

Figure 11. Path planning with appearance of unexpected U-shaped trap. Robot initially plans path (blue line) in empty static map (**top-left**) and starts to travel on it (**top-right**). When blockage is within sensor range, a bypass path is computed by a red line (**bottom-left**). The path is smoothed (green line) and connects with the final gradient-descent path (blue line) using CtG (**bottom-right**).

In accordance with the algorithm, a 5 by 5 window of cells is examined around the stuck cell (the yellow cell in Figure 11 (bottom-left) where the robot's initial gradient-descent path is blocked by the obstacle) to find the lowest CtG value below the current one (temporary intermediate goal cell marked by green in Figure 11 (bottom-left)). As explained earlier, such a cell is considered as being in an area where the CtG values are unaffected by the new blockage. If in some other case, the chosen 5 by 5 search window size does not produce a suitable target, because the cell with a minimum CtG value smaller than the current value lies within the trap zone, the search can be repeated using a larger 7 by 7 window and so on. This will eventually yield a temporary bypass goal outside the trap. The A* bypass path to this cell is also shown in Figure 11 (bottom-left) in red. The smoothed version obtained by the bilinear gradient interpolation in green, the post-bypass path segment based on the gradient descent, and the overall composite path between the start and goal points is shown in Figure 11 (bottom-right). Note that the path follows the computed bypass only until the leave point where the old CtG values and its interpolated gradient-descent path can be followed again.

6. Discussion

Although bilinear interpolation and the calculation of gradients from a discrete grid are well-established in image processing, their direct application to a discrete APF can lead to several problems. At the points where the cell is connected to its neighbour, discontinuities can occur in the gradients, making the use of a gradient descent problematic. This can lead to undesirable zigzag paths when following the direction of the gradient descent. This problem is much more pronounced near obstacles. The potential of occupied cells is not defined by the CtG assignment and can be considered infinite, as the cell should not be part of a path to the goal. This can lead to problems with local minima near obstacles where the direction of the gradient descent could change by more than 90°.

To interpolate the potential appropriately, one could also use some other higher-order interpolation technique, such as the bicubic interpolation [15,36]. The advantage of this technique would be a smooth gradient transition at the cell border because it uses third-order polynomials for the interpolation, which are continuous up to the second derivative (C^2). However, bicubic interpolation requires the use of a 4 by 4 neighbourhood, which becomes problematic near obstacles or in narrow corridors because the occupied cells have infinite (undefined) potential values. These occupied cells need special treatment before they can be used in the interpolation. Therefore, bicubic interpolation brings its own problems, as it requires a neighbourhood of 16 cells for the interpolation, in contrast to bilinear interpolation, which only requires 4 cells.

An additional problem that plagues bicubic interpolation is the occurrence of anomalies such as surface oscillations and the possibility of local minima near obstacles. Third-order polynomials have a continuous gradient which, while fitting the equidistant cell centres near obstacles (e.g., obstacle corners), causes oscillations in between (a common problem in the interpolation where the fit is perfect at the data point but could be oscillatory in between, known as the Runge phenomenon).

In image processing, the gradient is normally computed with the convolution of the image with the gradient operator. There are various gradient operators, such as one-dimensional operators (e.g., $[1,\ -1]$, $[1,\ 0\ -1]$) and Robert's cross or the Prewitt, Sobel and Scharr operators [37], which are more robust to noise. Some of these filters introduce gradient shifting and/or smoothing. In our case, averaging is not required, as the APF inherently does not contain noise. The proposed method for calculating the APF gradient is therefore different from the gradient methods used in image processing because it is designed in a way that the gradient in the cell centres always points towards the neighbouring cell with the lowest potential, or the gradient is zero if the current cell has the lowest potential. This ensures that the interpolated gradient points towards the cells with the lowest potential, regardless of the potential magnitude in the cell neighbourhood (without an undesired gradient shift and averaging). Therefore, an optimum smooth path from the start to the goal can be obtained, because the obtained path accurately follows the bottom of the valley that is defined by the APF.

The calculation of the potential field and the gradient in the free space away from the obstacles is straightforward, but in the vicinity of the obstacles, special care is needed to determine the appropriate potential and gradient, as presented in the paper. In the cells surrounding the obstacles, the gradient is calculated from the potential field of the neighbouring cells. If some of these adjacent cells are occupied by obstacles, the potential in these cells may be different depending on which side of the obstacle the gradient is calculated—this occurs in the case of thin or diagonally touching obstacles. Therefore, the batch calculation of the APF can only be performed for the cells that do not touch the obstacles. Moreover, to determine the optimal path, the proposed approach can calculate the gradients online only for the cells that are surrounding the tip of the path while it is being generated. One could resample the grid to double/quadruple the resolution of the map to alleviate the problems encountered near thin obstacles. However, this would require more computational resources (by a factor of four in the case of a double resolution) to calculate the APF and its gradient. However, because the gradient is interpolated, smooth paths are obtained even if the resolution of the map is low.

The proposed bilinear interpolation is applied to the path planning in a static and dynamic environment. A simple model for combining the global and local path planning that also derives from the original potential fields concept is used. Its key aspects are as follows. A method is needed for multiple missions that could potentially require navigation to the same destination in the environment. A static APF is therefore interpolated based on the pre-calculated CtG values for the cells navigating the path to the goal. A global plan is created based on these CtG values using a gradient descent on the static APF. Along its path, local map changes in the environment can be detected in various ways. A bypass strategy is formulated that enables the robot to find and evaluate a temporary bypass path through the use of the A* algorithm. A case-based decision is made whether to take the alternate path. Obtaining the diversionary path is relatively fast compared to regenerating CtG values for the entire environment. The benefits will be proportionately even greater for larger and less constricted environments.

In summary, the proposed approach introduces the following contributions/modifications in APF-based path planning: a small required neighbourhood (2×2, compared to other interpolations), an easier treatment of the occupied cells in the interpolation, and no anomalies that could result in local minima or the oscillating direction of the gradient-descent path near obstacles. The basic bilinear interpolation has a discontinuous gradient between cells, which we take into account by our proposed additional interpolation for gradients. Standard image processing techniques usually apply a batch calculation to the entire image, which simplifies the algorithmic flow in an obstacle-free area. However, additional special care is required to determine the appropriate potential and gradient near the obstacles. The proposed approach reduces the computational effort as the interpolation is only performed on cells along the path and not for the entire environment as is usually the case with batch image processing algorithms.

Due to the applied interpolation in the potential function, good quality paths are obtained even at a rough grid resolution of the environment. These contribute to the computational and memory requirement efficiency of the approach.

7. Conclusions

This work proposes a new approach to construct a navigation function in a variant of artificial potential fields (APF) that can be applied to navigation, path planning and the control of a mobile robot. The navigation function guarantees safe guidance to a goal without local minima in concave traps, which is a common problem with APFs. Optimality and convergence are inherited from an optimal grid-based search which results in a discrete APF.

To obtain a smooth navigation function and the associated gradient-descent driving direction, we apply a bilinear interpolation with several novel extensions that allow an efficient application to path planning. First, we propose a reconstruction of the discrete potential for the neighbourhood of cells used in the interpolation that belong to obstacles or are outside the environment. Second, we introduce an additional interpolation of the gradient-descent directions from the estimated discrete gradients of the interpolated cells. This leads to a smooth, optimal and collision-safe path or navigation towards the goal. Third, the proposed interpolation approach can be performed online and is computationally efficient as it only interpolates the discrete cell potentials along the planned path.

Several path planning results are provided to illustrate the performance of the navigation function. To illustrate the application in dynamic environments, we propose a simple strategy to combine global and local path planning to bypass detected dynamic changes. The strategy enables a robot to find and evaluate a temporary bypass path around newly detected obstacles through the use of the A* algorithm.

Author Contributions: Conceptualisation, methodology and software, M.K., A.Z. and G.K.; writing—original draft preparation, M.K.; writing—review and editing, M.K., A.Z. and G.K. All authors have read and agreed to the published version of the manuscript.

Funding: This work was partly founded by a US Fulbright Scholar grant to Mohan Krishnan and partly by the Slovenian Research Agency under Grants P2-0219 and L2-3168.

Institutional Review Board Statement: Not applicable.

Informed Consent Statement: Not applicable.

Data Availability Statement: Not applicable.

Conflicts of Interest: The authors declare no conflict of interest.

References

1. LaValle, S.M. *Planning Algorithms*; Cambridge University Press: Cambridge, UK, 2006; p. 1007.
2. Stentz, A. Optimal and efficient path planning for partially-known environments. In Proceedings of the 1994 IEEE International Conference on Robotics and Automation, San Diego, CA, USA, 8–13 May 1994; Volume 4; pp. 3310–3317. [CrossRef]
3. Stentz, A. The Focussed D* Algorithm for Real-Time Replanning. In Proceedings of the International Joint Conference on Artificial Intelligence, Montreal, QC, Canada, 20–25 August 1995.
4. Koenig, S.; Likhachev, M. Fast replanning for navigation in unknown terrain. *IEEE Trans. Robot.* **2005**, *21*, 354–363. [CrossRef]
5. Philippsen, R.; Siegwart, R. An Interpolated Dynamic Navigation Function. In Proceedings of the 2005 IEEE International Conference on Robotics and Automation, Barcelona, Spain, 18–22 April 2005; pp. 3782–3789. [CrossRef]
6. Karaman, S.; Frazzoli, E. Sampling-based algorithms for optimal motion planning. *Int. J. Robot. Res.* **2011**, *30*, 846–894. [CrossRef]
7. Sánchez-Ibáñez, J.R.; Pérez-del Pulgar, C.J.; García-Cerezo, A. Path Planning for Autonomous Mobile Robots: A Review. *Sensors* **2021**, *21*, 7898. [CrossRef] [PubMed]
8. Khatib, O. Real-Time Obstacle Avoidance for Manipulators and Mobile Robots. *Int. J. Robot. Res.* **1986**, *5*, 90–98. [CrossRef]
9. Bhattacharya, P.; Gavrilova, M.L. Roadmap-based path planning-using the voronoi diagram for a clearance-based shortest path. *IEEE Robot. Autom. Mag.* **2008**, *15*, 58–66. [CrossRef]
10. Liu, Y.H.; Arimoto, S. Path Planning Using a Tangent Graph for Mobile Robots Among Polygonal and Curved Obstacles: Communication. *Int. J. Robot. Res.* **1992**, *11*, 376–382. [CrossRef]
11. Savkin, A.V.; Hoy, M. Reactive and the shortest path navigation of a wheeled mobile robot in cluttered environments. *Robotica* **2013**, *31*, 323–330. [CrossRef]
12. Eapen, N.A. Path planning of a mobile robot among curved obstacles through tangent drawing and trapezoidal decomposition. *Eng. Sci. Technol. Int. J.* **2021**, *24*, 1415–1427.[CrossRef]
13. Dijkstra, E.W. A Note on Two Problems in Connexion with Graphs. *Numer. Math.* **1959**, *1*, 269–271. [CrossRef]
14. Fragapane, G.; de Koster, R.; Sgarbossa, F.; Strandhagen, J.O. Planning and control of autonomous mobile robots for intralogistics: Literature review and research agenda. *Eur. J. Oper. Res.* **2021**, *294*, 405–426. [CrossRef]
15. Klančar, G.; Seder, M. Coordinated Multi-Robotic Vehicles Navigation and Control in Shop Floor Automation. *Sensors* **2022**, *22*, 1455. [CrossRef] [PubMed]
16. Mali, A.D.; Motion Planning in Computer Games. In *Encyclopedia of Computer Graphics and Games*; Lee, N., Ed.; Springer International Publishing: Cham, Switzerland, 2018; pp. 1–6. [CrossRef]
17. Banik, S.; Bopardikar, S.D. Attack-Resilient Path Planning Using Dynamic Games With Stopping States. *IEEE Trans. Robot.* **2022**, *38*, 25–41. [CrossRef]
18. Hart, P.E.; Nilsson, N.J.; Raphael, B. A Formal Basis for the Heuristic Determination of Minimum Cost Paths. *IEEE Trans. Syst. Sci. Cybern.* **1968**, *4*, 100–107. [CrossRef]
19. Choset, H.; Lynch, K.; Hutchinson, S.; Kantor, G.; Burgard, W.; Kavraki, L.; Thrun, S. *Principles of Robot Motion: Theory, Algorithms, and Implementations*; MIT Press: Cambridge, MA, USA, 2005; p. 603.
20. Ratering, S.; Gini, M. Robot navigation in a known environment with unknown moving obstacles. *Auton. Robot.* **1995**, *1*, 149–165. [CrossRef]
21. Wang, L.C.; Yong, L.S.; Ang, M. Hybrid of global path planning and local navigation implemented on a mobile robot in indoor environment. In Proceedings of the IEEE Internatinal Symposium on Intelligent Control, Vancouver, BC, Canada, 30 October 2002; pp. 821–826. [CrossRef]
22. Azmi, M.Z.; Ito, T. Artificial Potential Field with Discrete Map Transformation for Feasible Indoor Path Planning. *Appl. Sci.* **2020**, *10*, 8987. [CrossRef]
23. Lazarowska, A. Comparison of Discrete Artificial Potential Field Algorithm and Wave-Front Algorithm for Autonomous Ship Trajectory Planning. *IEEE Access* **2020**, *8*, 221013–221026. [CrossRef]
24. Amiryan, J.; Jamzad, M. Adaptive Motion Planning with Artificial Potential Fields Using a Prior Path. *arXiv* **2020**, arXiv:2005.04191.
25. Mandava, R.; Bondada, S.; Vundavilli, P. An Optimized Path Planning for the Mobile Robot Using Potential Field Method and PSO Algorithm. In *Soft Computing for Problem Solving*; Advances in Intelligent Systems and Computing; Springer: Singapore, 2019; Volume 2, pp. 139–150. [CrossRef]
26. Lyu, H.; Yin, Y. Fast Path Planning for Autonomous Ships in Restricted Waters. *Appl. Sci.* **2018**, *8*, 2592. [CrossRef]
27. Park, M.; Lee, M.C. A new technique to escape local minimum in artificial potential field based path planning. *KSME Int. J.* **2003**, *17*, 1876–1885. [CrossRef]
28. Guerra, M.; Efimov, D.; Zheng, G.; Perruquetti, W. Avoiding Local Minima in the Potential Field Method using Input-to-State Stability. *Control Eng. Pract.* **2016**, *55*, 174–184. [CrossRef]

29. Cheng, C.; Chen, Y. A Neural Network based Mobile Robot Navigation Approach using Reinforcement Learning Parameter Tuning Mechanism. In Proceedings of the 2021 China Automation Congress (CAC), Beijing, China, 22–24 October 2021; pp. 2600–2605. [CrossRef]
30. Porta-Garcia, M.; Montiel Ross, O.; Castillo, O.; Sepúlveda, R.; Melin, P. Path planning for autonomous mobile robot navigation with ant colony optimization and fuzzy cost function evaluation. *Appl. Soft Comput.* **2009**, *9*, 1102–1110. [CrossRef]
31. Tarokh, M. Path planning of rovers using fuzzy logic and genetic algorithm. In Proceedings of the World Automation Conference ISORA-026, Maui, HI, USA, 11–16 June 2000; pp. 1–7.
32. Đakulović, M.; Sprunk, C.; Spinello, L.; Petrovic, I.; Burgard, W. Efficient navigation for anyshape holonomic mobile robots in dynamic environments. In Proceedings of the 2013 IEEE/RSJ International Conference on Intelligent Robots and Systems, Tokyo, Japan, 3–7 November 2013; pp. 2644–2649. [CrossRef]
33. Karaman, S.; Frazzoli, E. Optimal kinodynamic motion planning using incremental sampling-based methods. In Proceedings of the 49th IEEE Conference on Decision and Control (CDC), Atlanta, GA, USA, 15–17 December 2010; pp. 7681–7687.
34. Press, W.H.; Teukolsky, S.A.; Vetterling, W.T.; Flannery, B.P. *Numerical Recipes in C: The Art of Scientific Computing*, 2nd ed.; ACM: New York, NY, USA, 1992; pp. 123–128.
35. Bilinear Interpolation. Available online: https://en.wikipedia.org/wiki/Bilinear_interpolation (accessed on 15 February 2022).
36. Keys, R. Cubic convolution interpolation for digital image processing. *IEEE Trans. Acoust. Speech Signal Process.* **1981**, *29*, 1153–1160. [CrossRef]
37. Forsyth, D.; Ponce, J. *Computer Vision: A Modern Approach*; Prentice Hall: Hoboken, NJ, USA, 2011.

sensors

MDPI

Article

Real-Time Short-Term Pedestrian Trajectory Prediction Based on Gait Biomechanics

Leticia González *, Antonio M. López, Juan C. Álvarez and Diego Álvarez

Electrical Engineering Department, Campus of Gijon, University of Oviedo, 33204 Gijón, Spain
* Correspondence: leti@dieecs.com; Tel.: +34-985-181-994

Abstract: The short-term prediction of a person's trajectory during normal walking becomes necessary in many environments shared by humans and robots. Physics-based approaches based on Newton's laws of motion seem best suited for short-term predictions, but the intrinsic properties of human walking conflict with the foundations of the basic kinematical models compromising their performance. In this paper, we propose a short-time prediction method based on gait biomechanics for real-time applications. This method relays on a single biomechanical variable, and it has a low computational burden, turning it into a feasible solution to implement in low-cost portable devices. We evaluate its performance from an experimental benchmark where several subjects walked steadily over straight and curved paths. With this approach, the results indicate a performance good enough to be applicable to a wide range of human–robot interaction applications.

Keywords: motion trajectory prediction; kinematical models; gait biomechanics

Citation: González, L.; López, A.M.; Álvarez, J.C.; Álvarez, D. Real-Time Short-Term Pedestrian Trajectory Prediction Based on Gait Biomechanics. *Sensors* **2022**, *22*, 5828. https://doi.org/10.3390/s22155828

Academic Editor: Biswanath Samanta

Received: 1 July 2022
Accepted: 1 August 2022
Published: 4 August 2022

1. Introduction

Human motion trajectory prediction (HMTP) is a critical technology in applications where people share their workspace with autonomous moving machines. That is needed, for example, in collaborative robotics for obstacle avoidance [1,2], in automatic driving assistance systems for safety assurance [3,4], in prostheses or exoskeletons for better performance [5], or in virtual reality to improve the sensation of immersion perceived by the user [6]. The strong interest in all these application fields explains the exponential growth of scientific communications devoted to this problem in the last few years [7].

A basic instance of the problem of HMTP can be described as how to estimate the future location of a specific mark in the body of a walking human within a given short temporal window (see Figure 1). With that prediction, the possibilities of enhancing intelligent human–robot collaborative environments increase. Robots can plan their motion to adjust their actions better for more efficient and safe collaboration with humans [8]. The prediction is based on information coming from the monitorisation of the human with environmental sensors, or sensors mounted on the robot, or wearables [9–11]. A viable real-time prediction of human trajectory must consider the sensors and signals that will be available and whether it will be possible to achieve the reactivity or fast response that the application demands.

As will be discussed below, there are different proposals in the literature for short-range anticipation of a walker's position in real-time. All of them differ in both the amount of sensory information that needs to be provided and the type of information processing needed to solve the problem. In this paper, we propose a method with the following characteristics: (1) it is designed for close-proximity applications, with prediction windows around 1 s, (2) it relays on a single biomechanical variable that can be measured with inexpensive wearable sensors, and (3) it has a low computational load and is therefore feasible to implement in low-cost portable devices.

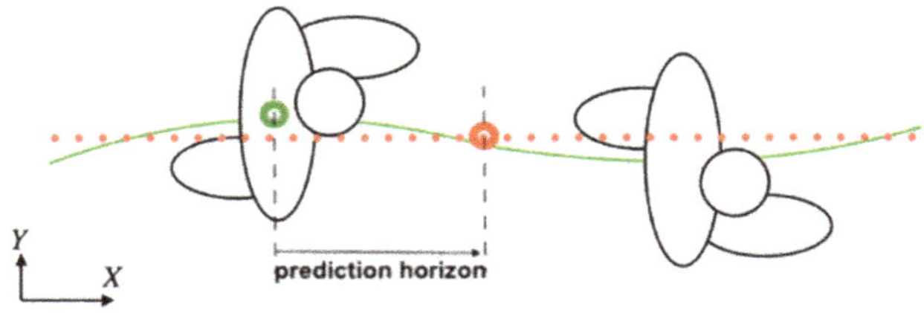

Figure 1. An instance of the problem of human motion trajectory prediction: the location of a specific landmark in the walking human body (green dot) is projected to its future location (red dot) within a given prediction horizon. The overall person's mark movement (solid green line) results in an estimated walk trajectory (red dots line).

In the following section, we will present the relevant state-of-the-art. After describing the method details in Section 3, we evaluate its performance from an experimental benchmark where several subjects walked steadily over straight and curved paths, Section 4. We show that the main error source of the estimations has a specific biomechanical source: the left-to-right fluctuations of the position of the subject in the direction of the displacement induced by the alternation of the step. This effect is adequately compensated in our algorithm by introducing a sensor-based biomechanical compensation. The results in Section 5 indicate a performance good enough to be applicable to a wide range of human–robot interaction applications.

2. Related Work

Following the general taxonomy of HMTP tools proposed in [7], physics-based approaches based on Newton's laws of motion seem best suited for real-time sensor-based short-term predictions because they operate within a reactive sense–predict scheme, avoiding intermediate processing time. For that reason, together with simplicity, we have ruled out methods that involved a higher level of cognition, such as learning based on data [12] or planning based on reasoning about motion intent [13].

Physics-based estimators, however, being useful to predict the motion of vehicles, fall short of capturing the quite complex dynamics of human walking. This has been addressed by resorting to a blending of a mixture of multiple dynamic models, but no performance improvement was found [14].

A similar strategy is to combine dynamic models with other learning or planning algorithms, even if, in principle, it goes against the speed of response of dynamic methods. A learning approach in [15] conveys promising results, but its performance on ambulatory motions such as walking has not been evaluated.

A third approach is to combine dynamic models with other information coming from the target agent himself. In [16], the short-term trajectory prediction is improved by tracking the user's head in the context of virtual reality applications. Head orientation anticipates the trajectory, but it is also heavily influenced by other ambient-related impulses or distractions [17]. The authors suggest the use of eye trackers to overcome this problem, but this technology is too expensive to be used in more general intelligent environments. Our work is framed in this line but seeks to improve the prediction with the measurement of biomechanical signals that are easily implementable in practice.

3. Formulation of the Prediction Method

For the short-term prediction of the pedestrian, we will adopt a kinematic model and a sensor-based biomechanical compensation method as described below.

3.1. Kinematical Trajectory Prediction Models

Kinematical models are a good approach for the short-term prediction of the position of moving objects [18,19]. They are derived by applying the Newton laws from an initial position (x, y) (see Figure 2), considering a set of initial conditions defined by the orientation of the displacement ϕ, the linear velocity v, the angular velocity w and the tangential acceleration a_t as follows:

$$\dot{x}(t) = v(t)\cos(\phi(t)) \quad (1)$$

$$\dot{y}(t) = v(t)sen(\phi(t)) \quad (2)$$

$$\dot{\phi}(t) = w(t) \quad (3)$$

$$\dot{v}(t) = a_t(t) \quad (4)$$

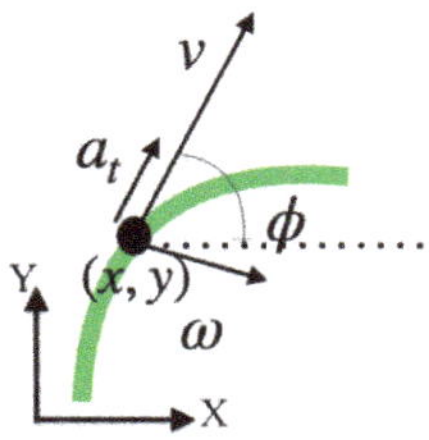

Figure 2. Geometry of motion for the 2-dimensional kinematical model of a moving object.

Depending on the assumptions taken about the translational and rotational velocities and the accelerations of the moving object, several particularisations can be formulated [18] (see Figure 3):

(i) Constant Velocity (CV) model ($w = a = 0$),
(ii) Constant Acceleration (CA) model ($w = 0, a = \text{constant} \neq 0$),
(iii) Constant Turn Rate and Velocity (CTRV) model ($w = \text{constant} \neq 0, a = 0$) and
(iv) Constant Turn Rate and Acceleration (CTRA) model ($w = constant \neq 0$, $a = constant \neq 0$).

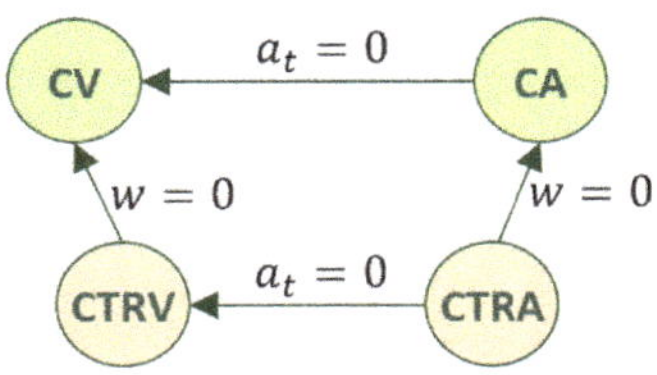

Figure 3. Relationship between common movement patterns defined from simplifying assumptions over the general model: CV (straight displacement at Constant Velocity), CA (straight displacement at Constant Acceleration), CTRV (curved displacement, at constant displacement and rotational velocities), and CTRA (curved displacement, at constant rotational velocity and constant accelerated displacement).

Defining the model state space vector as $[x, y, \phi, v, a, w]\prime$, which includes the velocity vector $[v, \phi]\prime$, we can obtain the discrete version of the kinematical Equations (1)–(4). Assuming that in the time interval Δt, the rotational velocity and the translational acceleration are constant (CTRA model) and results in:

$$\begin{bmatrix} x_{k+1} \\ y_{k+1} \\ \phi_{k+1} \\ v_{k+1} \\ a_{k+1} \\ w_{k+1} \end{bmatrix}_{CTRA} = \begin{bmatrix} x_k \\ y_k \\ \phi_k \\ v_k \\ a_k \\ w_k \end{bmatrix} + \begin{bmatrix} \frac{a}{w_k^2}\cos(\phi_k + w_k\Delta t) - \frac{a}{w_k^2}\cos(\phi_k) + \frac{v_k + a\Delta t}{w_k}\sin(\phi_k + w_k\Delta t) - \frac{v_k}{w_k}\sin(\phi_k) \\ \frac{a}{w_k^2}\sin(\phi_k + w_k\Delta t) - \frac{a}{w_k^2}\sin(\phi_k) - \frac{v_k + a\Delta t}{w_k}\cos(\phi_k + w_k\Delta t) + \frac{v_k}{w_k}\cos(\phi_k) \\ w_k \cdot \Delta t \\ a_k \cdot \Delta t \\ 0 \\ 0 \end{bmatrix} \quad (5)$$

From this discrete CTRA model, the equations of the other three can be calculated by substituting their respective values of translational and rotational velocities and the accelerations:

$$\begin{bmatrix} x_{k+1} \\ y_{k+1} \\ \phi_{k+1} \\ v_{k+1} \\ w_{k+1} \end{bmatrix}_{CTRV} = \begin{bmatrix} x_k \\ y_k \\ \phi_k \\ v_k \\ w_k \end{bmatrix} + \begin{bmatrix} \frac{v_k}{w_k}\sin(\phi_k + w_k\Delta t) - \frac{v_k}{w_k}\sin(\phi_k) \\ -\frac{v_k}{w_k}\cos(\phi_k + w_k\Delta t) + \frac{v_k}{w_k}\cos(\phi_k) \\ w_k \cdot \Delta t \\ 0 \\ 0 \end{bmatrix} \tag{6}$$

$$\begin{bmatrix} x_{k+1} \\ y_{k+1} \\ \phi_{k+1} \\ v_{k+1} \\ a_{k+1} \end{bmatrix}_{CA} = \begin{bmatrix} x_k \\ y_k \\ \phi_k \\ v_k \\ a_k \end{bmatrix} + \begin{bmatrix} \left(v_k\Delta t + \frac{a}{2}\Delta t^2\right)\cos(\phi_k) \\ \left(v_k\Delta t + \frac{a}{2}\Delta t^2\right)\sin(\phi_k) \\ 0 \\ a_k \cdot \Delta t \\ 0 \end{bmatrix} \tag{7}$$

$$\begin{bmatrix} x_{k+1} \\ y_{k+1} \\ \phi_{k+1} \\ v_{k+1} \end{bmatrix}_{CV} = \begin{bmatrix} x_k \\ y_k \\ \phi_k \\ v_k \end{bmatrix} + \begin{bmatrix} v_k\cos(\phi_k)\Delta t \\ v_k\sin(\phi_k)\Delta t \\ 0 \\ 0 \end{bmatrix} \tag{8}$$

At every sampling time, k, the model state vector $[x_k, y_k, \phi_k, v_k, a_k, w_k]$, must be computed to apply it to the model. The actual orientation of the displacement (ϕ_k^r) and the translational (v_k^r) are estimated from the position (x_k^r, y_k^r):

$$\phi_k^r = \tan^{-1}\left(\left(y_k^r - y_{k-1}^r\right)/\left(x_k^r - x_{k-1}^r\right)\right) \tag{9}$$

$$v_k^r = \mathrm{sqrt}\left(\left(x_k^r - x_{k-1}^r\right)^2 + \left(y_k^r - x_{k-1}^r\right)^2\right)/\Delta t \tag{10}$$

The rotational velocities (w_k^r) and translational acceleration (a_k^r) of the displacement are estimated from (ϕ_k^r, v_k^r):

$$a_k^r = \left(v_k^r - v_{k-1}^r\right)/\Delta t \tag{11}$$

$$w_k^r = \left(\phi_k^r - \phi_{k-1}^r\right)/\Delta t \tag{12}$$

In the following, we will refer to the sequence $(x_k^r,\ y_k^r,\ \phi_k^r,\ v_k^r,\ a_k^r,\ w_k^r)$ as the raw trajectory since its variables are obtained recursively directly from the kinematic models, without any other consideration of the biomechanics of human walking. As discussed above, the predictive power of these equations alone can be expected to be unsatisfactory.

3.2. Offline Compensations

To correct the predictable limitations of the model discussed above, we will resort to studying the effects induced in it by gait biomechanics. It is known that in normal walking, the pelvis moves from side to side once per cycle, the trunk being over each leg to maintain balance. Similarly, the forward movement is not constant, and it produces variations in velocity according to the phase of the step. The resulting velocity and orientation signals corresponding to the pelvis have a sinusoidal shape [20,21].

To include that fact in the model, we propose offline filtering of the whole raw trajectory signals (x_k^r, y_k^r) to remove their waviness. We applied a polynomial regression, defining the offline estimated trajectories $\left(x_k^b, y_k^b\right)$ for each of the experiments.

From the corrected position $\left(x_k^b, y_k^b\right)$, we calculated the orientation of the displacement (ϕ_k^b), the translational and rotational velocities (v_k^b and w_k^b, respectively), and translational acceleration (a_k^b) of the displacement with previous Equations (9)–(12). By removing the undulations from the positions $\left(x_k^b, y_k^b\right)$, the undulations will, in turn, be removed from the signals $\left(\phi_k^b, v_k^b, a_k^b, w_k^b\right)$.

In the following, we will refer to the sequence $\left(x_k^b,\ y_k^b,\ \phi_k^b,\ v_k^b,\ a_k^b,\ w_k^b\right)$ as the offline trajectory since its variables cannot be computed in real-time conditions. However, they could be taken as an instrumental way to test the utility of the proposed compensation.

3.3. Real-Time Sensor-Based Biomechanical Compensations

Assuming that the previous compensation will produce better estimations, it has the disadvantage of requiring filtering that induces time lags. We propose a real-time alternative to the offline compensation, which modifies the translational velocity and orientation signals $\left[v_k^r,\ \phi_k^r\right]$ considering previous knowledge about gait biomechanics.

The global velocity of the centre-of-mass velocity of the body, v_k^r, is an oscillatory signal whose average value coincides approximately with the value it takes in the initial and final contacts of the feet, as described in gait studies [22]. Therefore, the displacement velocity can be better computed from signals taken from the initial contact of the ipsilateral foot (Figure 4(A)) to the subsequent final contact of the contralateral foot (Figure 4(B)), and then to the subsequent initial contact of the contralateral foot (Figure 4(C)).

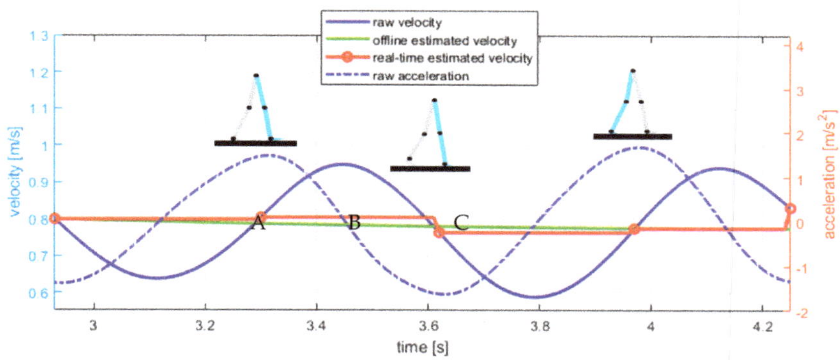

Figure 4. Forward real-time estimated velocity (red line) and offline estimated velocity (green line). The estimation of the forward velocity is addressed by holding the actual value of the raw velocity (continuous blue line) at the initial contact of the ipsilateral foot until the final contact of the contralateral foot and vice–versa. Initial and final contacts are detected from local maxima and minima of the derivative of the velocity (dashed blue line).

This way, we can fix the raw displacement velocity, v_k^r (Equation (10)), measured between initial and final contact foot events. We will refer to this sequence, $\widetilde{v}_k$, as the real-time estimated velocity. Notice that the initial and final foot contacts can be detected from local maximums and minimums of the derivative of the raw velocity [23,24], so this correction does not require specific new sensors. The acceleration in these segments, $\widetilde{a}_k$, can also be corrected and computed as the increment of velocity between consecutive events.

Regarding the orientation of the displacement ϕ_k, it is known in gait biomechanics that the body trunk and the pelvis orientation evolve in counter-phase [21] (see Figure 5). Therefore, the raw position signal, ϕ_k^r, can be corrected by measuring the pelvis orientation ϕ_k^h. A proportional average can cancel the oscillations [25] and return the basic signal trend, $\widetilde{\phi}_k$, closer to the actual forward orientation:

$$\widetilde{\phi}_k = \left(K \cdot \phi_k^h + (1 - K)\ \phi_k^r\right) \tag{13}$$

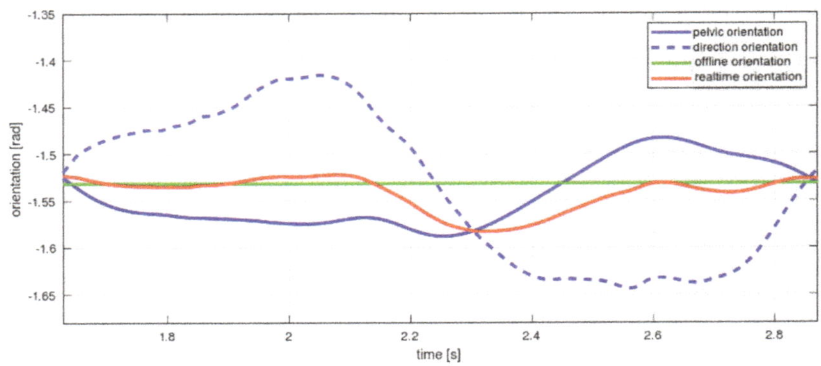

Figure 5. Orientation estimation (red line) and offline estimated orientation (green line). The real-time estimation is based on a proportional combination of the counter–phase signals pelvic orientation (continuous blue line) and the orientation estimated from the sampled raw spatial position of the subject (dashed blue line).

The calibration gain K has to be experimentally calculated for each subject to minimise the least square error between the raw and the baseline trajectories. From the estimated orientation, ϕ_k, we can compute the corresponding rotational velocity:

$$\widetilde{w}_k = (\widetilde{\phi}_k - \widetilde{\phi}_{k-1}\ )\,/\,\Delta t \tag{14}$$

From the point of view of the real-time implementation, we have introduced the need for a sensor that measures hip orientation. As in the previous case of trunk velocity, it is possible to make this estimate with an IMU sensor placed on the hip [26].

In the following, we will refer to the sequence $\left(x_k^r,\ y_k^r,\ \phi_k,\ \widetilde{v}_k,\ \widetilde{a}_k,\ \widetilde{w}_k\right)$ as the estimated trajectory. This sequence defines the initial conditions for the application of the kinematical models in real time to forecast the position of the walking subject. From the discussion above, it is expected that this model will produce a better trajectory estimation than the raw trajectory and close to the offline trajectory.

In the following, we will test this proposed method to evaluate and compare the performance of the raw, offline estimated and real-time estimated trajectories.

4. Evaluation

4.1. Experimental Setup

For the evaluation of the models, an experimental benchmark was designed involving five adult subjects aged between 21 and 52. Two types of trajectories, *straight* and *curved*, were defined (see Figure 6). For the straight trajectories, participants walked in a straight line 5.4 m in length. Participants were instructed to perform a normal gait while maintaining a constant walking velocity. This trajectory was repeated five times per individual. For the curved trajectories, participants walk in circles around a central point, 1.5 m radius, again five repetitions. They were instructed to keep the turning radius as constant as possible with the help of a guide painted on the floor, increasing the speed as the experiment progressed.

An Optitrack system composed of 10 Flex3 cameras was used to monitor the experiments with a sampling frequency of 100 Hz. Calibration of the system was performed following the recommendations of the manufacturer, and the nominal residual errors achieved were 1.4 mm (mean). Following [27] (see Figure 7), we used the bidimensional raw position $(x_k,\ y_k)$ of the centroid of the rigid body formed from five markers placed around the waist as the subject position. Pelvis orientation, ϕ_k^h, was estimated from the actual orientation of the rigid body from the defined reference system. A 3rd-order low-pass

Butterworth filter with a cut-off frequency of 6 Hz was applied during the acquisition to remove frequential components above those pertinent for the human gait.

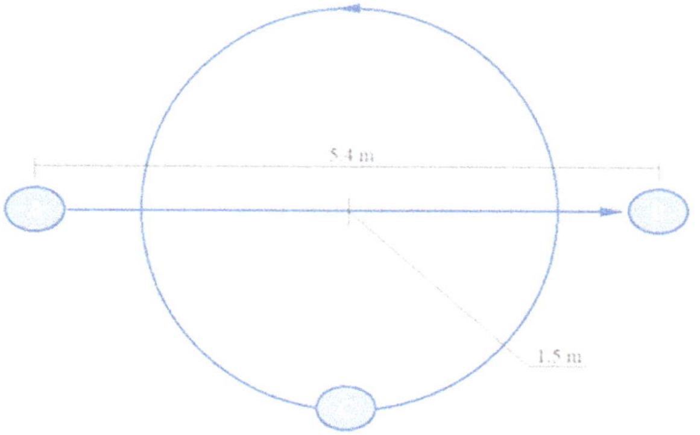

Figure 6. Experiments involved displacements over straight (5.4 m length) and circular trajectories (1.5 m radius). Straight trajectories ran from point A to B. For curved trajectories, participants describe a circle starting and ending at point C.

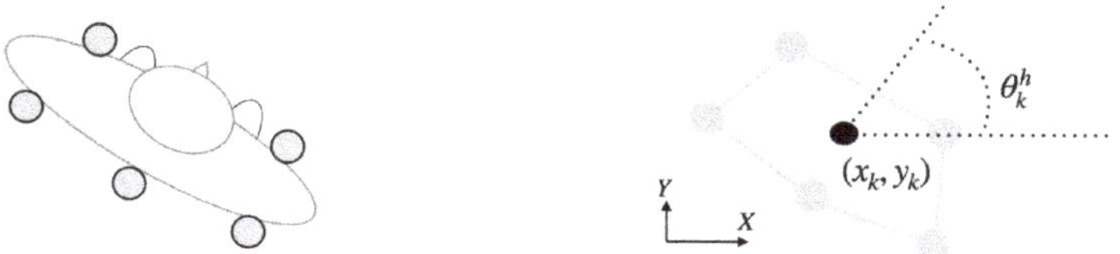

Figure 7. Five reflective markers were placed around the waist of the subject (**left**) for the estimation of the 2-dimensional spatial position of the body (x_k, y_k) and the orientation $\left(\theta_k^h\right)$ of the pelvis (**right**).

4.2. Model Application and Error Analysis

Data from the curved experiment was divided into five trajectory segments corresponding to each of the individual turns. This way, we defined for each subject five straight trajectory segments and five curved trajectory segments (each corresponding to a whole turn). To form the offline estimated trajectories, we use linear regression for straight segments and for the curved segments, a grade 8 polynomial regression.

The four prediction models (CV, CA, CTRV, CTRA) were then sequentially applied to the sequence $(x_k^r, y_k^r, \widetilde{\phi}_k, \widetilde{v}_k, \widetilde{a}_k, \widetilde{w}_k)$ starting at every sample from $k = 1$ to $k = N - 100$, being N the length of the current trajectory data. Each sample is predicted to have a horizon of 1 s (100 samples) (see Figure 8). For this purpose, the models are applied recursively by introducing as input the output of the previous iteration until the 1-s horizon is raised.

Similarly, these models are applied directly to the signals without removing the wavelets $(x_k^r, y_k^r, \phi_k^r, v_k^r, a_k^r, w_k^r)$ and on the post-processed signal with offline compensation $\left(x_k^b, y_k^b, \phi_k^b, v_k^b, a_k^b, w_k^b\right)$.

For every application of a model starting at sample k, the prediction error sequence e_k was defined as the sequence of Euclidean distances between the predicted $(\hat{x}_{k...k+99}, \hat{y}_{k...k+99})$ and actual position values $\left(x_{k...k+99}^r, y_{k...k+99}^r\right)$ of the raw trajectory.

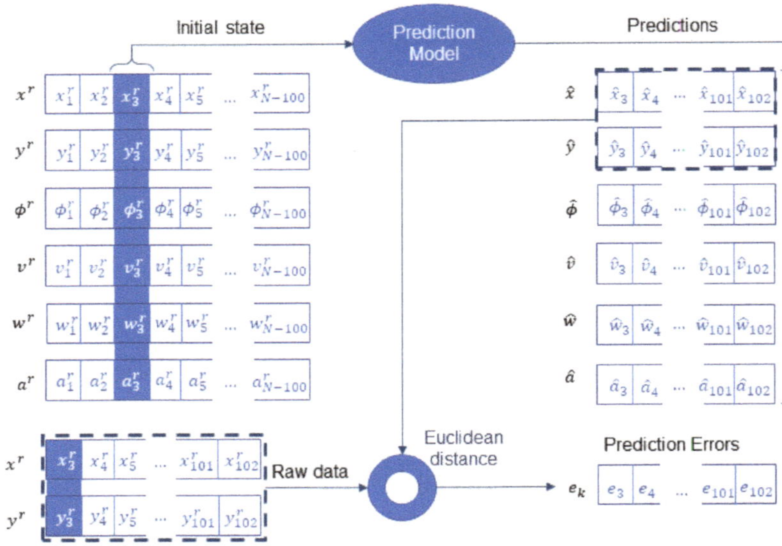

Figure 8. The figure describes the application of a prediction model starting at the third sample ($k = 3$) of a raw trajectory $(x_k^r, y_k^r, \phi_k^r, v_k^r, a_k^r, w_k^r)$. Predictions are sequentially extended for one second (100 samples, $k = 3 \ldots 102$) using the initial state vector highlighted in the blue column (**left**). The prediction error sequence is then defined from the Euclidean distance of predicted and actual positions of the subject ($k = 3 \ldots 102$).

5. Results

Figure 9 shows the Root Mean Square of the error sequences (predictions from one sample to one hundred samples, i.e., 0.01 s to 1 s) for each model (CV, CA, CTRV, CTRA) for the raw trajectories (top), offline estimated trajectories (centre), and real-time estimated trajectories (bottom).

We have found that prediction errors present an exponential growth with time, characteristic of the usual drift present in estimations based on integral procedures, with different growth rates for the different models. RMS prediction error at 1-s length is the biggest for predictions from the raw trajectories, with RMS error of predictions from offline estimated trajectories the lowest in general.

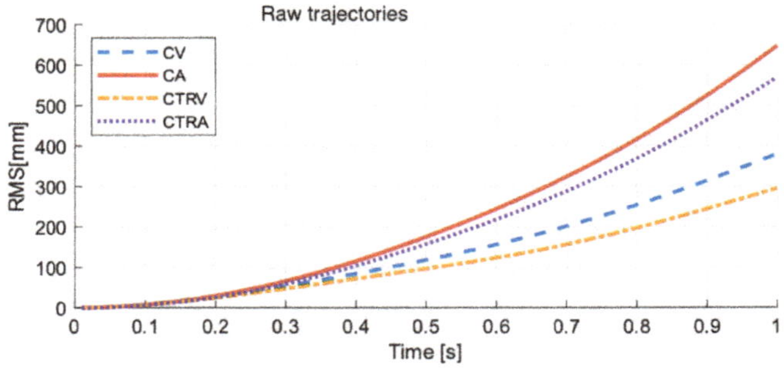

Figure 9. *Cont.*

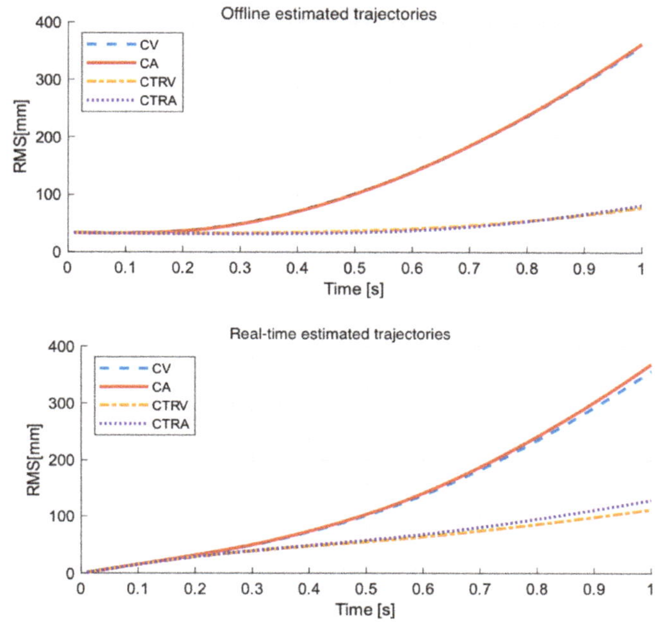

Figure 9. RMS prediction errors up to a prediction horizon of 1 s from the raw (**top**), offline estimated (**centre**), and real-time estimated (**bottom**) trajectories.

Table 1 contains the RMS error at a one-second prediction length from each trajectory (raw, offline estimated, and real-time estimated) and each prediction model. In general, as could be expected, curved models (CTRV, CTRA) perform better for curved trajectories, and straight models (CV, CA) perform better for straight trajectories. We found an exception in predictions from offline estimated trajectories, where there was no difference between the four prediction models for the straight trajectories. The translational acceleration and rotational velocity estimated for these trajectories were almost zero, as planned with the design of the experiment, which implies that all models adopted the behaviour of the CV model. Results also show that accelerated models perform in general worse than non-accelerated models, especially in the raw trajectories. In aggregated terms, the CTRV performs the best in all situations. Considering the type of path (straight, curved), the CTRV still performs the best except for prediction for straight real-time estimated trajectories, where straight models (CV, CA) perform better than the CTRV model.

To further analyse the performance of the prediction models, a multiway analysis of the variance (ANOVA) was carried out using MATLAB® R2020b from MathWorks to check if the mean prediction at a horizon of prediction of 1 s was affected by the intervening factors: the subject performing the test, the type of path (straight or curved), and the model applied for the prediction (CV, CA, CTRV, CTRA). The subject factor was treated as a random effect, while the type of path and model factors were treated as a fixed effect.

The results showed that mean prediction errors were not affected by the subject performing the test (p-value > 0.01) for the raw, offline estimated, and real-time estimated trajectories. On the contrary, the type of path and the prediction model significantly affected the mean prediction error for the raw, offline estimated, and real-time estimated trajectories (p-value < 0.01). A multiple comparison test was then carried out to analyse significant differences in the mean prediction error for pairs of type of trajectory/model.

Table 1. RMS error [mm] of the prediction on the theoretical offline estimated, raw and real-time estimated trajectories at a one-second prediction horizon. Results are shown for each prediction model and each type of trajectory (straight/curved). Aggregated values show the RMS prediction error considering straight and curved trajectories together.

		CV	CA	CTRV	CTRA
Offline estimated	Aggregated	360.83	363.80	78.15	82.83
	Curved	445.13	448.82	92.1	98.17
	Straights	40.12	40.12	40.12	40.12
Raw	Aggregated	379.52	646.16	294.27	568.33
	Curved	454.5	718.52	305.17	579.13
	Straights	160.28	480.28	272.44	547.31
Real-time estimated	Aggregated	357.5	369.04	113.33	130.59
	Curved	434.66	447.78	106.2	127.94
	Straights	110.2	120.29	125.73	135.44

For the offline estimated trajectories, we found four groups with mean prediction errors that were not significantly different among them and were significantly different from errors from the other three groups. These groups, expressed in ascending order or the averaged prediction error, were:

- Group 1: CV and CA in straight segments.
- Group 2: CTRV and CTRA in straight segments.
- Group 2: CTRV and CTRA in curved segments.
- Group 3: CV and CA in curved segments.

For the real-time estimated trajectories, we identified two different groups:

- Group 1: CV in straight segments; CTRV in straight and curved segments.
- Group 2: CA in straight segments; CTRA in straight and curved segments.

Regarding the raw trajectories, the test showed that the mean prediction error was significantly different for all type of trajectory/model pairs.

6. Discussion

6.1. Experiments Design

Experiments were designed to evaluate the performance of the proposed model during normal walking and, therefore, they comprised walks in straight and curved trajectories. We designed experiments for the extreme cases, straight trajectories at a constant velocity and circular trajectories with increasing velocity of displacement. The prediction results of the proposed model are compared with those obtained by applying the kinematic models directly to the raw data and to the post-processed data to be able to quantify the improvement achieved.

6.2. Impact of Velocity and Orientation on Predictions

Kinematical models based on the well-known Newton Laws have been used in the state-of-the-art as a general approach to forecasting the position of general moving objects. For instance, in [28], the authors perform a comparative study of several dynamical models (CV, CA, CTRV) and a multi-model algorithm combining such basic models. Particularly in terms of single motion models, they observe benefits for the CV model in some scenes and CTRV in others, making complex models inefficient. Similarly, we have observed that these models have usually been applied as a complement to other advanced techniques [15]. For our proposal prediction method, we chose to use the four models describing four types of trajectories where the occurrence of translational accelerations and angular velocities are combined.

The results of applying these models to our raw data confirm how the left-to-right fluctuations of the position of the subject in the direction of the displacement induced by the alternation of the step affect the performance of the basic kinematical models. Prediction on the raw data has the worst performance. On these, we can distinguish two effects. Firstly, the RMS of the accelerated models is significantly higher than that of the non-accelerated models for both curves and straight lines, in contrast to the offline prediction. This is the result of the influence of using the sinusoidal velocity in the kinematic models. A similar effect can also be found for straight versus curved models on straight trajectories. While in the offline prediction, all models have the same mean error, with the raw data, the CTRV model outperforms the RMS of the CV and the CTRA of the CA. This is due to the oscillation of the orientation, which is picked up by the curvilinear models.

6.3. *Impact of Basic Kinematic Models on Predictions*

The best model to apply in a general situation is the CTRV model. It shows the best prediction errors in aggregated terms. We have also found that the use of accelerated models supposes a concern. On the one side, we have found that for offline curved trajectories, curved models, accelerated and non-accelerated (CTRV, CTRA) performed equally. This finding was unexpected, as subjects were told to accelerate during the development of the curved experiments to analyse the performance of accelerated models (CA, CTRA) compared to non-accelerated models (CV, CTRV). However, we found that the acceleration estimated from the curved baseline trajectories (a_k^b) was on average 0.04 ± 0.09 m/s^2, coherent with reference data for normal walking reported in the literature [9]. This value was possibly too small to make a difference in performance between the two groups of models, as confirmed by the multiple comparison test. Therefore, this finding supports the idea that accelerated models do not provide a real improved performance over non-accelerated models in the context of predicting position during human walking. Moreover, the use of the acceleration in the models may suppose an issue, as the estimated prediction seems to present a higher error to this parameter, even with the velocity compensation.

In general terms, for the reasons stated above, the CTRV performs the best for unconstrained trajectories. We provide a statistical analysis that reveals the model most efficient for each type of path. In case the type of path is known in advance, the taxonomy of models may help to choose the most appropriate for a given application framework.

6.4. *Performance of Real-Time Sensor-Based Compensation Method*

The increased performance of predictions from offline filtered trajectories compared to predictions from raw trajectories confirms the necessity of a technique to remove the fluctuations from the raw signals before applying the prediction kinematical models. For this work, we have made batch filtering considering the whole raw signals to define the offline filtered trajectories. However, in a real application scenario of prediction of the subject position, such undulations should be removed in real-time, considering only the signal sampled to the actual moment. Future positions of the subject would not be available, and processing like the one performed to define the baseline trajectories would not be possible.

Classical frequential filters would be an option to remove the gait-induced undulations found in the sampled signals. In certain situations, these filters could perform properly. For instance, if the trajectory is essentially straight or the velocity is essentially constant, a low pass filter of a few seconds in length could be adequate. However, they may present some drawbacks, as the low step frequencies of human gait (between 0.74 and 1.3 two-step/s [29]) would require slow filters that would introduce a delay too high for a short-time prediction (a four-pole Butterworth filter with a cutting frequency of 1 Hz is expected to add a delay about 0.5 s [30]). So, for changing orientations and eventually changing velocities, these filters could not be very efficient. Recent results [25] confirm that this approach may lead to erroneous results and that a different approach is needed for a general solution to the problem.

In this paper, we propose to address the elimination of the undulations in the orientation and the velocity from the consideration of the biomechanical characteristics of human gait. We aimed to provide a general approach applicable to unconstrained trajectories mixing straight and curved paths and eventually different linear and rotational velocities and accelerations. With this approach, we show how prediction errors are considerably reduced from predictions from raw trajectories, making them comparable to the prediction from baseline estimations. Anyway, there is still room for improvement. Regarding the use of the acceleration in the models, it is true that the biomechanical inspired corrections made over the acceleration reduce the neglecting effect they present of the prediction from the raw trajectories. However, as acceleration makes a big impact on error and acceleration in human walking is low, the use of accelerated models may not be very profitable.

Comparing the results with those of other works, we find results in the same range of values. An example is the case of [15], where the proposed prediction method achieves an average error between 100 and 200 mm for predictions at a horizon of 0.5 s. The result obtained by individually applying a CV model (called as Velocity-Based Position Projection Method) gives a mean error of 250 mm. In our case, the RMS for the four kinematic models at a prediction horizon of 0.5 s is between 50 and 100 mm, as opposed to an RMS of ≈120 mm by the CV model applied to the raw data.

7. Conclusions

In a human–robot collaborative space, the short-term prediction of a person's walking position becomes necessary from a real and perceived safety point of view. In this paper, we propose a human trajectory prediction method for real-time application based on the biomechanical characteristics of human gait. This approach can be applied with inexpensive wearable sensors, and it has a low computational load. Experiments were designed to evaluate the performance of the proposed method during normal walking, comprising straight and curved paths. The results confirm how the left-to-right fluctuations of the position of the subject in the direction of the displacement induced by the alternation of the step affect the performance of the basic kinematical models to predict by themselves. We have found that if such undulations are removed, prediction performance would be improved and propose a prediction method that compensates the body swing by correcting velocity and orientation based on the initial and final contacts of the feet and the pelvic motion. The results show that, as acceleration makes a big impact on error and acceleration in human walking is low, the use of accelerated models may not be very profitable. Despite this, the performance of the proposed prediction method improves the use of basic kinematical models and produces results compatible with real-time applications.

Author Contributions: Conceptualization, L.G. and A.M.L.; methodology, L.G. and A.M.L.; software, L.G.; validation, J.C.Á., D.Á. and A.M.L.; investigation, L.G., A.M.L., J.C.Á. and D.Á.; data curation, L.G.; writing—original draft preparation, L.G. and A.M.L.; writing—review and editing, L.G. and J.C.Á.; visualization, L.G.; supervision, J.C.Á., A.M.L. and D.Á.; project administration, A.M.L.; funding acquisition, J.C.Á., A.M.L. and D.Á. All authors have read and agreed to the published version of the manuscript.

Funding: This research received no external funding.

Institutional Review Board Statement: Not applicable.

Informed Consent Statement: Not applicable.

Data Availability Statement: The data presented in this study are available on request from the corresponding authors.

Conflicts of Interest: The authors declare no conflict of interest.

References

1. Unhelkar, V.V.; Lasota, P.A.; Tyroller, Q.; Buhai, R.-D.; Marceau, L.; Deml, B.; Shah, J.A. Human-Aware Robotic Assistant for Collaborative Assembly: Integrating Human Motion Prediction with Planning in Time. *IEEE Robot. Autom. Lett.* **2018**, *3*, 2394–2401. [CrossRef]

2. Kruse, T.; Pandey, A.K.; Alami, R.; Kirsch, A. Human-Aware Robot Navigation: A Survey. *Robot. Auton.* **2013**, *61*, 1726–1743. [CrossRef]
3. Kooij, J.F.P.; Flohr, F.; Pool, E.A.I.; Gavrila, D.M. Context-Based Path Prediction for Targets with Switching Dynamics. *Int. J. Comput. Vis.* **2019**, *127*, 239–262. [CrossRef]
4. Koehler, S.; Goldhammer, M.; Bauer, S.; Zecha, S.; Doll, K.; Brunsmann, U.; Dietmayer, K. Stationary Detection of the Pedestrian? S Intention at Intersections. *IEEE Intell. Transp. Syst. Mag.* **2013**, *5*, 87–99. [CrossRef]
5. Zhang, F.; Liu, M.; Huang, H. Effects of Locomotion Mode Recognition Errors on Volitional Control of Powered Above-Knee Prostheses. *IEEE Trans. Neural Syst. Rehabil. Eng.* **2015**, *23*, 64–72. [CrossRef] [PubMed]
6. Schwind, V.; Halbhuber, D.; Fehle, J.; Sasse, J.; Pfaffelhuber, A.; Tögel, C.; Dietz, J.; Henze, N. The Effects of Full-Body Avatar Movement Predictions in Virtual Reality Using Neural Networks. In Proceedings of the ACM Symposium on Virtual Reality Software and Technology, VRST, Virtual Event, 1–4 November 2020; Association for Computing Machinery: New York, NY, USA, 2020.
7. Rudenko, A.; Palmieri, L.; Herman, M.; Kitani, K.M.; Gavrila, D.M.; Arras, K.O. Human Motion Trajectory Prediction: A Survey. *Int. J. Robot. Res.* **2020**, *39*, 895–935. [CrossRef]
8. Karim, B.; Tan, Q.; Alvarez, J.C. Towards Fog-Based HiTLCPS for Human Robot Interactions in Smart Lab: Use Cases and Architecture Overview. In Proceedings of the 15th International Conference on Soft Computing Models in Industrial and Environmental Applications (SOCO 2020), Burgos, Spain, 16–18 September 2020; Herrero, Á., Cambra, C., Urda, D., Sedano, J., Quintián, H., Corchado, E., Eds.; Springer International Publishing: Cham, Switzerland, 2021; pp. 590–599.
9. González, L.; Álvarez, J.C.; López, A.M.; Álvarez, D. Metrological Evaluation of Human–Robot Collaborative Environments Based on Optical Motion Capture Systems. *Sensors* **2021**, *21*, 3748. [CrossRef] [PubMed]
10. Álvarez, D.; Alvarez, J.C.; González, R.C.; López, A.M. Ambulatory Human Upper Limb Joint Motion Monitoring. In Proceedings of the 2012 IEEE I2MTC—International Instrumentation and Measurement Technology Conference, Graz, Austria, 13–16 May 2012; pp. 15–19.
11. Alvarez, J.C.; Álvarez, D.; López, A.M. Accelerometry-Based Distance Estimation for Ambulatory Human Motion Analysis. *Sensors* **2018**, *18*, 4441. [CrossRef] [PubMed]
12. Knoedler, L.; Salmi, C.; Zhu, H.; Brito, B.; Alonso-Mora, J. Improving Pedestrian Prediction Models with Self-Supervised Continual Learning. *IEEE Robot. Autom. Lett.* **2022**, *7*, 4781–4788. [CrossRef]
13. Bajcsy, A.; Bansal, S.; Ratner, E.; Tomlin, C.J.; Dragan, A.D. A Robust Control Framework for Human Motion Prediction. *IEEE Robot. Autom. Lett.* **2021**, *6*, 24–31. [CrossRef]
14. Schneider, N.; Gavrila, D.M. Pedestrian Path Prediction with Recursive Bayesian Filters: A Comparative Study. In *German Conference on Pattern Recognition*; Springer: Berlin/Heidelberg, Germany, 2013; pp. 174–183. [CrossRef]
15. Lasota, P.A.; Shah, J.A. A Multiple-Predictor Approach to Human Motion Prediction. In Proceedings of the IEEE International Conference on Robotics and Automation, Singapore, 29 May–3 June 2017; pp. 2300–2307.
16. Nescher, T.; Kunz, A. Using Head Tracking Data for Robust Short Term Path Prediction of Human Locomotion. In *Transactions on Computational Science XVIII*; LNCS 7848; Springer: Berlin/Heidelberg, Germany, 2013; pp. 172–191.
17. López, A.M.; Alvarez, J.C.; Álvarez, D. Walking Turn Prediction from Upper Body Kinematics: A Systematic Review with Implications for Human-Robot Interaction. *Appl. Sci.* **2019**, *9*, 361. [CrossRef]
18. Li, X.R.; Jilkov, V.P. Survey of Maneuvering Target Tracking. Part I: Dynamic Models. *IEEE Trans. Aerosp. Electron. Syst.* **2003**, *39*, 1333–1364.
19. Best, R.A.; Norton, J.P. A New Model and Efficient Tracker for a Target with Curvilinear Motion. *IEEE Trans. Aerosp. Electron. Syst.* **1997**, *33*, 1030–1037. [CrossRef]
20. Whittle, M.W. Gait Analysis. In *The Soft Tissues*; Butterworth-Heinemann: Oxford, UK, 1993.
21. Perry, J. *Gait Analysis Normal and Pathological Function*; SLACK Inc.: West Deptford, NJ, USA, 2010; Volume 159, ISBN 9781556421921.
22. Jian, Y.; Winter, D.A.; Ishac, M.G.; Gilchrist, L. Trajectory of the Body COG and COP during Initiation and Termination of Gait. *Gait Posture* **1993**, *1*, 9–22. [CrossRef]
23. Zijlstra, W.; Hof, A.L. Displacement of the Pelvis during Human Walking: Experimental Data and Model Predictions. *Gait Posture* **1997**, *6*, 249–262. [CrossRef]
24. Alvarez, J.C.; Alvarez, D.; López, A.; González, R.C. Pedestrian Navigation Based on a Waist-Worn Inertial Sensor. *Sensors* **2012**, *12*, 10536–10549. [CrossRef] [PubMed]
25. Gonzalez, L.; Lopez, A.M.; Alvarez, D.; Alvarez, J.C. Evaluation of Filtering Methods for the Prediction of Human Position during Walking by Means of Kinematical Models. In Proceedings of the Conference Record—IEEE Instrumentation and Measurement Technology Conference, Glasgow, UK, 17–20 May 2021; Institute of Electrical and Electronics Engineers Inc.: Piscataway, NJ, USA, 2021.
26. Bolink, S.A.A.N.; Naisas, H.; Senden, R.; Essers, H.; Heyligers, I.C.; Meijer, K.; Grimm, B. Validity of an Inertial Measurement Unit to Assess Pelvic Orientation Angles during Gait, Sit-Stand Transfers and Step-up Transfers: Comparison with an Optoelectronic Motion Capture System. *Med. Eng. Phys.* **2016**, *38*, 225–231. [CrossRef] [PubMed]
27. Bernardin, D.; Kadone, H.; Bennequin, D.; Sugar, T.; Zaoui, M.; Berthoz, A. Gaze Anticipation during Human Locomotion. *Exp. Brain Res.* **2012**, *223*, 65–78. [CrossRef] [PubMed]

28. Schneider, N.; Gavrila, D.M. Pedestrian Path Prediction with Recursive Bayesian Filters: A Comparative Study. In *Proceedings of the Lecture Notes in Computer Science (Including Subseries Lecture Notes in Artificial Intelligence and Lecture Notes in Bioinformatics)*; Springer: Berlin, Germany, 2013; Volume 8142 LNCS, pp. 174–183.
29. Oberg, T.; Karsznia, A.; Oberg, K. Basic Gait Parameters: Reference Data for Normal Subjects, 10–79 Years of Age. *J. Rehabil. Res. Dev.* **1993**, *30*, 210–223. [PubMed]
30. Manal, K.; Rose, W. A General Solution for the Time Delay Introduced by a Low-Pass Butterworth Digital Filter: An Application to Musculoskeletal Modeling. *J. Biomech.* **2007**, *40*, 678–681. [CrossRef] [PubMed]

sensors

MDPI

Article

An Adaptive Prescribed Performance Tracking Motion Control Methodology for Robotic Manipulators with Global Finite-Time Stability

Anh Tuan Vo †, Thanh Nguyen Truong † and Hee-Jun Kang *,†

Department of Electrical, Electronic and Computer Engineering, University of Ulsan, Ulsan 44610, Korea
* Correspondence: hjkang@ulsan.ac.kr; Tel.: +82-52-259-2207
† These authors contributed equally to this work.

Abstract: In this paper, the problem of an APPTMC for manipulators is investigated. During the robot's operation, the error states should be kept within an outlined range to ensure a steady-state and dynamic attitude. Firstly, we propose the modified PPFs. Afterward, a series of transformed errors is used to convert "constrained" systems into equivalent "unconstrained" ones, to facilitate control design. The modified PPFs ensure position tracking errors are managed in a pre-designed performance domain. Especially, the SSE boundaries will be symmetrical to zero, so when the transformed error is zero, the tracking error will be as well. Secondly, a modified NISMS based on the transformed errors allows for determining the highest acceptable range of the tracking errors in the steady-state, finite-time convergence index, and singularity elimination. Thirdly, a fixed-time USOSMO is proposed to directly estimate the lumped uncertainty. Fourthly, an ASTwCL is applied to deal with observer output errors and chattering. Finally, an observer-based-control solution is synthesized from the above techniques to achieve PCP in the sense of finite-time Lyapunov stability. In addition, the precision, robustness, as well as harmful chattering reduction of the proposed APPTMC are improved significantly. The Lyapunov theory is used to analyze the stability of closed-loop systems. Throughout simulations, the proposed PPTMC has been shown to perform well and be effective.

Keywords: Uniform Second-Order Sliding Mode Observer; Prescribed Performance Control; robot manipulators; finite-time Stability

Citation: Vo, A.T.; Truong, T.N.; Kang, H.-J. An Adaptive Prescribed Performance Tracking Motion Control Methodology for Robotic Manipulators with Global Finite-Time Stability. *Sensors* **2022**, *22*, 7834. https://doi.org/10.3390/s22207834

Academic Editors: Gregor Klancar, Marija Seder and Sašo Blažič

Received: 20 September 2022
Accepted: 12 October 2022
Published: 15 October 2022

1. Introduction

Increasing performance requirements are put into practice with a wide range of the robot's applications [1] such as fire prevention, medical support, industrial assembly, etc. However, some general problems of mechanical systems the dynamical uncertainties such as state constraints, frictions, high nonlinearity, parametric variations, etc., are unavoidable in reality [2]. They can be also exterior disturbances leading to the robot system may perform poorly in transient and steady-state states, causing instability in the robot's operation. Moreover, system uncertainties have highly complicated dynamics since their dynamics are influenced by the state of the system, its derivatives, and its inputs. Thus, it remains an open problem to determine an effective compensation method for system uncertainties in robot manipulators' trajectory tracking control. Under the influence of time-varying disturbances, the traditional PID controllers [3,4] have difficulty in maintaining accurate tracking. Therefore, a few more advanced controllers such as the modified PID control [5,6], Sliding Mode Control (SMC) [7–9], Computed Torque Control (CTC) [10], Back-stepping Control Method (BsCM) [11], Adaptive Control Method (ACM) [12], and so on, have been widely used in control design to reduce the effects of system uncertainty. SMC is most used by the control community due to its robustness, accuracy, and ease of implementation. However, unknown terms must be suppressed by the SMC's switching terms to ensure the existence of the sliding surface-reaching motion, leading to large chattering [13]. Moreover, it is unfortunate that most of these methods, including SMC, can only asymptotically converge to the neighborhood equilibrium points.

To obtain effective anti-disturbance ability and high tracking accuracy for robot systems with complicated dynamics and external disturbances, there are a lot of disturbance rejection control methods in the literature such as Sliding Mode Observer-based Control Method (SMO-CM) [14–18], Time-Delay Estimation-based Control Method (TDE-CM) [9,19], Disturbance Observer-based Control Method (DO-CM) [20], Active Disturbance Rejection Control Method (ADRCM) [21], and so on. In addition to removing the unreasonable assumption as H2 norm-bounded assumption [22], the SMO-CMs possess the robust control performance of the SMC methods. Using the SMC in conjunction with an observer, its switching part with a small sliding gain can compensate for the estimation error of Disturbance Observer (DO) along with minimizing chattering. This has prompted SMO-CM studies to become increasingly popular. Despite the fact that the SMO-CMs can offer powerful performance for controlled uncertain systems, most SMO-CMs employ asymptotical stability theory for their design. Therefore, those schemes only achieve asymptotical convergence. In control systems, fast/finite-time/fixed-time convergence is an important performance property. Finite-time/fixed-time convergence differs from asymptotic convergence in that the system states converge to zero in a finite amount of time or in fixed time. Therefore, the Finite-Time Control Method (FnTCM) [23,24] or the Fixed-Time Control Method (FxTCM) [17,25] could be achieved a better convergence rate and tracking precision.

Recently, a series of SMC with finite-time/fixed-time convergence have been introduced along with the expansion of FnTCM and FxTCM theory, such as Integral SMC (ISMC) [26,27], Terminal SMC (TSMC) [28,29], Non-singular TSMC (NTSMC) [30,31], Fast TSMC (FTSMC) [29,32,33], Fast NTSMC (FNTSMC) [34,35], and so on. Therefore, the Finite-Time Disturbance Observers (FnTDOs) or Fixed-Time Disturbance Observers (FxTDOs) have been developed such as Second-Order Sliding Mode Observer (SOSMO) [16,36], Uniform SOSMO (USOSMO) [37,38], or Third-Order Sliding Mode Observer (TOSMO) [14,39,40]. It can be seen from a comparison between FnTDO and FxTDO that under the same observer's gains, FnTDO cannot achieve a similar fast convergence performance as FxTDO. With the FxTDO, system states and estimation errors have uniform convergence time, and their upper bounds are not affected by the system's initial condition. The FxTDO is therefore a good candidate for handling unknown components. In addition, a combination of the FxTDO and the SMC also can minimize the effects of the chattering, as mentioned above.

In stabilization and tracking problems, transient performance is an important index for controlled systems that we need to concentrate on it. Though all of the conventional control methodologies can manipulate the state error variables to a residual set with an unknown size, it is not guaranteed convergence of trajectory states within a small maximum overshoot and maintained the steady states in a predefined boundary because of the lack of suitable techniques. The concept of the Prescribed Performance Control (PPC) was first proposed in [41] for satisfying transient behavior. That means both transient performance and steady-state performance are guaranteed with the following conditions: (1) tracking errors are limited to a small residual set; (2) the convergence rate is not less than a predetermined constant; (3) the maximum overshoot is limited to a predetermined space. Most current PPC studies [41–43] used a single Prescribed Performance Function (PPF) to generate boundaries of specific performance. For example, ref. [41] used a PPF $P(t)$ to determine the operating space in which $P(t)$ is prescribed as the upper boundary and $-NP(t)$ $(0 < N < 1)$ is prescribed as the lower boundary. This method has some drawbacks, as follows: the operating domain of specified performance be scaled down over a specified static error value because the lower boundary will be N times smaller than the upper boundary. In the steady state, these two boundaries will not be symmetrical about each other through zero if a ratio of PPF is used to create the lower boundary. Therefore, the transformed error can be converged to zero but the tracking error differs from zero.

This situation presents a real challenge in choosing an Error Transformation Function (ETF). In addition, some ETFs [44–46] have a singularity problem, which negatively affects the operation of the real system.

Inspired by the mentioned investigation, we propose an Adaptive Prescribed Performance Tracking Motion Control (APPTMC) for robotic manipulators with global finite-time stability. Our achievements include:

- the proposed PPFs ensure position tracking errors are managed in a pre-designed performance domain. Especially, the Steady-State Error (SSE) boundaries will be symmetrical to zero, so when the transformed error is zero, the tracking error will be as well;
- a fixed-time USOSMO is proposed to directly estimate the lumped uncertainty;
- in addition to determining the highest acceptable range of tracking errors at the steady state, the modified Non-singular Integral Sliding Mode Surface (NISMS) can also eliminate singularities and achieve finite-time convergence;
- the Adaptive Super-twisting Control Law (ASTwCL) is applied to deal with observer output errors and chattering. In this way, the control design clears the upper boundary requirement of all uncertainty.
- the proposed APPTMC ensures the effective reduction of harmful chattering behaviors by active compensations;
- guarantees prescribed performance in the sense of finite-time Lyapunov stability;
- the effectiveness of the APPTMC has been fully confirmed through simulations.

Following is a summary of the rest of the article. Section 2 describes the related preliminaries and mathematical formulas for robot dynamics. Throughout Section 3, the USOSMO design and the APPTMC design are presented along with their combination to solve the tracking control problems. A discussion of innovative features is presented in Section 4 through simulation examples on a 3-Degrees of Freedom (DOF) robot manipulator. As a result of this research, we draw some important conclusions and look ahead to future research directions in Section 5.

A list of nomenclature is provided in Table 1 for the reader's convenience. In addition, some other physical symbols will be fully defined in the paper.

Table 1. List of nomenclature.

Description	Notation
the real n-dimensional space	$\mathbb{R}^n$
the set of m by n real matrices	$\mathbb{R}^{n\times m}$
the transpose of	$\cdot^T$
Euclidean norm of	$\|\cdot\|$
absolute value of	$\lvert\cdot\rvert$
vector of joint angular acceleration	$\ddot{p} \in \mathbb{R}^{n\times 1}$
vector of joint angular velocity	$\dot{p} \in \mathbb{R}^{n\times 1}$
vector of joint angular position	$p \in \mathbb{R}^{n\times 1}$
vector of system state	$z = [z_1, z_2]^T = [p, \dot{p}]^T \in \mathbb{R}^{n\times 1}$
vector of tracking error	$z_e = \left[z_{e_1}^T, z_{e_2}^T\right]^T \in \mathbb{R}^{2n\times 1}$
vector of the desired trajectory	$z_d \in \mathbb{R}^{n\times 1}$
vector of NISMS	$s \in \mathbb{R}^{n\times 1}$
the first-order derivative of x	$\dot{x}$
the second-order derivative of x	$\ddot{x}$
Euler's number	$\mathbf{e}$

2. Problem Statement

2.1. Dynamic Modeling of Robotic Manipulators

Dynamic modeling of an n-DOF robot manipulator is described as [2]:

$$H(p)\ddot{p} + C(p,\dot{p})\dot{p} + g(p) + F_r(\dot{p}) = \tau - \tau_d, \tag{1}$$

where $H(p) = H_0(p) + \delta H(p) \in \mathbb{R}^{n\times n}$ is an inertial matrix that is nonsingular. $C(p,\dot{p}) = C_0(p,\dot{p}) + \delta C(p,\dot{p}) \in \mathbb{R}^{n\times n}$ represent Centripetal and Coriolis matrix and $g(p) = g_0(p) + \delta g(p) \in \mathbb{R}^{n\times 1}$ is gravity vector. $H_0(p) \in \mathbb{R}^{n\times n}, C_0(p,\dot{p}) \in \mathbb{R}^{n\times n}$, and $g_0(p) \in \mathbb{R}^{n\times 1}$ symbolize the computed dynamic function of $H(p), C(p,\dot{p})$, and $g(p)$, respectively. $\delta H(p) \in \mathbb{R}^{n\times n}$, $\delta C(p,\dot{p}) \in \mathbb{R}^{n\times n}$, and $\delta g(p) \in \mathbb{R}^{n\times 1}$ symbolize undefined dynamic function of $H(p)$, $C(p,\dot{p})$, and $g(p)$, respectively. Friction forces, external disturbances, and control torques are represented by the vectors $F_r(\dot{p}) \in \mathbb{R}^{n\times 1}$, $\tau_d \in \mathbb{R}^{n\times 1}$, and $\tau \in \mathbb{R}^{n\times 1}$, respectively.

Let $z = [z_1, z_2]^T = [p, \dot{p}]^T$ and $u = \tau$. then, the robot dynamics (1) can be described in form of the second-order state-space formula:

$$\begin{cases} \dot{z}_1 = z_2 \\ \dot{z}_2 = J(z)u + W(z) - \Delta(z,\delta,\tau_d) \end{cases}, \tag{2}$$

where $J(z) = H_0^{-1}(p)$, $W(z) = -H_0^{-1}(p)(C_0(p,\dot{p})\dot{p} + g_0(p))$ stands for the calculable or measurable terms and $\Delta(z,\delta,\tau_d) = H_0^{-1}(p)(F_r(\dot{p}) + \delta H(p)\ddot{p} + \delta C(p,\dot{p})\dot{p} + \delta g(p) + \tau_d)$ stands for the lumped unknown terms.

Let $z_e = \left[z_{e_1}^T, z_{e_2}^T\right]^T = \left[[z_1 - z_d]^T, [z_2 - \dot{z}_d]^T\right]^T$. So, Equation (2) is rewritten as:

$$\begin{cases} \dot{z}_{e_1} = z_{e_2} \\ \dot{z}_{e_2} = J(z)u + W(z) - \Delta(z,\delta,\tau_d) - \ddot{z}_d \end{cases}, \tag{3}$$

For improvements in the overall control performance, our article develops an APPTMC with global finite-time stability for robots that ensures transient performance and Prescribed Control Performance (PCP) within the prescribed domain.

A subsection below discusses mathematical statements, assumptions, lemmas, and definitions that will confirm the stability and convergence of the APPTMC.

2.2. Related Definitions and Lemmas

Some notations are described as follows: $[z]^0 = \text{sign}(z) = \begin{cases} 1 \,\text{if}\, z > 0 \\ 0 \,\text{if}\, z = 0 \\ -1 \,\text{otherwise} \end{cases}$ and $[z]^\phi = |z|^\phi \text{sign}(z)$ with $\phi > 0$.

Assumption 1. *Suppose that the desired trajectory z_d and their higher order time derivatives are continuous and bounded.*

Assumption 2. *Suppose that $\left|\dot{\Delta}_i(z,\delta,\tau_d)\right| \le \bar{\Delta}_i$, in which $\bar{\Delta}_i > 0$ is a predefined positive constant, $i = 1, \cdots, n$.*

Consider the differential formula:

$$\dot{z} = f(z(t)), f(0) = 0, z(0) = z_0, z \in \mathbb{D} \tag{4}$$

where f: $\mathbb{D} \to \mathbb{R}^n$ is continuous.

Definition 1 ([47]). *It is defined that Equation (4)'s origin point is global finite time stable if the following two conditions are met: (1) Equation (4) is globally asymptotically stable; (2) any solution*

$z(z_0,t)$ approach to the origin point at some finite time moments, i.e., $z(z_0,t) = 0, \forall t \geq T(z_0)$, where $T(z_0)$ presents the settling-time function.

Lemma 1 ([37]). *Consider the following dynamic system:*

$$\begin{cases} \dot{q}_0 = -\Pi_1 \Psi_{q_0} + q_1 \\ \dot{q}_1 = -\Pi_2 \Psi_{q_1} - \dot{\Delta} \end{cases} \tag{5}$$

where Ψ_{q_0} and Ψ_{q_1} are given by:

$$\begin{cases} \Psi_{q_0} = [q_0]^{\frac{1}{2}} + A[q_0]^{\frac{3}{2}} \\ \Psi_{q_1} = \frac{1}{2}[q_0]^0 + 2Aq_0 + \frac{3}{2}A^2[q_0]^2 \end{cases}$$

If $A > 0$, $|\dot{\Delta}| \leq \Delta_{\max}$, $\Delta_{\max} > 0$ is a predefined positive constant, and Π_1 and Π_2 are selected in the set below:

$$\Pi = \left\{ (\Pi_1, \Pi_2) \in \mathbb{R}^2 \middle| 0 < \Pi_1 \leq 2\sqrt{\Delta_{\max}}, \Pi_2 > \frac{\Pi_1^2}{4} + \frac{4\Delta_{\max}^2}{\Pi_1^2} \right\} \cup \left\{ (\Pi_1, \Pi_2) \in \mathbb{R}^2 \middle| \Pi_1 > 2\sqrt{\Delta_{\max}}, \Pi_2 > 2\Delta_{\max} \right\}.$$

Then $q_0 = 0$ and $q_1 = 0$ can be achieved in fixed time T_0 [37].

Lemma 2 ([48]). *Consider the differential formula with the following origin:*

$$\left[Q^{(j)}\right]^{\frac{\beta}{h-j}} + \lambda_{j-1}\left\{ \left[Q^{(j-1)}\right]^{\frac{\beta}{h-j+1}} + \ldots + \lambda_2 \left[[\ddot{Q}]^{\frac{\beta}{h-2}} + \lambda_1 \left([\dot{Q}]^{\frac{\beta}{h-1}} + \lambda_0 [Q]^{\frac{\beta}{h}} \right) \right] \ldots \right\} = 0, \tag{6}$$

If β is a positive scalar, $h \geq 2$ is an integer, and λ_k, $(k = 0, \ldots, h-1)$ are chosen sufficiently large then, Equation (6) is finite-time stable for each $j = 1, \ldots, h-1$.

Lemma 3 ([49]). *Consider the system:*

$$\begin{cases} \dot{\varpi} = -\nu_1(t)[\varpi]^{1/2} - \nu_2(t)\varpi + \gamma \\ \dot{\gamma} = -\nu_3(t)[\varpi]^0 - \nu_4(t)\varpi + \chi(t) \end{cases}. \tag{7}$$

Suppose that $|\chi(t)| \leq \delta_\chi$ with unknown scalar $\delta_\chi \geq 0$. The time-varying gains $\nu_m(t)$, $(m = 1,2,3,4)$ are obtained by:

$$\begin{aligned} &\nu_1(t) = \nu_{10}\sqrt{\rho_0(t)}; \nu_3(t) = \nu_{30}\rho_0(t); \\ &\nu_2(t) = \nu_{20}\rho_0(t); \nu_4(t) = \nu_{40}\rho_0^2(t), \end{aligned} \tag{8}$$

where positive constants ν_{m0} that satisfy the condition: $4\nu_{30}\nu_{40} \geq (8\nu_{30} + 9\nu_{10}^2)\nu_{20}^2$. $\rho_0(t)$ is a positive function and is tuned by the below adaptive law:

$$\dot{\rho}_0(t) = \begin{cases} \varepsilon \text{ if } |\varpi| \geq \delta_\varpi \\ 0 \text{ otherwise} \end{cases}, \tag{9}$$

where $\varepsilon, \delta_\varpi$ is arbitrary positive scalar.

Thus, the states in Equation (7) converge towards the origin within a finite amount of time.

3. Development of the Proposed Strategy

3.1. Design of an USOSMO

This subsection designs a USOSMO to estimate directly all uncertain terms. For bounded uncertain terms, the developed observer converges exactly in finite time, and also with a convergence time that is uniformly bounded for all initial conditions.

Using Equation (2), the observer is designed as follows:

$$\begin{cases} \tilde{z}_2 = z_2 - \hat{z}_2 \\ \dot{\hat{z}}_2 = J(z)u + W(z) - \hat{\Delta} + \theta_1 \Psi_1(\tilde{z}_2) \\ \dot{\hat{\Delta}} = -\theta_2 \Psi_2(\tilde{z}_2) \end{cases} \tag{10}$$

where $\Psi_1(\tilde{z}_2)$ and $\Psi_2(\tilde{z}_2)$ are selected as:

$$\begin{cases} \Psi_1(\tilde{z}_2) = [\tilde{z}_2]^{\frac{1}{2}} + \alpha[\tilde{z}_2]^{\frac{3}{2}} \\ \Psi_2(\tilde{z}_2) = \frac{1}{2}[\tilde{z}_2]^0 + 2\alpha\tilde{z}_2 + \frac{3}{2}\alpha^2[\tilde{z}_2]^2 \end{cases} \tag{11}$$

z_2 has an approximate value of $\hat{z}_2$. θ_1, θ_2, and α represent user-designed parameters of observer. θ_1 and θ_2 are selected respectively with Π_1 and Π_2 in the set as stated in Lemma 1.

The following theorem describes the design procedure of the observer.

Theorem 1. *The proposed observer's estimate errors will converge towards zero in a fixed time regardless of the initial conditions and of bounded uncertain terms* $\Delta(z, \delta, \tau_d)$.

Proof of Theorem 1. The proposed observer's estimate errors can be rewritten in the below expression.

$$\begin{cases} \tilde{z}_2 = z_2 - \hat{z}_2 \\ \tilde{\Delta} = \hat{\Delta} - \Delta \end{cases} \tag{12}$$

Taking time derivative of Equation (12) and using Equation (10) yields

$$\begin{cases} \dot{\tilde{z}}_2 = -\theta_1 \Psi_1(\tilde{z}_2) + \tilde{\Delta} \\ \dot{\tilde{\Delta}} = -\theta_2 \Psi_2(\tilde{z}_2) - \dot{\Delta} \end{cases} \tag{13}$$

where $\tilde{\Delta}$ represents the estimation error of the lumped uncertainty.

According to Lemma 1, it is concluded that the differentiator (13) is uniformly exact convergent, $\tilde{z}_2 = 0$ and $\tilde{\Delta} = 0$ are achieved in fixed time T_0 regardless of the initial conditions and of bounded uncertain terms. For the sake of brevity, the definition of T_0 could be found in the study [37]. T_0 was defined in Equation (12), as an upper bound for the convergence time of any trajectory of Equation (3) in the study [37].

This proof is completed. □

Remark 1. *Comparing with some recently proposed observers such as [16,36,39] we found that all three observers achieve only finite time convergence i.e., the convergence time of the observer depends on the initial condition whereas the proposed observer achieves uniform convergence in fixed time. In addition, refs. [16,36] require a measured value of the acceleration, which is not usually available, ref. [39] is known as a TOSMO and the feature of this observer is slow convergence. Therefore, the proposed observer can improve some shortcomings of the three observers.*

3.2. Design of the PPC

Based on the theory of the PPC, the tracking error z_e is constrained to the following domain:

$$-P_l(t) < z_e \text{sign}(z_e(0)) < P_u(t) \tag{14}$$

where $z_e(0)$ is the initial error, the PPFs are $P_u(t) = (P_0 - P_\infty)\mathbf{e}^{-rt} + P_\infty$ and $P_l(t) = (P_1 - P_\infty)\mathbf{e}^{-rt} + P_\infty$, and the $P_u(t)$ and $P_l(t)$ are defined as: $P_u(t)$ and $P_l(t) : R_+ \to R_+$ are smoothly, positive, and decreasing functions which respectively satisfying $\lim\limits_{t\to\infty} P_u(t) =$

$P_\infty > 0$, $\lim_{t \to \infty} P_l(t) = P_\infty > 0$. $P_0 > |z_e(0)| > 0, P_0 \geqslant P_1 \geqslant P_\infty$, r are design constants to adjust the specified performance domain.

Different from the existing PPC studies [41–46], two separate PPFs including $P_u(t)$ and $P_l(t)$ are proposed to manage the tracking errors in our paper. When the sign of the initial error changes, the lower and upper bounds will be reversed through the signum function. $P_u(t)$ and $P_l(t)$ represent upper and lower bounds for the performance domain, respectively. The upper boundary $P_u(t)$ sets the maximum allowable tracking error z_e at steady-state and limits the convergence rate while the lower boundary $P_l(t)$ sets the allowable maximum boundary of the overshoot and limits the allowable maximum size of the SSE z_e at the lower boundary. Because both PPFs are set the same boundary of the control error at a steady state lead to the specified performance space is increased compared to the classical PPC. Furthermore, the SSE boundaries will be symmetrical to zero, so when the transformed error is zero, the tracking error will be as well. Using the above proposal, ETFs can be designed more easily. The designed ETF does not suffer from singularity issues. Figure 1 shows the description of the prescribed performance definition that is proposed in our paper.

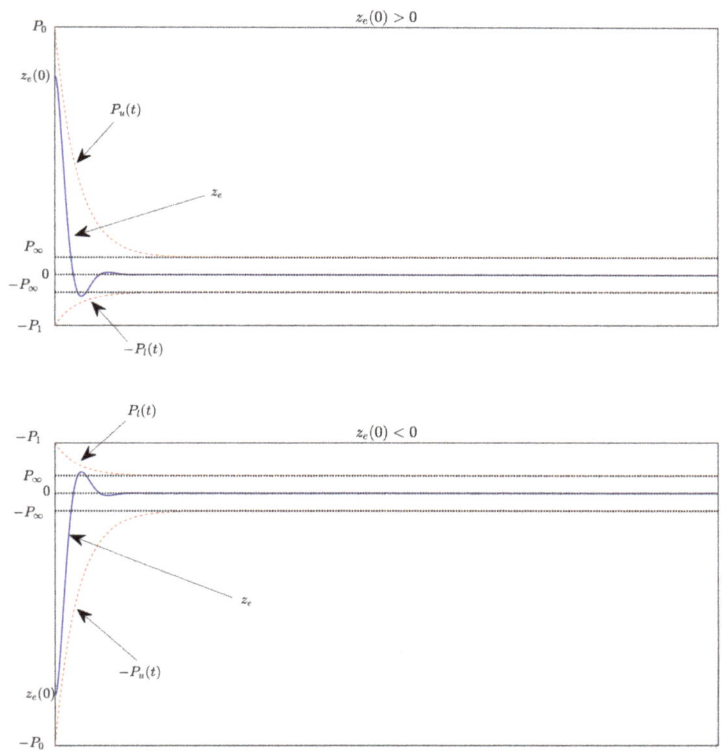

Figure 1. Description of the prescribed performance definition.

Remark 2. *It is prescribed that the allowable maximum size of tracking steady state error z_e is P_∞, that its maximum overshoot must be smaller than P_1, and that convergence rate of z_e depends on the decreasing rate of $P_u(t)$ adjusted by r. The output trajectory of the system is determined by the appropriate selection of $P_u(t)$ and $P_l(t)$.*

The constrained error dynamics are converted to their equivalent unconstrained dynamics by the following ETF:

$$z_{e_1} = P(t)T(\varrho_1) \tag{15}$$

where ϱ_1 is a transformed error, $T(\varrho_1)$ is an ETF, and

$$P(t) = \begin{cases} P_u(t) & \text{if } \operatorname{sign}(z_e.z_e(0)) > 0 \\ P_l(t) & \text{if } \operatorname{sign}(z_e.z_e(0)) < 0 \end{cases}.$$

$T(\varrho_1)$ has the properties:

- it is a smooth and strictly increasing function;
- $-1 < T(\varrho_1) < 1$;
- $T(\varrho_1) = 0$ if $\varrho_1 = 0$;
- $\begin{cases} \lim\limits_{\varrho_1 \to -\infty} T(\varrho_1) = -1 \\ \lim\limits_{\varrho_1 \to +\infty} T(\varrho_1) = 1 \end{cases}.$

Considering all possible scenarios, as follows:

If $z_e(0) > 0$ and $z_e > 0$ then $0 \leqslant T(\varrho_1) < 1$ and $P_u(t) > 0$. Hence, $0 \leqslant P_u(t)T(\varrho_1) < P_u(t)$; If $z_e(0) > 0$ and $z_e < 0$ then $-1 < T(\varrho_1) \leqslant 0$ and $P_l(t) > 0$. Hence, $-P_l(t) < P_l(t)T(\varrho_1) \leqslant 0$. It is concluded that whenever $z_e(0) > 0$, then $-P_l(t) < z_e < P_u(t)$.

If $z_e(0) < 0$ and $z_e < 0$ then $-P_u(t) < P_u(t)T(\varrho_1) < 0$. If $z_e(0) < 0$ and $z_e > 0$ then $0 < P_l(t)T(\varrho_1) < P_l(t)$. It is concluded that whenever $z_e(0) < 0$ then $-P_u(t) < z_e < P_l(t)$

Consequently, Equation (14) can be obtained fully which means the tracking error behavior will be prescribed over transient and steady-state scenarios.

The ETF in Equation (15) is proposed as

$$T(\varrho_1) = \frac{2}{\pi}\arctan(\varrho_1) \tag{16}$$

As a result, the transformed error ϱ_1 is given by:

$$\varrho_1 = \tan\left(\frac{\pi z_{e_1}}{2P(t)}\right) \tag{17}$$

Calculating the first-order derivative of $\arctan(\varrho_1)$ with respect to time obtains

$$(\arctan(\varrho_1))' = \frac{\dot{\varrho}_1}{1+\varrho_1^2} \tag{18}$$

Using Equations (16) and (18), the first-order derivative of z_{e_1} is

$$\begin{aligned} \dot{z}_{e_1} &= \dot{P}(t)T(\varrho_1) + P(t)\dot{T}(\varrho_1) \\ &= \dot{P}(t)\frac{2}{\pi}\arctan(\varrho_1) + P(t)\frac{2}{\pi}\frac{\dot{\varrho}_1}{1+\varrho_1^2} \end{aligned} \tag{19}$$

where $\dot{P}(t) = \begin{cases} \dot{P}_u(t) & \text{if } \operatorname{sign}(z_e.z_e(0)) > 0 \\ \dot{P}_l(t) & \text{if } \operatorname{sign}(z_e.z_e(0)) < 0 \end{cases}.$

Therefore, the first-order derivative of ϱ_1 is derived from Equation (19):

$$\dot{\varrho}_1 = \frac{\pi(1+\varrho_1^2)}{2P(t)}\left(\dot{z}_{e_1} - \frac{2\dot{P}(t)}{\pi}\arctan(\varrho_1)\right) \tag{20}$$

Calculating the second-order derivative of $\arctan(\varrho_1)$ with respect to time obtains

$$(\arctan(\varrho_1))'' = \frac{\ddot{\varrho}_1(1+\varrho_1^2) - 2\varrho_1\dot{\varrho}_1^2}{(1+\varrho_1^2)^2} \tag{21}$$

Using Equations (16), (18), and (21), the second-order derivative of z_{e_1} is

$$\begin{aligned}\ddot{z}_{e_1} &= \left(\dot{P}(t)T(\varrho_1) + P(t)\dot{T}(\varrho_1)\right)' \\ &= \ddot{P}(t)T(\varrho_1) + 2\dot{P}(t)\dot{T}(\varrho_1) + P(t)\ddot{T}(\varrho_1) \\ &= \frac{2}{\pi}\left(\ddot{P}(t)\arctan(\varrho_1) + \frac{2\dot{P}(t)\dot{\varrho}_1}{1+\varrho_1^2} - \frac{2P(t)\varrho_1\dot{\varrho}_1^2}{(1+\varrho_1^2)^2}\right) + \frac{2P(t)}{\pi}\frac{\ddot{\varrho}_1}{(1+\varrho_1^2)}\end{aligned} \tag{22}$$

where $\ddot{P}(t) = \begin{cases}\ddot{P}_u(t) \text{ if } \operatorname{sign}(z_e.z_e(0)) > 0 \\ \ddot{P}_l(t) \text{ if } \operatorname{sign}(z_e.z_e(0)) < 0\end{cases}$.

Therefore, the second-order derivative of ϱ_1 is derived from Equation (22):

$$\ddot{\varrho}_1 = \frac{\pi(1+\varrho_1^2)}{2P(t)}\left(\ddot{z}_{e_1} - \frac{2}{\pi}\left(\ddot{P}(t)\arctan(\varrho_1) + \frac{2\dot{P}(t)\dot{\varrho}_1}{1+\varrho_1^2} - \frac{2P(t)\varrho_1\dot{\varrho}_1^2}{(1+\varrho_1^2)^2}\right)\right) \tag{23}$$

with $\frac{\pi(1+e^2)}{2P(t)} > 0$.

Referring Equations (3) and (23), the robot dynamics can be presented in unconstrained dynamics:

$$\begin{cases}\dot{\varrho}_1 = \varrho_2 \\ \dot{\varrho}_2 = \Theta(J(z)u + W(z) - \Delta(z,\delta,\tau_d) - \ddot{z}_d - \bar{P})\end{cases}. \tag{24}$$

where $\Theta = \frac{\pi(1+e^2)}{2P(t)} > 0$ and $\bar{P} = \frac{2}{\pi}\left(\ddot{P}(t)\arctan(\varrho_1) + \frac{2\dot{P}(t)\dot{\varrho}_1}{1+\varrho_1^2} - \frac{2P(t)\varrho_1\dot{\varrho}_1^2}{(1+\varrho_1^2)^2}\right)$.

3.3. Design of NISMS

A modified NISMS is proposed to control the transformed errors to be skated on its surface in finite time, as follows:

$$s = \varrho_2 - \varrho_2(0) + \int_0^t \left[\sigma_1\left([\varrho_2]^{\frac{\beta}{h-1}} + \sigma_0[\varrho_1]^{\frac{\beta}{h}}\right)\right]^{\frac{h-2}{\beta}} d\iota, \tag{25}$$

where ι is the variable according to time, σ_0 and σ_1 are design constants. Due to its integral form, the proposed NISMS does not have any singularity issues.

If $s = 0$ and $\dot{s} = 0$, then the proposed system is in sliding mode. Equation (25) provides the following results:

$$\dot{\varrho}_2 = -\left[\sigma_1\left([\varrho_2]^{\frac{\beta}{h-1}} + \sigma_0[\varrho_1]^{\frac{\beta}{h}}\right)\right]^{\frac{h-2}{\beta}}. \tag{26}$$

Then, Equation (26) can be presented in the following form:

$$\begin{cases}\dot{\varrho}_1 = \varrho_2 \\ [\ddot{\varrho}_1]^{\frac{\beta}{h-2}} + \sigma_1\left([\varrho_2]^{\frac{\beta}{h-1}} + \sigma_0\varrho_1\right) = 0\end{cases}. \tag{27}$$

With $\beta = h = 3$ and $j = 2$, Equation (27) can be obtained the results as Equation (6); According to Lemma 2, for any initial states ϱ_0, the states $\varrho(t)$ of the system (27) will approach the origin within a finite period. Therefore, for any initial states $z_e(0)$, the tracking errors $z_e(t)$ will also converge to its origin within a finite period.

Remark 3. *As a result of designing the NISMS (25), the second-order sliding mode for s variable, i.e., $s = \dot{s} = 0$ leads to a third-order sliding mode of $\varrho_1(t)$ variable, i.e., $\varrho_1 = \varrho_2 = \dot{\varrho}_2 = 0, (r = 3)$. Therefore, the proposed controller can achieve 3-sliding accuracy even when measurement noise or sampling effects are existing [50].*

3.4. Proposed Controller Design

This subsection presents the process of the strategy being synthesized and its stability proof.

Calculating the first-order derivative of s and noting the dynamics (24) yields:

$$\dot{s} = \Theta(J(z)u + W(z) - \Delta(z, \delta, \tau_d) - \ddot{z}_d - \bar{P}) + \left[\sigma_1\left([\varrho_2]^{\frac{\beta}{h-1}} + \sigma_0[\varrho_1]^{\frac{\beta}{h}}\right)\right]^{\frac{h-2}{\beta}} \tag{28}$$

The proposed strategy is designed with the control torques as follows:

$$u = -J^{-1}\Theta^{-1}(z)(u_0 + u_{ob} + u_{astw}), \tag{29}$$

where the term u_0 is designed as:

$$u_0 = \Theta(W(z) - \ddot{z}_d - \bar{P}) + \left[\sigma_1\left([\varrho_2]^{\frac{\beta}{h-1}} + \sigma_0[\varrho_1]^{\frac{\beta}{h}}\right)\right]^{\frac{h-2}{\beta}},$$

the term u_{ob} is obtained from the observer's output as

$$u_{ob} = -\Theta\hat{\Delta},$$

and the reaching term u_{astw} is designed according to Lemma 2, as follows:

$$u_{astw} = \nu_1(t)[s]^{\frac{1}{2}} + \nu_2(t)s + \int_0^t \left[\nu_3(t)[s]^0 + \nu_4(t)s\right]d\iota.$$

Figure 2 illustrates the control system's block diagram.

The below theorem summarizes the control design process.

Theorem 2. *For the unconstrained system of the robot system, the sliding mode motions, $s = 0$, $\varrho_1 = 0$, and $z_{e1} = 0$, will take place in finite-time if the control torque (29) is designed based on the observer's output (10), the proposed NISMS (25), and Lemma 3.*

Proof of Theorem 2. Applying the control torque (29) to dynamic (28) obtains

$$\begin{aligned} \dot{s} &= \Theta\tilde{\Delta} - u_{astw} \\ &= \Theta\tilde{\Delta} - \nu_1(t)[s]^{\frac{1}{2}} - \nu_2(t)s - \int_0^t \left[\nu_3(t)[s]^0 + \nu_4(t)s\right]d\iota \end{aligned}. \tag{30}$$

Dynamic (30) can be represented by:

$$\begin{cases} \dot{s} = -\nu_1(t)[s]^{\frac{1}{2}} - \nu_2(t)s + \gamma \\ \dot{\gamma} = -\nu_3(t)[s]^0 - \nu_4(t)s + \dot{\Theta}\dot{\tilde{\Delta}} \end{cases}. \tag{31}$$

where $\gamma = -\int_0^t \left[\nu_3(t)[s]^0 + \nu_4(t)s\right]d\iota + \Theta\tilde{\Delta}$. Suppose that $\left|\dot{\Theta}\dot{\tilde{\Delta}}\right|$ is bounded by $\left|\dot{\Theta}\dot{\tilde{\Delta}}\right| < K$ which is a Lipschitz continuous function according to time, $K > 0$.

According to Lemma 3, the convergence of Equation (31) is finite time. Therefore, $s = 0$ and $\gamma = 0$ will be achieved within a finite amount of time. □

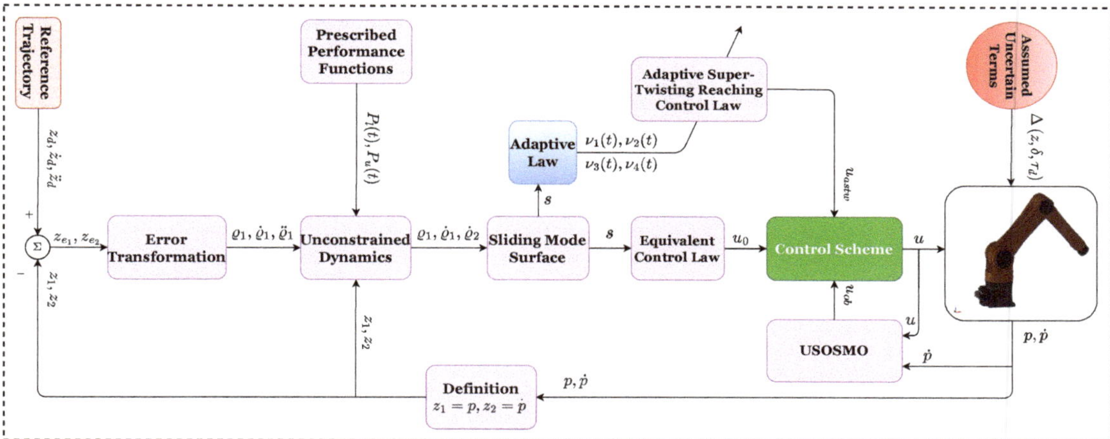

Figure 2. Algorithm diagram for the proposed control procedure.

4. Simulations

The performance of the trajectory tracking motion control is simulated in this section to show the effectiveness of the APPTMC. Simulations were performed in MATLAB/SIMULINK environment to evaluate aspects including maximum overshoot, convergence index, transient response, and SSEs. In addition, approximation ability, chattering reduction, accuracy, and robustness of the control proposal also are considered thoroughly via comparison to other equivalent solutions including the SMC [7], the TSMC [29] and the FTSMC [29]. All controllers are applied to a 3-DOF robotic manipulator to investigate their effectiveness. The dynamic mathematics and kinematic design of this robot are derived from studies [2,51]. The system parameters of the robot are selected from [15,25]. In the studies [15,25], we describe in detail how the robot system was built using MATLAB/SIMULINK, and SOLIDWORKS software. In MATLAB/SIMULINK, the differential equations are solved using Euler's method with a sampling time of $t_s = 10^{-3}$.

4.1. Configuration of the Robot System and Control Parameter Selection

The basic design parameters of the robot system including the length and weight of links, the center of mass, and inertia are reported in Table 2. A geometric representation of the robot model is shown in Figure 3.

Assigning a trajectory to the robot's end-effector is the robot's primary objective:

$$\begin{cases} X = 0.85 - 0.01t \\ Y = 0.2 + 0.2\sin(0.5t) \quad (\text{m}). \\ Z = 0.7 + 0.2\cos(0.5t) \end{cases} \tag{32}$$

To evaluate the robustness and the effectiveness of the developed scheme in presence of uncertain terms including calculated-dynamical errors, frictions, and exterior disturbances, they are assumed in Table 3.

Figure 3. Geometric representation of the robot model.

Table 2. Basic design parameters of a 3-DOF robot system.

Description	Link 1	Link 2	Link 3
Link Length (m)	$l_1 = 0.25$	$l_2 = 0.7$	$l_3 = 0.6$
Link Weight (kg)	$m_1 = 33.429$	$m_2 = 34.129$	$m_3 = 15.612$
Center of Mass (mm)	$l_{c1x} = 0$ $l_{c1y} = 0$ $l_{c1z} = -0.7461$	$l_{c2x} = 0.3477$ $l_{c2y} = 0$ $l_{c2z} = 0$	$l_{c3x} = 0.3142$ $l_{c3y} = 0$ $l_{c3z} = 0$
Inertia (kg.m^2)	$I_{1xx} = 0.7486$ $I_{1yy} = 0.5518$ $I_{1zz} = 0.5570$	$I_{2xx} = 0.3080$ $I_{2yy} = 2.4655$ $I_{2zz} = 2.3938$	$I_{3xx} = 0.0446$ $I_{3yy} = 0.7092$ $I_{3zz} = 0.7207$

Table 3. Assumed Uncertain Terms.

Type of the Assumed Uncertainty	Functions
Calculated-Dynamical Errors	$\delta H(p) = 0.2H(p)$ $\delta C(p, \dot{p}) = 0.2C(p, \dot{p})$ $\delta g(p) = 0.2g(p)$
Frictions $F_r(\dot{p})$ (N. m)	$F_{r1}(\dot{p}) = 0.1\text{sign}(\dot{p}_1) + 2\dot{p}_1$ $F_{r2}(\dot{p}) = 0.1\text{sign}(\dot{p}_2) + 2\dot{p}_2$ $F_{r3}(\dot{p}) = 0.1\text{sign}(\dot{p}_3) + 2\dot{p}_3$
Exterior Disturbances τ_d (N. m)	$\tau_{d1} = 4\sin(t)$ $\tau_{d2} = 5\sin(t)$ $\tau_{d3} = 6\sin(t)$

Following is a specific guide to choosing the control parameters.

Remark 4. *The parameters of the proposed sliding surface including $\beta, h, j, \sigma_0, \sigma_1$ are chosen according to Lemma 2. The parameters of the term u_{astw} including v_1, v_2, v_3 and v_4 are chosen according to Lemma 3. The parameters of the observer including θ_1, θ_2 are chosen based on the set, as stated in Lemma 1 while α is chosen to be greater than zero. The parameters of the PPF including P_0, P_1, P_∞, r are chosen to specify preset performance, as mentioned in Remark 1.*

Each controller's parameters are selected to optimize performance within its capabilities. Accordingly, Table 4 provides the control parameters selected for each algorithm.

Table 4. Control parameter selection for the proposed scheme.

Description	Symbol	Value
USOSMO (10)	$\theta_1, \theta_2, \alpha$	$10, 60, 2\sqrt{30}$
PPF (14)	P_0, P_1, P_∞, r	$0.023, 0.006, 0.0015, 3$
NISMS (25)	$\beta, h, j, \sigma_0, \sigma_1$	$3, 3, 2, 50, 10$
Proposed Control Law (29)	$\varepsilon, \nu_{10}, \nu_{20}, \nu_{30}, \nu_{40}$	$3, 2, 6, 10, 100$

4.2. Simulation Results and Discussion

We first investigate the efficiency and approximation of the proposed observer. We compare the estimation accuracy of the proposed FxTDO (USOSMO) with that of the FnTDO (TOSMO) [39]. The description of performance estimation from the FnTDO and the proposed FxTDO can be found in Figure 4. The estimated errors of the two observers are also plotted in Figure 5 to facilitate comparisons between them. According to Figures 4 and 5, both observers seem to achieve the same good accuracy. However, the proposed observer provides much faster convergence than the FnTDO. The convergence of the FnTDO was achieved in finite time, thus, the FnTDO depended on the initial value. In contrast, the proposed FxTDO provided fixed-time uniform convergence of the estimation errors. The displayed advantages of the proposed observer have a major contribution to improving overall control performance for robot manipulators.

We will then investigate the simulation results in terms of regulatory issues and tracking issues. Based on the results displayed in Figures 6–8, we analyze the regulation problem.

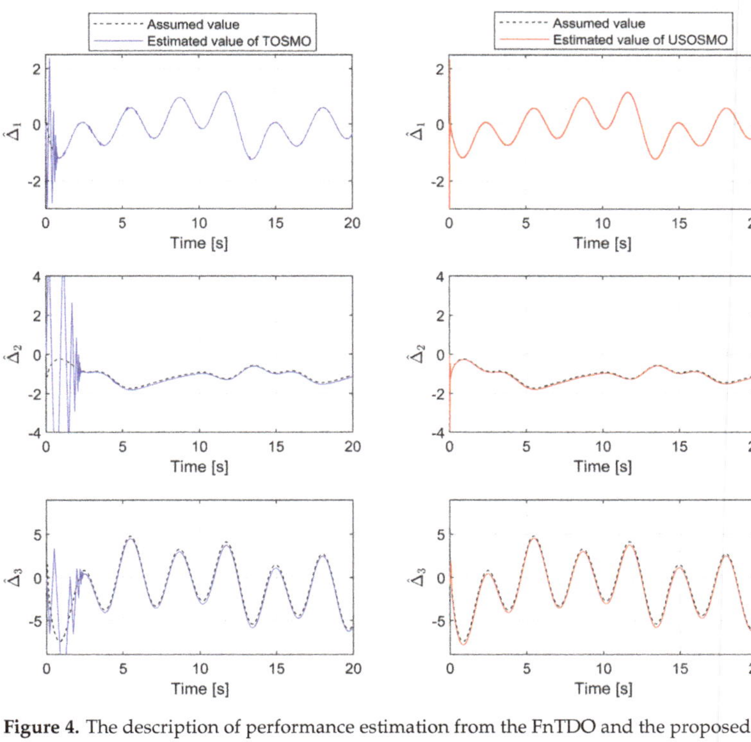

Figure 4. The description of performance estimation from the FnTDO and the proposed FxTDO.

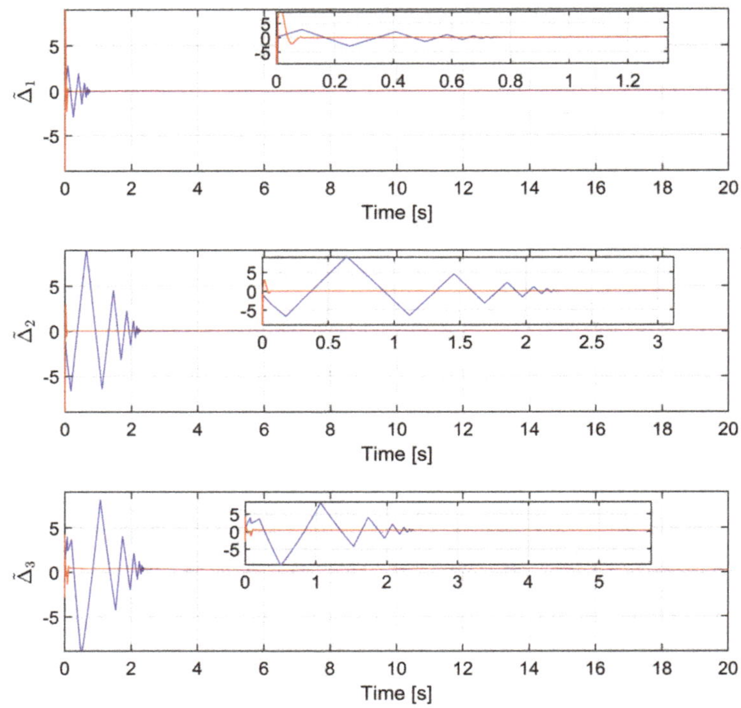

Figure 5. The comparison of the estimated errors between the FnTDO and the proposed FxTDO.

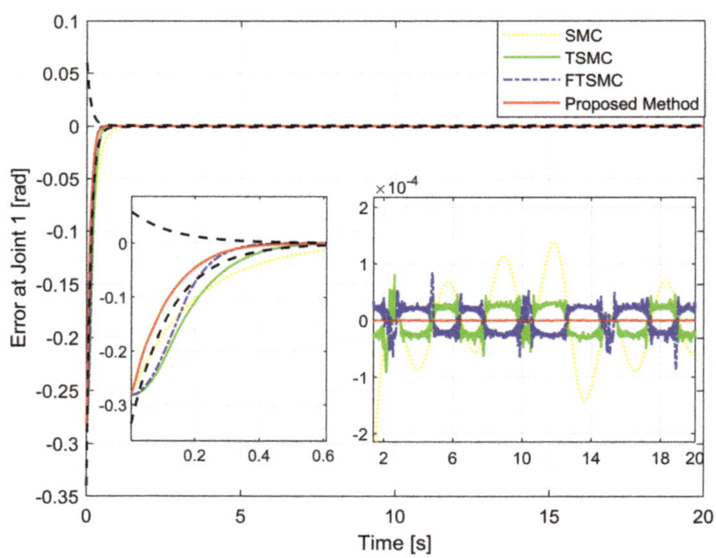

Figure 6. Tracking error of the first joint versus the desired trajectory.

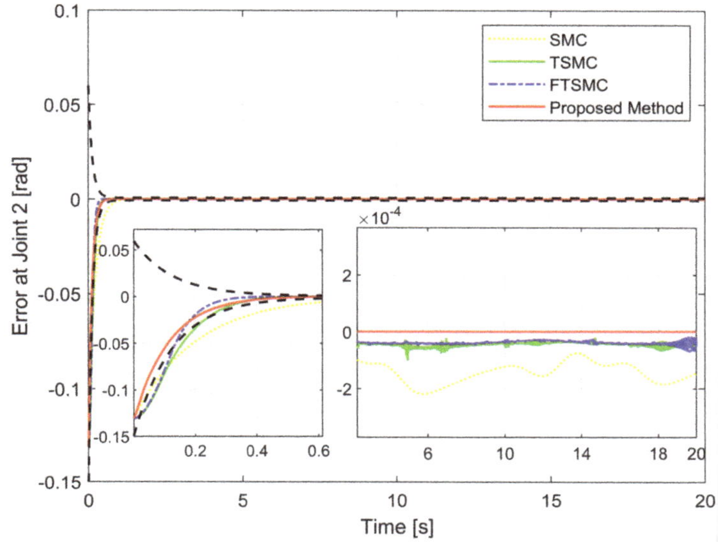

Figure 7. Tracking error of the second joint versus the desired trajectory.

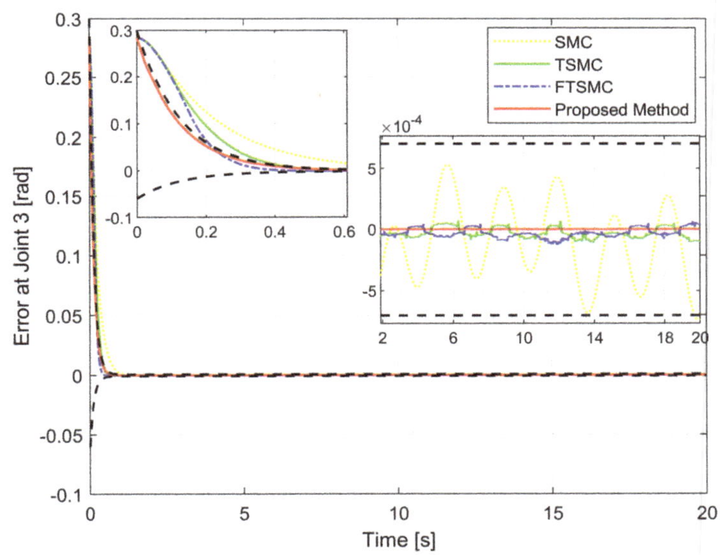

Figure 8. Tracking error of the third joint versus the desired trajectory.

For a fair investigation, the system states are considered with the same initial conditions. We investigate two terms in the approach stage (from the 0th second to the 0.6th second), including convergence rate and maximum overshoot, and find that the proposed strategy fulfills these both performance indices with a prescribed performance defined by Equation (14). By adjusting the design parameters including P_0, P_1, P_∞, and r we can control the output trajectory of the system within a predefined performance domain as described in Remark 2. However, the zoomed-in portions of Figures 6–8 clearly show that none of the other three methods satisfy both of the above performance indices.

Consider the trajectory tracking problems when controlling the robotic arm to follow the desired trajectory, as stated in Equation (32). Tracking accuracy and control performance can be evaluated by analyzing SSEs after the convergence period to equilibrium. Therefore, the time used to calculate the SSE can be calculated from the 2nd to 20th seconds through the Roots-Mean-Square Method (RMSM) as introduced below.

$$\begin{aligned} E_X &= \sqrt{\tfrac{1}{S}\sum_{i=1}^{S}|(X_{ri}-X_i)|^2}; E_Y = \sqrt{\tfrac{1}{S}\sum_{i=1}^{S}|(Y_{ri}-Y_i)|^2}; E_Z = \sqrt{\tfrac{1}{S}\sum_{i=1}^{S}|(Z_{ri}-Z_i)|^2}; \\ E_1 &= \sqrt{\tfrac{1}{S}\sum_{i=1}^{S}|(p_{r1i}-p_{1i})|^2}; E_2 = \sqrt{\tfrac{1}{S}\sum_{i=1}^{S}|(p_{r2i}-p_{2i})|^2}; E_3 = \sqrt{\tfrac{1}{S}\sum_{i=1}^{S}|(p_{r3i}-p_{3i})|^2}, \end{aligned} \tag{33}$$

where S denotes the number of the calculated samples. Roots-Mean-Square Errors (RMSEs) for joint 1, joint 2, and joint 3 are E_1, E_2, and E_3, respectively. RMSEs for X axis,Y axis, and Z axis are E_X E_Y, and E_Z respectively. $[X_i, Y_i, Z_i]^T$ denotes the actual position and $[X_{ri}, Y_{ri}, Z_{ri}]^T$ denotes the reference position at time index i. $[p_{1i}, p_{2i}, p_{3i}]^T$ denotes the actual joint angle and $[p_{r1i}, p_{r2i}, p_{r3i}]^T$ denotes the reference joint angle at time index i.

Figure 9 depicts the trajectory of the effective point of the robot arm separately controlled by four different methods. It is generally possible to control the robotic arm using each of the four methods to complete orbital tracking well. According to Figures 6–8, tracking errors are compared between the real robot trajectory and the reference trajectory at each joint. Based on Figure 10, the end effector's position and the reference trajectory are compared in terms of X-axis, Y-axis, and Z-axis errors. Using RMSE levels for joint errors, X-axis, Y-axis, and Z-axis errors, tracking accuracy was evaluated. The results pointed in Figures 6–8, 10, and Table 5 show that the proposed strategy has obtained the highest tracking accuracy and the smallest steady-state errors. Overall, both controllers including TSMC, and FTSMC have proven their effectiveness in trajectory tracking when they could provide relatively high tracking accuracy. Their SSEs can be within predetermined performance boundaries while the SSEs of the SMC sometimes cross performance boundaries.

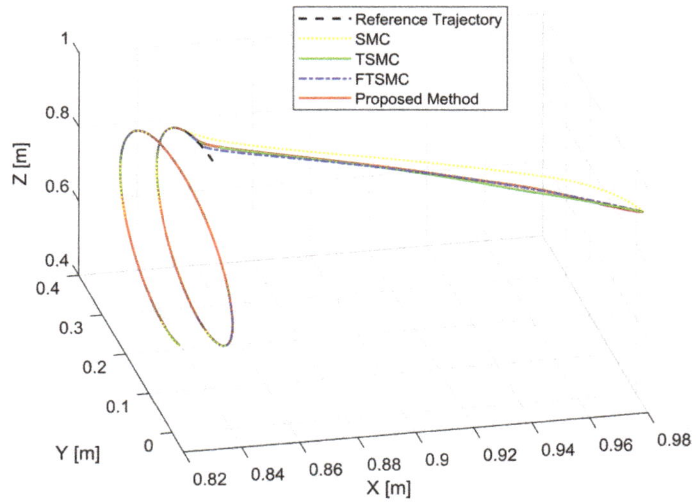

Figure 9. The real trajectories under all controllers versus the desired trajectory.

Table 5. RMSEs via four Control Strategies.

Control System	E_X	E_Y	E_Z	E_1	E_2	E_3
SMC [7]	1.1565×10^{-4}	8.4785×10^{-5}	2.1955×10^{-4}	6.6134×10^{-5}	1.4889×10^{-4}	3.3847×10^{-4}
TSMC [29]	1.4363×10^{-5}	2.4533×10^{-5}	5.8271×10^{-5}	2.5713×10^{-5}	4.6512×10^{-5}	5.1967×10^{-5}
FTSMC [29]	1.3054×10^{-5}	2.2247×10^{-5}	5.2373×10^{-5}	2.3968×10^{-5}	3.9330×10^{-5}	5.0069×10^{-5}
Proposed Controller	1.2158×10^{-7}	2.9631×10^{-7}	2.2370×10^{-7}	3.4814×10^{-7}	2.3686×10^{-7}	1.9566×10^{-7}

Figure 11 shows the control torque provided by the four different control schemes. The proposed scheme achieved smoother control torques for the robot as a result of estimating uncertainty terms from observers and using the ASTwCL for the reaching phase, as well as robustness that allowed it to cope with the effects of uncertain elements and preserve tracking precision despite uncertain components. As a result of the application of a high-frequency reaching control law, the three remaining control schemes produced control torques with harmful chattering phenomena. Although those control schemes still guarantee robustness as well as provide a good level of tracking performance. In reality, chattering may result in arm vibrations, moving parts in actuators, mechanical abrasions, and even heat generation in the controlled systems [13,52]. Therefore, chattering should be removed/reduced its effects.

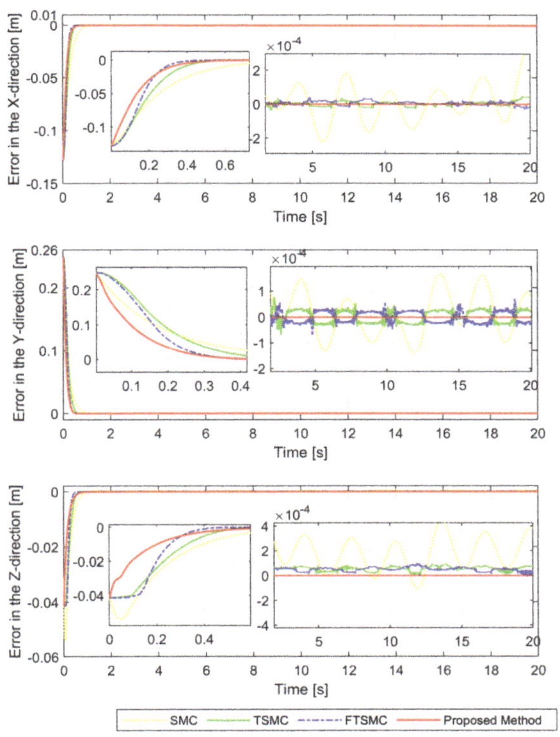

Figure 10. X-axis, Y-axis, and Z-axis error comparisons between the position of the end effector and the reference trajectory.

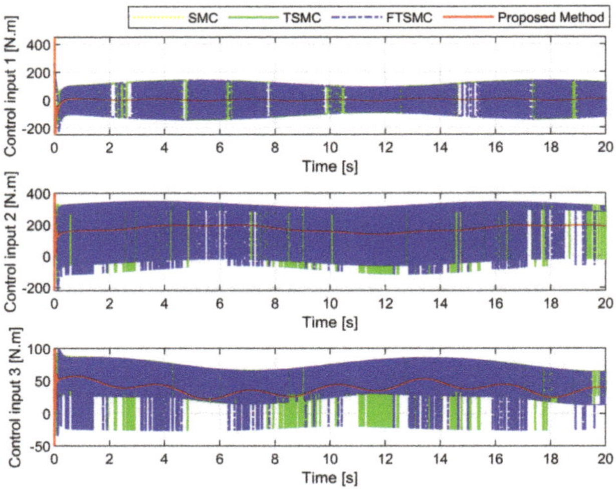

Figure 11. The control torque of the four different strategies.

To prove the universality of the algorithm, the robot manipulator is controlled to follow a different trajectory. This trajectory tracking performance of the robot is presented in Figure 12. Through the obtained simulation results, we observed that they have the results as those of the first example. Therefore, to avoid repeated analysis, we only present briefly the tracking control performance as shown in Figure 12.

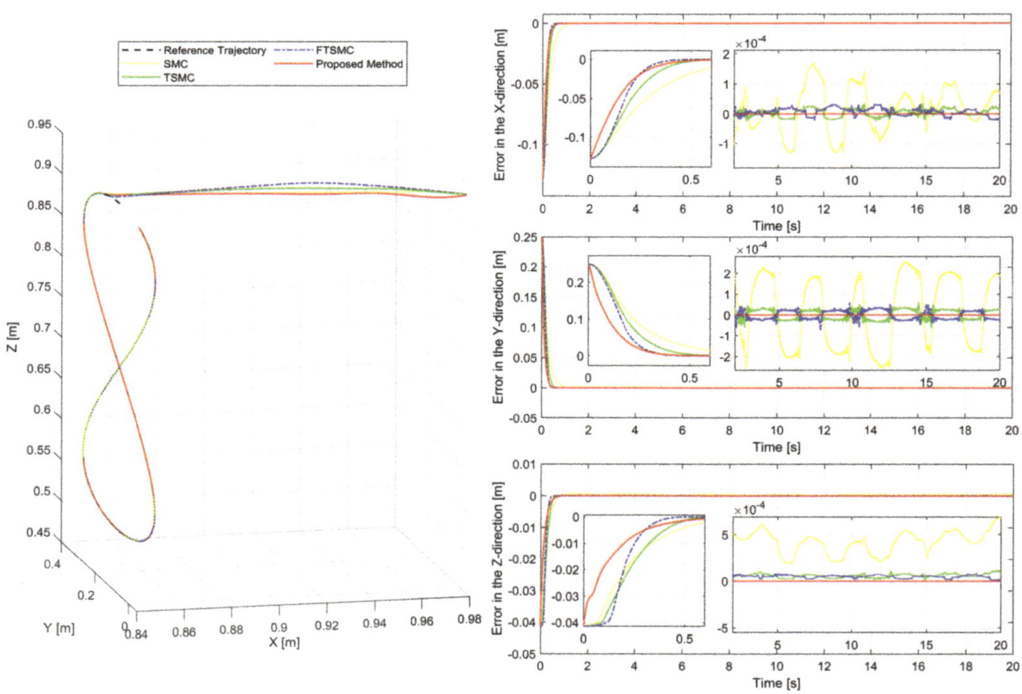

Figure 12. Performance of the control system in tracking another trajectory.

5. Conclusions

The proposed APPTMC with the capability of obtaining prescribed performance has been presented to solve the tracking control problem of robot manipulators under the influence of disturbances and dynamical uncertainties. The modified PPFs have been proposed to manipulate position tracking errors in a pre-designed performance domain. Especially, the SSE boundaries will be symmetrical to zero with the modified PPFs, so when the transformed error is zero, the tracking error will be as well. A new NISMS based on the transformed errors allows knowing the allowable maximum size of the control errors in the steady-state, finite-time convergence speed, and singularity elimination. A fixed-time USOSMO was proposed to directly estimate the lumped uncertainty. The integration of the designed USOSMO, the suggested sliding mode surface based on the transformed errors, and the transformed errors formed an APPTMC for robotic manipulators with global finite-time stability. The developed control solution provided prescribed performance, chattering reduction ability, and robustness in coping with the effects of uncertain elements. The stability of the whole closed-loop system of the tracking control method has been carried out by Lyapunov theory. The effectiveness and robustness of the proposed method have been fully confirmed through numerical simulations.

We examined the robot system in our paper with matched uncertain terms, including dynamic uncertainties, external disturbances, and frictions. Therefore, we plan to extend the consideration of time-varying mismatched as well as time-varying matched uncertainties to robot systems in the future.

Author Contributions: Conceptualization, methodology, validation, writing—original draft preparation, and writing—review and editing, A.T.V.; software, visualization, and resources, T.N.T.; supervision, funding acquisition, and project administration, H.-J.K.; formal analysis, investigation, and data curation, T.N.T. and H.-J.K. All authors have read and agreed to the published version of the manuscript.

Funding: This research was funded by the Ministry of Education (NRF-2019R1D1A3A03103528).

Institutional Review Board Statement: Not applicable.

Informed Consent Statement: Not applicable.

Data Availability Statement: The data sets generated and/or analyzed during the current study are available from the corresponding author on reasonable request.

Acknowledgments: This research was supported by the Basic Science Research Program through the National Research Foundation of Korea (NRF) funded by the Ministry of Education (NRF-2019R1D1A3A03103528).

Conflicts of Interest: The authors declare no conflict of interest.

Abbreviations

The following abbreviations are used in this manuscript:

CTC	Computed Torque Control
ACM	Adaptive Control Method
BsCM	Back-stepping Control Method
SMC	Sliding Mode Control
ISMC	Integral Sliding Mode Control
SSE	Steady-State Error
SOSMC	Second-Order Sliding Mode Control
TSMC	Terminal Sliding Mode Control
NTSMC	Non-singular Terminal Sliding Mode Control
FTSMC	Fast Terminal Sliding Mode Control
FNTSMC	Fast Non-singular Terminal Sliding Mode Control
NISMS	Nonsingular Integral Sliding Mode Surfac

FnTCM	Finite-Time Control Method
FxTCM	Fixed-Time Control Method
DO	Disturbance Observer
FnTDO	Finite-Time Disturbance Observer
FxTDO	Fixed-Time Disturbance Observer
SOSMO	Second-Order Sliding Mode Observer
USOSMO	Uniform Second-Order Sliding Mode Observer
TOSMO	Third-Order Sliding Mode Observer
ASTwCL	Adaptive Super-twisting Control Law
PPC	Prescribed Performance Control
PCP	Prescribed Control Performance
PPF	Prescribed Performance Function
ETF	Error Transformation Function
DOF	Degrees of Freedom
RMSM	Roots-Mean-Square Method
RMSE	Roots-Mean-Square Error
SMO-CM	Sliding Mode Observer-based Control Method
TDE-CM	Time-Delay Estimation-based Control Method
DO-CM	Disturbance Observer-based Control Method
ADRCM	Active Disturbance Rejection Control Method
APPTMC	Adaptive Prescribed Performance Tracking Motion Control

References

1. Appleton, E.; Williams, D.J. *Industrial Robot Applications*; Springer Science & Business Media: New York, NY, USA, 2012.
2. Craig, J.J. *Introduction to Robotics: Mechanics and Control*; Pearson Educacion: Madrid, Spain, 2005.
3. Kelly, R. A tuning procedure for stable PID control of robot manipulators. *Robotica* **1995**, *13*, 141–148. [CrossRef]
4. Pervozvanski, A.A.; Freidovich, L.B. Robust stabilization of robotic manipulators by PID controllers. *Dyn. Control* **1999**, *9*, 203–222. [CrossRef]
5. Cervantes, I.; Alvarez-Ramirez, J. On the PID tracking control of robot manipulators. *Syst. Control Lett.* **2001**, *42*, 37–46. [CrossRef]
6. Parra-Vega, V.; Arimoto, S.; Liu, Y.H.; Hirzinger, G.; Akella, P. Dynamic sliding PID control for tracking of robot manipulators: Theory and experiments. *IEEE Trans. Robot. Autom.* **2003**, *19*, 967–976. [CrossRef]
7. Edwards, C.; Colet, E.F.; Fridman, L.; Colet, E.F.; Fridman, L.M. *Advances in Variable Structure and Sliding Mode Control*; Springer: Berlin/Heidelberg, Germany, 2006; Volume 334.
8. Su, C.Y.; Leung, T.P. A sliding mode controller with bound estimation for robot manipulators. *IEEE Trans. Robot. Autom.* **1993**, *9*, 208–214. [CrossRef]
9. Baek, J.; Jin, M.; Han, S. A new adaptive sliding-mode control scheme for application to robot manipulators. *IEEE Trans. Ind. Electron.* **2016**, *63*, 3628–3637. [CrossRef]
10. Shang, W.; Cong, S. Nonlinear computed torque control for a high-speed planar parallel manipulator. *Mechatronics* **2009**, *19*, 987–992. [CrossRef]
11. Truong, T.N.; Vo, A.T.; Kang, H.J. A backstepping global fast terminal sliding mode control for trajectory tracking control of industrial robotic manipulators. *IEEE Access* **2021**, *9*, 31921–31931. [CrossRef]
12. Slotine, J.J.E.; Li, W. Composite adaptive control of robot manipulators. *Automatica* **1989**, *25*, 509–519. [CrossRef]
13. Utkin, V.; Lee, H. Chattering problem in sliding mode control systems. In Proceedings of the International Workshop on Variable Structure Systems, 2006. VSS'06, Alghero, Sardinia, 5–7 June 2006; pp. 346–350.
14. Nguyen, V.C.; Vo, A.T.; Kang, H.J. A non-singular fast terminal sliding mode control based on third-order sliding mode observer for a class of second-order uncertain nonlinear systems and its application to robot manipulators. *IEEE Access* **2020**, *8*, 78109–78120. [CrossRef]
15. Vo, A.T.; Truong, T.N.; Kang, H.J. A Novel Tracking Control Algorithm With Finite-Time Disturbance Observer for a Class of Second-Order Nonlinear Systems and its Applications. *IEEE Access* **2021**, *9*, 31373–31389. [CrossRef]
16. Pan, H.; Zhang, G.; Ouyang, H.; Mei, L. Novel Fixed-Time Nonsingular Fast Terminal Sliding Mode Control for Second-Order Uncertain Systems Based on Adaptive Disturbance Observer. *IEEE Access* **2020**, *8*, 126615–126627. doi: [CrossRef]
17. Van, M.; Ceglarek, D. Robust fault tolerant control of robot manipulators with global fixed-time convergence. *J. Frankl. Inst.* **2020**, *358*, 699–722. [CrossRef]
18. Rojsiraphisal, T.; Mobayen, S.; Asad, J.H.; Vu, M.T.; Chang, A.; Puangmalai, J. Fast terminal sliding control of underactuated robotic systems based on disturbance observer with experimental validation. *Mathematics* **2021**, *9*, 1935. [CrossRef]
19. Ba, D.X. A Fast Adaptive Time-delay-estimation Sliding Mode Control for Robot Manipulators. *Adv. Sci. Technol. Eng. Syst. J.* **2020**, *5*, 904–911. [CrossRef]

20. Vo, A.T.; Kang, H. A Novel Fault-Tolerant Control Method for Robot Manipulators Based on Non-Singular Fast Terminal Sliding Mode Control and Disturbance Observer. *IEEE Access* **2020**, *8*, 109388–109400. [CrossRef]
21. Huang, Y.; Xue, W. Active disturbance rejection control: Methodology and theoretical analysis. *ISA Trans.* **2014**, *53*, 963–976. [CrossRef]
22. Chang, J.L. Dynamic output integral sliding-mode control with disturbance attenuation. *IEEE Trans. Autom. Control* **2009**, *54*, 2653–2658. [CrossRef]
23. Truong, T.N.; Vo, A.T.; Kang, H.J.; Van, M. A Novel Active Fault-Tolerant Tracking Control for Robot Manipulators with Finite-Time Stability. *Sensors* **2021**, *21*, 8101. [CrossRef]
24. Van, M.; Ge, S.S.; Ren, H. Finite time fault tolerant control for robot manipulators using time delay estimation and continuous nonsingular fast terminal sliding mode control. *IEEE Trans. Cybern.* **2016**, *47*, 1681–1693. [CrossRef]
25. Vo, A.T.; Truong, T.N.; Kang, H.J.; Van, M. A Robust Observer-Based Control Strategy for n-DOF Uncertain Robot Manipulators with Fixed-Time Stability. *Sensors* **2021**, *21*, 7084. [CrossRef] [PubMed]
26. Rubagotti, M.; Estrada, A.; Castaños, F.; Ferrara, A.; Fridman, L. Integral sliding mode control for nonlinear systems with matched and unmatched perturbations. *IEEE Trans. Autom. Control* **2011**, *56*, 2699–2704. [CrossRef]
27. Pan, Y.; Yang, C.; Pan, L.; Yu, H. Integral sliding mode control: Performance, modification, and improvement. *IEEE Trans. Ind. Inform.* **2017**, *14*, 3087–3096. [CrossRef]
28. Mu, C.; He, H. Dynamic behavior of terminal sliding mode control. *IEEE Trans. Ind. Electron.* **2017**, *65*, 3480–3490. [CrossRef]
29. Yu, S.; Yu, X.; Shirinzadeh, B.; Man, Z. Continuous finite-time control for robotic manipulators with terminal sliding mode. *Automatica* **2005**, *41*, 1957–1964. doi: [CrossRef]
30. Rabiee, H.; Ataei, M.; Ekramian, M. Continuous nonsingular terminal sliding mode control based on adaptive sliding mode disturbance observer for uncertain nonlinear systems. *Automatica* **2019**, *109*, 108515. [CrossRef]
31. Vo, A.T.; Kang, H.J. An Adaptive Terminal Sliding Mode Control for Robot Manipulators With Non-Singular Terminal Sliding Surface Variables. *IEEE Access* **2019**, *7*, 8701–8712. [CrossRef]
32. Doan, Q.V.; Vo, A.T.; Le, T.D.; Kang, H.J.; Nguyen, N.H.A. A novel fast terminal sliding mode tracking control methodology for robot manipulators. *Appl. Sci.* **2020**, *10*, 3010. [CrossRef]
33. Labbadi, M.; Cherkaoui, M. Robust adaptive backstepping fast terminal sliding mode controller for uncertain quadrotor UAV. *Aerosp. Sci. Technol.* **2019**, *93*, 105306. [CrossRef]
34. Nojavanzadeh, D.; Badamchizadeh, M. Adaptive fractional-order non-singular fast terminal sliding mode control for robot manipulators. *IET Control Theory Appl.* **2016**, *10*, 1565–1572. [CrossRef]
35. Van, M. An enhanced robust fault tolerant control based on an adaptive fuzzy PID-nonsingular fast terminal sliding mode control for uncertain nonlinear systems. *IEEE/ASME Trans. Mechatron.* **2018**, *23*, 1362–1371. [CrossRef]
36. Van, M.; Do, V.T.; Khyam, M.O.; Do, X.P. Tracking control of uncertain surface vessels with global finite-time convergence. *Ocean Eng.* **2021**, *241*, 109974. [CrossRef]
37. Cruz-Zavala, E.; Moreno, J.A.; Fridman, L.M. Uniform robust exact differentiator. *IEEE Trans. Autom. Control* **2011**, *56*, 2727–2733. [CrossRef]
38. Wang, Y.; Chen, M. Fixed-time Disturbance Observer-based Sliding Mode Control for Mismatched Uncertain Systems. *Int. J. Control Autom. Syst.* **2022**, *20*,2792–2804. [CrossRef]
39. Nguyen, V.C.; Vo, A.T.; Kang, H.J. A finite-time fault-tolerant control using non-singular fast terminal sliding mode control and third-order sliding mode observer for robotic manipulators. *IEEE Access* **2021**, *9*, 31225–31235. [CrossRef]
40. Van, M.; Franciosa, P.; Ceglarek, D. Fault diagnosis and fault-tolerant control of uncertain robot manipulators using high-order sliding mode. *Math. Probl. Eng.* **2016**, *2016*, 7926280. [CrossRef]
41. Bechlioulis, C.P.; Rovithakis, G.A. Robust adaptive control of feedback linearizable MIMO nonlinear systems with prescribed performance. *IEEE Trans. Autom. Control* **2008**, *53*, 2090–2099. [CrossRef]
42. Tran, X.T.; Oh, H. Prescribed performance adaptive finite-time control for uncertain horizontal platform systems. *ISA Trans.* **2020**, *103*, 122–130. [CrossRef]
43. Huang, H.; He, W.; Li, J.; Xu, B.; Yang, C.; Zhang, W. Disturbance observer-based fault-tolerant control for robotic systems with guaranteed prescribed performance. *IEEE Trans. Cybern.* **2020**, *52*, 772–783. [CrossRef]
44. Yang, P.; Su, Y. Proximate fixed-time prescribed performance tracking control of uncertain robot manipulators. *IEEE/ASME Trans. Mechatron.* **2021**, *27*, 3275–3285. [CrossRef]
45. Wang, S.; Na, J.; Chen, Q. Adaptive predefined performance sliding mode control of motor driving systems with disturbances. *IEEE Trans. Energy Convers.* **2020**, *36*, 1931–1939. [CrossRef]
46. Ding, Y.; Wang, Y.; Chen, B. Fault-tolerant control of an aerial manipulator with guaranteed tracking performance. *Int. J. Robust Nonlinear Control* **2022**, *32*, 960–986. [CrossRef]
47. Zuo, Z. Non-singular fixed-time terminal sliding mode control of non-linear systems. *IET Control Theory Appl.* **2015**, *9*, 545–552. [CrossRef]
48. Ding, S.; Levant, A.; Li, S. Simple homogeneous sliding-mode controller. *Automatica* **2016**, *67*, 22–32. [CrossRef]
49. Laghrouche, S.; Liu, J.; Ahmed, F.S.; Harmouche, M.; Wack, M. Adaptive second-order sliding mode observer-based fault reconstruction for PEM fuel cell air-feed system. *IEEE Trans. Control Syst. Technol.* **2014**, *23*, 1098–1109. [CrossRef]
50. Levant, A. Higher-order sliding modes, differentiation and output-feedback control. *Int. J. Control* **2003**, *76*, 924–941. [CrossRef]

51. Niku, S.B. *Introduction to Robotics: Analysis, Control, Applications*; John Wiley & Sons: Hoboken, NJ, USA, 2020.
52. Levant, A. Chattering analysis. *IEEE Trans. Autom. Control* **2010**, *55*, 1380–1389. [CrossRef]

sensors

MDPI

Article

Visual Servoing Approach to Autonomous UAV Landing on a Moving Vehicle

Azarakhsh Keipour [1,*,†], Guilherme A. S. Pereira [2], Rogerio Bonatti [3], Rohit Garg [4], Puru Rastogi [5], Geetesh Dubey [4] and Sebastian Scherer [6]

1 Robotics AI, Amazon, Arlington, VA 22202, USA
2 Department of Mechanical and Aerospace Engineering, West Virginia University, Morgantown, WV 26506, USA
3 Autonomous Systems Group, Microsoft, Redmond, WA 98052, USA
4 Gather AI, Pittsburgh, PA 15217, USA
5 Mowito, Bengaluru 560091, India
6 Robotics Institute, Carnegie Mellon University, Pittsburgh, PA 15213, USA
* Correspondence: keipour@gmail.com; Tel.: +1-412-273-2712
† The publication was written prior to A. Keipour joining Amazon. During the realization of this work, he was affiliated with the Robotics Institute at Carnegie Mellon University.

Abstract: Many aerial robotic applications require the ability to land on moving platforms, such as delivery trucks and marine research boats. We present a method to autonomously land an Unmanned Aerial Vehicle on a moving vehicle. A visual servoing controller approaches the ground vehicle using velocity commands calculated directly in image space. The control laws generate velocity commands in all three dimensions, eliminating the need for a separate height controller. The method has shown the ability to approach and land on the moving deck in simulation, indoor and outdoor environments, and compared to the other available methods, it has provided the fastest landing approach. Unlike many existing methods for landing on fast-moving platforms, this method does not rely on additional external setups, such as RTK, motion capture system, ground station, offboard processing, or communication with the vehicle, and it requires only the minimal set of hardware and localization sensors. The videos and source codes are also provided.

Keywords: visual servoing; autonomous landing; monocular vision; aerial robotics

Citation: Keipour, A.; Pereira, G.A.S.; Bonatti, R.; Garg, R.; Rastogi, P.; Dubey, G.; Scherer, S. Visual Servoing Approach to Autonomous UAV Landing on a Moving Vehicle. *Sensors* **2022**, *22*, 6549. https://doi.org/10.3390/s22176549

Academic Editors: Gregor Klancar, Marija Seder and Sašo Blažič

Received: 17 July 2022
Accepted: 25 August 2022
Published: 30 August 2022

Publisher's Note: MDPI stays neutral with regard to jurisdictional claims in published maps and institutional affiliations.

1. Introduction

The recent advances in Unmanned Aerial Vehicles (UAVs) have allowed innovative applications ranging from package delivery to infrastructure inspection, early fire detection, and cinematography [1–3]. While many of these applications require landing on the ground and static platforms, the ability to land on dynamic platforms is essential for some other applications. A typical example of a real-world scenario is an autonomous landing on a ship deck [4] or maritime Search and Rescue operations [5].

The autonomous landing of Unmanned Aerial Vehicles (UAVs) on known patterns has been an active area of research for several years [6–12]. Some of the key challenges of the problem include dealing with environmental conditions, such as changes in light and wind, and robust detection of the landing zone. The subsequent maneuver in trying to land also needs to take care of the potential ground effects at the proximity of the landing surface.

The Mohamed Bin Zayed International Robotics Challenge (MBZIRC) is a set of real-world robotics challenges happening every few years [13]. Challenge 1 took place in March 2017, focusing on landing UAVs on a moving platform. In this challenge, there was a ground vehicle (truck) moving on an 8-shaped road in a 90 × 60 m arena with a predefined speed of 15 km/h (4.17 m/s). On top of the truck, at 1.5 m height, there was a flat horizontal ferromagnetic deck with a predefined 1.5 × 1.5 m pattern (Figure 1) printed on top of it [14].

The goal was to land the UAV on this deck autonomously. To make the challenge realistic, no communication between the UAV and the ground vehicle was allowed, no precise state estimation sensors such as RTK or Motion Capture were provided, and finally, the location of the vehicle was not given to the UAV, requiring a visual detection of the pattern, which could result in false and imperfect detections among other issues.

Figure 1. Landing zone pattern utilized at the MBZIRC competition [14].

This paper describes our approach to landing on moving platforms, which uses only a single monocular camera, a point Lidar, and an onboard computer. The method is based on a visual-servoing controller and has been successfully tested for landing on moving ground vehicles at speeds of up to 15 km/h (4.17 m/s).

Our contributions include proposing a method that does not depend on a special environment setup (e.g., IR markers and communication channels), can work with the minimal UAV sensor setup, only requires a monocular camera, has minimum processing requirements (e.g., a lightweight onboard computer), and approaches the landing platform with high certainty. We also provide our source codes and simulation environment to assist with future developments. The videos of our experiments and all the codes are available at http://theairlab.org/landing-on-vehicle.

The paper is organized as follows: Section 2 reviews the relevant work and state of the art; Section 3 explains our method for autonomous landing; Section 4 describes the architecture and the experiments performed and discusses the results from simulation, indoor and outdoor tests.

2. Background and Relevant Research

A review of the existing research reveals several approaches to solving this problem. One of the earliest vision-based approaches introduced an image moment-based method to land a scale helicopter on a static landing pad that has a distinct geometric shape [15]. This work further extends to the helicopter landing on a moving target [6]. However, a limitation of the approach is that it tracks the target in a single dimension, and the image data are processed offboard.

A more recent work tries to land a quadcopter on top of a ground vehicle traveling at high speed using an AprilTag fiducial marker for landing pad detection [16]. They devise a combination of Proportional Navigation (PN) guidance and Proportional-Derivative (PD) control to execute the task. The UAV has landed on a target moving at up to 50 km/h (13.89 m/s). However, the PN guidance strategy used to approach the target has to rely on the wireless transmission of GPS and IMU measurements from the ground vehicle to the UAV, which is impractical in most applications. In addition, two cameras are used for the final approach to detect the AprilTag from far and near.

Visual servoing control is another approach that can be applied to this problem. In [17], it is used to generate a velocity reference command to the lower level controller of a quadrotor, guiding the vehicle to a target landing pad moving at 0.07 m/s. It is computationally cheaper to compute the control signals in image space than in 3D space. However, this work relies on offboard computing as well as a VICON motion capture system for accurate position feedback during its patrol to search for the target. Another closed-loop approach for landing was proposed in [18]. In this paper, the authors presented a landing vector field. Unlike visual servoing, the method's main advantage is that the vector field can enforce the shape of the vehicle's trajectory during landing. In addition, different from visual servoing, the 3D localization of the UAV is necessary.

Some other works emphasize the design of a landing pad that is robust to detection, which simplifies the task by eliminating the detection uncertainty [19–21]. In [7], a Wii (IR) camera is used to track a T-shaped 3D pattern of infrared lights mounted on the vehicle, and, in [22], an IR beacon is placed on the landing pad. Reliable detection enables a ground-based system to estimate the position of the UAV relative to the landing target. Xing et al. [21] propose a notched ring with a square landmark inside to enable robust detection of the landing zone, then it uses a minimum-jerk trajectory to land on the detected pattern. Finally, the method presented in [23] localizes the vehicle using two nested ArUco markers [24] in an illuminated landing pad, allowing landing during the night using visual servoing.

Optical flow is another technique used to provide visual feedback for guiding the UAV. Ruffier and Franceschini [25] developed an autopilot that uses a ventral optic flow regulator in one of its feedback loops for controlling lift (or altitude). The UAV can land on a platform that moves along two axes. The method in [26] also uses the optical flow information obtained from a textured landing target but does not attempt to reconstruct velocity or distance to the goal. Instead, the approach chooses to control the UAV within the image-based paradigm.

The method in [27] transfers all the computing power to the ground station eliminating the need for an onboard computer. They only report results for very slow vehicle speeds (10–15 cm/s), and the method cannot extend to higher speeds due to communication delays.

Several approaches were introduced for the MBZIRC challenge. The method used by [10] tries to follow the platform until the pattern detection rate is high enough for landing. Then, it continues to follow the platform on the horizontal plane while slowly decreasing the altitude of the UAV. A Lidar (Light Detection and Ranging sensor) is used to determine if the UAV should land on the platform or not. The landing is performed by a fast descent followed by switching off the motors. The state estimation uses monocular VI-Sensor data fused with GPS, IMU, and RTK data.

Researchers from the University of Catania implemented a system that detects and tracks the target pattern using a Tracking-Learning-Detection-based method integrated with the Circle Hough Transform to find the precise location of the landing zone. Then, a Kalman filter is used to estimate the vehicle's trajectory for the UAV to follow and approach it for landing [28,29].

Baca et al. [30] took advantage of a SuperFisheye high-resolution monocular camera with a high FPS rate for pattern detection. They used adaptive thresholding to improve the method's robustness to the light intensity, followed by undistorting the image. Then, the circle and the cross are detected to find the landing pattern in the frame. After applying a Kalman filter on the detected coordinates to estimate and predict the location of the ground vehicle, a model-predictive controller (MPC) is devised to generate the reference trajectory for the UAV in real-time, which is tracked by a nonlinear feedback controller to approach and land on the ground vehicle.

Researchers at the University of Bonn have achieved the fastest approach for landing among the other successful runs in the MBZIRC competition (measured from the time of detection to the successful landing). They used two cameras for the landing pattern detection, a Nonlinear Model Predictive Control (MPC) for time-optimal trajectory generation and control, and a separate proportional controller for yaw [11].

Tzoumanikas et al. [31] used an RGB-D camera for visual-inertial state estimation. After the initial detection, the UAV flies in the vehicle's direction until it reaches the velocity and position close to the ground vehicle. Then, the UAV starts descending toward the target until a certain altitude, when it will continue descending in an open-loop manner. During the landing, the UAV uses pattern detection feedback only to decide between aborting the mission or not and not for correcting controller commands.

University of Zurich researchers have introduced a system in [32] that first finds the quadrangle of the pattern and then searches for the ellipse or the cross, validating the detection using RANSAC. Then, the platform's position is estimated from its relative position to the UAV, and an optimal landing trajectory is generated to land the UAV on the vehicle.

An ultimate goal in robotics' real-world applications is to achieve good results while reducing the robot's costs and the amount of the prior setup. Dependence on the additional hardware or external hardware (e.g., motion capture systems, RTK, multiple cameras) is costly and not always feasible in real-world applications. In contrast with other available methods, our approach uses only a single monocular camera (with comparatively low resolution and low frame rate), a point Lidar, and an onboard computer to land on the moving ground vehicle at high speeds (tested at 15 km/h (4.17 m/s)) using a visual-servoing controller. The approach does not depend on accurate localization sensors (e.g., RTK or motion capture systems) and can work with the state estimation provided by a commercial UAV platform (obtained only using GPS, IMU, and barometer data). Our real-time elliptic pattern detection and tracking method can track the landing deck in challenging environment conditions (e.g., changes in lighting) with a high frequency [33].

3. Materials and Methods

This section explains our approach to sensing the vehicle passing below the UAV and then landing the quadrotor on the vehicle. Our system was developed in the MBZIRC Challenge 1 context, discussed in Section 1. Therefore, our primary goal was to land on a known platform (see Figure 1) fixed on top of a vehicle moving at a fixed speed.

3.1. General Strategy

In our approach, we assume that the drone is hovering above the point where the moving vehicle is expected to pass (which can be any point on the road if the vehicle is in a loop). In addition, we assume that the direction and the speed of the moving vehicle are known, which can be previously obtained, for example, from some consecutive deck pattern detections. We further assume that the vehicle is almost moving in a straight line during the few seconds when the UAV attempts to land.

A time-optimal trajectory to landing is theoretically possible, while in practice, there is a variation in the speed of the moving vehicle (which is driven by a human driver), and there are delays and errors in state estimation. Therefore, an approach with constant feedback to correct the trajectory would work better than an open-loop landing using an optimal trajectory. The method we chose is to continually measure the target's position, size, and orientation directly in the image space and translate it to velocity commands for the UAV using visual servoing. This approach is described in Section 3.3.

Given the assumptions mentioned above, the defined strategy is as follows:

1. The quadrotor hovers at the point at the height of h meters until it senses the passing ground vehicle.
2. As soon as the ground vehicle is sensed, the UAV starts flying in the direction of movement with a predefined speed slightly higher than the vehicle speed to compensate for the distance between the UAV and the vehicle. This gap results from the delays in processing the sensors and dynamics of the UAV in getting up to the vehicle speed.
3. After the initial acceleration, the UAV flies with a feed-forwarding speed equal to the ground vehicle speed. Then, the visual servoing controller tries to decrease the remaining gap between the UAV and the target.
4. After successfully landing, magnets at the bottom of the UAV legs stick to the metal platform, and the propellers are shut down.

Figure 2 depicts the steps of the described strategy. The following subsections explain the main components of the landing strategy in greater detail.

We used a path tracker similar to the one proposed by [34] to fly to the hover point and to fly the straight lines.

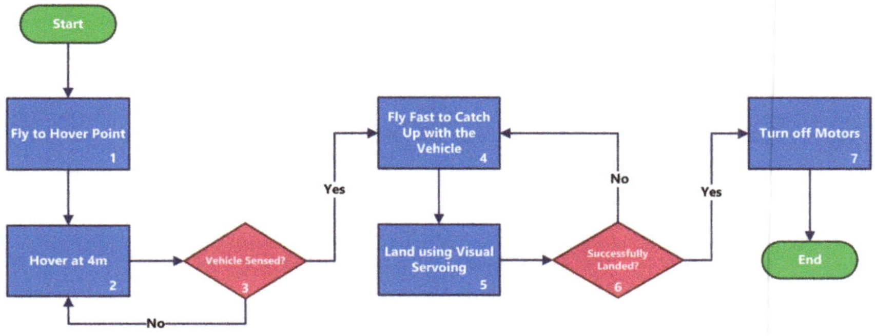

Figure 2. Our strategy for landing on a moving vehicle. The vehicle hovers above the road until it senses the passing vehicle. Then, it accelerates to catch up with the vehicle and tries to land using the visual servoing method. As soon as the landing is detected, all the motors are turned off.

3.2. Sensing the Passing Vehicle

We considered two strategies for detecting the passing vehicle below the UAV when it is hovering above the road on the vehicle's path: the camera or using laser sensors. Due to the reliability of our pattern detection method, using the camera gives more reliable results. The chosen hovering altitude can range from a few to tens of meters, where the pattern detection is reliable. The lower altitude allows for a faster approach to landing, while the higher altitude results in more flight time required to catch up with the vehicle (box 4 in Figure 2).

The strategy of sensing the vehicle using a camera worked well in our tests; however, for higher vehicle speeds, we developed a laser-based solution to avoid the delays introduced by the image processing and to get an even lower reaction time. We used three point-laser sensors on the bottom of the UAV (as described in Section 4.1) to detect the vehicle passing below it. If the distance measured by any of the three lasers has a sudden drop of more than a certain threshold within a specified period, we assume that the vehicle is passing below the quadrotor, and the landing system is triggered.

The choice of the number of the lasers and the ideal height of the hovering depend on the road's width and the vehicle's width and height. For our tests, the road was 3 m wide, the vehicle was 1.5 m wide, and its height was 1.5 m. In order to capture this vehicle, the center laser points directly downwards, and the other two lasers are angled at approximately 30 degrees on each side. With this configuration, at the altitude of 4 m, two additional laser rays will hit the ground at a distance of 1.44 m from the center laser. When the moving vehicle (with a width and height of 1.5 m) passes on the road below the quadrotor, at least one of the three lasers will sense the change in the measured height. Algorithm 1 illustrates the method.

Algorithm 1: Approach for sensing the passing vehicle using three point lasers.

```
function DetectPassingVehicle
    ▷ Keep reading the sensors until the vehicle is sensed
    while true do
        ▷ Read the laser measurements
        dist_l, dist_c, dist_r ← ReadLasers()
        ▷ Check if we have a sudden decrease in the measured distance
        ▷ thr is set to a large number smaller than the vehicle's height
        if (last_l - dist_l) > thr or (last_c - dist_c) > thr or (last_r - dist_r) > thr then
            return true
        else
            last_l ← dist_l, last_c ← dist_c, last_r ← dist_r
        end if
    end while
end function
```

Figure 3 shows examples of a laser triggering after sudden measurement changes.

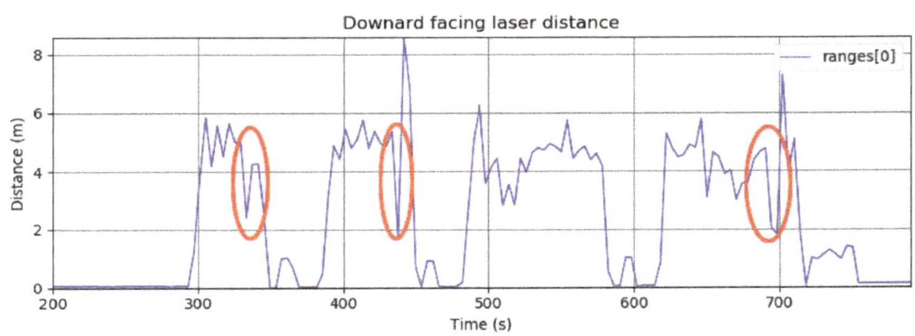

Figure 3. Plot of laser distance measurements during a flight in our trials. The passing vehicle is detected in several moments, indicated by the red ellipses. Detection by this laser fails in the third trial due to vehicle misalignment with the laser direction.

3.3. *Landing Using Visual Servoing*

After the vehicle is detected (using the lasers or the visual pattern detection), the next task is trying to land on the defined pattern on the deck of the moving ground vehicle. In our strategy, the UAV follows and lands the moving deck using visual servoing. Visual servoing consists of techniques that use information extracted from visual data to control the robot's motion. Assuming that the camera is attached to the drone through a gimbal, the vision configuration is called an eye-in-hand configuration. We used an Image-Based Visual Servoing scheme (IBVS), in which the error signal is estimated directly based on the 2D features of the target, which we considered to be the center of the target ellipse and the corners of its circumscribed rectangle. The error is then computed as the difference (in pixels) between the features' positions when the UAV is right above the target, at the height of about 50 cm, and the current features' positions given by the deck detection algorithm (discussed in Section 3.6). Since the robot operates in task space coordinates, there must be a mapping between the changes in image feature parameters and the robot's position. This mapping is applied using the image Jacobian, also known as the Interaction Matrix [35].

3.3.1. Interaction Matrix

If τ represents the task space and F represents the feature parameter space, then the image Jacobian L_v is the transformation from the tangent space of τ at r to the tangent space of F at f, where r and f are vectors in task space and feature parameter space, respectively. This relationship could be represented as:

$$\dot{f} = L_v(r)\dot{r}, \tag{1}$$

which implies

$$L_v(r) = \left[\frac{\partial f}{\partial r}\right]. \tag{2}$$

In practice, $\dot{r}$ needs to be calculated given the value of $\dot{f}$ and the interaction matrix at r. Depending on the size and rank of the interaction matrix, different approaches can be used to calculate the inverse or pseudo-inverse of the interaction matrix, which gives the required $\dot{r}$ [36].

3.3.2. Calculating Quadrotor's Velocity

The developed visual servoing controller for this project assumes the presence of five 2D image features: the center of the detected ellipse $\mathbf{s}_c$ and the four corners of the rectangle that circumscribes the detected elliptic target. Since the gimbal has faster dynamics than the quadrotor, two separate controllers were developed. The first controller is a simple proportional controller for the gimbal pitch angle to keep the deck in the center of the image:

$$\delta = k\|\mathbf{s}_c^\star - \mathbf{s}_c\|, \tag{3}$$

where k is a positive gain, $\mathbf{s}_c^\star$ is the desired location of the deck center in the image, and $\|\cdot\|$ stands for the Euclidean distance. Since the gimbal is not mounted at the center of the quadrotor, position $\mathbf{s}_c^\star$ has an offset to compensate for this.

The second control problem, which relies on image features $\mathbf{s} = [x, y]^T$, is a traditional visual servoing problem, which is solved using the following control law:

$$\mathbf{v} = -\lambda\left(\mathbf{L}_s{}^c\mathbf{V}_b{}^b\mathbf{J}_b\right)^+(\mathbf{s}^\star - \mathbf{s}) + \mathbf{v}_{\text{ff}} \tag{4}$$

where $\mathbf{v}$ is the quadrotor velocity vector in the body control frame, λ is a positive gain, $\mathbf{L}_s$ is the interaction matrix, ${}^c\mathbf{V}_b$ is a matrix that transforms velocities from the camera frame to the body frame, ${}^b\mathbf{J}_b$ is the robot Jacobian, $()^+$ is the pseudo-inverse operator, $(\mathbf{s}^\star - \mathbf{s})$ is the error computed in the feature space and $\mathbf{v}_{\text{ff}}$ is a feed-forward term obtained by transforming the truck velocity to the body reference frame [37]. In our case, the error $(\mathbf{s}^\star - \mathbf{s})$ is minimized using only linear velocities and yaw rate. Therefore, $\mathbf{v} = [v_x, v_y, v_z, \omega]^T$, which forces the Jacobian matrix to be written as:

$${}^b\mathbf{J}_b = \begin{bmatrix} 1 & 0 & 0 & 0 \\ 0 & 1 & 0 & 0 \\ 0 & 0 & 1 & 0 \\ 0 & 0 & 0 & 0 \\ 0 & 0 & 0 & 0 \\ 0 & 0 & 0 & 1 \end{bmatrix}. \tag{5}$$

This matrix is responsible for the transformation from the general six-dimensional velocity space to a four-dimensional space that the UAV can indeed follow.

The interaction matrix $\mathbf{L}_s(\mathbf{s}^\star, Z^\star)$ in our method is constant, and is computed at the target location using image features $(\mathbf{s}^\star = [x^\star, y^\star]^T)$ and the distance of the camera to the deck $(Z^\star)$ as [35]:

$$\mathbf{L}_s = \begin{bmatrix} -1/Z^\star & 0 & x^\star/Z^\star & x^\star y^\star & -(1 + x^{\star 2}) & y^\star \\ 0 & -1/Z^\star & y^\star/Z^\star & 1 - y^{\star 2} & -x^\star y^\star & -x^\star \end{bmatrix}. \tag{6}$$

It is essential to mention that, ideally, the interaction matrix should be computed online using the current features and height information. However, in our approach, we use a constant matrix to be more robust regarding the errors related to the UAV position estimation and information necessary to compute Z. For the same reason, the distance from the deck is not used as part of the feature, as is done in some visual servoing approaches [37]. In addition, a constant matrix causes the system to be less sensitive to eventual spurious detection of the target. On the other hand, since the camera is mounted on a gimbal, matrix ${}^c\mathbf{V}_b$ must be computed online as:

$$ {}^c\mathbf{V}_b = \begin{bmatrix} {}^c\mathbf{R}_b & [{}^c\mathbf{t}_b]_\times {}^c\mathbf{R}_b \\ \mathbf{0}_{3\times 3} & {}^c\mathbf{R}_b \end{bmatrix} \tag{7} $$

where ${}^c\mathbf{R}_b$ is the rotation matrix between the image and the robot body, which is a function of the gimbal angles, ${}^c\mathbf{t}_b$ is the corresponding constant translation vector, and $[\mathbf{t}]_\times$ is the skew-symmetric matrix related to $\mathbf{t}$.

Since we are dealing with a moving deck, the feed-forward term $\mathbf{v}_{\text{ff}}$ in Equation (4) is necessary to reduce the error $(\mathbf{s}^\star - \mathbf{s})$ to zero [38]. In our case, $\mathbf{v}_{\text{ff}}$ originates from the ground vehicle velocity vector, which is a vector tangent to the moving vehicle's path. Thus, the precise estimation of this vector would require the localization of the vehicle with respect to the track, which is a challenging task. To make such an estimation more robust to sensor noise, we relaxed this problem to compute $\mathbf{v}_{\text{ff}}$ only on the straight line segments of the track, where we assume the ground vehicle speed vector to be constant for several seconds. This restriction causes our UAV to land only on such segments.

All the visual servoing control laws in Equations (1)–(7) are implemented using the Visual Servoing Platform (ViSP) [37] version 3.0.1, which provides a suitable data structure for the problem and efficiently computes the required pseudo-inverse matrix.

3.4. *Initial Quadrotor Acceleration*

As described in Section 3.1, the quadrotor will hover until it senses the ground vehicle passing below it. Due to the relatively slow dynamics of the quadrotor controller, it takes some time to accelerate to the ground vehicle's speed. By this time, the deck is already out of the camera's sight, and the visual servoing procedure cannot be used for landing.

To compensate for the delay in reaching the ground vehicle's speed and to gain sight of the deck again, an initial period of high acceleration is set right after sensing the passing ground vehicle. In this period, the desired speed is set to a value higher than the vehicle's speed until the quadrotor compensates for the created gap with the vehicle. After this time, the quadrotor switches to the visual servoing procedure described in Section 3.3 with a feed-forward velocity lowered to the current ground vehicle's speed.

We tested two different criteria for switching from the higher-speed flight to the visual servoing (box 5 in Figure 2) procedure: vision-based and timing-based. In the vision-based approach, the switch to visual servoing happens when the pattern on the moving vehicle is detected again. In the timing-based approach, depending on the processing and dynamics delays of the sensing and considering the known speed of the vehicle and the set speed for the UAV, it is possible to calculate a fixed time t_a that is enough for catching up with the vehicle. After this time, the UAV can start the visual servoing landing.

3.5. *Final Steps and Landing*

When the UAV gets too close to the ground vehicle, the camera can no longer see the pattern. To prevent the UAV from stopping when the target detector (described in Section 3.6) is unable to detect the deck during landing, the UAV is still commanded to move with the known vehicle's speed for a few sampling periods. If using one of its point lasers, the UAV detects that it has the correct height to land, it increases its downward vertical velocity and activates the drone landing procedure, which shuts down the propellers.

Note that the "blind" approach can only work due to the low delays between losing the pattern and landing (generally 30–70 ms). This approach may fail if the ground vehicle aggressively changes its direction or velocity. However, for many applications, the vehicle's motion (e.g., a ship or a delivery truck) can be reasonably assumed stable for such a short time.

3.6. Detection and Tracking of the Deck

Figure 1 shows the target pattern on top of the deck. There are some challenges in the detection of this pattern from the camera frames, including:

- This algorithm is used for landing the quadrotor on a moving vehicle. Therefore, it should work online (with a frequency greater than 10 Hz) on a resource-limited onboard computer.
- The shape details cannot be seen in the video frames when the quadrotor is flying far from the deck.
- The shape of the target is transformed by a projective distortion, which occurs when the shape is seen from different points of view.
- There is a wide range of illumination conditions (e.g., cloudy, sunny, morning, evening).
- Due to the reflection of the light (e.g., from sources like the sun or bulbs), the target may not always be seen in all the frames, even when the camera is close to the vehicle.
- In some frames, there may be shadows on the target shape (e.g., the shadow of the quadrotor or trees).
- In some frames, only a part of the target shape may be seen.

Considering these challenges and the shape of the pattern, different participants of the MBZIRC challenge chose different approaches for detection. For example, in [10], the authors implemented a quadrilateral detector for detecting the pattern from far distances and a cross detector for close-range situations. The method in [11] uses the line and circular Hough-transform algorithms to calculate a confidence score for the pattern for the initial detection and then uses the rectangular area around the pattern for the tracking. In [39], a Convolutional Neural Network is developed to detect the elliptic pattern, which was trained with over 5000 images collected from the pattern moving at a maximum of 15 km/h (4.17 m/s) at various heights. In [40] the cross and the circle are detected for far images, and only cross detection is used for the closer frames. The method uses the [41] method for ellipse detection. In [31], the outer square is detected first, and then the detection is verified by a template matching algorithm.

In our work, we developed a novel method to overcome the problem challenges mentioned above, which detects and tracks the pattern by exploiting the structural properties of the shape without the need for any training [33]. More specifically, the deck detector system detects and tracks the circular shape (seen as an ellipse due to the projective transformation) on top of the deck, while ignoring the cross in the middle and then verifying that the detected ellipse belongs to the deck pattern. The developed real-time ellipse detection method can also detect the ellipses with partial occlusion or the ellipses that are exceeding the image boundaries.

Figure 4 shows results for the detection of the deck in some sample frames. A more detailed description of the methods, database, and results are provided in [33].

(a) (b)

Figure 4. *Cont.*

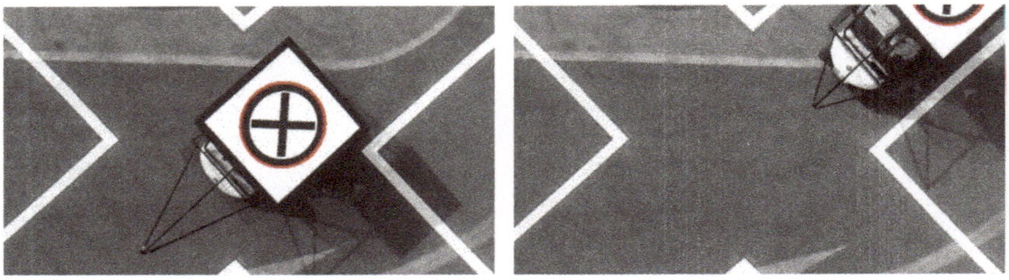

(c) (d)

Figure 4. (**a–d**) Result of the deck detection and tracking algorithm on sample frames. The red ellipse indicates the detected pattern on the deck of the moving vehicle.

4. Experiments and Results

This section outlines the architecture of our proposed landing method from both hardware and software perspectives and presents the results obtained from the tests in different conditions: simulation, indoor testing, and outdoor testing. Some videos of experiments and sequences shown in this section are available at http://theairlab.org/landing-on-vehicle.

4.1. Hardware

The aerial platform chosen for this project is a DJI Matrice 100 quadrotor. This UAV is programmable, can carry additional sensors, is fast, provides velocity commands, and is accurate enough for this problem. Using an off-the-shelf vs. home-built platform vastly increased the speed of the development, reducing the time needed for dealing with hardware bugs. The selected platform has a GPS module for state estimation and is additionally equipped with an ARM-based DJI Manifold onboard computer (based on NVIDIA Jetson TK1 computer), a DJI Zenmuse X3 Gimbal and Camera system, a set of three SF30 altimeters, and four permanent magnets at the bottom of legs. One altimeter is pointed straight down to measure the current height of the flight with respect to the ground. The other two altimeters are mounted on the two sides of the quadrotor, pointing down with approximately 30 degrees outward skew to measure the sudden changes in the height on the sides of the quadrotor for the detection of the truck passing next to the quadrotor. The camera is mounted on a three-axis gimbal and outputs grayscale images at a frequency of 30 Hz with a resolution of 640×360 pixels.

A wireless adapter allows communication between the robot computer and the ground station for development and monitoring. For indoor tests, a set of propeller guards and a DJI Guidance System, a vision-based system able to provide indoor localization, velocity estimation, and obstacle avoidance, are used to provide velocity estimation and improve the safety of the testing. The guards and the Guidance system are removed for outdoor tests to reduce the weight and increase the robustness to the wind. Figure 5 shows the picture of the robot.

Figure 5. The robot developed for our project. The figure shows our DJI Matrice 100 quadrotor equipped with the indoor testing parts (DJI Guidance System and the propeller guards), as well as the DJI Zenmuse X3 Gimbal and Camera system, the SF30 altimeter, and the DJI Manifold onboard computer.

The main characteristics of the robot are summarized in Table 1.

Table 1. The main characteristics and parameters of the UAV.

Parameter	**Value**
Maximum time of flight	19 min
Maximum horizontal speed	17 m/s (61 km/h)
Maximum vertical speed	4 m/s (14.4 km/h)
External diameter (with propeller guards)	1.2 m
Height	0.46 m
Processor	Quad-core ARM CORTEX-A15
RAM Memory	2 GB
Maximum robot speed	8.33 m/s (30 km/h)

4.2. *Software*

The robot's software is developed using the Robot Operating System (ROS) to achieve a modular structure. The modularity allowed the team members to work on different parts of the software independent of each other and reduced the debugging time. DJI's Onboard SDK is used to interact with the quadrotor's controller. The software is constructed in a way that it can control both the simulator and the actual robot with just a few modifications. Figure 6 shows a general view of the system architecture.

In Figure 6, the main block of our architecture (implemented as a ROS node), called "Mission Control", dictates the robot's behavior, informing the other blocks of the current task (mission status) using a ROS Parameter. The Mission Control block also generates trajectories for the robot when the current mission mode requires the robot to fly to a different position. To dictate the robot's behavior, Mission Control relies on the robot's odometry information and the robot's distance to the ground or target.

The other blocks in Figure 6 are:

- Deck tracking—detects the deck target and provides its position and orientation in the image reference frame;
- Visual Servoing—controls the UAV to track and approach the deck;
- Trajectory controller—provides velocity commands to the robot so it can follow the trajectories generated by the Mission Control node;
- Mux—selects the velocities to be sent to the quadrotor, depending on the mission status.

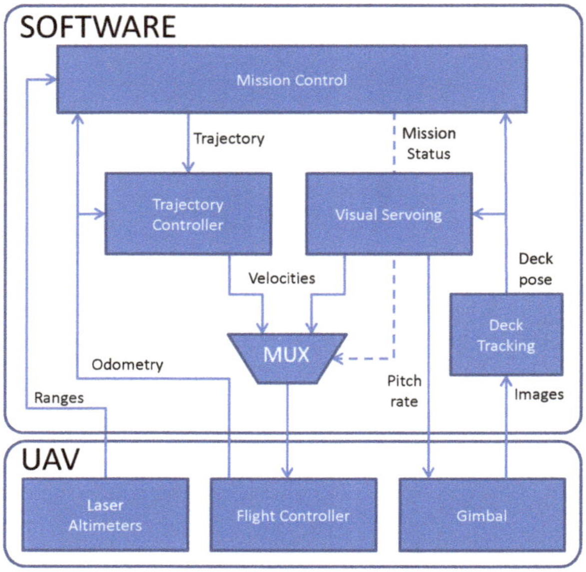

Figure 6. System architecture used in the development of the system.

The implementation of the project and the datasets used for our tests are provided as open-source for public use on our website.

4.3. Test Results

Before testing the landing outdoors, we performed several indoor experiments to develop and validate our visual servoing approach. We performed experiments both with a static and moving deck using this setup at speeds up to 2 m/s. Figure 7 shows snapshots of one of our experiments. A video of the continuous sequence of experiments can be accessed from our website.

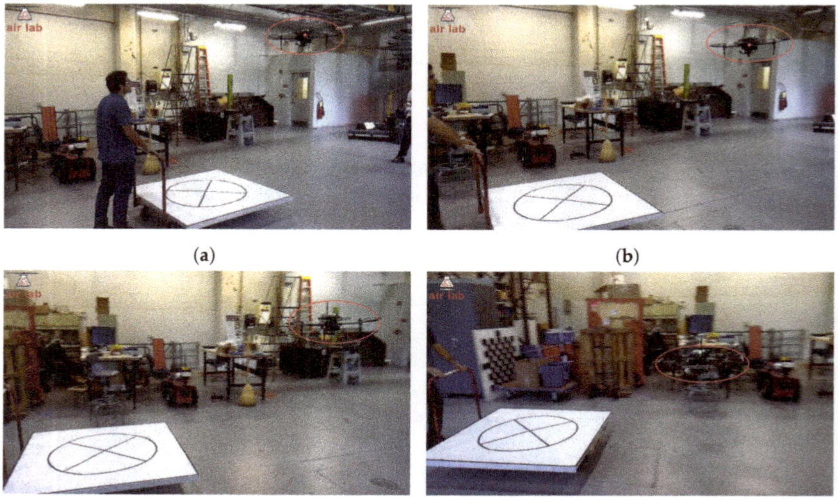

(a) (b)

(c) (d)

Figure 7. *Cont.*

(e) (f)

Figure 7. (**a–f**) Screenshots from a video sequence showing our quadrotor landing on a moving platform. In this experiment, the quadrotor is additionally equipped with a DJI Guidance sensor for safe operation in the indoor environment.

Out of 19 recorded trials to land indoors on a platform moving at a speed of 5–10 km/h (1.39–2.78 m/s), there were 17 successful landings on the platform. One failure was due to loss of detection of the pattern, which happened midway to landing and the other failure was due to loss of the pattern at the last stage, which resulted in trying to land behind the deck.

We developed a simulation environment using Gazebo [42]. It has a ground vehicle (truck) of 1.5 m in height with the target pattern on top of it. The truck's motion is controlled by a ROS node that is parameterized by the truck speed and the direction of movement.

Our UAV is simulated using the Hector Quadrotor [43], which provides a velocity-based controller similar to the one provided by DJI's ROS SDK. The simulated quadrotor is also equipped with a (non-gimbaled) camera and a height sensor. Since the simulator has no gimbal, we fixed the camera on 45°. The odometry of the drone provides position estimates in the same message type provided by DJI. This similarity allows testing the entire software in simulation before actually flying the UAV. Therefore, the same software could be used both in simulation and on real hardware, provided that some parameters related to target detection were changed. Due to the kinematic nature of our controller, it is possible to run the trajectory controller with unchanged parameters.

Figure 8 shows a snapshot of the simulator when the quadrotor is about to land on the moving deck. The upper-left corner of the figure shows the UAV's view of the deck.

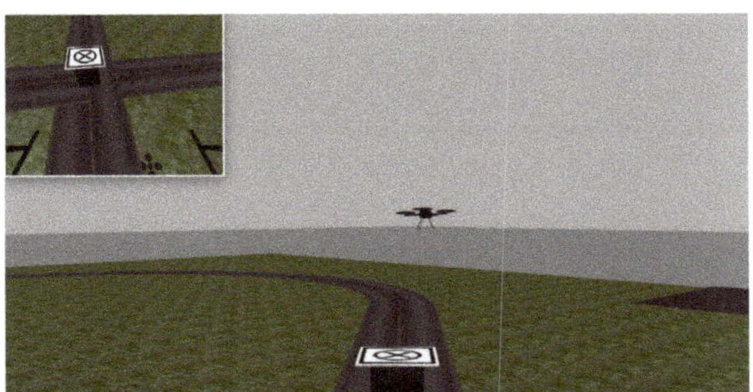

Figure 8. A snapshot of the Gazebo simulation. The quadrotor's camera view is shown in the top left corner.

To test the system in the real scenarios, a Toro MDE eWorkman electric vehicle was modified to support the pattern at the height of 1.5 m (Figure 9). A ferromagnetic deck with a painted landing pattern was attached to the top as the landing zone for the quadrotor.

Figure 9. The Toro Workman ground vehicle used for our tests.

The ground vehicle was manually driven with an approximate speed of 15 km/h (4.17 m/s) measured by a GPS device. Visual servoing gains were empirically calibrated, and the robustness of the approach to landing was evaluated.

Although we did not collect detailed statistics on the landing procedure, there were several successful landings on the vehicle at the speed of 15 km/h (4.17 m/s). It is hard to compare our method's success rate with the other methods, as none of the publications from the MBZIRC challenge have reported their success rates and have only reported limited results from the attempts during the challenge trials.

Figure 10 shows a successful autonomous landing of the quadrotor on the moving vehicle used in our experiments at 15 km/h (4.17 m/s) speed.

Out of 22 recorded trials, visual servoing was successful in bringing the UAV to the deck in 19 cases. All three failure cases were due to the extreme sun reflection where the pattern was no longer seen in the frame; therefore, the visual servoing lost track of the pattern, and the mission was aborted. The cases where the switching from initial acceleration to visual servoing happened too soon or too late (which results in the loss of the target even at the beginning) were excluded from the analysis, as they were irrelevant to the study of the visual servoing performance. The results show the successful development of our visual servoing approach in bringing the UAV to the vicinity of the moving vehicle.

The time from the first detection of the platform to the stable landing varied between different runs from 5.8 to 6.5 s, which is less than the reported times of approaches claiming to be the fastest methods (7.65 s reported by [31] and 7.8 s reported by [11]). This timing can also be seen in the accompanying videos on our website.

(a)

(b)

Figure 10. *Cont.*

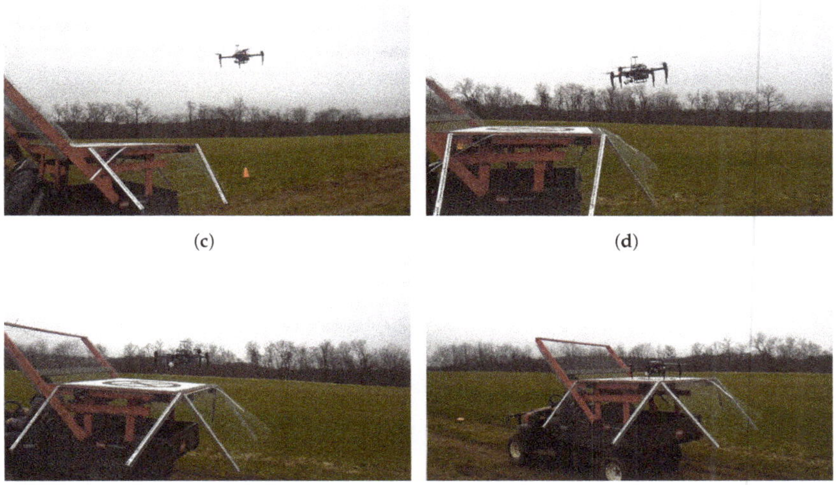

(**c**) (**d**)

(**e**) (**f**)

Figure 10. (**a–f**) Screenshots from a video sequence showing our quadrotor landing on the moving vehicle at 15 km/h (4.17 m/s) speed. The video is available on our website.

5. Conclusions

This paper presented our approach to landing an autonomous UAV on a moving vehicle. With a few modifications, such as close-range landing zone detection and vehicle turn estimation, the proposed methods can work for a generalized autonomous landing scenario in a real-world application.

The visual servoing controller showed the ability to approach and land on the moving deck in simulation, indoor environments, and outdoor environments. The controller can work with a different set of localization sensors like GPS and DJI Guidance and does not rely on very precise sensors (e.g., motion capture system).

The visual servoing controller showed promising results in reliably approaching the moving vehicle. A potential way to improve the landing efficiency further and tackle the loss of visual target problem is to divide the landing task into two subtasks: approaching the vehicle and landing from a short distance. After reaching the vehicle, the visual servoing approach can switch to a short-range landing algorithm. We believe that, by extending this approach, it can be used in a wide range of applications.

The robustness of the approach to occasional false positives in tracking the target and nondeterministic target detection rate has allowed us to use monocular vision with a real-time general-purpose ellipse detection method developed for this work (see [33]), further reducing the overall cost and weight of the system. However, as discussed in Section 4.3, this has caused the UAV to stop the landing procedure due to the sun reflection in 13.6% of our outdoor tests. If this compromise cannot be accepted in an application, a more reliable target tracking method (such as infrared or radio markers) can be devised, or another attempt should be made at finding and approaching the vehicle.

While the approach has been tested for a range of vehicle velocities up to 15 km/h, using the estimated vehicle speed as a feed-forwarding velocity means that the maximum ground vehicle's speed for UAV landing is mainly limited by the maximum speed of the UAV platform. However, the other potentially limiting factor can be the target detection rate and the gust factor, which at higher speeds can result in higher errors between two detections with insufficient time to correct the course.

Author Contributions: Conceptualization, A.K., G.A.S.P. and R.B.; methodology, A.K. and G.A.S.P.; software, A.K., G.A.S.P., R.B., R.G., P.R. and G.D.; validation, A.K., G.A.S.P. and S.S.; data curation, A.K., R.B., R.G., P.R. and G.D.; writing—original draft preparation, A.K., G.A.S.P. and R.B.; writing—review and editing, A.K., G.A.S.P., R.G., G.D., P.R. and S.S.; supervision, G.A.S.P. and S.S.; project administration, S.S.; funding acquisition, S.S. All authors have read and agreed to the published version of the manuscript.

Funding: The project was sponsored by Carnegie Mellon University Robotics Institute and Mohamed Bin Zayed International Robotics Challenge. During the realization of this work, Guilherme A.S. Pereira was supported by UFMG and CNPq/Brazil.

Institutional Review Board Statement: Not applicable.

Informed Consent Statement: Not applicable.

Data Availability Statement: The data presented in this study are openly available at http://theairlab.org/landing-on-vehicle.

Acknowledgments: The authors want to thank Nikhil Baheti, Miaolei He, Zihan (Atlas) Yu, Koushil Sreenath, and Near Earth Autonomy for their support and help in this project.

Conflicts of Interest: The authors declare no conflict of interest.

Abbreviations

The following abbreviations are used in this manuscript:

UAV	Unmanned Aerial Vehicle
GPS	Global Positioning System
IMU	Inertial Measurement Unit
MBZIRC	Mohamed Bin Zayed International Robotics Challenge
RTK	Real-Time kinematic positioning
MPC	Model Predictive Control
RANSAC	Random Sample Consensus
RGB-D	Red, Green, Blue and Depth
MUX	Multiplexer

References

1. Bonatti, R.; Ho, C.; Wang, W.; Choudhury, S.; Scherer, S. Towards a Robust Aerial Cinematography Platform: Localizing and Tracking Moving Targets in Unstructured Environments. In Proceedings of the 2019 IEEE/RSJ International Conference on Intelligent Robots and Systems (IROS), Macau, China, 3–8 November 2019; pp. 229–236. [CrossRef]
2. Keipour, A.; Mousaei, M.; Ashley, A.T.; Scherer, S. Integration of Fully-Actuated Multirotors into Real-World Applications. *arXiv* **2020**, arXiv:2011.06666.
3. Keipour, A. Physical Interaction and Manipulation of the Environment Using Aerial Robots. Ph.D. Thesis, Carnegie Mellon University, Pittsburgh, PA, USA, 2022.
4. Arora, S.; Jain, S.; Scherer, S.; Nuske, S.; Chamberlain, L.; Singh, S. Infrastructure-free shipdeck tracking for autonomous landing. In Proceedings of the 2013 IEEE International Conference on Robotics and Automation, Karlsruhe, Germany, 6–10 May 2013; pp. 323–330. [CrossRef]
5. Almeshal, A.; Alenezi, M. A Vision-Based Neural Network Controller for the Autonomous Landing of a Quadrotor on Moving Targets. *Robotics* **2018**, *7*, 71. [CrossRef]
6. Saripalli, S.; Sukhatme, G.S. Landing on a Moving Target Using an Autonomous Helicopter. In *Field and Service Robotics: Recent Advances in Research and Applications*; Springer: Berlin/Heidelberg, Germany, 2006; pp. 277–286. [CrossRef]
7. Wenzel, K.E.; Masselli, A.; Zell, A. Automatic Take Off, Tracking and Landing of a Miniature UAV on a Moving Carrier Vehicle. *J. Intell. Robot. Syst.* **2011**, *61*, 221–238. [CrossRef]
8. Gautam, A.; Sujit, P.B.; Saripalli, S. A survey of autonomous landing techniques for UAVs. In Proceedings of the 2014 International Conference on Unmanned Aircraft Systems (ICUAS), Orlando, FL, USA, 27–30 May 2014; pp. 1210–1218. [CrossRef]
9. Kim, J.; Jung, Y.; Lee, D.; Shim, D.H. Outdoor autonomous landing on a moving platform for quadrotors using an omnidirectional camera. In Proceedings of the 2014 International Conference on Unmanned Aircraft Systems (ICUAS), Orlando, FL, USA, 27–30 May 2014; IEEE: Piscataway, NJ, USA, 2014; pp. 1243–1252. [CrossRef]
10. Bahnemann, R.; Pantic, M.; Popović, M.; Schindler, D.; Tranzatto, M.; Kamel, M.; Grimm, M.; Widauer, J.; Siegwart, R.; Nieto, J. The ETH-MAV Team in the MBZ International Robotics Challenge. *J. Field Robot.* **2019**, *36*, 78–103. [CrossRef]

11. Beul, M.; Nieuwenhuisen, M.; Quenzel, J.; Rosu, R.A.; Horn, J.; Pavlichenko, D.; Houben, S.; Behnke, S. Team NimbRo at MBZIRC 2017: Fast landing on a moving target and treasure hunting with a team of micro aerial vehicles. *J. Field Robot.* **2019**, *36*, 204–229. [CrossRef]
12. Alarcón, F.; García, M.; Maza, I.; Viguria, A.; Ollero, A. A Precise and GNSS-Free Landing System on Moving Platforms for Rotary-Wing UAVs. *Sensors* **2019**, *19*, 886. [CrossRef]
13. Bhattacharya, A.; Gandhi, A.; Merkle, L.; Tiwari, R.; Warrior, K.; Winata, S.; Saba, A.; Zhang, K.; Kroemer, O.; Scherer, S. Mission-level Robustness with Rapidly-deployed, Autonomous Aerial Vehicles by Carnegie Mellon Team Tartan at MBZIRC 2020. *Field Robot.* **2022**, *2*, 172–200. [CrossRef]
14. MBZIRC. *MBZIRC Challenge Description*, 2nd ed.; Khalifa University: Abu Dhabi, United Arab Emirates, 2015. Available online: www.mbzirc.com (accessed on 24 August 2022).
15. Saripalli, S.; Sukhatme, G.S.; Montgomery, J.F. An Experimental Study of the Autonomous Helicopter Landing Problem. In *Experimental Robotics VIII*; Springer: Berlin/Heidelberg, Germany, 2003; pp. 466–475. [CrossRef]
16. Borowczyk, A.; Nguyen, D.T.; Phu-Van Nguyen, A.; Nguyen, D.Q.; Saussié, D.; Le Ny, J. Autonomous Landing of a Quadcopter on a High-Speed Ground Vehicle. *J. Guid. Control Dyn.* **2017**, *40*, 1–8. [CrossRef]
17. Lee, D.; Ryan, T.; Kim, H.J. Autonomous landing of a VTOL UAV on a moving platform using image-based visual servoing. In Proceedings of the 2012 IEEE International Conference on Robotics and Automation, Saint Paul, MN, USA, 14–18 May 2012; pp. 971–976. [CrossRef]
18. Gonçalves, V.M.; McLaughlin, R.; Pereira, G.A.S. Precise Landing of Autonomous Aerial Vehicles Using Vector Fields. *IEEE Robot. Autom. Lett.* **2020**, *5*, 4337–4344. [CrossRef]
19. Lange, S.; Sunderhauf, N.; Protzel, P. A vision based onboard approach for landing and position control of an autonomous multirotor UAV in GPS-denied environments. In Proceedings of the 2009 International Conference on Advanced Robotics, Munich, Germany, 22–26 June 2009; pp. 1–6.
20. Merz, T.; Duranti, S.; Conte, G. Autonomous Landing of an Unmanned Helicopter based on Vision and Inertial Sensing. In *Experimental Robotics IX: The 9th International Symposium on Experimental Robotics*; Springer: Berlin/Heidelberg, Germany, 2006; pp. 343–352. [CrossRef]
21. Xing, B.Y.; Pan, F.; Feng, X.X.; Li, W.X.; Gao, Q. Autonomous Landing of a Micro Aerial Vehicle on a Moving Platform Using a Composite Landmark. *Int. J. Aerosp. Eng.* **2019**, *2019*, 4723869. [CrossRef]
22. Xuan-Mung, N.; Hong, S.K.; Nguyen, N.P.; Ha, L.N.N.T.; Le, T.L. Autonomous Quadcopter Precision Landing Onto a Heaving Platform: New Method and Experiment. *IEEE Access* **2020**, *8*, 167192–167202. [CrossRef]
23. Wynn, J.S.; McLain, T.W. Visual Servoing for Multirotor Precision Landing in Daylight and After-Dark Conditions. In Proceedings of the 2019 International Conference on Unmanned Aircraft Systems (ICUAS), Atlanta, GA, USA, 11–14 June 2019; pp. 1242–1248.
24. Romero-Ramirez, F.; Muñoz-Salinas, R.; Medina-Carnicer, R. Speeded Up Detection of Squared Fiducial Markers. *Image Vis. Comput.* **2018**, *76*, 38–47. [CrossRef]
25. Ruffier, F.; Franceschini, N. Optic Flow Regulation in Unsteady Environments: A Tethered MAV Achieves Terrain Following and Targeted Landing Over a Moving Platform. *J. Intell. Robot. Syst.* **2015**, *79*, 275–293. [CrossRef]
26. Herissé, B.; Hamel, T.; Mahony, R.; Russotto, F.X. Landing a VTOL Unmanned Aerial Vehicle on a Moving Platform Using Optical Flow. *IEEE Trans. Robot.* **2012**, *28*, 77–89. [CrossRef]
27. Chang, C.W.; Lo, L.Y.; Cheung, H.C.; Feng, Y.; Yang, A.S.; Wen, C.Y.; Zhou, W. Proactive Guidance for Accurate UAV Landing on a Dynamic Platform: A Visual-Inertial Approach. *Sensors* **2022**, *22*, 404. [CrossRef] [PubMed]
28. Cantelli, L.; Guastella, D.; Melita, C.D.; Muscato, G.; Battiato, S.; D'Urso, F.; Farinella, G.M.; Ortis, A.; Santoro, C. Autonomous Landing of a UAV on a Moving Vehicle for the MBZIRC. In *Human-Centric Robotics, Proceedings of the CLAWAR 2017: 20th International Conference on Climbing and Walking Robots and the Support Technologies for Mobile Machines, Porto, Portugal, 11–13 September 2017*; World Scientific: Singapore, 2017; pp. 197–204. [CrossRef]
29. Battiato, S.; Cantelli, L.; D'Urso, F.; Farinella, G.M.; Guarnera, L.; Guastella, D.; Melita, C.D.; Muscato, G.; Ortis, A.; Ragusa, F.; et al. A System for Autonomous Landing of a UAV on a Moving Vehicle. In *Image Analysis and Processing, Proceedings of the International Conference on Image Analysis and Processing, Catania, Italy, 11–15 September 2017*; Battiato, S., Gallo, G., Schettini, R., Stanco, F., Eds.; Springer International Publishing: Cham, Switzerland, 2017; pp. 129–139.
30. Baca, T.; Stepan, P.; Spurny, V.; Hert, D.; Penicka, R.; Saska, M.; Thomas, J.; Loianno, G.; Kumar, V. Autonomous landing on a moving vehicle with an unmanned aerial vehicle. *J. Field Robot.* **2019**, *36*, 874–891. [CrossRef]
31. Tzoumanikas, D.; Li, W.; Grimm, M.; Zhang, K.; Kovac, M.; Leutenegger, S. Fully autonomous micro air vehicle flight and landing on a moving target using visual–inertial estimation and model-predictive control. *J. Field Robot.* **2019**, *36*, 49–77. [CrossRef]
32. Falanga, D.; Zanchettin, A.; Simovic, A.; Delmerico, J.; Scaramuzza, D. Vision-based autonomous quadrotor landing on a moving platform. In Proceedings of the 2017 IEEE International Symposium on Safety, Security and Rescue Robotics (SSRR), Shanghai, China, 11–13 October 2017; IEEE: Piscataway, NJ, USA, 2017; pp. 200–207. [CrossRef]
33. Keipour, A.; Pereira, G.A.; Scherer, S. Real-Time Ellipse Detection for Robotics Applications. *IEEE Robot. Autom. Lett.* **2021**, *6*, 7009–7016. [CrossRef]
34. Hoffmann, G.M.; Waslander, S.L.; Tomlin, C.J. Quadrotor helicopter trajectory tracking control. In Proceedings of the AIAA Guidance, Navigation and Control Conference and Exhibit, Honolulu, HI, USA, 18–21 August 2008; pp. 1–14.

35. Espiau, B.; Chaumette, F.; Rives, P. A new approach to visual servoing in robotics. *IEEE Trans. Robot. Autom.* **1992**, *8*, 313–326. [CrossRef]
36. Hutchinson, S.; Hager, G.D.; Corke, P.I. A tutorial on visual servo control. *IEEE Trans. Robot. Autom.* **1996**, *12*, 651–670. [CrossRef]
37. Marchand, É.; Spindler, F.; Chaumette, F. ViSP for visual servoing: A generic software platform with a wide class of robot control skills. *IEEE Robot. Autom. Mag.* **2005**, *12*, 40–52. [CrossRef]
38. Corke, P.I.; Good, M.C. Dynamic effects in visual closed-loop systems. *IEEE Trans. Robot. Autom.* **1996**, *12*, 671–683. [CrossRef]
39. Jin, R.; Owais, H.M.; Lin, D.; Song, T.; Yuan, Y. Ellipse proposal and convolutional neural network discriminant for autonomous landing marker detection. *J. Field Robot.* **2019**, *36*, 6–16. [CrossRef]
40. Li, Z.; Meng, C.; Zhou, F.; Ding, X.; Wang, X.; Zhang, H.; Guo, P.; Meng, X. Fast vision-based autonomous detection of moving cooperative target for unmanned aerial vehicle landing. *J. Field Robot.* **2019**, *36*, 34–48. [CrossRef]
41. Fornaciari, M.; Prati, A.; Cucchiara, R. A fast and effective ellipse detector for embedded vision applications. *Pattern Recognit.* **2014**, *47*, 3693–3708. [CrossRef]
42. Koenig, N.; Howard, A. Design and use paradigms for Gazebo, an open-source multi-robot simulator. In Proceedings of the IEEE/RSJ International Conference on Intelligent Robots and Systems (IROS), Sendai, Japan, 28 September–2 October 2004; Volume 3, pp. 2149–2154.
43. Meyer, J.; Sendobry, A.; Kohlbrecher, S.; Klingauf, U.; von Stryk, O. Comprehensive Simulation of Quadrotor UAVs Using ROS and Gazebo. In *Simulation, Modeling, and Programming for Autonomous Robots, Proceedings of the Third International Conference, SIMPAR 2012, Tsukuba, Japan, 5–8 November 2012*; Noda, I., Ando, N., Brugali, D., Kuffner, J.J., Eds.; Springer: Berlin/Heidelberg, Germany, 2012; pp. 400–411.

sensors

MDPI

Article

Modeling and Control of a Spherical Robot in the CoppeliaSim Simulator

Guelis Montenegro [1,†], Roberto Chacón [2,†], Ernesto Fabregas [3], Gonzalo Garcia [4], Karla Schröder [2], Alberto Marroquín [2], Sebastián Dormido-Canto [3] and Gonzalo Farias [2,*]

1 Departamento de Electrotecnia e Informática, Universidad Técnica Federico Santa María, Av. Federico Santa María 6090, Viña del Mar 2520001, Chile
2 Escuela de Ingeniería Eléctrica, Pontificia Universidad Católica de Valparaíso, Av. Brasil 2147, Valparaíso 2362804, Chile
3 Departamento de Informática y Automática, Universidad Nacional de Educación a Distancia, Juan del Rosal 16, 28040 Madrid, Spain
4 Ocean and Mechanical Engineering Department, Florida Atlantic University, 777 Glades Road EW 190, Boca Raton, FL 33431, USA
* Correspondence: gonzalo.farias@pucv.cl; Tel.: +56-32-2273-673
† These authors contributed equally to this work.

Abstract: This article presents the development of a model of a spherical robot that rolls to move and has a single point of support with the surface. The model was developed in the CoppeliaSim simulator, which is a versatile tool for implementing this kind of experience. The model was tested under several scenarios and control goals (i.e., position control, path-following and formation control) with control strategies such as reinforcement learning, and Villela and IPC algorithms. The results of these approaches were compared using performance indexes to analyze the performance of the model under different scenarios. The model and examples with different control scenarios are available online.

Keywords: spherical robot; model; simulation; CoppeliaSim (V-REP)

Citation: Montenegro, G.; Chacón, R.; Fabregas, E.; Garcia, G.; Schröder, K.; Marroquín, A.; Dormido-Canto, S.; Farias, G. Modeling and Control of a Spherical Robot in the CoppeliaSim Simulator. *Sensors* **2022**, *22*, 6020. https://doi.org/10.3390/s22166020

Academic Editors: Sašo Blažič, Gregor Klancar and Marija Seder

Received: 1 July 2022
Accepted: 10 August 2022
Published: 12 August 2022

1. Introduction

The field of robotics is very extensive with respect to robot design, as it is necessary to investigate and analyse the different types of mechanisms that a robot could eventually integrate. Mobile robots can move around their environment using different mechanisms, such as wheels, caterpillars and others [1–4]. Mobile robots include robots with a spherical shape, that roll to move, like a football. This type of robot represents a particular challenge because it must roll with all of its components inside [5,6]. Examples of this kind of robot can be found in the literature with different driving/steering mechanisms, including single wheels [7,8], pendulums [9–11], and omnidirectional wheel mechanisms [12].

It is well-known that, currently, advanced robots are very expensive and can be exposed to damage during laboratory experimentation. For this reason, simulators are very important in this field. Virtual laboratories using these simulators offer significant benefits for robotics education [13]. Using such virtual laboratories, students can test and gain an understanding of concepts that are not easy to follow in the classroom, at any time and pace, and from anywhere [14,15]. They can also design and test control strategies before implementing them in an actual robot in the laboratory without risk of damage to the physical device.

Currently, there are many simulators for different areas of robotics. For example, Argos [16], Webots [17], RFCSIM [18], and CoppeliaSim (formerly V-REP) [19], to mention only those most used. Some of these platforms have licenses that can be used free of charge for educational purposes. They have competitive functionalities with various components that can interact with each other and can be programmed with different programming languages.

The CoppeliaSim simulator deserves special attention for being one of the most widely used for pedagogical purposes today. This simulator provides a versatile and scalable framework for creating 3D simulations in a relatively short period of time. CoppeliaSim has an integrated development environment (IDE) that is based on a distributed and scripted architecture—each scene object can have a built-in script attached, all operating at the same time, in the form of threads. CoppeliaSim contains a wealth of examples, robot models, sensors, and actuators to create and interact with a virtual world at run-time. New models can be designed and added to CoppeliaSim to implement custom-designed simulation experiments [20,21]. CoppeliaSim has an extensive list of mobile robots, among which there is no spherical robot. Therefore, it would be interesting to add a model of a spherical robot so that it would be available and could be used by the community for experiments.

In our laboratory, we have recently developed a spherical mobile robot that can be easily reproduced with a 3D-printer and some basic electronic components. Our idea was to develop and test the simulation model for this robot, as we did in previous investigations [14,15] with a model of the Khepera IV robot in the CoppeliaSim simulator. In the model, the physical properties and components, such as mass, dimensions, and other variables were carefully taken into account. This model was tested with several control strategies under different scenarios. The results obtained with the model were very similar to those obtained with the actual Khepera IV robot with the same control algorithms and experimental conditions [22]. We expect the resulting new model to be very similar to the physical robot, based on our previous results obtained with the Khepera IV robot.

This paper presents the development and testing of a model of a spherical robot whose movement is based on an internal pendulum. The robot consists of a spherical-shaped cover that protects the pendulum and the internal circuitry and allows it to roll to move from one place to another. This type of morphology has a fundamental advantage in that there is no possibility of the robot tipping over, which gives it certain stability in movement. At the same time, it has certain disadvantages with respect to sliding on the surface and difficulties with the presence of obstacles or irregularities in the terrain [9,12,23].

The main contribution of this article is the modelling and control of a non-linear model of a spherical robot that does not exist in the CoppeliaSim simulator. This is a challenging task due to the complexity of the spherical robot model. The developed model is controlled under different scenarios with several control algorithms implemented by the authors in previous studies, including Villela [24], IPC (integral proportional controller) [25], and reinforcement learning (RL) [26,27]. The experiments undertaken to test the robot model included investigation of position control, path-following and formation control. As a result of this work, the model and some examples are available online for use by the community that works with mobile robots.

The article is organised as follows: Section 2 describes the model of the spherical robot and its design and implementation in CoppeliaSim. Section 3 describes the control laws and experiments implemented with the spherical robot. Section 4 shows the tests performed on position control, path-following and formation control. Finally, Section 5 presents the conclusions and discusses future work.

2. Spherical Robot Model

This section presents the details of the mathematical model of the spherical robot in different situations.

2.1. Robot Description

A spherical robot is a mobile machine that has a spherical casing with a single contact with the surface on which it rests. The housing not only allows the robot to be in a balanced state, but also enables the robot to roll from one place to another without sliding. Figure 1 shows a picture of the actual robot in the laboratory.

Figure 1. Actual spherical robot.

For practical purposes, the mechanism that moves the robot and maintains balance is based on a pendulum and three motors. One motor operates the pendulum and modifies the centre of gravity of the robot; the other motors are connected to the casing-sphere, which allows its movement using rotation. The design of this robot is based on [28]. Figure 2 shows a schematic view of the robot, where the right side is the front view and the left side is the lateral view. The components are as follows: (1) casing sphere, a 3D-printed shell which is the body of the robot; (2) bearings, the joins of the case with the shaft; (3) pendulum, which is a stick that allows the robot to turn; (4) counterweights, which are two body masses to improve the stability of the robot; (5) circuitry box, is a small space that contains all the electronic components; (6) shaft (fixed axle), which is the join between the circuit box and the case; and (7) articulation, which is the join between the pendulum and the shaft.

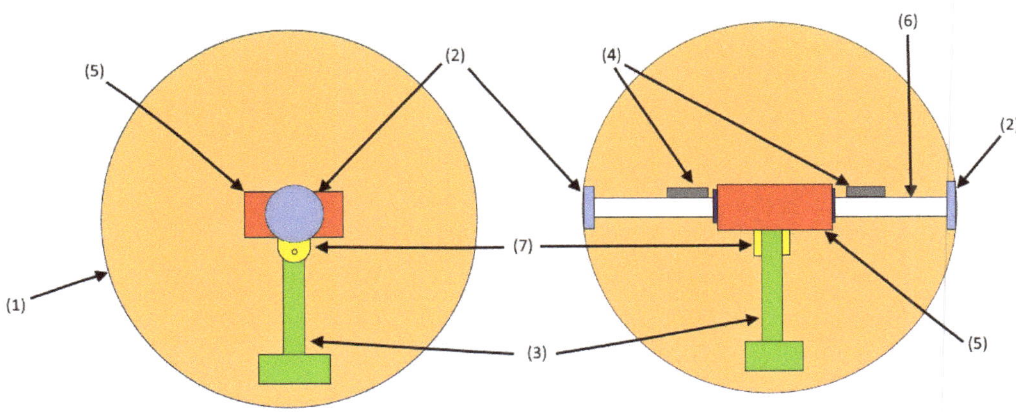

Figure 2. Details of the spherical robot.

2.2. Horizontal Motion Equations

A simplified model of the robot can be seen in Figure 3, which is based on [28]. The model considers moments of inertia, the radius of the sphere, masses of the elements that make up the robot, angles, and the direction of linear velocity.

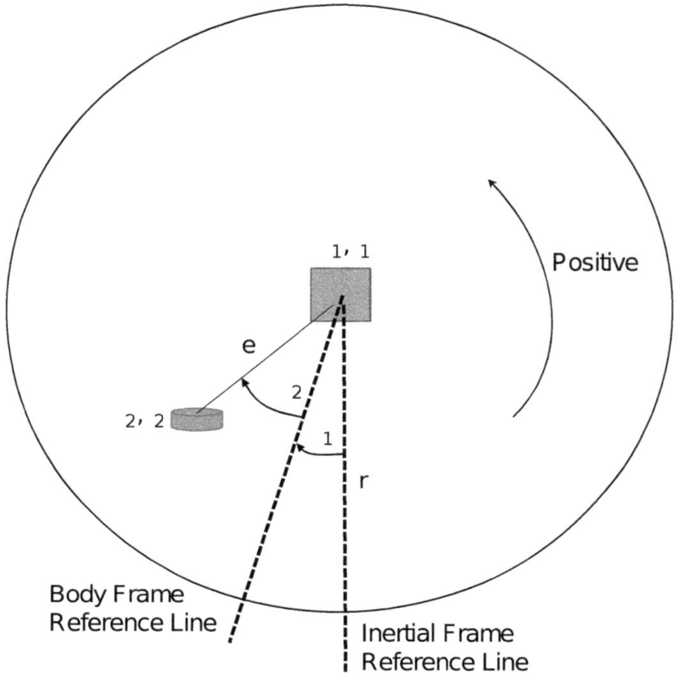

Figure 3. Simplified scheme of the horizontal movement of the robot (1 rotation angle of the ball, 2 rotation angle of the pendulum with respect to the ball).

In this model, it is considered that the casing is always in contact with the surface, which allows the robot to roll without slipping. The dynamic model of the spherical robot for a horizontal movement considers the balance of potential energy, kinetic energy and rotational energy, respectively, as can be seen in Equations (1)–(3):

$$\begin{cases} U_1 = 0 \\ U_2 = -M_2 g e \cos(\theta_1 + \theta_2) \end{cases} \tag{1}$$

$$\begin{cases} K_1 = \frac{1}{2} M_1 (r w_1)^2 \\ K_2 = \frac{1}{2} M_2 \Big\{ (r\omega_1 - e\cos(\theta_1 + \theta_2)(\omega_1 + \omega_2))^2 \end{cases} \tag{2}$$

$$\begin{cases} T_1 = \frac{1}{2} J_1 \omega_1^2 \\ T_2 = \frac{1}{2} J_2 (\omega_1 + \omega_2)^2 \end{cases} \tag{3}$$

where the main variables are the following:
U_1: is the potential energy of the spherical housing with respect to the height of its centroid.
U_2: is the potential energy of the pendulum with respect to the height of the centroid of the spherical casing.
K_1: is the kinetic energy of the spherical shell.
K_2: is the kinetic energy of the pendulum.

T_1: is the rotational energy of the spherical shell.
T_2: is the rotational energy of the pendulum.
e: is the distance between the centroid of the spherical shell and the pendulum.
M_1: is the mass of the spherical shell.
M_2: is the mass of the pendulum.

The Lagrange equations are calculated as follows:

$$L = K_1 + K_2 + T_1 + T_2 - U_1 - U_2 \tag{4}$$

$$\frac{d}{dt}\left(\frac{\partial L}{\partial \omega_1}\right) - \frac{\partial L}{\partial \theta_1} = -T + T_f \tag{5}$$

$$\frac{d}{dt}\left(\frac{\partial L}{\partial \omega_2}\right) - \frac{\partial L}{\partial \theta_2} = T \tag{6}$$

where t is the independent variable of time, T is the torque applied between the casing and the circuitry, and T_f is the torque that appears with the friction force that occurs between the casing and the ground. Solving the Equations (5) and (6), we obtain:

$$T = a_1(J_2 - M_2 re\cos(\theta_1 + \theta_2) + M_2 e^2) + a_2(J_2 + M_2 e^2) + M_2 ge\sin(\theta_1 + \theta_2) \tag{7}$$

This last equation is useful to determine which motor may be used for the construction of the spherical robot and to consider its design.

2.3. Robot Turning Equations

The motion that allows changing the direction of the spherical robot is based on the mass of the pendulum using a CMG (control moment gyroscope). The calculation of this motion is based on a torsion pendulum, as shown in Figure 4.

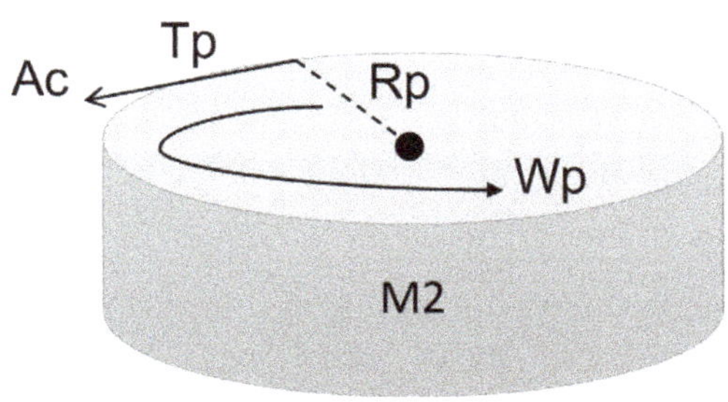

Figure 4. Pendulum schema reaction wheel.

Where T_p is the torque of the pendulum; R_p is the radius of the pendulum; ω_p is the angular velocity of the pendulum; A_c is the swing acceleration of the pendulum; and M_2 is the mass of the pendulum. The equation of the rotational motion of the pendulum is as follows:

$$T_p = 4R_p M_2 A_c \tag{8}$$

Figure 5 shows an angle θ corresponding to the angular position of a spherical mobile robot, whose orientation is controlled by varying the speed of a reaction wheel. The angular velocity of the reaction wheel, relative to the spherical mobile robot, can be varied by varying the voltage applied to the electric motor. This means that, if the motor rotates clockwise, the spherical robot will orient itself in the opposite direction. The effect is achieved by analysing the angular momentum at the axis of rotation; as the speed of the wheel varies, the speed of the mobile robot will also start to vary so that the momentum remains constant.

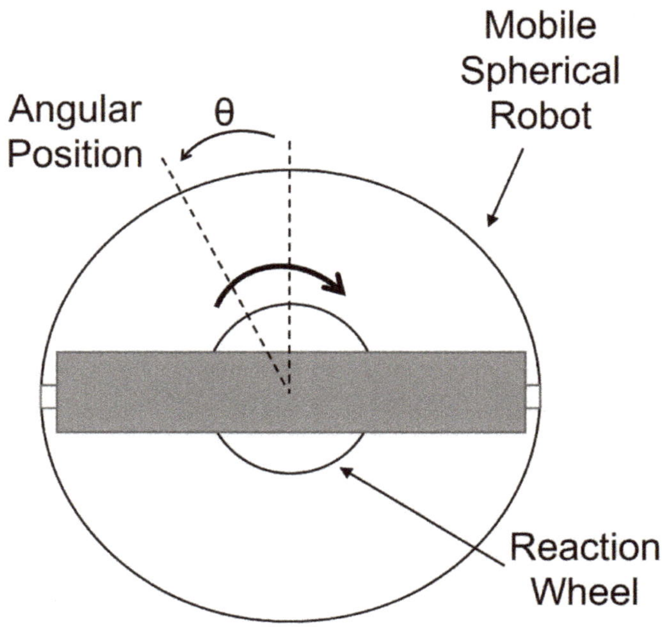

Figure 5. Top view of the robot for rotational movement.

The equations governing this phenomenon are obtained after analysing the momentum of the spherical robot and the reaction wheel around the axis of rotation. As shown below:

$$\dot{\theta} = \omega \tag{9}$$

$$\dot{\omega} = -\frac{B_1}{J_1}\omega + \frac{B_2}{J_1}\Omega - \frac{1}{J_1}\tau_m \tag{10}$$

$$\dot{\Omega} = -\frac{B_2}{J_{eq}}\Omega + \frac{1}{J_{eq}}\tau_m \tag{11}$$

where the main variables are the following: θ is the angle of the spherical robot casing (robot angle); ω is the angular velocity of the spherical robot; Ω is the angular velocity of the pendulum wheel; J_1 is the moment of inertia of the spherical robot casing; J_2 is the moment of inertia of the pendulum wheel of the spherical robot; and J_{eq} is the equivalent moment of inertia where: $\frac{1}{J_{eq}} = \frac{1}{J_1} + \frac{1}{J_2}$.

2.4. Building the Model of the Robot

The parts of the spherical robot were designed using the 3D design software Autodesk Fusion 360 [29] based on the actual robot shown in Figure 2. These parts were then imported into the CoppeliaSim simulator [30] working environment and the robot was assembled manually. Figure 6 shows the process of building the robot.

As can be seen, on the right side, the imported parts of the robot are shown (i.e., housing, pendulum motor, housing motors). On the left side, the result of the robot assembly is shown. The total diameter of the robot is 18 centimetres. Note that the motors and internal elements were built in the same software.

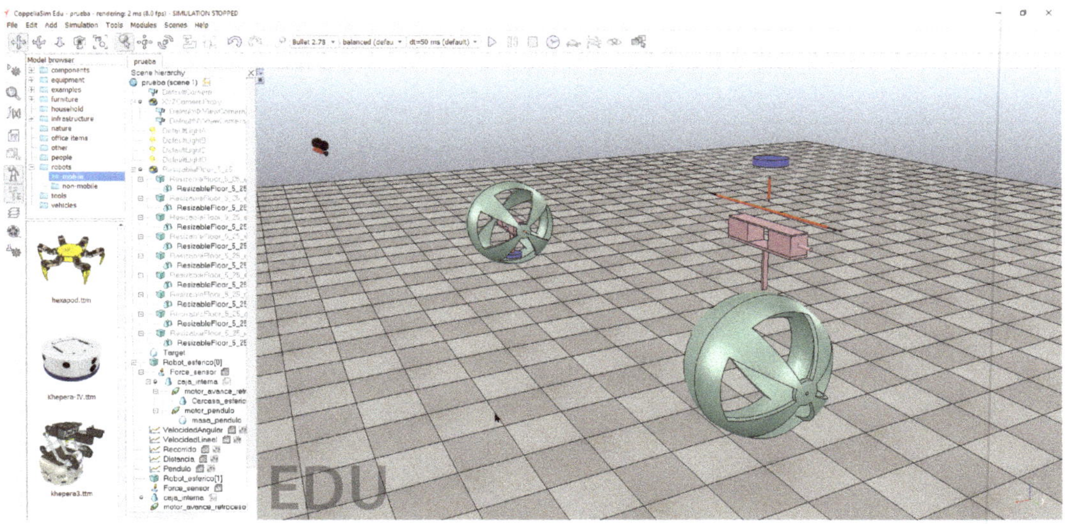

Figure 6. Assembling the robot in CoppeliaSim.

3. Experiments with the Spherical Robot

In this section, some tests/experiments implemented with the robot are presented and commented on.

3.1. Position Control

This experiment consisted of getting the robot to move from one point C (current position) to another T_p (target point) in the most efficient way possible, which implies that it does so by following the shortest path to the destination point. Note that this robot can rotate without displacement, which means that its model is holonomic. To add complexity to the experiment, we imposed non-holonomic constraints on the model. This means that the robot has to move in order to rotate (it cannot rotate about its own position). Figure 7 shows a representation of this experiment.

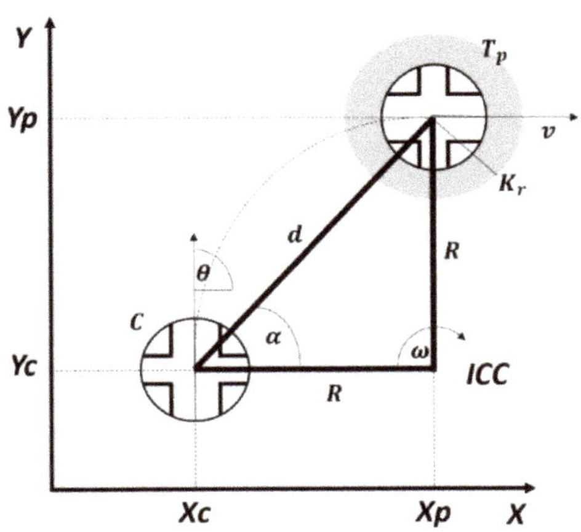

Figure 7. Position control experiment.

As can be seen, the variables involved in this experiment were, on one hand, the pose $C(x, y, \theta)$, which includes the position (x,y) and the orientation angle (θ) of the robot; on the other hand, the distance (d) and the angle (α) at which the target point is located. These variables are calculated as follows:

$$d = \sqrt{(y_p - y_c)^2 + (x_p - x_c)^2} \quad (12)$$

$$\alpha = \tan^{-1}\left(\frac{y_p - y_c}{x_p - x_c}\right) \quad (13)$$

The control law is calculated using the angular error as input ($\alpha_e = \theta - \alpha$) and obtaining as outputs the linear velocity (v) and the angular velocity of the robot (ω). Then the corresponding values for the housing motor and pendulum are calculated for the robot to move with these angular and linear velocities. As a result, the robot is positioned in a new pose C (x,y,θ), which is used to recalculate the values described above. Figure 8 shows the block diagram of the control loop for this experiment. Where the block Compute implements Equations (12) and (13); while the Control Law block can be implemented in different ways, with artificial intelligence or conventional control law approaches, as is explained in the next subsections.

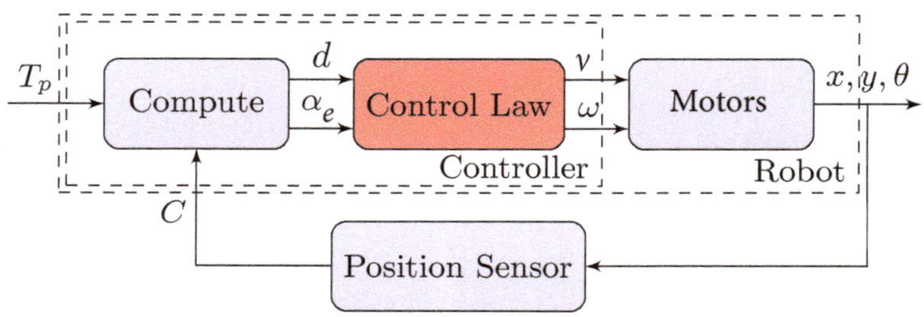

Figure 8. Position control block diagram.

In the case of the spherical robot, the linear velocity (v) is applied as a voltage to the servomotors of Figure 2, which allows movement of the pendulum backwards or forwards to change the centre of gravity of the robot and to make the robot move in one of those directions. The angular velocity (ω) is applied to the DC motor of the pendulum, which allows the robot to rotate clockwise or counter-clockwise.

In addition to this experiment, the model was tested with other approaches: (1) path-following and (2) formation control. In the first case, the robot must control its position by following a trajectory received as a reference. In the second experiment, more robots were added to the scenario. One of them acts as the leader and the rest as followers. The followers use the position of the leader to make a formation around the leader.

3.2. Reinforcement Learning Approach

Reinforcement learning is able to provide an optimal solution despite the complexity of the system. The system learns by acting on the environment while operating in real-time. This is an advantage of traditional optimization methods that rely on a mathematical model and are tuned backwards in time [31,32].

The reinforcement learning approach for this research, called Q-learning, is based on solving the Bellman equation, and the principle of optimality. This technique allows an optimal learning process to be carried out during regular operation, based on the robot's dynamics, and continuous-time signals. In the limit, the Q-matrix captures a discretized version of the optimal action-state combination in terms of the highest long-term reward.

Given a system described by the dynamics $x_{k+1} = f(x_k, u_k)$ and a reward function $\sigma(x_k, u_k)$, where x_k is the state of the system, and $u_k = \pi(x_k)$ the control policy, a long-term reward can be defined by Equation (14):

$$\sum_{k=0}^{\infty} \gamma^k \sigma(x_k, u_k) = \sum_{k=0}^{\infty} \gamma^k \sigma(x_k, \pi(x_k)) \tag{14}$$

where $0 < \gamma < 1$ is a discount factor required to penalize future rewards and to ensure convergence of the summation. This expression represents the discounted accumulated rewards starting from the current state x_0 and the application of the policy π.

To apply Bellman's optimality principle, the previous long-term reward expression (14) is redefined in terms of the function $Q(x_k, u_k)$, called action-value, which allows for the splitting of the reward assignment into two consecutive steps. This action-value function conveys the long-term reward by the contribution of the immediate reward due to applying an arbitrary action u_k while in the state x_k, and by the discounted accumulated reward continuing with the control policy π. This is as shown in (15) starting from x_0:

$$Q^{\pi}(x_0, u_0) = \sigma(x_0, u_0) + \sum_{k=1}^{\infty} \gamma^k \sigma(x_k, \pi(x_k)) = \sigma(x_0, u_0) + \gamma \sum_{k=0}^{\infty} \gamma^k \sigma(x_{k+1}, \pi(x_{k+1})) \tag{15}$$

The optimal value is obtained by maximizing the future rewards; using the optimal policy defined by π^*, a recursive equation is obtained:

$$Q^*(x_k, u_k) = \sigma(x_k, u_k) + \gamma \max_{\mu} Q^*(x_{k+1}, \mu) \tag{16}$$

This equation captures the optimal principle by stating that future optimal control actions are not specified by past optimal values, but, instead, only by the current state. The major advance in these calculations is the viability of forward-in-time learning, as opposed to a standard optimal search performed backwards-in-time. This method is also known as Q-learning. From (16), the following recursive equation can be devised that asymptotically converges to the fixed manifold Q^* [33,34]:

$$Q^{i+1}(x_k, u_k) = Q^i(x_k, u_k) + \alpha(\sigma(x_k, u_k) + \gamma \max_{\mu} Q^i(x_{k+1}, \mu) - Q^i(x_k, u_k)). \tag{17}$$

The term $\sigma(x_k, u_k) + \gamma \max_{\mu} Q^i(x_{k+1}, \mu) - Q^i(x_k, u_k)$ is typically labeled temporal difference $TD^i(x_k, u_k)$, or error between the target value $\sigma(x_k, u_k) + \gamma \max_{\mu} x_{k+1}, \mu$ and the current value $Q^{i+1}(x_k, u_k)$, with $0 < \alpha < 1$ a learning rate. The expression (17) resembles a gradient descend numerical search. Another interpretation of (17) is the structure of a low-pass filter, by rearranging it as $Q^{i+1}(x_k, u_k) = \alpha TD^i(x_k, u_k) + (1 - \alpha)Q^i(x_k, u_k)$. The learning rate α, or numerical search step size, establishes the effect of new information overriding previous information. A small value will reduce the rate of learning, while a larger value will rely more heavily on new data, despite what was previously learned.

3.3. Control Laws: Villela and IPC Approaches

As was mentioned before, the Control Law block of Figure 9 can be implemented with traditional control laws or with a machine learning approach. In this subsection, we show both control laws that will later be implemented in the robot. For example, the Villela control law [24], named after its author, was used previously with different kinds of robots with good results [15,18,22]. It calculates the linear velocity (v) and the angular velocity (ω) of the robot, as shown in Equations (18).

$$v = \begin{cases} v_{max} & if|d| > k_r \\ d\left(\frac{v_{max}}{k_r}\right) & if|d| \leq k_r \end{cases} \quad (18)$$

$$\omega = \omega_{max}\sin(\alpha_e)$$

where v_{max} is the maximum linear velocity, k_r is the radius of a docking area (around the target point) and ω_{max} is the maximum angular velocity of the robot.

Based on the Villela control law, in a previous study, we developed what we term an integral proportional controller IPC [25], which was compared with the Villela algorithm and was found to produce better results. The controller implements the velocities as follows in Equation (19):

$$\begin{aligned} v &= \min\{K_v p(\alpha_e)d, v_{\max}\} \\ \omega &= K_p \sin(\alpha_e) + K_i \int_0^t \alpha_e dt \end{aligned} \quad (19)$$

where $p(\alpha_e) = 1 - |\alpha_e|/\pi$, for $\alpha_e \in [-\pi, \pi]$ and K_v, K_p, and K_i are tuning parameters of the control law. We tested this control law with a differential wheeled mobile robot, so it is challenging to implement this controller with a spherical robot.

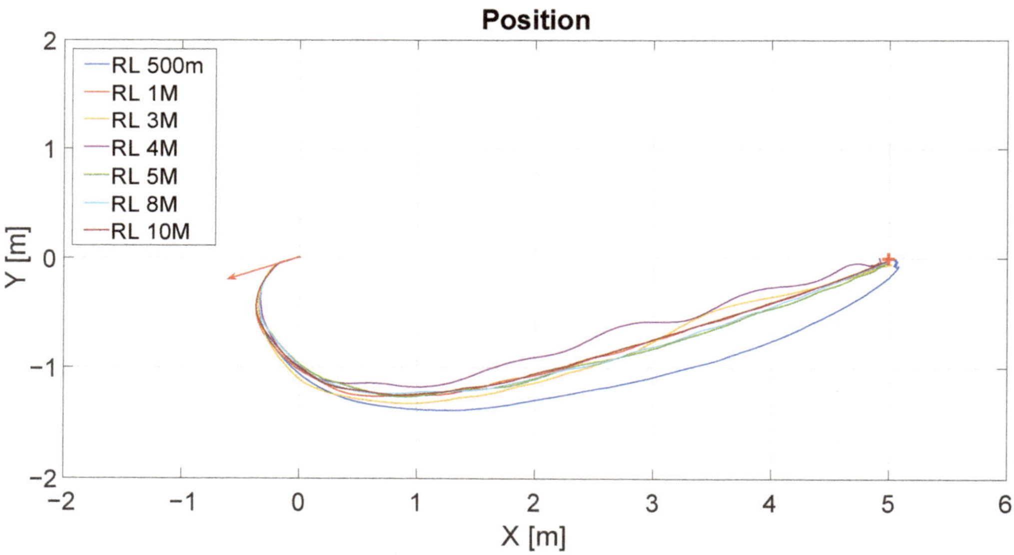

Figure 9. Obtained trajectories for different values of iteration in RL algorithm.

4. Results

In this section, the results of the implementation of the control experiments with the developed model are presented.

4.1. Reinforcement Learning Results

This subsection shows the simulation results for different tests in several iterations to build the Q-matrix. The Q-matrix was obtained in MATLAB during the learning stage and exported to Python (Spyder-Anaconda IDE). The CoppeliaSim software was connected to Spyder via remote API, which ensured it was compatible with Python programming. The experiments were performed with the CoppeliaSim simulator using the developed spherical robot model.

In the learning stage, the algorithm builds the Q-matrix to learn how to reach the destination point. To this end, the angle error (α_e) is used to obtain the angular velocity (ω) in order to control the position of the robot. Note that, initially, the linear velocity (ν) is kept constant at its maximum value until the robot reaches the docking area.

The Q-matrix is composed of the sets (state, action), where the state is the angle error (α_e), and the action is the angular velocity (ω). The criterion for obtaining the rewards of the Q-matrix is to penalize significant changes in the angle error and small changes in the angular velocity of the pendulum. In this case, the matrix Q has a size of 126×41, where 126 states are generated linearly spaced between $-\pi$ and π, and 41 actions linearly spaced between $-\pi/2$ and $\pi/2$. The array is made up of initial reward values. These initial values are adjusted according to the number of iterations of the algorithm based on a learning rate, a discount rate and a coefficient of relationship between exploration and use. They explore and use values to allow the robot to explore the space to complete knowledge of it and later use that knowledge.

Figure 9 shows the results of the position control experiment for different iterations of the RL algorithm (RL 500 m-500.000 iterations, RL 1M-1.000.000 iterations, and so on). The lines describe the trajectories followed by the robot for each value of iteration. The initial position of the robot is represented by the base of the red arrow at (0; 0), and the target point is represented by the red cross located at (5; 0). The direction of the arrow represents the initial orientation of the robot.

Figure 10 shows the distance to the destination point for these experiments. The y-axis represents the distance in meters and the x-axis represents the time in seconds. As can be seen, for all experiences, the time to arrive at the destination was similar, around 14 s. This would be expected given the similarity between the trajectories shown in the figure above.

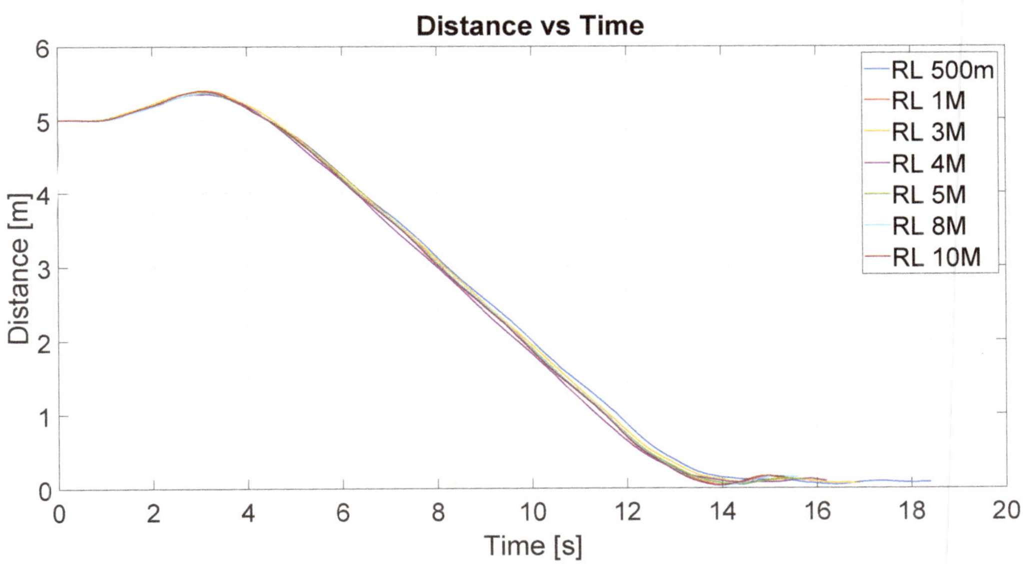

Figure 10. Distance vs. time of all experiences/iterations.

The quality of each control algorithm can be evaluated using performance indexes. These indexes use the integral of the error, which is, in our case, the distance to the target point. The performance indexes considered in this work are the following: (1) integral square error (ISE), (2) integral absolute error (IAE), (3) integral time squared error (ITSE), and (4) integral time absolute error (ITAE). Note that the last two also include the time in the analysis [35]. Table 1 shows the performance indexes to compare the results of each algorithm. Note that all the indexes showed similar results, which is logical in view of the above results. The best performance was shown by the RL4M algorithm. For that reason, we selected this algorithm to compare with the other approaches.

Table 1. Performance indexes for each algorithm. In bold, the better case.

Index	RL 500 m	RL 1M	RL 3M	RL4M	RL 5M	RL 8M	RL 10M
IAE	46.81	45.84	46.25	**45.19**	45.67	45.66	45.52
ISE	198.29	195.68	197.05	**191.79**	194.27	194.47	194.23
ITSE	804.22	781.07	791.03	**755.07**	775.13	776.11	771.11
ITAE	234.15	222.85	227.18	**217.66**	221.73	221.49	219.67

4.2. Comparison between Different Approaches (RL, Villela and IPC)

To establish a basis for comparison of the results of the different control algorithms with the spherical robot, in addition to the RL, the Villela and the IPC algorithms were selected. In both algorithms, the parameters were selected based on our previous experience with the implementation of these experiments with the Khepera IV robot (see for example [25,27]). In the Villela algorithm, the parameters were the following: $V_{max} = 1$ and $\omega_{max} = \pi/2$. For the IPC algorithm, the parameters were the following: $K_v = 0.15$, $K_p = 1.5$, $K_i = 0.000001$, $V_{max} = 1$ and $\omega_{max} = \pi/2$.

Figure 11 shows the results of the position control experiment for the different algorithms (Villela, IPC, and RL). As in the previous case, the red arrow represents the initial orientation of the robot and the red cross represents the target point. The lines describe the trajectories followed by the robot for each control law. The initial position of the robot is represented by the base of the arrow at (0; 0), and the target point located at (5; 0).

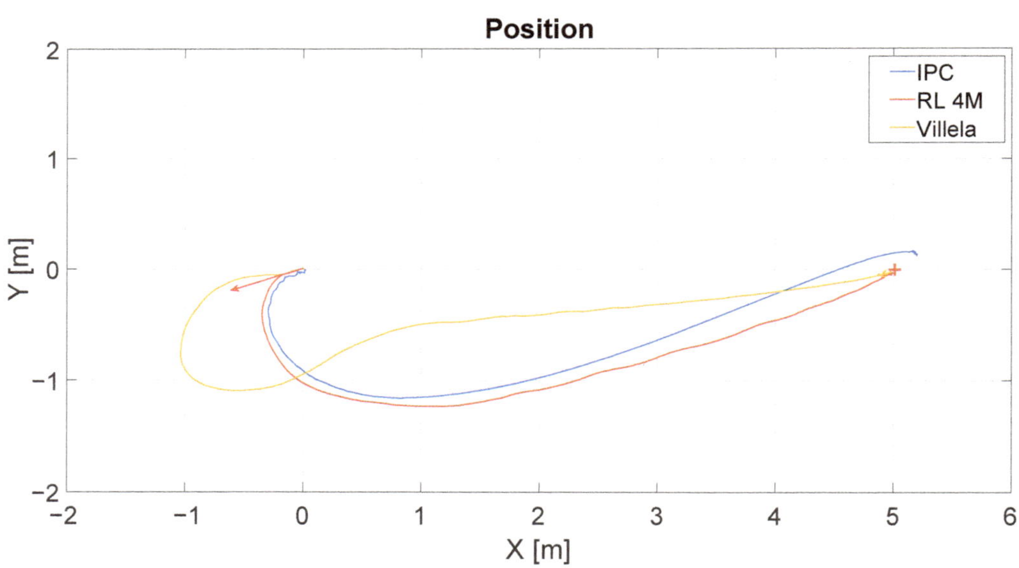

Figure 11. Obtained trajectories for each control algorithm (Villela, IPC and RL4M).

As can be seen, the IPC and RL4M algorithms describe similar trajectories, while Villela's approach shows the worst trajectory. In order to provide a better comparison, we can analyze the graph of distance vs. time. Figure 12 shows the distance to the destination point for these experiments. The y-axis represents the distance in meters and the x-axis represents the time in seconds.

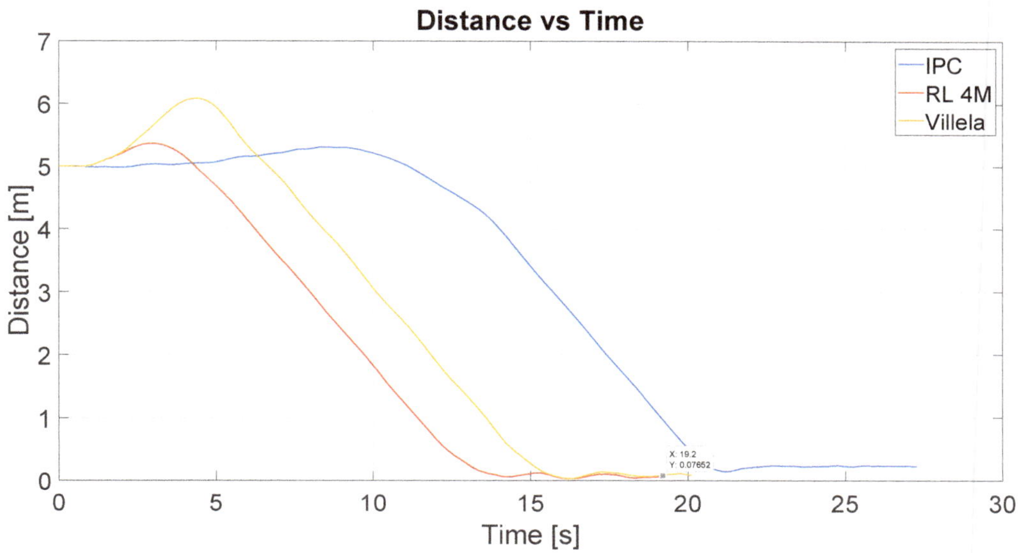

Figure 12. Distance vs. time of all experiences (RL, Villela and IPC).

As can be seen, the better performance was demonstrated by the RL4M algorithm, which took around 14 s to reach the destination point. Note that, in the previous figure, it appears that IPC had a similar trajectory to RL4M, but when the time is taken into account in the analysis, the differences are clearer. With the IPC algorithm, the robot took more than 20 s to reach the destination point. So the trajectory was similar but took more time and the performance was the worst of all, while RL4M showed the best behaviour.

Table 2 shows the performance indexes for all algorithms. As can be seen, as was expected, the best performance was shown by the RL4M algorithm, which confirms the previous results.

Table 2. Performance indexes for each algorithm in the position control experiment. In bold, the better case.

Index	Villela	RL4M	IPC
IAE	58.59	**45.39**	85.28
ISE	273.33	**191.93**	388.68
ITSE	1.334.0	**754.55**	2.985.5
ITAE	338.50	**220.96**	743.39

4.3. Path Following

To test the control strategies in a different scenario, we implemented a path-following example [36–38]. This experiment is widely known in the field of mobile robot control because it is used to demonstrate the behaviour of the implemented control algorithm. It consists of "dynamic" position control of the robot in which the reference point constantly changes to describe a trajectory by joining all points. The result is that the robot follows the points one by one to create the trajectory. Figure 13 shows the implementation of this experiment with the spherical robot in CoppeliaSim.

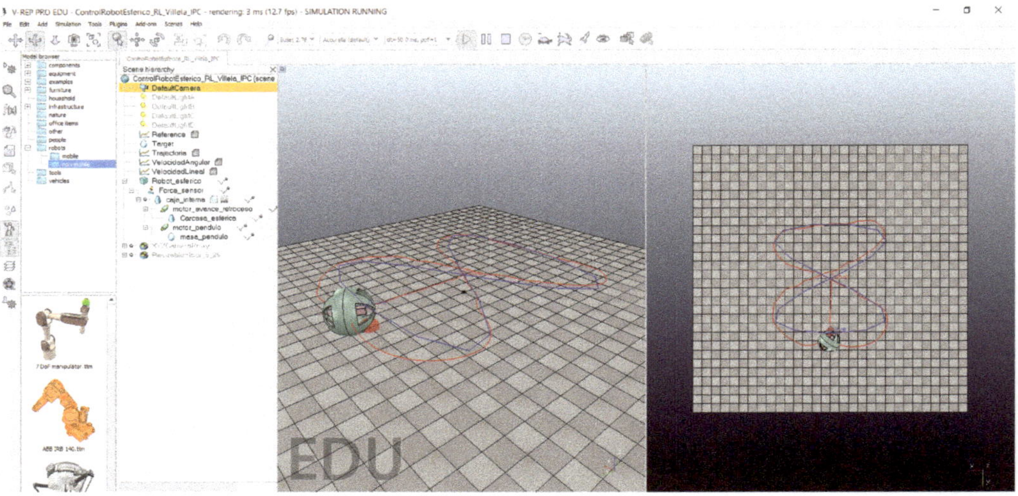

Figure 13. Path following of a Lissajous figure.

Figure 14 shows the trajectories described by the robot for different control algorithms: RL4M (red line), Villela (green line) and IPC (violet line). The dashed line represents the trajectory that the robot receives as a reference. As can be seen, the robot follows the trajectory with different behaviours for all the algorithms.

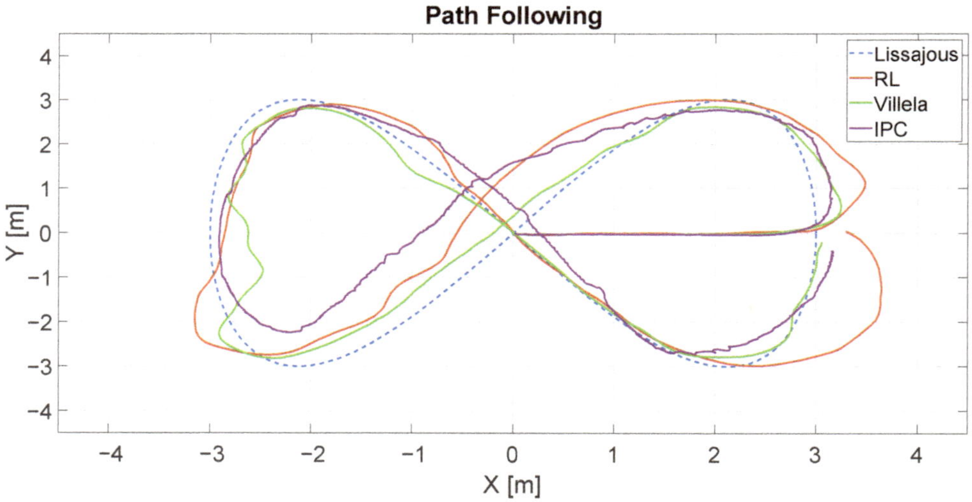

Figure 14. Path following example for Villela, IPC and RL4M.

At first glance, it appears that the best performance was shown by the Villela algorithm. To perform a better comparison, we calculated the performance indexes for all algorithms. Table 3 shows these results.

As can be seen, the lowest values in all indexes were for the Villela algorithm, which means that, for this algorithm, the robot followed the trajectory better.

Table 3. Performance indexes for each algorithm in the path-following experiment. In bold, the better case.

Index	RL4M	Villela	IPC
IAE	87.79	**85.12**	212.07
ISE	162.91	**162.12**	384.11
ITSE	4.18×10^3	$\mathbf{3.78 \times 10^3}$	2.34×10^4
ITAE	2.25×10^3	$\mathbf{2.02 \times 10^3}$	1.32×10^4

4.4. Multi-Agent Formation Control

This experiment was based on [22,39] and consisted of making a formation in a cooperative and decentralized way. One robot acted as the leader and the rest as followers. The positions of the follower's robots were controlled as in the previous experiment. To make a formation, the followers' robots have to reach a position using the leader position as the reference. Equation (20) shows how the velocity of the leader robot is calculated as a function of its own position error (E_{pm}) and the followers' errors in the formation (E_f).

$$v_m(t) = K_p E_{pm}(t) - K_f E_f(t) \tag{20}$$

The values of K_p and K_f are manually adjusted to control the influence of each error in the velocity of the leader robot. If $K_f = 0$, the errors of the followers in the formation are not taken into account, and the control is made in a non-cooperative way because the leader robot does not consider the errors of the followers. Equation (21) shows how the formation error is calculated.

$$E_f(t) = \sum_{i=1}^{N} E_{pi}(t) \tag{21}$$

Figure 15 shows this experiment in the CoppeliaSim simulator for the RL algorithm. For the leader robot, the reference is the target point at the left of the image (red semi-sphere) and, for the followers, the target points are their positions in the formation. In this case, the followers use the position of the leader to make a triangular formation around it. Both followers are situated at a fixed 4 m from the leader and 30° and −30° behind it, respectively.

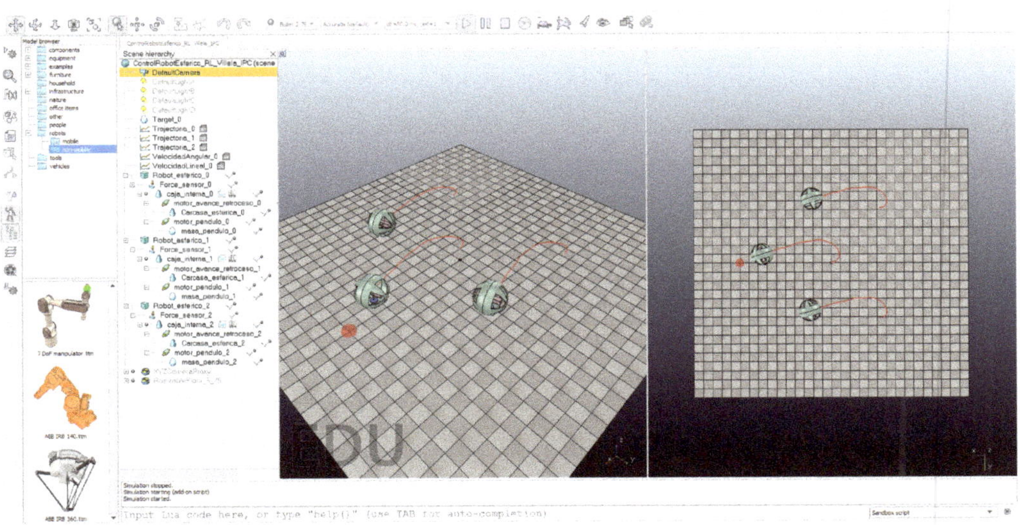

Figure 15. Formation control experiment.

As can be seen, initially, the robots make a triangle using the leader robot as a reference. At the end of the experiment, the followers maintain the formation around the leader robot. Figure 16 shows the data for this experiment. The blue small circle represents the initial position of the leader robot and the green cross represents the destination point. The blue line represents the trajectory described by the leader robot and the red and orange lines represent the trajectories described by the following robots. As can be seen, the robots maintain the formation during the experiment.

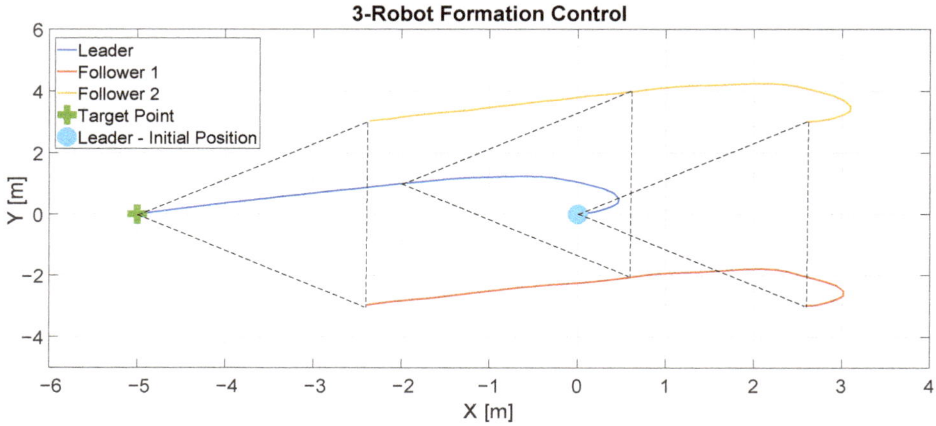

Figure 16. Results of the formation control experiment.

Figure 17 shows the results of this experiment for all the algorithms. The y-axis represents the distance travelled by the robots. The leader robot is represented by the blue lines and they show the distance from the robot to its target point. The followers are represented by the red and orange lines, which show the distance between each follower and the leader.

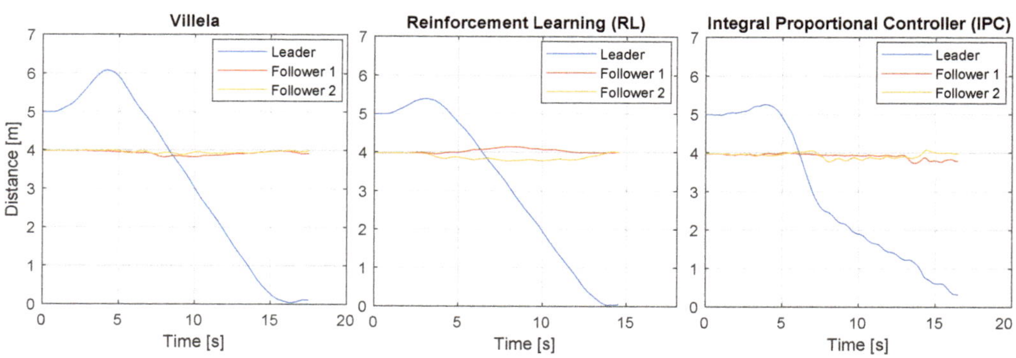

Figure 17. Positions of the robots for all algorithms.

As can be seen, at the beginning, the leader robot moves away from the target because, initially, the target is at its back. After a few seconds, the leader reaches the destination point, while the followers maintain a constant distance to the leader (4 m), which means that the formation is maintained during the experience.

By simple visual inspection, it can be observed that the RL algorithm showed better performance because the leader robot reached the destination point in less time and travelled the shortest distance. However, to be on the safe side, we calculated the performance indexes for each experiment to establish a more accurate comparison. Tables 4–6 show the performance indexes for each algorithm and robot in each experiment.

Table 4. Performance indexes for the Villela algorithm in the formation control experiment.

Robot	IAE	ISE	ITSE	ITAE
Leader	57.97	271.45	1.32×10^3	332.37
Follower 1	68.41	268.24	2.34×10^3	599.01
Follower 2	68.90	272.06	2.38×10^3	603.86
Sum	195.28	811.76	6049.30	1535.24

Table 5. Performance indexes for the RL4M algorithm in the formation control experiment.

Robot	IAE	ISE	ITSE	ITAE
Leader	46.00	197.80	802.10	224.79
Follower 1	58.41	235.35	1.72×10^3	427.47
Follower 2	56.12	217.32	1.57×10^3	408.16
Sum	**160.53**	**650.48**	**4103.30**	**1060.42**

Table 6. Performance indexes for the IPC algorithm in the formation control experiment.

Robot	IAE	ISE	ITSE	ITAE
Leader	95.52	377.80	3.11×10^3	1.04×10^3
Follower 1	129.12	506.93	8.24×10^3	2.11×10^3
Follower 2	129.44	509.42	8.38×10^3	2.13×10^3
Sum	354.09	1394.17	19,747.70	5297.60

All experiences were generated with the same initial conditions, and only the control algorithm was changed in each case. We can then compare the results using Equation (21), which is shown in the row Sum in Tables 4–6. As can be seen, the least values in all cases for the Sum row were observed for the RL experiment, which confirms that the better performance was produced by this algorithm.

5. Conclusions

This article presents the design and implementation of a model of an actual spherical robot, the method of movement of which is based on an internal pendulum. The design of the model was developed using the 3D-design and modeling software, Autodesk Fusion 360. The model was incorporated piece-by-piece into the CoppeliaSim simulator where the hardware was assembled; the position control strategy was programmed in the LUA and Python programming languages to verify its operation. Different experiments, concerning position control, path-following, and multi-robot formation control were performed. The results obtained with the different control laws and experiments showed that the design and implementation of the robot model were satisfactory since its behaviour was similar to that previously obtained with a differential model of the Khepera IV robot. Future work will include performing these experiments with the actual robot in the platform previously implemented in our laboratory [22]. This is a challenging task due to the complexity of obtaining the absolute position of the robot in the platform.

Author Contributions: Conceptualization, G.M., R.C. and G.F.; methodology, G.M. and G.F.; software, G.M., G.G. and R.C.; validation, E.F., K.S. and A.M.; formal analysis, G.M.; investigation, E.F. and G.M.; resources, G.F. and S.D.-C.; data curation, G.G.; writing—original draft preparation, E.F., G.M. and G.G.; writing—review and editing, G.M., E.F., G.F. and S.D.-C.; visualization, E.F. and G.M.; supervision, G.F. and S.D.-C.; project administration, G.F.; funding acquisition, G.F. and S.D.-C. All authors have read and agreed to the published version of the manuscript.

Funding: This research was supported, in part, by the Chilean Research and Development Agency (ANID) under Project FONDECYT 1191188, The National University of Distance Education under Project 2021V/-TAJOV/00, and the Ministry of Science and Innovation of Spain under Project PID2019-108377RB-C32.

Institutional Review Board Statement: Not applicable.

Informed Consent Statement: Not applicable.

Data Availability Statement: Not applicable.

Conflicts of Interest: The authors declare no conflict of interest.

References

1. Jones, J.; Seiger, B.; Flynn, A. *Mobile Robots: Inspiration to Implementation*; CRC Press: Boca Raton, FL, USA , 1998.
2. Siegwart, R.; Nourbakhsh, I.R.; Scaramuzza, D. *Introduction to Autonomous Mobile Robots*; MIT Press: Cambridge, MA, USA, 2011.
3. Fankhauser, P.; Gwerder, C. Modeling and Control of a Ballbot. Ph.D. Thesis, Eidgenössische Technische Hochschule Zürich, Zürich, Switzerland, 2010.
4. Rubio, F.; Valero, F.; Llopis-Albert, C. A review of mobile robots: Concepts, methods, theoretical framework, and applications. *Int. J. Adv. Robot. Syst.* **2019**, *16*. [CrossRef]
5. Chen, W.H.; Chen, C.P.; Tsai, J.S.; Yang, J.; Lin, P.C. Design and implementation of a ball-driven omnidirectional spherical robot. *Mech. Mach. Theory* **2013**, *68*, 35–48. [CrossRef]
6. Cook, G.; Zhang, F. *Mobile Robots: Navigation, Control and Sensing, Surface Robots and AUVs*; John Wiley & Sons: Hoboken, NY, USA, 2020.
7. Bujňák, M.; Pirník, R.; Rástočný, K.; Janota, A.; Nemec, D.; Kuchár, P.; Tichý, T.; Łukasik, Z. Spherical Robots for Special Purposes: A Review on Current Possibilities. *Sensors* **2022**, *22*, 1413. [CrossRef] [PubMed]
8. Alexey, B.; Alexander, K.; Yury, K.; Anton, K. Stabilization of the motion of a spherical robot using feedbacks. *Appl. Math. Model.* **2019**, *69*, 583–592. [CrossRef]
9. Kilin, A.; Pivovarova, E.; Ivanova, T.B. Spherical robot of combined type: Dynamics and control. *Regul. Chaotic Dyn.* **2015**, *20*, 716–728. [CrossRef]
10. Asiri, S.; Khademianzadeh, F.; Monadjemi, A.; Moallem, P. The design and development of a dynamic model of a low-power consumption, two-pendulum spherical robot. *IEEE/Asme Trans. Mechatron.* **2019**, *24*, 2406–2415. [CrossRef]
11. Ma, L.; Sun, H.; Song, J. Fractional-order adaptive integral hierarchical sliding mode control method for high-speed linear motion of spherical robot. *IEEE Access* **2020**, *8*, 66243–66256. [CrossRef]
12. Karavaev, Y.L.; Kilin, A.A. The dynamics and control of a spherical robot with an internal omniwheel platform. *Regul. Chaotic Dyn.* **2015**, *20*, 134–152. [CrossRef]
13. Galán, D.; Fabregas, E.; Garcia, G.; Sáenz, J.; Farias, G.; Dormido-Canto, S.; Dormido, S. Online virtual control laboratory of mobile robots. *IFAC-PapersOnLine* **2018**, *51*, 316–321. [CrossRef]
14. Peralta, E.; Fabregas, E.; Farias, G.; Vargas, H.; Dormido, S. Development of a Khepera IV Library for the V-REP Simulator. In Proceedings of the 11th IFAC Symposium on Advances in Control Education ACE, Bratislava, Slovakia, 1–3 June 2016; Volume 49, pp. 81–86. [CrossRef]
15. Farias, G.; Fabregas, E.; Peralta, E.; Torres, E.; Dormido, S. A Khepera IV library for robotic control education using V-REP. *IFAC-PapersOnLine* **2017**, *50*, 9150–9155. [CrossRef]
16. Pinciroli, C.; Trianni, V.; O'Grady, R.; Pini, G.; Brutschy, A.; Brambilla, M.; Mathews, N.; Ferrante, E.; Di Caro, G.; Ducatelle, F.; et al. ARGoS: A modular, parallel, multi-engine simulator for multi-robot systems. *Swarm Intell.* **2012**, *6*, 271–295. [CrossRef]
17. Michel, O. Webots: Symbiosis between virtual and real mobile robots. In Proceedings of the International Conference on Virtual Worlds, Paris, France, 1–3 July 1998; Springer: Berlin/Heidelberg, Germany, 1998; pp. 254–263.
18. Fabregas, E.; Farias, G.; Dormido-Canto, S.; Dormido, S. RFCSIM simulador interactivo de robótica móvil para control de formación con evitación de obstáculos. In Proceedings of the XVI Congreso Latinoamericano de Control Automático, Cancún, México, 14–17 October 2014.
19. Rohmer, E.; Singh, S.P.; Freese, M. V-REP: A versatile and scalable robot simulation framework. In Proceedings of the 2013 IEEE/RSJ International Conference on Intelligent Robots and Systems, Tokyo, Japan, 3–7 November 2013; IEEE: Piscataway, NY, USA, 2013; pp. 1321–1326. [CrossRef]
20. Fabregas, E.; Farias, G.; Peralta, E.; Vargas, H.; Dormido, S. Teaching control in mobile robotics with V-REP and a Khepera IV library. In Proceedings of the 2016 IEEE Conference on Control Applications (CCA), Buenos Aires, Argentina, 19–22 September 2016; pp. 821–826. [CrossRef]
21. Farias, G.; Torres, E.; Fabregas, E.; Vargas, H.; Dormido-Canto, S.; Dormido, S. Navigation control of the Khepera IV model with OpenCV in V-REP simulator. In Proceedings of the 2018 IEEE International Conference on Automation/XXIII Congress of the Chilean Association of Automatic Control (ICA-ACCA), Concepcion, Chile, 17–19 October 2018; IEEE: Piscataway, NY, USA, 2018; pp. 1–6.

22. Farias, G.; Fabregas, E.; Peralta, E.; Vargas, H.; Dormido-Canto, S.; Dormido, S. Development of an easy-to-use multi-agent platform for teaching mobile robotics. *IEEE Access* **2019**, *7*, 55885–55897. [CrossRef]
23. Zhan, Q.; Cai, Y.; Yan, C. Design, analysis and experiments of an omni-directional spherical robot. In Proceedings of the 2011 IEEE International Conference on Robotics and Automation, Shanghai, China, 9–13 May 2011; IEEE: Piscataway, NY, USA, 2011; pp. 4921–4926.
24. Villela, V.J.G.; Parkin, R.; Parra, M.L.; González, J.M.D.; Liho, M.J.G. A wheeled mobile robot with obstacle avoidance capability. *Inge. Mecánica. Tecnol. y Desarro.* **2004**, *1*, 159–166.
25. Fabregas, E.; Farias, G.; Aranda-Escolástico, E.; Garcia, G.; Chaos, D.; Dormido-Canto, S.; Bencomo, S.D. Simulation and experimental results of a new control strategy for point stabilization of nonholonomic mobile robots. *IEEE Trans. Ind. Electron.* **2019**, *67*, 6679–6687. [CrossRef]
26. Farias, G.; Garcia, G.; Montenegro, G.; Fabregas, E.; Dormido-Canto, S.; Dormido, S. Position control of a mobile robot using reinforcement learning. *IFAC-PapersOnLine* **2020**, *53*, 17393–17398. [CrossRef]
27. Farias, G.; Garcia, G.; Montenegro, G.; Fabregas, E.; Dormido-Canto, S.; Dormido, S. Reinforcement Learning for Position Control Problem of a Mobile Robot. *IEEE Access* **2020**, *8*, 152941–152951. [CrossRef]
28. Nagai, M. Control System for a Spherical Robot. Master's Thesis, Luleå University of Technology, Luleå, Sweden, 2008.
29. Autodesk Fusion 360. Available online: https://www.autodesk.com/products/fusion-360/overview (accessed on 30 Septemble 2021).
30. CoppeliaSim. Available online: https://www.coppeliarobotics.com/downloads (accessed on 30 Septemble 2021).
31. Sutton, R.S.; Barto, A.G. *Reinforcement Learning: An Introduction*; MIT Press: Cambridge, MA, USA, 2018.
32. Sumanas, M.; Petronis, A.; Bucinskas, V.; Dzedzickis, A.; Virzonis, D.; Morkvenaite-Vilkonciene, I. Deep Q-Learning in Robotics: Improvement of Accuracy and Repeatability. *Sensors* **2022**, *22*, 3911. [CrossRef]
33. Watkins, C.J.C.H. Learning from delayed rewards. Ph.D. Thesis, King's College, Cambridge United Kingdom, Cambridge, UK, 1989.
34. Watkins, C.J.C.H.; Dayan, P. Technical Note: Q -Learning. *Mach. Learn.* **1992**, *8*, 279–292. [CrossRef]
35. Sultan, A.J. Optimal load frequency control in a single area power system based Genetic Algorithm. *Int. J. Sci. Eng. Res.* **2014**, *5*, 2196–2200. [CrossRef]
36. Coelho, P.; Nunes, U. Path-following control of mobile robots in presence of uncertainties. *IEEE Trans. Robot.* **2005**, *21*, 252–261. [CrossRef]
37. Lapierre, L.; Soetanto, D. Nonlinear path-following control of an AUV. *Ocean. Eng.* **2007**, *34*, 1734–1744. [CrossRef]
38. Rayguru, M.; Elara, M.; Ramalingam, B.; Viraj, J.M.M.A.; Bhagya, P.S.S.M. A Path Tracking Strategy for Car Like Robots with Sensor Unpredictability and Measurement Errors. *Sensors* **2020**, *20*, 3077. [CrossRef]
39. Lawton, J.R.; Beard, R.W.; Young, B.J. A decentralized approach to formation maneuvers. *IEEE Trans. Robot. Autom.* **2003**, *19*, 933–941. [CrossRef]

sensors

MDPI

Article

Coordinated Multi-Robotic Vehicles Navigation and Control in Shop Floor Automation

Gregor Klančar [1,*] and Marija Seder [2]

[1] Faculty of Electrical Engineering, University of Ljubljana , Tržaška 25, SI-1000 Ljubljana, Slovenia
[2] Faculty of Electrical Engineering and Computing, University of Zagreb, Unska 3, 10000 Zagreb, Croatia; marija.seder@fer.hr
* Correspondence: gregor.klancar@fe.uni-lj.si

Abstract: In this paper, we propose a global navigation function applied to model predictive control (MPC) for autonomous mobile robots, with application to warehouse automation. The approach considers static and dynamic obstacles and generates smooth, collision-free trajectories. The navigation function is based on a potential field derived from an E* graph search algorithm on a discrete occupancy grid and by bicubic interpolation. It has convergent behavior from anywhere to the target and is computed in advance to increase computational efficiency. The novel optimization strategy used in MPC combines a discrete set of velocity candidates with randomly perturbed candidates from particle swarm optimization. Adaptive horizon length is used to improve performance. The efficiency of the proposed approaches is validated using simulations and experimental results.

Keywords: navigation; model predictive control; path planing; mobile robots; warehouse automation

Citation: Klančar, G.; Seder, M. Coordinated Multi-Robotic Vehicles Navigation and Control in Shop Floor Automation. *Sensors* **2022**, *22*, 1455. https://doi.org/10.3390/s22041455

Academic Editor: Jesús Ureña

Received: 16 January 2022
Accepted: 31 January 2022
Published: 14 February 2022

1. Introduction

Mobile robotic platforms have found numerous applications over the past decade, a significant portion of which can be attributed to intralogistics and transportation in modern manufacturing, warehousing, and utility markets [1–3]. Although numerous sites have been automated by either automated guided vehicles (AGV) or autonomous mobile robots (AMR) with impressive deployments, the market is expected to grow by about 30% over the next five years [4,5].

Since the first AGV was built in 1953 [3], AGVs have evolved into today's solution, which is the standard in automating internal logistics. Typically, AGVs move along predefined paths and can only deliver to fixed points along the path. This makes these transportation systems simpler and more robust. Since movement is limited to fixed paths, the complexity of path planning and coordinating multiple AGVs is reduced. Nevertheless, planning collision-proof safe paths for a group of AGVs and creating and optimizing schedules to achieve better performance (higher throughput and less likely occurrence of conflicts) remains a challenging task [6]. Path planning is usually solved using graph-based search algorithms such as A*-based search, where optimal approaches [7,8] are feasible for a smaller number of vehicles since the computational complexity is exponential with the number of vehicles. The coordination overhead in multi-AGV systems is further reduced by suboptimal approaches, where the problem is decoupled from finding individual vehicles and conflicts are resolved by assigning traffic rules, priorities, or distributed multi-agent negotiations [6,9,10].

AMRs (unlike AGVs) are more flexible (in terms of their navigation capabilities and the services they can provide) and can move freely in dynamic environments where they locate, navigate, and act autonomously [4]. Free space is mapped based on knowledge of static obstacles, and dynamic obstacles are avoided using sensors. Since movement is not restricted to predefined paths but is possible in the continuum of obstacle-free space, the complexity of path planning must be reduced. A common approach is to discretize the environment into cells of equal size and use grid-based path planning [11,12]. Since pathfinding usually examines only 4 or 8 neighbourhood directions, the paths obtained are not smooth.

Another approach is to apply a discrete set of motion primitives or actions that a vehicle can apply to advance to new locations. The motion primitives can be Bezier curves [13,14], clothoids [15], or other smooth curvature curves [16–18]. This usually results in smoother paths that a vehicle can easily follow. Different optimization strategies can be used to select suitable motion primitives. In high-dimensional spaces, randomised planners such as the Rapid Exploring Random Tree (RRT*) and the kinodynamic RRT* are popular choices [3,19]. A state lattice graph can be constructed from a discrete set of motion primitives that have smooth curvature transitions in the joints [20,21]. Graph search algorithms such as A* [22,23], D* [11,24], and E* [25] can be used to find the final path. The computational complexity can be addressed by hybrid search approaches such as Hybrid A* or HE * [14,26], where a computationally efficient discrete graph-based search is applied to obtain the heuristics for more efficient construction and search of the lattice graph, where the motion primitives form the edges of the graph.

Potential field-planning methods are also popular, where the potential function for online navigation can be used to guide the search or control algorithm. The goal with minimum potential value can be achieved by simply following the direction of the steepest descent of the potential field. A common problem of the potential field is local minima in which the robot may be trapped. Several approaches have been proposed to avoid local minima. Concave obstacles can be simply modelled as convex [27] in the environment map, or an adaptive potential field can be generated using multiple points of attraction instead of just one in the target [28]. It is also possible to modify the potential field in unstable equilibrium by introducing perturbations into the field [29] or adding virtual obstacles to repel the robot from the local minima [30]. The potential field can also be interpolated from a discrete cost map obtained from an optimal grid-based search [31,32]. Relying only on the reactive behavior of a potential field may result in unwanted oscillations in the presence of obstacles where alternating repulsive and attractive fields may cause approaching and moving-away behavior [33]. Therefore, prediction capabilities are needed to achieve more deliberative actions where planning and control are combined in receding dynamic window approaches or trajectory roll-out algorithms considering convergent navigation function, such as in [31,34–36]. Here, the obtained performance depends on a control law and uses an objective function, which needs to incorporate mapped static obstacles and sensed dynamic obstacles to find feasible optimal trajectories in a prediction horizon.

Moving obstacle avoidance is required for efficient multiple vehicle navigation. Coordinated motion of multiple vehicles can be dealled by assigning traffic rules in decentralised manner as in [10] or by combining a centralized supervisor, which detects collisions and assignes priorities for decentralised planner and scheduling for collision avoidance [37]. The decentralised decoupled approach is proposed in [18], where vehicles first plan optimal paths independently; then, conflict resolution is performed based on a priority scheme. In [38], a model predictive scheme is proposed where local deviations from the existing reference path are optimized considering collision avoidance with static and moving obstacles. In [39], an integration of the focused D* graph search algorithm for path planning and the dynamic window algorithm for generating admissible robot trajectories around the planned global path is proposed.

In this paper, the main contributions are the following. We propose a global navigation function applied in model predictive control to safely navigate the vehicle to the goal destination. The navigation function depends on a potential field for the environment layout and the driving direction. The potential function for a known target is computed in advance by an E* graph search algorithm on a discrete rectangular grid. A smooth surface with arbitrary potential values and slope directions is obtained by bicubic interpolation. It allows navigation from any location to the target and can be precomputed for any known target to which AMR must deliver.

Constrained Model Predictive Control (MPC) is a method of finding optimal trajectories given the proposed navigation function and constraints on robot kinematics, maximum velocities and accelerations, convergent behavior, and the coordination of multiple robots. MPC combines local motion planning and control in the presence of static and dynamic obstacles.

Adaptive horizon length is introduced in MPC to improve performance in terms of safety and achieved curve optimality.

A novel optimization strategy for MPC is proposed that combines a discrete set of command velocities proposed in [36] with randomly perturbed particle swarm optimization candidates. This approach extends the navigation of a single robot [32] to multiple robots.

Coordinated navigation in the presence of multiple vehicles is obtained locally as a constraint in the MPC objective function. The approach assumes that cooperative vehicles share their planned trajectories within the prediction horizon. For non-cooperative objects, the motion trajectories must be estimated from measurements.

The performance of the proposed approaches is illustrated by several examples.

2. Vehicle Autonomous Navigation and Control

For an example of a simple production layout, see Figure 1, where robots typically need to transport material between known, fixed locations. Suppose a destination is the dropoff point shown in Figure 1, to which robots must deliver material from several other locations, such as pickup point, workstations, and storage aria. A navigation function can be created to guide the robot safely from any starting point in the environment to the desired destination.

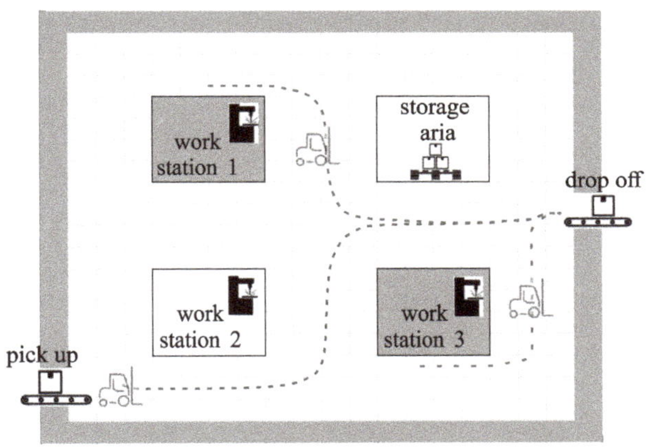

Figure 1. Production layout with multiple known and fixed delivery points and defined free corridors. Shown are robot delivery paths from three starting locations to the same destination. The paths can be effectively determined by a single navigation function, as shown in Figure 6.

Basic idea of applied navigation and control diagram is illustrated in Figure 2.

In Figure 1, three paths are shown that are automatically determined based on the derived navigation function shown in Figure 6, which is interpolated at runtime from a stored discrete potential field of the layout, as shown in Figure 3. Other navigation functions are determined for other desired frequent destinations such as pickup belts, workstations, or battery charging arias. Such navigation functions can be computed in advance if the pickup and drop-off locations are fixed and the robots can move in predefined corridors. This leads to a computationally efficient approach with high-quality trajectories that takes static obstacles into account during design and can also be extended to include detected dynamic obstacles. Further details of the navigation function and control algorithm are presented below.

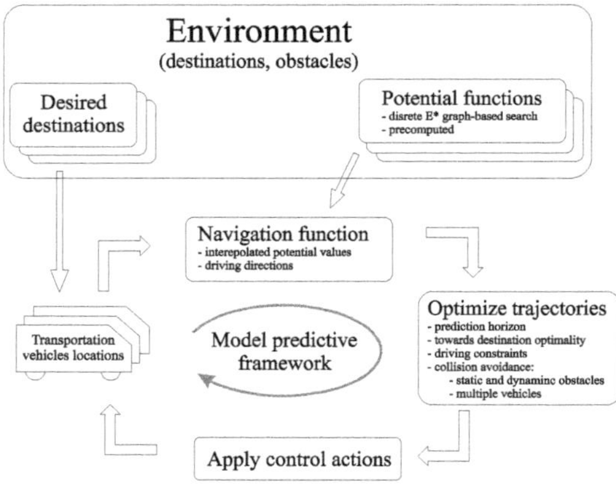

Figure 2. Basic idea of navigation and control of multiple vehicles based on the proposed navigation function and model predictive control.

2.1. Concept of Navigation

The navigation function $N(x, y, \varphi)$ is used to drive a wheeled robot with position x, y and orientation φ) safely between obstacles to the goal. A control algorithm therefore steers the robot to locations where $N(x, y, \varphi)$ decreases. The unimodal potential function is a good choice for $N(x, y, \varphi)$ because it has a single minimum ($N(x, y, \varphi) = 0$) at the goal and no local minima where the control algorithm might get stuck. Additionally, $N(x, y, \varphi)$ must have the highest values at the obstacles. A graph search algorithm such as D* for dynamic environments can be used to obtain such a potential function $U(x, y)$ as shown in Figure 3. The value of $U(x, y)$ represents the distance to the target cell, which is computed as the sum of the distances between cells (d_c) along the path. Such a search is computationally efficient since it is performed on a discrete grid of the environment but is not suitable for a control algorithm since $U(x, y)$ is constant for any robot position within a discrete cell. Therefore, the grid-based navigation function must be modified to obtain a unique value for each position within a cell that retains the property of a single minimum [31]. In the following, we propose a bicubic interpolation approach.

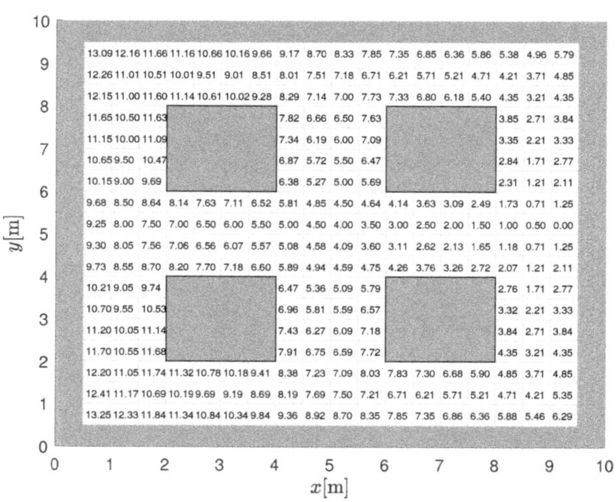

Figure 3. 3D view of the discrete potential function from obtained a grid-based search with 0.5 m resolution, a target position at $x = 9.25$ m, $y = 5.25$ m, and occupied cells with $U(x, y) = \infty$ (grey cells).

2.2. Bicubic Interpolation

To obtain a smooth interpolated potential $P(x,y)$ from a discrete potential $U(x,y)$, bicubic interpolation [40] is used. For a given arbitrary position $[x,y]^T$ within a cell **M**, an interpolated potential is calculated based on a 4×4 cell neighborhood, as shown in Figure 4. Depending on the quadrant of cell **M** in which the point $[x,y]^T$ is located, an appropriate four-cell neighborhood is determined whose centers form a square, as shown in Figure 4 shown by a dashed line.

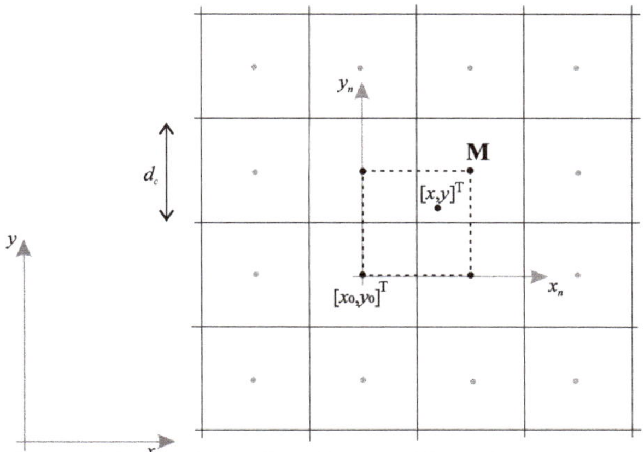

Figure 4. Selection of the cell neighbourhood for the bicubic interpolation of the potential field based on a point $[x,y]^T$ within the cell **M**.

The normalized coordinates $x_n, y_n \in [0,1]$ are defined as $x_n = \frac{x-x_0}{d_c}$, $y_n = \frac{y-y_0}{d_c}$, where the origin $[x_0, y_0]$ is defined by the lower left corner of the dashed square and d_c is the cell size. The interpolated and discrete potential in normalized coordinates are expressed as $P_n(x_n,y_n) = P(x,y)$ and $U_n(x_n,y_n) = U(x,y)$, respectively. Define the potential and estimated partial derivatives for the four adjacent cell centers (corners of a dashed square in Figure 4)

$$p_{rc} = U_n(x_n,y_n)\Big|_{x_n=r,\ y_n=c}$$

$$f_{xrc} = \frac{\partial P_n}{\partial x_n}\Big|_{x_n=r,\ y_n=c} \approx \frac{U_n(r+1,c) - U_n(r-1,c)}{2}$$

$$f_{yrc} = \frac{\partial P_n}{\partial y_n}\Big|_{x_n=r,\ y_n=c} \approx \frac{U_n(r,c+1) - U_n(r,c-1)}{2}$$

$$f_{xyrc} = \frac{\partial^2 P_n}{\partial x_n \partial y_n}\Big|_{x_n=r,\ y_n=c} \approx \frac{U_n(r+1,c+1) - U_n(r-1,c+1) - U_n(r+1,c-1) - U_n(r-1,c-1)}{4},$$

where $r,c \in \{0,1\}$ and $U_n(x_n,y_n) = U(x,y)$.

For a given arbitrary position, the interpolated potential is then defined by bicubic interpolation as follows

$$P_n(x_n,y_n) = \begin{bmatrix} 1 & x_n & x_n^2 & x_n^3 \end{bmatrix} \mathbf{A} \begin{bmatrix} 1 & y_n & y_n^2 & y_n^3 \end{bmatrix}^T, \tag{1}$$

where the matrix of coefficients is

$$\mathbf{A} = \begin{bmatrix} 1 & 0 & 0 & 0 \\ 0 & 0 & 1 & 0 \\ -3 & 3 & -2 & -1 \\ 2 & -2 & 1 & 1 \end{bmatrix} \begin{bmatrix} p_{00} & p_{01} & f_{y00} & f_{y00} \\ p_{10} & p_{11} & f_{y10} & f_{y11} \\ f_{x00} & f_{x01} & f_{xy00} & f_{xy01} \\ f_{x10} & f_{x11} & f_{xy10} & f_{xy11} \end{bmatrix} \begin{bmatrix} 1 & 0 & -3 & 2 \\ 0 & 0 & 3 & -2 \\ 0 & 1 & -2 & 1 \\ 0 & 0 & -1 & 1 \end{bmatrix}.$$

And negative gradient of $P(x,y)$ in $[x,y]^T$ is computed in a closed-form as

$$\begin{aligned} \vec{\mathbf{g}}(x,y) &= -\nabla P(x,y) = -\left[\frac{\partial P(x,y)}{\partial x}, \frac{\partial P(x,y)}{\partial y}\right]^T = \\ &= -\frac{1}{d_c}\left[\frac{\partial P(x_n,y_n)}{\partial x_n}, \frac{\partial P(x_n,y_n)}{\partial y_n}\right]^T \end{aligned} . \tag{2}$$

For non-holonomic robots (with the kinematics given in 6), the final navigation function depends on the interpolated potential $P(x,y)$ and on the robot orientation φ

$$\begin{aligned} N(x,y,\varphi) &= P(x,y) + \xi e(\varphi) \\ e(\varphi) &= \min_{k=\{0,1,-1\}} |\angle \vec{\mathbf{g}}(x,y) - \varphi + 2k\pi|, \end{aligned} \tag{3}$$

where $e(\varphi)$ is the absolute orientation error, $\xi > 0$, and $\angle \vec{\mathbf{g}}(x,y)$ is the orientation of the negative gradient .

In Figure 5, the interpolated potential function $P(x,y)$ of the free space and the centrally located target is shown for discrete potentials obtained by A* and E* grid-based searches. Since A* uses 4 and 8 neighbourhood connections, respectively, the gradients of $P(x,y)$ remain multiples of 90° and 45°, respectively (see contours of the same potential in Figure 5). The E* [25] is a dynamic path planning algorithm that can approximate continuous gradients (and contours). It uses 4 neighbour connectivity (such as A* or D*), but instead of one, two parent nodes in orthogonal directions are used to get a better cost-to-goal estimate for each cell. Using the discrete potential field obtained from the E* algorithm, the interpolated potential function (1) is smoother with arbitrary direction of the negative gradient (2).

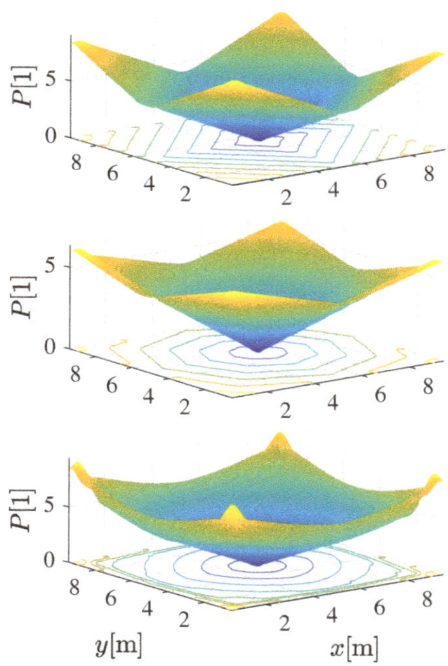

Figure 5. Interpolated potential function $P(x,y)$ based on the discrete potential obtained by A* with 4 and 8 neighbour cells connections and by E*. The obtained gradient is in the directions of multiples of 90° or 45° when 4 or 8 neighbourhood connections are used in A*. While E* can have arbitrary directions, which can be seen from the contours of equal potentials orthogonal to the gradient direction.

An example of an occupied space and its computed interpolated smooth navigation function is given in Figure 6. Additionally, three paths are drawn from different starting points following the negative gradient towards the target with the lowest potential. The obtained paths are orthogonal to the contours of the same potential (see the lower part of Figure 6).

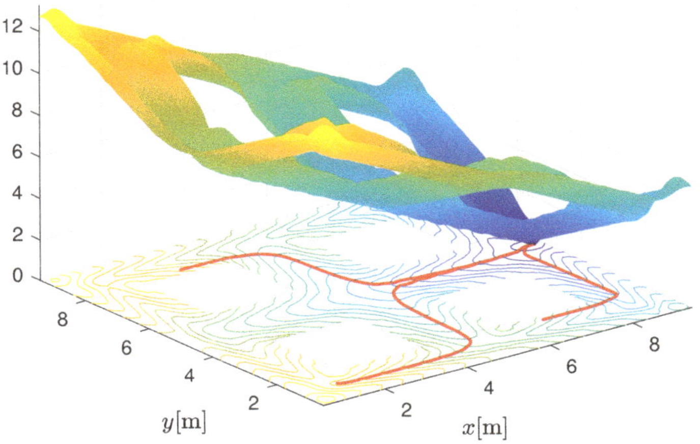

Figure 6. 3D view of the interpolated navigation function $N(x, y, \varphi)$. For clarity, $e(\varphi)$ is set to zero in Equation (3). Three paths are drawn from different starting points, following the negative gradient towards the goal location with the lowest potential value.

3. Coordinated Model Predictive Control

The proposed interpolated potential function Equation (1) allows the simple application of control of a single robot to safely navigate from anywhere to the target while automatically avoiding obstacles. The vehicle only needs to follow the given negative gradient direction Equation (2), and since the gradient has soft transitions (see e.g., Figure 3), feasible trajectories result. Such an approach lacks predictive capabilities and assumes a static environment without any other vehicles.

Therefore, the control behavior is defined as follows. The simple gradient-following reactive behavior is improved by incorporating prediction, so that the current control action also depends on the future states of the vehicles. The navigation function already includes knowledge of a static map in which the space occupied by obstacles has infinite potential. However, observed dynamic obstacles (cooperative obstacles such as other transport vehicles or non-cooperative obstacles such as humans or forklifts controlled by humans, etc.) are not included in the navigation function as this would require constant replanning. Dynamic obstacles are observed by sensors (laser range finder, camera, etc.), and their movement is estimated in the prediction horizon of the controller. A feasible trajectory is determined in the prediction horizon that is consistent with the navigation function, does not conflict with other vehicles or other detected obstacles, and is within the kinematic and dynamic constraints of the vehicle.

3.1. Control Definition

Model predictive control (MPC) is defined as an optimization problem with constraints. Optimal controls $\mathbf{u}(t) = [v(t), \omega(t)]^T$ are found for a differential robot over a prediction horizon h that minimize the objective function J at the current robot state $\mathbf{s}(t) = [x, y, \varphi]^T$

$$
\begin{aligned}
J(\mathbf{s}(t)) \quad &= \min_{\mathbf{u}(i-1)} \textstyle\sum_{i=1}^{h} \left(N(\mathbf{s}(i)) + \mathbf{u}^T(i-1)\mathbf{R}\mathbf{u}(i-1)\right) \\
&\text{subject to:} \\
&\text{- free configuration space (Equation (5))} \\
&\text{- driving constraints (Equation (7))} \\
&\text{- convergent behavior (Equation (11))} \\
&\text{- collision-free with dynamic obstacles (Equation (12))}
\end{aligned}
\tag{4}
$$

where v and ω are translational and angular velocity, i is a shorthand notation for time $t + iT_s$, T_s is the sampling time, and $\mathbf{R}$ is the weighting matrix. Note that MPC considers the robot's motion model and the environment model, where the target and static obstacles are already considered in the navigation function. However, due to kinematic constraints, collisions with static obstacles may still occur. Therefore, a valid trajectory in the horizon must lie in the free configuration space of the static map Q_{free}

$$
\mathbf{s}(i) \in Q_{free} \quad , \quad i = 1, \ldots, h. \tag{5}
$$

Similarly, collisions with dynamic obstacles are considered. For more details, see the Section 3.1.4. The future state of the robot $\mathbf{s}(i) = [x(i), y(i), \varphi(i)]^T$ is predicted using differential drive kinematics

$$
\begin{aligned}
x(i+1) &= x(i) + v(i)T_s \cos\left(\varphi(i) + \tfrac{\omega(i)T_s}{2}\right) \\
y(i+1) &= y(i) + v(i)T_s \sin\left(\varphi(i) + \tfrac{\omega(i)T_s}{2}\right) \\
\varphi(i+1) &= \varphi(i) + \omega(i)T_s.
\end{aligned}
\tag{6}
$$

3.1.1. Driving Constraints

Control actions are constrained by maximum velocities and accelerations by

$$
\begin{aligned}
0 \leq v(i) \leq v_{\max} \quad &, \quad |\omega(i)| \leq \omega_{\max} \\
\tfrac{|v(i)-v(i-1)|}{T_s} \leq a_{\max} \quad &, \quad \tfrac{|\omega(i)-\omega(i-1)|}{T_s} \leq \alpha_{\max},
\end{aligned}
\tag{7}
$$

where $v_{\max}$, $\omega_{\max}$, $a_{\max}$, and $\alpha_{\max}$ are maximum allowable translational and rotational velocities and accelerations.

3.1.2. Length of the Horizon

During the horizon, let the robot travel on an arc, where $v(i)/\omega(i)$ is its radius. A constant arc in the horizon is convenient because it reduces the computational cost of the MPC problem since it only requires the optimization of two parameters. The choice of horizon length affects the driving performance, safety, and computational cost of MPC.

The minimum horizon length ($h = h_{min}$) is chosen so that the robot travelling at maximum speed can safely decelerate to a stop at the end of the prediction horizon

$$
h_{min} = \left\lceil \max\left(\frac{v_{\max}}{a_{\max}T_s}, \frac{w_{\max}}{\alpha_{\max}T_s}\right) \right\rceil + 1. \tag{8}
$$

This prevents the worst case collision with the maximum robot speed and the newly observed (static) object at the end of the prediction horizon. The control $\mathbf{u}(i), i \in 0,1,\ldots,h-1$ in the horizon therefore decreases linearly to zero at the end of the horizon, as follows

$$\mathbf{u}(i) = \begin{cases} \mathbf{u}(0) & , \quad 0 \le i \le h-1-N_{dec} \\ \mathbf{u}(0)\frac{h-1-i}{N_{dec}} & , \quad h-1-N_{dec} < i \le h-1. \end{cases} \tag{9}$$

where the required number of deceleration samples at the end of the horizon is $N_{dec} = \left\lceil \max\left(\frac{v(0)}{a_{\max}T_s}, \frac{w(0)}{\alpha_{\max}T_s}\right)\right\rceil$ at the current robot velocity $\mathbf{u}(0) = [v(0), \omega(0)]^T$.

Choosing a larger horizon ($h > h_{min}$) also increases safety for moving objects because the motion of the obstacle is predicted early enough to find alternative trajectories that efficiently avoid the collision. However, too large a horizon increases the computational cost and may lead to worse trajectories in free space due to the averaging effect since a longer constrained trajectory does not fit as optimally on the surface of the navigation function.

As a compromise, we choose a larger horizon ($h > h_{min}$) and allow the robot to stop even before the end of the horizon (e.g. at $h_{stop} \le h$) and keep the remaining number of samples $h - h_{stop}$ still. This effectively makes the horizon variable (with variable stopping time), and since all optimised trajectories have the same number of samples (h), their objective functions in Equation (11) are still comparable. The velocity profile in the horizon is then

$$\mathbf{u}(i) = \begin{cases} \mathbf{u}(0) & , \quad 0 \le i \le h_{stop}-1-N_{dec} \\ \mathbf{u}(0)\frac{h_{stop}-1-i}{N_{dec}} & , \quad h_{stop}-1-N_{dec} < i < h_{stop} \\ [0,0]^T & , \quad h_{stop} \le i \le h-1 \end{cases} \tag{10}$$

where the candidates for h_{stop} are selected according to the previous optimal curve by evaluating four possibilities $h_{stop} \to h_{stop} + \{0,-1,-2,+1\}$ that need to be in range $(N_{dec} + 1) \le h_{stop} \le h$. Initial value is set to $h_{stop} = h_{min}$.

Optimal control sequence $\mathbf{u}(0), \mathbf{u}(1), \ldots, \mathbf{u}(h-1)$, which minimizes Equation (4), defines the best feasible future trajectory, and its first control action is applied to the robot in the current time. In the next time sample, the procedure repeats.

3.1.3. Convergent Behavior

To ensure convergent behavior of the MPC control, the summands $V(\mathbf{s}(i)) = N(\mathbf{s}(i)) + \mathbf{u}^T(i-1)\mathbf{R}\mathbf{u}(i-1)$ in the criteria Equation (4) will have to decrease in the horizon. This follows from the convergence constraint in Equation (4)

$$N(\mathbf{s}(i)) \ge N(\mathbf{s}(h)), \; i = 1,\ldots,h. \tag{11}$$

In the worst case, if all the candidate control actions in Equation (4) result in trajectories that violate the convergence constraint Equation (11), the robot can still choose the optimal trajectory from the previous control step, shifted by one sample. Since the trajectory slows down at the end of the horizon (see Equations (9) and (10)), the robot will start slowing down earlier. This can happen if some dynamic (non-cooperative) objects block its path.

3.1.4. Preventing Conflicts with Dynamic Obstacles

The environment may contain dynamic obstacles that can be treated as cooperative objects (e.g., other robots) and non-cooperative objects (e.g., forklifts operated by humans). Cooperative objects are assumed to have intentions and trajectories known to the robot for at least a prediction horizon h. The intentions of non-cooperative objects can be estimated from sensor observations (e.g., laser range scans) of their past movements by estimating their velocities and predicting the most likely trajectories in the horizon.

For a given control $\mathbf{u}(i)$ in Equation (4), the robot trajectory $\mathbf{s}(i)$ is collision-safe (CS) if it does not collide with any moving object trajectory (static obstacles are already considered in Equation (5)). Let $o \in \mathcal{O}$ denote all other moving objects from a set $\mathcal{O}$, and let $\mathbf{s}_o(i) = [x_o(i), y_o(i), \varphi_o(i)]^T$ be the location of an object at horizon prediction time $i \in 1, \ldots, h$

$$CS \implies \nexists o \in \mathcal{O} : (||\mathbf{s}(i) - \mathbf{s}_o(i)|| < d_{safe}) \; and \; |\varphi(i) - \arctan(\tfrac{y_o(i)-y(i)}{x_o(i)-x(i)})| < \varphi_{safe}, \quad (12)$$

d_{safe} is the required safety distance between the robot and an object o, and φ_{safe} is the range of angular deviation from the robot's forward motion that can lead to a collision, $\arctan(\cdot)$ is the four-quadrant inverse tangent version. Any robot trajectory that is not collision safe to moving objects is rejected in Equation (4).

The same procedure is followed for all cooperative robots. After one robot determines optimal controls (and trajectory in the horizon), the other robots can adapt by finding collision-proof trajectories to the previous robot. To prevent chattering behavior (where two robots can switch between different optimal controls while avoiding collisions), the selection of optimal controls (trajectories) is done sequentially, which is natural if all robots are controlled from a central computer. If the first robot determines an optimal trajectory that avoids all other robots (taking into account the predicted trajectories of the others), the next robot will adapt by finding collision-safe paths (e.g., swerving to the other side or slowing down). This will automatically lead to consensus since the current predicted trajectories are inherited and known to the others in the same sample. When robots plan and control autonomously, if there is a possibility of collision, they need to only negotiate the order in which they compute their trajectories. Alternatively, they can consider priorities when they are assigned (e.g., for priorities for transportation tasks), as in [10] and [18]. In this way, the first robot computes the optimal trajectory in the prediction horizon, which takes into account the previous trajectories of the others. Additional traffic rules (e.g., swerving to the right for head-on collisions) can also be used [9].

The proposed navigation approach is computationally efficient since the navigation function is precomputed for a static environment, and collisions with dynamic obstacles in the control (Equation (4)) are avoided at runtime.

Note that local minima can still occur (but this is unlikely in practise) if another robot (the second robot or the moving object) approaches another robot (the first robot) exactly from the direction opposite to the negative gradient of the first robot's navigation function. This also means that the second robot uses a different navigation function, to a different destination whose gradient is in the opposite direction to that of the first robot. When this happens, both robots may slow down as this can be cheaper (according to the MPC control cost (Equation (4)) than driving with the increased values of the navigation functions while avoiding a collision.

In this particular case, it may be a good choice to perform a dynamic replanning of the navigation functions with the detected possible collision position of the other robot. In this way, no slowdown or local minima can occur since the navigation function also considers moving obstacles. This replanning requires additional computation time and should therefore be performed incrementally using the dynamic E* algorithm [25]. Moreover, a relatively fine grid should be chosen (e.g., at most half the robot size) to reduce the discretization error when the predicted collision location is snapped to a grid. Therefore, replanning can only occur if an object is in the collision with the robot in the predicted horizon. Note that the presented MPC approach remains the same if replaned navigation function is used in (Equation (4)), where the objects contained in the navigation function will no longer appear as dynamic collision constraints (Equation (12).

4. Optimisation Strategy in MPC Control

In the following, we propose a novel optimization strategy for solving (Equation (4)) that combines optimization with a fixed set of control action candidates and particle swarm optimization.

4.1. Fixed Candidate Optimization

To reduce the computational cost, Fixed Candidate Optimization (FCO) is introduced in [36] to solve the MPC problem. Given the current velocities $\mathbf{u}_c = [v_c, \omega_c]^T$ (applied to the robot at time $t - T_s$), a set of possible discrete accelerations $a_c \in \{-a_{max}, 0, a_{max}\}$, $\alpha_c \in \{-\alpha_{max}, 0, \alpha_{max}\}$ is defined to produce a set of 9 candidate velocities for optimization

$$\mathbf{u}(t) = \mathbf{u}_c + T_s[a_c, \alpha_c]^T \tag{13}$$

constrained by Equation (7).

The main strengths of the proposed MPC with FCO are the low computational complexity and the generation of near-optimal trajectories with a guaranteed convergence, as shown in [36]. However, the obtained velocity profile contains higher noise due to the coarse set of possible accelerations.

4.2. Particle Swarm Optimization

Particle swarm optimization (PSO) uses a stochastic strategy with a swarm of randomly perturbed particles to find a solution. Applying PSO to the MPC problem yields arbitrary velocities $\mathbf{u}(t)$ sampled from a continuum and constrained by Equation (7). Each particle k is parameterized by a parameter vector $\mathbf{p}_k = [v_k, \omega_k]^T$ defining its velocities and an increment vector $\Delta\mathbf{p}_k$ defining the change in velocities. During MPC optimization, the population of all particles is iteratively updated and validated according to the objective function (4). Each particle keeps track of its parameters and remembers its best previously achieved parameter $\mathbf{pB}_k$, along with its associated objective function $J_k = f(\mathbf{pB}_k)$, where f is the function that is minimized in Equation (4). During optimization, the global best parameter vector of the entire swarm $\mathbf{gB}$ is also remembered.

In each sample time, the particles are iteratively updated according to the following rules

$$\begin{aligned} \Delta\mathbf{p}_k &\leftarrow \gamma\Delta\mathbf{p}_k + c_1\mathbf{rand}(0,1)\cdot(\mathbf{pB}_k - \mathbf{p}_k) + c_2\mathbf{rand}(0,1)\cdot(\mathbf{gB} - \mathbf{p}_k) \\ \mathbf{p}_k &\leftarrow \mathbf{p}_k + \Delta\mathbf{p}_k, \end{aligned} \tag{14}$$

where $\gamma > 0$ is the inertia factor, $c_1 > 0$ is the self-cognitive constant, $c_2 > 0$ is the social constant, and $\mathbf{rand}(0,1)$ is a vector of uniformly distributed values in the range $[0,1]$. At the end of the optimization, the best parameter is applied to the robot $\mathbf{u}(t) = \mathbf{gB}(t)$.

MPC with PSO produce smoother velocity profiles and can find better solutions since no velocity discretization is used. However, the computational complexity becomes much higher (compared to MPC with FCO) due to multiple required iterations with more particles. Due to the random nature of PSO, both the solution and the convergence of the search are not guaranteed.

4.3. Combined Deterministic-Stochastic Optimization

The main idea is to combine FCO and PSO in the so-called combined deterministic-stochastic optimization (CDS) and to exploit the advantages of both algorithms to generate trajectories with a smooth speed profile, with guaranteed convergence and low computational complexity.

CDS is a modified PSO algorithm (shown in Algorithm 1) that executes $K_F = 9$ fixed particles and K_C changing particles in parallel. Fixed particles are initialized by Equation (13) and are not updated during optimization. These fixed particles provide good starting parameters that can be used by other changing particles through $\mathbf{gB}$ when iteratively updated through Equation (14). In this way, CDS provides a better (more optimal and smoother) or at least as good a solution as FCO itself. MPC with CDS is guaranteed to converge to the goal in a finite time from any unoccupied location in the environment where the goal is reachable ($N(x, y, \varphi) < \infty$). The algorithm is computationally efficient since the number of changing particles K_C in CDS can be much smaller than in the corresponding PSO.

Algorithm 1 Combined deterministic-stochastic optimization.

Require: List of particles $k = 1, \cdots, K$ where first $k = 1, \ldots, K_F$ are fixed particles.
  for each particle $k = 1, \cdots, K_F$ **do**
    Initialize $\mathbf{p}_k$ by Equation (13).
  end for
  for each particle $k = K_F + 1, \cdots, K$ **do**
    Randomly initialize $\mathbf{p}_k$, $\Delta\mathbf{p}_k = [0, 0]^T$, $\mathbf{pB}_k = \mathbf{p}_k$.
  end for
  $J_{\text{best}} = \infty$, $iter = 1$
  repeat
    for each particle $k = 1, \cdots, k$ **do**
      if $k > K_F \mid iter == 1$ **then**
        Compute objective J_k by (4) considering ramp down (Equation (9) and add penalty for Equation (7), Equation (12) violation.
        Set *convergent* condition by Equation (11).
        if $J_k < f(\mathbf{pB}_k)$ **then**
          $\mathbf{pB}_k = \mathbf{p}_k$
        end if
        if $f(\mathbf{pB}_k) < J_{\text{best}}$ & *convergent* **then**
          $\mathbf{gB} = \mathbf{pB}_k$, $J_{\text{best}} = f(\mathbf{gB})$
        end if
      end if
    end for
    for each particle $k = K_F + 1, \cdots, K$ **do**
      Update $\mathbf{p}_k$ and $\Delta\mathbf{p}_k$ by Equation (14) constrained by Equation (7).
    end for
  until $iter \leq MAXiter$

5. Simulation Results

5.1. Single Robot Navigation

The performance of a single robot in the environment from Figure 1 is first illustrated when the robot needs to transport products or materials from different starting locations to different destinations. A possible scenario could be that the robot has to deliver a semi-finished product to workstation 1 and then transport it to the dropoff location (see top left image in Figure 7).

The navigation functions are computed in advance for known locations, which minimizes the online computational overhead (no online planning is required) and makes the system robust to disturbances during control (e.g., deviations from the original desired path due to errors in robot location, control performance, or dynamic obstacles). Since only a discrete cost map needs to be stored to interpolate the navigation values from it, this is also not memory-intensive (20×20 cost-to-goal for the environment in Figure 1) with the cell size $d_c = 0.5$ m.

In Figure 7, the desired target is at $x = 3.1$ m, $y = 8.3$ m (e.g., workstation 1 in Figure 1) for red paths. The destination can be safely reached from any location considering the navigation function (the top right image in Figure 7) in the proposed MPC control. For a different desired destination (e.g., drop-off location at $x = 9.3$ m, $y = 5.1$ m in Figure 1), a different navigation function is used for all green paths (the bottom right image in Figure 7). The simulation results are obtained using the following parameters. The interpolation of the navigation function is performed on the grid-based search with a cell size resolution of $d_c = 0.5$ m. As the interpolation is applied, good navigation and control performance can be obtained even at coarse resolutions. Optimization in MPC is performed using a fixed horizon length $h = 14$ (with deceleration at the end, as shown in Equation (9)) and sample time $T_s = 0.1$ s, and by considering the constraints on velocities and accelerations $v_{max} = 1$ ms^{-1}, $\omega_{max} = 6$ s^{-1}, $a_{max} = 1$ ms^{-2}, and $\alpha_{max} = 6$ s^{-2}.

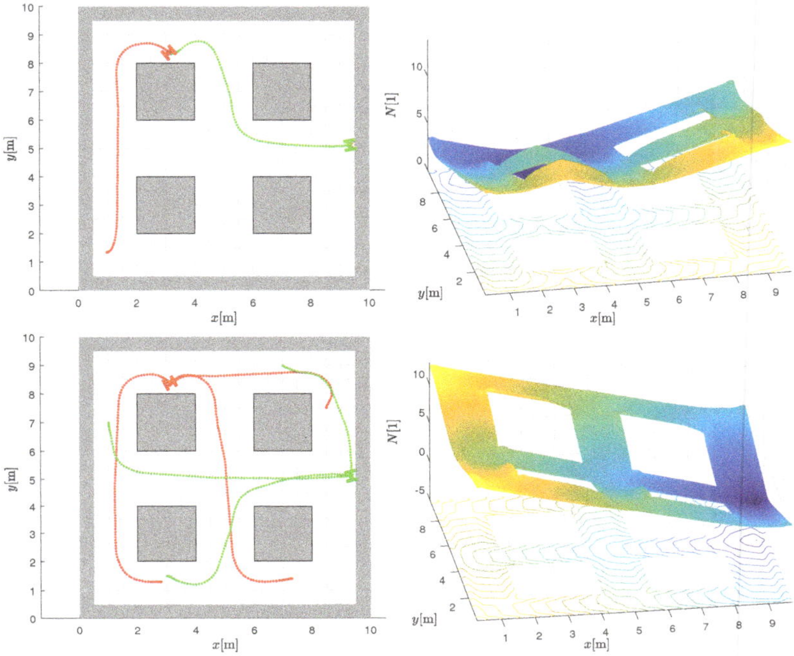

Figure 7. Examples of navigation and control of a single robot in two different destinations defined by the minimum values of the navigation functions in the right column. All red paths are obtained by the navigation function in the top right image, while the green paths are obtained by the navigation function in the bottom right image.

The performance of model predictive control with the proposed combined deterministic-stochastic optimization (MPC-CDS) is compared with the results of fixed candidate optimization (MPC-FCO) and particle swarm optimization (MPC-PSO). In addition, MPC-based algorithms are compared with the kinodynamic stable sparse RRT planning approach (SST) approach [19]. The results are shown in Figures 8 and 9 for the U-obstacle map, Maze map, and the Random-obstacle map. All MPC trajectories are computed for the bicubic interpolation function and also collected in Table 1. The results are compared in terms of obtained trajectories, velocity profiles, length of trajectory L, travel time t_{goal}, cumulative navigation A_N, and normalized computational efficiency E_{comp} (according to MPC-FCO). The computational complexity of MPC-FCO depends on its implementation. In our case, it allows for real-time operation with a refresh rate of at least 50 Hz on a 2.80 GHz Intel dual-core processor with C++ implementation.

The best performance of MPC is obtained by the PSO optimization approach (Figures 8 and 9 and Table 1), where the obtained trajectories are short and fast and the velocities have a smooth profile. However, the computational complexity of MPC-PSO is higher (than MPC-FCO or MPC-CDS) because it uses 25 particles and 20 iterations to optimize each control sample. Similar performance in terms of trajectory length and travel time is obtained with MPC-CDS, which uses nine fixed particles and only two changing particles. The results of CDS are a compromise between the quality of trajectories generated by PSO and the computational complexity of FCO. CDS produces smoother velocity profiles than FCO and requires much less computational effort than PSO. SST produces similarly long trajectories, sometimes shorter since it does not take into account safety costs around the obstacles, but with much slower velocity profiles due to the randomness of velocity selection during the search process. Unlike the MPC-based algorithms, the SST algorithm computes the entire trajectory to the goal before the robot begins execution.

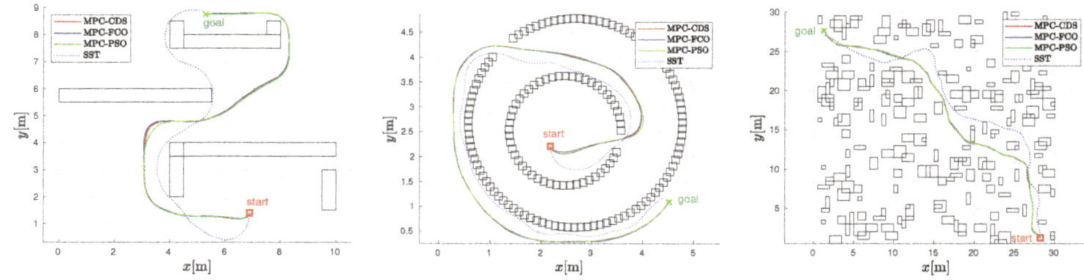

Figure 8. Trajectory comparison for the U-obstacle map (left), the Maze map (middle), the Random-obstacle map (right), the bicubic interpolated navigation, and the SST algorithm.

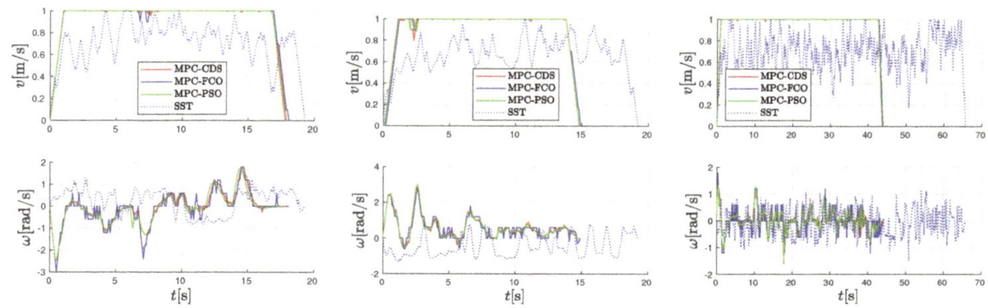

Figure 9. Comparison of velocity profiles for the U-obstacle map (left), the Maze map (middle), and the Random-obstacle map (right).

Table 1. Algorithms validation on three different maps.

	Alg.	L [m]	t_{goal} [s]	A_N [1]	E_{comp} [1]
U map	MPC-CDS	16.88	17.90	230.39	1.61
	MPC-FCO	16.95	18.10	230.84	1.00
	MPC-PSO	16.79	17.80	228.68	13.11
	SST	13.82	19.18	n/a	n/a
Maze map	MPC-CDS	13.63	14.80	111.30	1.73
	MPC-FCO	13.64	14.90	111.66	1.00
	MPC-PSO	13.65	14.80	111.73	13.63
	SST	12.87	19.09	n/a	n/a
Rnd map	MPC-CDS	42.69	43.70	1090.24	2.90
	MPC-FCO	42.78	43.90	1092.28	1.00
	MPC-PSO	42.59	43.60	1091.18	28.01
	SST	46.07	65.59	n/a	n/a

5.2. Multiple Robot Coordinated Navigation

Analysis of the selection of the horizon length in MPC performance is first explained. The minimum horizon length h_{min} (Equation (8)) is sufficient for navigation in static environment (obstacles mapped or unknown in the navigation function), but it may not be good enough for moving obstacles. For moving obstacles and $h > h_{min}$, safety and navigation performance increases because collision threats can be predicted early enough so that better avoidance routes can be found. The analysis of the varying horizon length (where the moment of deceleration can also occur before the end of the horizon, as defined in Equation (10)) for the navigation of two robots approaching a head-on collision and a cross collision ([9]) is shown in Figure 10 and Table 2. In a head-on collision (left image in Figure 10), the robots stop to avoid collision when the minimum horizon $h = h_{min} = 11$ is chosen, while the robots can safely navigate to the target for $h > h_{min}$. This scenario is more difficult than the cross-collision (right part of Figure 10) since the other robots move in the opposite direction of the negative gradient of the navigation function. A larger horizon can provide better collision avoidance but increases the computational cost of navigation (E_{comp} increases in Table 2). E_{comp} is the normalized computational load corresponding to the computational time at $h = h_{min}$ (where in the first line the value $E_{comp} = 1$ is normalized by the computational time of the ignored collision, since the robots do not reach the goals). A larger horizon can slightly reduce both the distance traveled (joint distance $\sum L$ in Table 2) and the travel time (joint travel time $\sum T$ in Table 2), which means that the robots do not need to wait or slow down to avoid a collision. Note that the improvement in travel time and distance is relatively small compared to the increased computational cost. Therefore, the main reason for increasing the horizon is safety and collision avoidance performance.

Table 2. Performance at variable horizon.

	h/h_{min}	$\sum L$ [m]	$\sum T$ [s]	E_{comp} [1]
head-on	11/11	/	/	1.00
	16/11	14.27	16.40	1.46
	22/11	14.26	16.40	2.05
cross	11/11	13.48	16.20	1.00
	16/11	13.38	15.90	1.63
	22/11	13.33	15.80	2.27

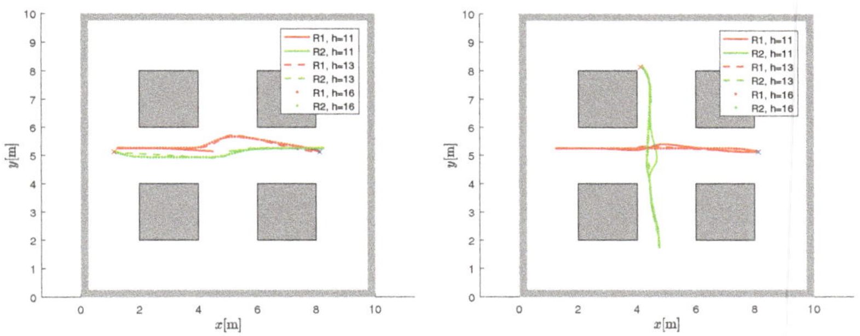

Figure 10. Analysis of horizon length in frontal collision avoidance (left) and cross collision avoidance (right). The target locations are indicated by a cross. A larger prediction horizon can find better trajectories than the minimum horizon $h_{min} = 11$.

Some other examples of coordinated navigation and control of multiple robots can be found in Figure 11, where the navigation results are shown without collision avoidance (left images, where the robots drive over each other) and with coordinated collision avoidance navigation (right images). The starting position of the i-th robot is marked with Ri, and its target position coincides with the final robot position. The occurrence of collisions is marked (left image in Figure 11) by ellipses Ci, where C1 is the first collision between robots 2 and 4; C2 between robots 2, 3 and 4; C3 between robots 1 and 3; and C4 between robots 4 and 5. The controller with coordinated predictive collision avoidance (right image in Figure 11) successfully avoids all collisions.

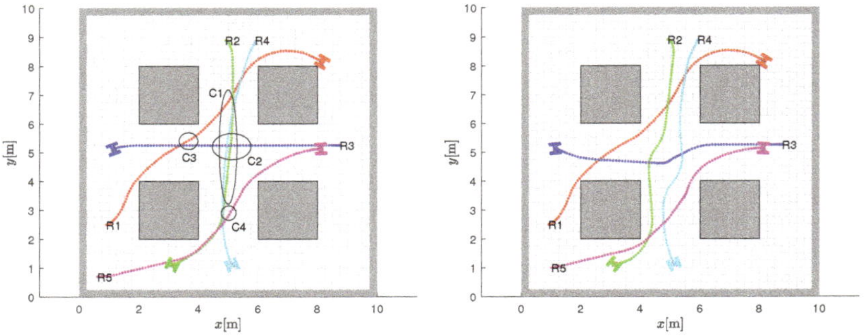

Figure 11. Coordinated collision avoidance. Obtained robot path without using collision avoidance in MPC (where robots drive over each other), with collision cases marked by ellipses (left image). Collision avoidance with prediction horizon $h = 20$ finds safe routes with similar travel times (right image).

Coordinated collision avoidance for symmetric initial locations and congested traffic in the centre of the map is shown in Figure 12. The obtained control with avoidance and prediction horizon $h = 15$ fails to navigate the robots to the destinations as the robots stop safely to avoid collisions (Figure 12, left image). Increasing the horizon to $h = 25$ results in safe trajectories to the targets (Figure 12, right image).

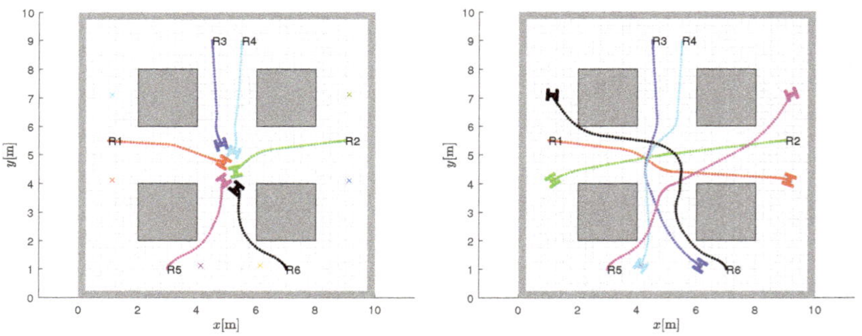

Figure 12. The obtained control with avoidance and prediction horizon $h = 15$ cannot steer the robots towards the destinations since the robots stop safely to prevent collision (left image). Increasing the horizon to $h = 25$ leads to safe trajectories towards the target locations (right image).

5.3. Experiments

Navigation is performed also in the real map shown in Figure 15 and 16 (floor plan of our laboratory). The map is an occupancy grid with 10 cm resolution created with the Sick LMS200 laser range finder. Four target locations $\mathbf{G}_{N1} = [2.4, 7]^T$, $\mathbf{G}_{N2} = [6, 6]^T$, $\mathbf{G}_{N3} = [10.5, 6]^T$, $\mathbf{G}_{N4} = [15, 4]^T$ (e.g., locations of workstations) are defined on the map. A robot can reach a desired goal through MPC control (Equation (4)) by following the navigation function (Equation (3)), which consists of an appropriate interpolated potential field. Figure 13 shows potential fields for the defined targets.

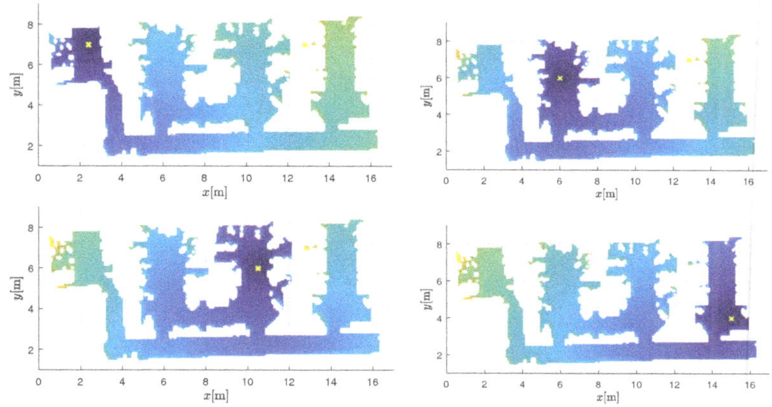

Figure 13. Four interpolated potential fields are used in the navigation functions $N_i(y, x, \varphi)$ $(i \in 1, \dots, 4)$ to find one of the desired target positions $\mathbf{G}_{N1} = [2.4, 7]^T$, $\mathbf{G}_{N2} = [6, 6]^T$, $\mathbf{G}_{N3} = [10.5, 6]^T$, and $\mathbf{G}_{N4} = [15, 4]^T$ from any initial position. The targets are located at the lowest value of the potential field (darkest region) and are marked by a cross.

In experiments, Roomba cleaning robots (Figure 14) are used to simulate transportation tasks between desired workstations. For localization, a camera is used to detect Aruco markers on the ceiling placed at known locations. The robots are controlled by a built-in Raspberry Pi, which sends velocity commands with an update frequency of 10 Hz ($Ts = 0.1$ s).

Figure 14. Roomba robots used to simulate transportation tasks in the laboratory layout from Figures 15 and 16. View of the robots (left) and closer robot view with integrated Raspberry Pi and camera (right).

During navigation, velocities and accelerations are constrained by $v_{max} = 0.45$ ms^{-1}, $\omega_{max} = 3$ s^{-1}, $a_{max} = 0.5$ ms^{-2}, and $\alpha_{max} = 3$ s^{-2}. To predict collision hazards with other robots, the horizon $h = 20 > h_{min}$ ($h_{min} = 11$ according to Equation (8)) is chosen and safety distance and angle are set to $d_{safe} = 0.35$ m and $\phi_{safe} = \pi/2$.

In Figure 15, the first robot uses the first navigation function (the upper left image in Figure 13) to get to the destination $\mathbf{G}_1 = \mathbf{G}_{N1}$. Similarly, the destination for the second robot is reached by the third navigation function ($\mathbf{G}_2 = \mathbf{G}_{N3}$) and the destination for the third robot is reached by the second navigation function ($\mathbf{G}_3 = \mathbf{G}_{N2}$).

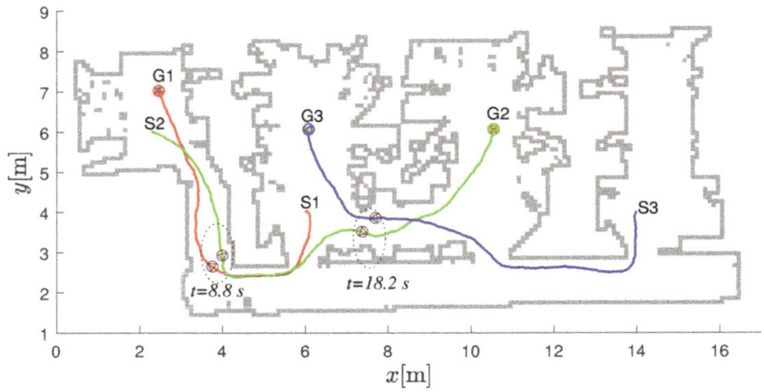

Figure 15. Robot paths in the laboratory layout. The start location of the i-th robot is denoted by $\mathbf{S}_i$ and the destination locations by $\mathbf{G}_i$. Safe navigation during collision avoidance is represented by dark gray circles belonging to passing robots at the same time (the left pair of circles belongs to time $t = 8.8$ s and the right pair to time $t = 18.2$ s).

In Figure 16, the first robot uses the second navigation function (the upper left image in Figure 13) to reach the destination $\mathbf{G}_1 = \mathbf{G}_{N2}$. Similarly, the destination for the second robot is reached by the fourth navigation function ($\mathbf{G}_2 = \mathbf{G}_{N4}$) and the destination for the third robot is reached by the third navigation function ($\mathbf{G}_3 = \mathbf{G}_{N3}$).

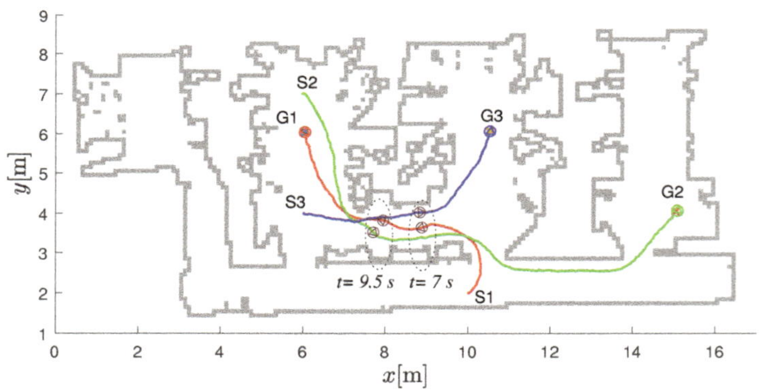

Figure 16. Robot paths in the laboratory layout. The start location of the i-th robot is denoted by $\mathbf{S}_i$ and the destination locations by $\mathbf{G}_i$. Safe navigation during collision avoidance is represented by dark gray circles belonging to the robots passing simultaneously (the right pair of circles belongs to time $t = 7$ s and the left pair to time $t = 9.5$ s).

6. Discussion

From simulations and experiments, the proposed interpolated navigation function combined with MPC computes collision-safe paths for a single vehicle that are near-optimal considering static obstacles. Moreover, the completeness of the system is guaranteed since the potential field does not contain local minima. The approach assumes that the global information about the system layout is known and static. This allows a discrete potential function (e.g., a distance-to-goal cost map) to be computed in advance and bicubic interpolation to be performed only at runtime to obtain a computationally efficient continuous estimate of the potential field values and their negative gradients. In manufacturing or similar applications, vehicles need to deliver cargo between several defined destinations. For each destination, a suitable navigation function can be precomputed, which increases the computational efficiency.

Good trajectories are also obtained when avoiding collisions with multiple robots. Other robots or moving objects are only considered locally within the prediction horizon. Choosing a minimum horizon (h_{min}) ensures that the robot navigates safely while moving past other robots and, in the worst case, stops to avoid a collision. Extending the prediction horizon (e.g., to $2h_{min}$ or more) allows the robots to navigate safely without unnecessary emergency stops to prevent collision. Since local information is considered, optimality is not guaranteed, although good results are obtained in practice. The robot could navigate into a narrow corridor (following the negative gradient) based on the static navigation function, although there is not enough space to avoid collision with another vehicle. MPC will try to follow the direction that reduces the potential while searching among the possible trajectories to avoid collision with another approaching vehicle. In the worst case, if there is not enough free space, the robots would safely stop before a collision occurs.

It would also be possible to globally take into account information about other moving objects and re-create the navigation function. This would require dynamic replanning of the discrete potential field, which is computationally more challenging. However, since the future positions of other objects are not known in advance (they can only be predicted within the sensor's field of view), the navigation function would need to be modified early enough for the robot to safely follow the modified negative gradient of the re-planned potential, which would prevent a collision. The same requirements apply to the length of the prediction horizon as for the static potential field (the horizon must be long enough according to h_{min}). Additionally, the resolution of the discrete grid would need to be fine (much smaller than the size of the robot) to allow for more accurate updating for detected moving obstacles, which is important for narrow passages. A finer resolution of the grid would further increase the computational complexity.

7. Conclusions

In this work, we proposed a novel navigation function obtained from a discrete graph search and smoothed by bicubic interpolation. The navigation function has no local minima and decreases monotonically in the direction of a target, allowing a mobile robot to safely navigate from an arbitrary initial configuration to a desired target. For environments where a set of desired targets is known and fixed, such as on the shop floor or in a warehouse, the appropriate navigation functions can be precomputed. This allows for computationally efficient navigation with rather modest memory requirements. The navigation function is coupled with model predictive control (MPC), which extends navigation to multiple robots and introduces variable horizon and combined stochastic and deterministic search in the optimization to improve performance. Coordination of multiple vehicles is solved locally in MPC as a constrained optimization problem where the cooperating vehicles must share their trajectories in the horizon, while for other objects the trajectories must be estimated from observations. The applicability of the proposed solutions is illustrated by several simulations and experiments. In the future, we will explore alternative approaches for interpolating navigation function. In addition, coordination overhead could be reduced by introducing traffic guidelines and a one-way option in the navigation function, which would improve performance and reduce coordination overhead in narrow corridor areas.

Author Contributions: Conceptualization, methodology, and software, G.K. and M.S.; writing—original draft preparation, G.K.; writing—review and editing, G.K. and M.S.; All authors have read and agreed to the published version of the manuscript.

Funding: This research was partially funded by Slovenian Research Agency and Epilog d.o.o. (research core funding no. P2-0219 and L2-3168) and partially by the Croatian Ministry of Science and Education through the European Regional Development Fund (DATACROSS) (grant KK.01.1.1.01.0009).

Institutional Review Board Statement: Not applicable.

Informed Consent Statement: Not applicable.

Data Availability Statement: Not applicable.

Acknowledgments: This work was partially supported by the Slovenian Research Agency and Epilog d.o.o. (research core funding no. P2-0219 and L2-3168) and partially by the Croatian Ministry of Science and Education through the European Regional Development Fund (DATACROSS) (grant KK.01.1.1.01.0009).

Conflicts of Interest: The authors declare no conflict of interest.

References

1. Jahn, U.; Heß, D.; Stampa, M.; Sutorma, A.; Röhrig, C.; Schulz, P.; Wolff, C. A taxonomy for mobile robots: Types, applications, capabilities, implementations, requirements, and challenges. *Robotics* **2020**, *9*,109. [CrossRef]
2. Gonzalez-Aguirre, J.A.; Osorio-Oliveros, R.; Rodríguez-Hernández, K.L.; Lizárraga-Iturralde, J.; Morales Menendez, R.; Ramírez-Mendoza, R.A.; Ramírez-Moreno, M.A.; Lozoya-Santos, J.d.J. Service Robots: Trends and Technology. *Appl. Sci.* **2021**, *11*,702. [CrossRef]
3. Oyekanlu, E.A.; Smith, A.C.; Thomas, W.P.; Mulroy, G.; Hitesh, D.; Ramsey, M.; Kuhn, D.J.; Mcghinnis, J.D.; Buonavita, S.C.; Looper, N.A.; et.al. A review of recent advances in automated guided vehicle technologies: Integration challenges and research areas for 5g-based smart manufacturing applications. *IEEE Access* **2020**, *8*, 202312–202353. [CrossRef]
4. Fragapane, G.; de Koster, R.; Sgarbossa, F.; Strandhagen, J.O. Planning and control of autonomous mobile robots for intralogistics: Literature review and research agenda. *Eur. J. Oper. Res.* **2021**, *294*, 405–426. [CrossRef]
5. Wood, L. Global AGV (Automated Guided Vehicles) and AMR (Autonomous Mobile Robots) Market Forecast to 2026: Contains Analysis of More Than 500 Players. 2021. Available online: https://www.globenewswire.com/news-release/2021/01/13/2157658/0/en/Global-AGV-Automated-Guided-Vehicles-and-AMR-Autonomous-Mobile-Robots-Market-Forecast-to-2026-Contains-Analysis-of-More-Than-500-Players.html (accessed on 15 January 2022)
6. Digani, V.; Sabattini, L.; Secchi, C.; Fantuzzi, C. Ensemble coordination approach in multi-agv systems applied to industrial warehouses. *IEEE Trans. Autom. Sci. Eng.* **2015**, *12*, 922–934. [CrossRef]
7. Sharon, G.; Stern, R.; Felner, A.; Sturtevant, N.R. Conflict-based search for optimal multi-agent pathfinding. *Artif. Intell.* **2015**, *219*, 40–66. [CrossRef]
8. Standley, T. Finding optimal solutions to cooperative pathfinding problems. In Proceedings of the AAAI Conference on Artificial Intelligence, Atlanta, GA, USA, 11–15 July 2010; Volume 24, pp. 173–178.
9. Zhang, Z.; Guo, Q.; Chen, J.; Yuan, P. Collision-free route planning for multiple agvs in an automated warehouse based on collision classification. *IEEE Access* **2018**, *6*, 26022–26035. [CrossRef]
10. Pallottino, L.; Scordio, V.G.; Bicchi, A.; Frazzoli, E. Decentralized cooperative policy for conflict resolution in multivehicle systems. *Trans. Rob.* **2007**, *23*, 1170–1183. [CrossRef]
11. Stentz, A. Optimal and efficient path planning for partially-known environments. In Proceedings of the IEEE International Conference on Robotics and Automation, San Diego, CA, USA, 8–13 May 1994; pp. 3310–3317.
12. Čikeš, M.; Đakulović, M.; Petrović, I. The path planning algorithms for a mobile robot based on the occupancy grid map of the environment – A comparative study. In Proceedings of the Information, Communication and Automation Technologies (ICAT), 2011 XXIII International Symposium on. IEEE, Sarajevo, Bosnia and Herzegovina, 27–29 October 2011; pp. 1–8.
13. Zdešar, A.; Škrjanc, I. Optimum velocity profile of multiple bernstein-bézier curves subject to constraints for mobile robots. *ACM Trans. Intell. Syst. Technol.* **2018**, *9*, 1–23. [CrossRef]
14. Klančar, G.; Seder, M.; Blažič, S.; Škrjanc, I.; Petrović, I. Drivable path planning using hybrid search algorithm based on e* and bernstein-bézier motion primitives. *IEEE Trans. Syst. Man, Cybern. Syst.* **2021**, *51*, 4868–4882. [CrossRef]
15. Gim, S.; Adouane, L.; Lee, S.; Dérutin, J.P. Clothoids composition method for smooth path generation of car-like vehicle navigation. *J. Intell. Robot. Syst. Theory Appl.* **2017**, *88*, 129–146. [CrossRef]
16. Klančar, G.; Blažič, S. Optimal Constant Acceleration Motion Primitives. *IEEE Trans. Veh. Technol.* **2019**, *68*, 8502–8511. [CrossRef]
17. Ghilardelli, F.; Gabriele, L.; Piazzi, A. Path generation using η4-splines for a truck and trailer vehicle. *IEEE Trans. Autom. Sci. Eng.* **2014**, *11*, 187–203. [CrossRef]
18. Draganjac, I.; Miklić, D.; Kovačić, Z.; Vasiljević, G.; Bogdan, S. Decentralized control of multi-agv systems in autonomous warehousing applications. *IEEE Trans. Autom. Sci. Eng.* **2016**, *13*, 1433–1447. [CrossRef]
19. Li, Y.; Littlefield, Z.; Bekris, K.E. Asymptotically optimal sampling-based kinodynamic planning. *Int. J. Robot. Res.* **2016**, *35*, 528–564. [CrossRef]
20. Pivtoraiko, M.; Knepper, R.; Kelly, A. Differentially constrained mobile robot motion planning in state lattices. *Robot. Auton. Syst.* **2009**, *26*, 308–333. [CrossRef]
21. Likhachev, M.; Ferguson, D. Planning long dynamically-feasible maneuvers for autonomous vehicles. *Int. J. Robot. Res.* **2009**, *28*, 933–945. [CrossRef]
22. Montemerlo, M.; Becker, J.; Bhat, S. ; Dahlkamp, H. ; Dolgov, D. ; Ettinger, S. ; Hähnel, D. ; Hilden, T. ; Hoffmann, G. ; Huhnke, B.; et al. Junior: The Stanford entry in the Urban Challenge. *J. Field Robot.* **2008**, *25*, 569–597. [CrossRef]
23. Wooden, D.; Malchano, M.; Blankespoor, K.; Howardy, A.; Rizzi, A.A.; Raibert, M. Autonomous navigation for BigDog. In Proceedings of the 2010 IEEE International Conference on Robotics and Automation, Anchorage, AK, USA, 3–8 May 2010; pp. 4736–4741. [CrossRef]

24. Bagnell, J.A.; Bradley, D.; Silver, D.; Sofman, B.; Stentz, A. Learning for autonomous navigation. *IEEE Robot. Autom. Mag.* **2010**, *17*, 74–84. [CrossRef]
25. Philippsen, R.; Siegwart, R. An interpolated dynamic navigation function. In Proceedings of the 2005 IEEE International Conference on Robotics and Automation (ICRA), Barcelona, Spain, 18–22 April **2005**, Volume 4, pp. 3782–3789.
26. Dolgov, D.; Thrun, S.; Montemerlo, M.; Diebel, J. Path planning for autonomous vehicles in unknown semi-structured environments. *Int. J. Robot. Res.* **2010**, *29*, 485–501. [CrossRef]
27. Lu, H.; Yin, Y. Fast path planning for autonomous ships in restricted waters. *Appl. Sci.* **2018**, *8*, 1–24. [CrossRef]
28. Cosio, F.A.; Castaneda, M.P. Autonomous robot navigation using adaptive potential fields. *Math. Comput. Model.* **2004**, *40*, 1141–1156. [CrossRef]
29. Guerra, M.; Efimov, D.; Zheng, G.; Perruquetti, W. Avoiding local minima in the potential field method using input-to-state stability. *Control. Eng. Pract.* **2016**, *55*, 174–184. [CrossRef]
30. Park, M.G.; Lee, M.C. A new technique to escape local minimum in artificial potential field based path planning. *Ksme Int. J.* **2003**, *17*, 1876–1885. [CrossRef]
31. Ogren, P.; Leonard, N.E. A convergent dynamic window approach to obstacle avoidance. *IEEE Trans. Robot.* **2005**, *21*, 188–195. [CrossRef]
32. Klančar, G.; Seder, M. Combined stochastic-deterministic predictive control using local-minima free navigation. In Proceedings of the 2021 IEEE/RSJ International Conference on Intelligent Robots and Systems (IROS), Prague, Czech Republic, 25–29 October, **2021**, pp. 5788–5793. [CrossRef]
33. Klančar, G.; Mušič, G.; Chen, H.; Seder, M. Wheeled robot navigation based on a unimodal potential function. In Proceedings of the 2019 International Conference on Computer, Information and Telecommunication Systems (CITS), Beijing, China, 28–30 August 2019, pp. 1–5. [CrossRef]
34. Borkowski, P. Computational mathematics in marine navigation. *Sci. J. Marit. Univ. Szczecin, Zesz. Nauk. Akad. Morskiej Szczecinie* **2010**, *21*, 20–27.
35. Kiss, D.; Tevesz, G. A receding horizon control approach to obstacle avoidance. In Proceedings of the 2011 6th IEEE International Symposium on Applied Computational Intelligence and Informatics (SACI), Timisoara, Romania, 19–21 May 2011, pp. 397–402.
36. Seder, M.; Baotić, M.; Petrović, I. Receding horizon control for convergent navigation of a differential drive mobile robot. *IEEE Trans. Control. Syst. Technol.* **2017**, *25*, 653–660. [CrossRef]
37. Demesure, G.; Defoort, M.; Bekrar, A.; Trentesaux, D.; Djemai, M. Decentralized motion planning and scheduling of AGVs in FMS. *IEEE Trans. Ind. Inf.* **2017**, *14*, 1744–1752. [CrossRef]
38. Brito, B.; Floor, B.; Ferranti, L.; Alonso-Mora, J. Model predictive contouring control for collision avoidance in unstructured dynamic environments. *IEEE Robot. Autom. Lett.* **2019**, *4*, 4459–4466. [CrossRef]
39. Seder, M.; Petrovic, I. Dynamic window based approach to mobile robot motion control in the presence of moving obstacles. In Proceedings of the 2007 IEEE International Conference on Robotics and Automation, Roma, Italy, 10–14 April 2007; pp. 1986–1991. [CrossRef]
40. Keys, R. Cubic convolution interpolation for digital image processing. *IEEE Trans. Acoust. Speech Signal Process.* **1981**, *29*, 1153–1160. [CrossRef]

MDPI
St. Alban-Anlage 66
4052 Basel
Switzerland
Tel. +41 61 683 77 34
Fax +41 61 302 89 18
www.mdpi.com

Sensors Editorial Office
E-mail: sensors@mdpi.com
www.mdpi.com/journal/sensors

www.ingramcontent.com/pod-product-compliance
Lightning Source LLC
Chambersburg PA
CBHW061557220926
51850CB00005B/361